THE BIRDWATCHER'S YEARBOOK
and
DIARY 2006

Designed and published by
Hilary Cromack

Edited by
David Cromack

BUCKINGHAM PRESS

in association with

SWAROVSKI
OPTIK

Published in 2005 by:
Buckingham Press
55 Thorpe Park Road, Peterborough
Cambridgeshire PE3 6LJ
United Kingdom

01733 561 739
e-mail: admin@buckinghampress.com

© Buckingham Press 2005

ISBN 0 9550339 0X
ISSN 0144-364 X

Cover image: Red-backed Shrikes by Dan Powell
Address: 4 Forth Close, Stubbington, Hampshire PO14 3SZ
E-mail: dan.powell@care4free.net

Black and white illustrations by Mabel Cheung
E-mail: chinita@talk21.com

Printed and bound in Great Britain by:
Cambridge University Press, Cambridge, UK

CONTENTS

CONTENTS

CONTENTS

PREFACE

DURING the course of 2005, research published by Target Group Index (TGI) indicated that no fewer than 2.85 million people in Britain were birdwatching on a regular basis. This is a staggering figure for what many consider a slightly nerdish minority activity, but the fact these newcomers have become involved, despite the image problem, speaks volumes for the deep joy wild birds can bring to our lives.

I think it is safe to say that Bill Oddie's TV programmes have played a leading role in getting new recruits into the activity, but where will these people turn for the information they need to progress as birdwatchers? As the Editor of *Bird Watching* magazine, I've taken the decision to add eight extra pages every issue to cater for the needs of the less experienced.

A number of websites also allow some interaction between the inexperienced and the expert birders but whenever I meet someone new to the hobby, I encourage them to seek out their local bird club or RSPB members group. These are often the best sources of information on sites, bird distribution in the area and general birding knowledge. Almost inevitably the first question is "Where can I find my local group?" and of course the answer is "They are all listed in *The Birdwatcher's Yearbook*." And to the best of my knowledge this is the only place where a nationwide listing of club contacts exists in such a handy format.

Naturally *The Yearbook* offers a lot more to all levels of birdwatcher – where else, for instance, can you get updated information on more than 400 nature reserves or tide-table information for months ahead?

Club committee members looking for new speakers for indoor meetings can pick out some new names from our regionalised directory of lecturers. And now that digital printing is so well entrenched, there is no reason why you cannot call one of the photographers in that section of *The Yearbook* and order a fabulous new image for the lounge wall, or maybe you'd prefer to talk to an artist about commissioning an original painting or sculpture.

The Yearbook's Log Chart section features the official BOU *British List* so that you can keep a record of your year's bird sightings. Last year we experimented with a different format (a shorter listing of commoner British species with the full array of tick columns, followed by a complete *British List* with just four entry columns) but subsequent feedback suggests that more readers prefer the original layout so we have reverted to it for this year.

I believe *The Yearbook* has remained successful because it has continually evolved to offer better value and more useful features, so if you have strong views on the Log Charts or any other aspect of the book, please do not hesitate to communicate them to me.

It would be remiss of me if I did not take this opportunity to thank Swarovski UK Ltd and its managing director John Brinkley for their ongoing sponsorship of this book and Buckingham Press' latest title *Best Birdwatching Sites in the Scottish Highlands*. Swarovski's financial input has allowed us to upgrade our products and yet keep price rises to an absolute minimum and it is just one way that the Austrian optical manufacturer supports the broader birdwatching community in this country.

Recognition should also be accorded to Buckingham Press publisher Hilary Cromack, who, despite her exalted title, had to bear the main responsibility for collecting, collating and checking the many thousands of pieces of information that make up each edition. She was grateful for the help she received from Alex Williams until the time came for her to go to university.

It was gratifying that the 2005 edition was a complete sell-out and if you have friends (beginners or experts) you think may appreciate what *The Birdwatcher's Yearbook* has to offer, do encourage them to order a copy as soon as they can to avoid possible disappointment.

David Cromack
EDITOR

Key contributors in this Edition

MABEL CHEUNG was originally from Newton Abbot in Devon, but is now based in Cardiff working for RSPB Cymru as Assistant Conservation Officer. Amongst her varied work history, highlights include working with sea turtle conservation charities in Costa Rica and BirdWatch Ireland. She works in pencil, ink pens, charcoal, pastels and acrylics, from life and/or photos and has contributed illustrations to the recently published *'Birds of Inishbofin Connemara'* by Tim Gordon.

RICHARD FACEY is a keen birder and photographer and is a regular contributor to Buckingham Press' quarterly magazine *Birds Illustrated*. He has a degree in zoology, with his main interests being bird behaviour and evolution. Richard currently works as a Community Project Officer with RSPB Cymru.

KEN HALL
Ken Hall has been an enthusiastic supporter of the Ligue pour la Protection des Oiseaux for many years and, along with his wife, is their UK representative. Together they run a website (www.kjhall.org.uk/lpo.htm) which gives publicity to the society's activities. He first came to prominence in the ornithological world with the publication of his book *Where to Watch Birds in Somerset, Gloucestershire and Wiltshire*.

GORDON HAMLETT, a freelance writer, is a regular contributor to *Bird Watching* magazine, both as a reviewer of books, DVDs and computer software and also as sub-editor of the *UK Bird Sightings* section. Gordon is the author of the acclaimed *Best Birdwatching Sites in the Highlands of Scotland* book (Buckingham Press) and in his spare time is a dedicated internet browser.

DAN POWELL is a well-established bird and wildlife artist who lives in the south of England. In 1995 he won the Artists for Nature Foundation award which resulted in a field trip to the Pyrennees and an exhibition in Barcelona. He has also won the British Birds' *Bird Illustrator of the Year* title.

Dan is best known for his fieldguide *The Dragonflies of Great Britain* that he wrote, designed and illustrated. His work has appeared in numerous wildlife books and publications. He can be contacted on: dan.powell@care4free.net

---

OUR SPECIAL THANKS go to Alex Williams and Derek Toomer for their help with putting this edition of the *Yearbook* together and Nick Williams for the use of his photographs. We would also like to thank all of our contacts in the various bird groups and clubs, bird reserves and national organisations as it would be impossible to produce this book without their continued support and assistance.

FEATURES

Lesser Spotted Woodpecker – one of several woodland species in decline. Drawn by Mabel Cheung

SCIENTIFIC DISCOVERIES IN 2005

(A REVIEW OF THE YEAR'S LITERATURE)

Keeping abreast of the world's ornithological publications is an almost impossible task. To ensure you don't miss anything interesting, we are pleased to present this digest of reports compiled by Richard Facey with illustrations by Mabel Cheung.

New species in Asia and South America

THOUGH the birding press constantly carries gloomy tidings about the imminent demise of many bird species, the last 12 months have seen some amazing discoveries and rediscoveries.

A century after ornithologists last visited the remote island of Calayan in the Philippines, a coalition of British and Filipino scientists came across an unusual rail which has been christened the Calayan Rail (*Gallirallus calayanensis*).

The Piding, as it is known locally, has dark plumage, bright red legs and like many island species is probably flightless. The discovery was made in lowland forest and though the rail does not seem to be in immediate danger, its restricted distribution and small population makes it potentially vulnerable to the effects of habitat destruction and fragmentation, as well as those posed by introduced species.

Another new species – this time an owl – was heard calling in a garden on one of the Togian Islands in central Sulawesi, Indonesia. Initially the birds were identified as Brown Hawk-owl (*N. scutulata), a* regular winter visitor to the region, but further investigations revealed that the birds were of a hitherto unknown species, now named as Togian Hawk-owl (*Ninox burhani*).

Calls similar to that made by the Togian Hawk-owl have also been heard on the islands of Malenge, Batudaka and Walea Bahi and this, along with information gathered from interviews with local people, has led researchers to believe that the new owl is fairly widespread and has a moderately sized population.

Switching continents, a new parrot has been discovered from the Amazon river basin. Named the Sulphur-breasted Parakeet (*Aratinga pintoi*), it closely resembles the juvenile of its close relative, the Sun Parakeet (*A. solstitialis*) and a case of mistaken identity meant its existence has been hidden for many years.

Museum specimens of 'Sun-parakeets' in Brazil were mainly mislabelled Sulphur-breasteds, while those in Europe and America were actually of the correct species. It wasn't until scientists compared the two that the mix-up was discovered.

The Sulphur-breasted Parakeet remains common around Monte Alegre in Brazil. However the region is unprotected, and it is feared that the creation of a new species will draw the attention of trappers, keen to fill international demand.

SCIENTIFIC DISCOVERIES IN 2005

Back in the limelight

TWO SPECIES and one subspecies also reappeared after several years of non-attendance from the register of the ornithological world.

A team from Asociacion Armonía (BirdLife in Bolivia), saw one and heard three Southern Helmeted Curassows (*Crax unicornis koepckeae*) in the Sira mountains of central Peru. While the species is known from central Bolivia, as well as central and east Peru, it is listed as Endangered as the species is believed to be declining very rapidly as a result of hunting and habitat destruction. This is the first time since 1969 that the endemic Peruvian race has been seen.

The Rusty-throated Wren-babbler (*Spelaeornis badeigularis*) was re-sighted in the Arunachal Pradesh region of India. The bird was originally lured from obscurity by using a tape-recording of its call – the recording had been made after it had responded to a recording of its near relative, the Rufous-throated Wren-babbler (*S. caudatus*).

Originally discovered in 1948, after being mist-netted near the Mishmi Hills, the Rusty-throated Wren-babbler has not been seen since – a gap of 58 years.

However, one other 'missing' species has generated far more press coverage than either of these two species – and seems destined to make headlines for many years to come.

The continued existence of the New World's Ivory-billed Woodpecker (*Campephilus principalis*) had been an ornithological enigma for some time – did it exist or did it not? That was the question. Well thanks to a number of sightings of at least one male bird, the answer appears to be yes.

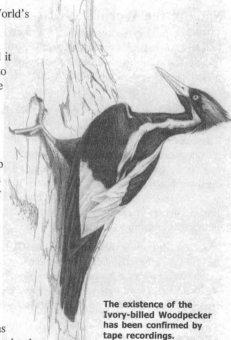

Proof that the Ivory-billed Woodpecker was still around in the Big Woods forest of the Mississippi River basin came from seven sightings made up until April 2005, all confined to 3km of a 41km search area. All sightings probably relate to the same individual. However more birds could easily be hiding, awaiting discovery, as the potential habitat in the region covers an expansive area of more than 220,000 hectares.

However, some scientists remained sceptical about the identification. The evidence came from video footage and photographs, which in some instances was blurry at best. The sceptics claimed it was simply

The existence of the Ivory-billed Woodpecker has been confirmed by tape recordings.

a case of mistaken identity with the more common Pileated Woodpecker (*Dryocopus pileatus*). The scene was set for a war of words in the scientific journals.

Then came a tape recordings made by the discovering team, which had not been analyised when the discovery was announced and this confirmed the existence of the Ivory-billed Woodpecker. After analysis of 17,000 hours of recordings, scientists heard the distinctive 'double-rapping', made as an Ivory-billed Woodpecker taps twice on the tree, which revealed the presence of two birds and the species nasal' *'Kent'* call.

There is a lot to do before the full size and status of the Ivory-billed Woodpecker population is known. But at least we can say it is back.

Sources:
Ivory-billed Woodpecker, Togian Hawk-owl, Southern Helmeted Curassow, Rusty-throated Wren-babbler - BirdLife International

Sulfur-breasted Parakeet – Silveria, Thadeo de Lima, Höfling. *The Auk,* volume 22, pages 292-305

Calayan Rail – Allen, Oliveros, Española, Broad and Gonzalez. *Forktail,* volume 20, pages 1-7

State of the World's Birds – more species in trouble

WE MAY be discovering species, but the world is in danger of losing a lot more. Exactly 2,000 species of bird – that's a fifth of the 9,775 known species – are in trouble according to BirdLife International's annual review of the world's birds for the 2005 IUCN Red List.

No fewer than 179 species have the dubious distinction of being classified as Critically Endangered. These species are on the highest level of threat and include such birds as the Azores Bullfinch (*Pyrrhula murina*), a species that has gone through a severe decline since the early 1990s. As native plants have been usurped by introduced species unpalatable to the finch, numbers have dropped to less than 300 individuals, making it Europe's rarest songbird.

Some of Europe's species are on the ICUN Red List for the first time. The European Roller *(Coracias garrulous)*, Krüper's Nuthatch (*Sitta krueperi*) and the Red Kite (*Milvus milvus*) are a trio that have moved from the Least Concern to Near Threatened category.

Alien species pose one of the greatest threats to biodiversity across the globe. Large numbers of bird species have already gone extinct in New Zealand and

What they mean...
The ICUN uses the following categories to classify species according to their conservation status

Critically Endangered
Facing an extremely high risk of extinction in the wild.

Endangered
Facing a very high risk of extinction

Vulnerable
Facing a high risk of extinction in the wild

Near Threatened
Close to qualifying as vulnerable

Least Concern
Abundant and widespread species. Those that do not qualify for any of the above categories

huge increases in introduced rat populations threaten two more of the archipelago's species. The rodents were responsible for the destruction and loss of two populations of Yellowhead (*Mohoua ochrocephala*), earning it an up-grade from Vulnerable to Endangered, while Malherbe's or Orange-fronted Parakeet (*Cyanoramphus malherbi*) reached the Critically Endangered list, as numbers are now measured in tens.

Every cloud has a silver lining and some species were down-listed, as the hard work of those involved in their conservation has begun to pay off. Five species which took a slide down the scale of conservation concern includes the brightly coloured Kirtland's Warbler (*Dendroica kirtlandii*).

The species, which breeds in Michigan and spends the winter in the Bahamas, reached a low of 167 individuals in the 1970s. Now, in 2005 the species' population stands at more than 1,200, and so it has been down-graded from Vulnerable to Near Threatened.

With 5,000 pairs totally restricted to Christmas island, Australia, Abbott's Booby (*Papasula abbotti*) was classified as Critically Endangered. The species' survival was threatened by the introduced yellow crazy ant, but recent efforts to control the six-legged interloper are paying dividends and the Booby is now classified as Endangered.

The Seychelles Magpie-robin (*Copsychus sechellarum*) is another species which has enjoyed the benefits of conservation efforts on its behalf. The species was known on at least six islands in the Seychelles but by 1965 numbers were as low as 12-15 individuals on Frégate.

However introduction of the species to predator-free islands and the provision of nestboxes and supplementary feeding, has lifted the species' population to more than 130 individuals on four islands, allowing it to be moved from Critically Endangered to Endangered.

Source: *State of the world's birds* – BirdLife International

Tough time for seabirds

THE BUSTLING seabird colonies that characterise Britain's offshore islands could becoming a whole lot quieter according to a report published, by the Joint Nature Conservation Committee (JNCC), in 2005.

The diminutive Kittiwake underwent tough times in Orkney in 2004

UK Seabirds in 2004 summarises the results of the Seabird Monitoring Programme at seabird colonies around Britain. The report presents the changes in number and breeding success of Guillemot (*Uria aalge*), Kittiwake (*Rissa tridactyla*) and Fulmar (*Fulmaris glacialis*).

The combined population of the three species accounts for almost 50% of the UK's seven million breeding seabirds. As they occupy different niches, studying these species gives scientists the best indication of what changes are occurring in our seas.

The results are not encouraging. Welsh Guillemot colonies have declined for the first time in 15 years and in 2004 a mere 28% of the pairs on Orkney successfully reared chicks, compared to the usual 75%. Breeding birds also did badly on Fair Isle and Shetland, where no chicks were raised and the population has fallen by 50% on the latter since 2000.

Orkney was also a poor place to be if you were a Kittiwake, as no chicks were reared at the study colonies. Though the species carries the distinction of being the UK's most abundant gull, the population has declined by 25% since 1986.

Food shortage are being blamed for these declines. The lesser sandeel is a small unassuming looking fish that forms the staple diet for many bird species but rising sea temperatures in the North Sea, resulting from climate change, are causing it problems. The plankton on which the sandeels feed is hatching later and thereby depriving larval sandeels of sufficient food. As a result, sandeels are either not surviving, growing to a large enough size or are being forced out of the reach of hunting seabirds.

Climate change has also begun to affect other seabirds, meaning that 2004 was the worst breeding season on record. Hundreds of Fulmars were reported washed up along the eastern coast of Britain, with similar reports coming from the continent. Post-mortems revealed that the birds had died from starvation – and 90% were female.

On the Farne Islands half the previous June's rainfall fell in a mere two days, resulting in the death of 1,000 chicks among the 2,000 strong breeding colony, though the 2005 breeding performance here was a welcome highspot in another generally dismal year.

Arctic Tern (*Sterna paradisaea*) colonies on Orkney and the south of Shetland failed to rear young, while Great Skuas (*Stercorarius skua*) have also suffered unprecedented breeding failures. Many turned to hunting seabirds or others of their own kind as alternative sources of prey, as fish stocks have diminished over the past two decades.

Seabirds also face competition for the ever-dwindling sandeel stocks. Industrial fisheries catch millions of tonnes of fish such as sandeel, sprats and anchovy to be turned into fish meal and oil, and concerns have been raised at how sustainable such an industry is. The North Sea sandeel fishery failed

Puffins – just one of several seabird species being hit by drops in the North Sea sandeel population.

SCIENTIFIC DISCOVERIES IN 2005

to meet 60% of criteria tested for in an RSPB investigation on the industry's sustainability.

Local inshore fisheries in the Shetlands voluntarily hung up their nets but the large, offshore fishery continued until a number of surveys during April and May 2005 showed that the North Sea's sandeel population was only 50% of the 300,000 million fish required to allow fishing to continue. The European Commission stepped in to close down sandeel fishing with immediate effect, but the decision may have come too late for many of our seabirds.

During 2005 many seabirds delayed their breeding, while others did not bother so the year may turn out to be the grimmest on record.

Source:
Seabirds – JNCC, (2004), UK Seabirds in 2004. also RSPB

Trends in British birds

RESULTS from the tenth year of the Breeding Bird Survey show a continuing trend of mixed fortunes for UK species, but with successes outweighing those in decline. In 2004, more than 2,000 volunteers visited 2,512 BBS squares and recorded 219 species – an average of 31 species per square.

Taking the UK as a whole, 23 of these species showed a decline in numbers, while 49 species have increased since the survey's inception in 1994. In Wales 20 species increased, while five have declined. The results from Scotland were that six species have fallen in number and 20 are on the up, while across the Irish Sea no species declined significantly in Northern Ireland, yet 11 species increased. In England 23 species decreased while 46 species increased.

The Red Kite (*Milvus milvus*) was recorded in 65 squares in 2004, compared to nine in 1994 but because, on average, the species was recorded on less than 40 squares since 1994 it was not possible for reliable trends to be produced. The same was true for Barn Owl (*Tyto alba*), Peregrine (*Falco peregrinus*) and Ring-necked Parakeet (*Psitticula krameri*). The latter species was recorded on 76 squares, compared to four in 1994.

Reintroduction schemes are clearly being successful in building up Red Kite numbers.

However, population trends were created for 100 species. Shelduck (*Tadorna tadorna*), Grey Partridge (*Perdix perdix*), Curlew (*Numenius arquata*), Turtle Dove (*Streptopelia turtur*), Yellow Wagtail (*Motacilla* flava) and Starling (*Sturnus vulgaris*) were among the species that suffered marked declines of between 25-50% on a national level. The Woodpigeon (*Columba palumbus*), Buzzard (*Buteo buteo*), Coot (*Fulica atra*), Canada Goose (*Branta canadensis*) and Great Spotted Woodpecker (*Dendrocopos major*) were among the 40 species that increased by more than 50%.

The BBS monitors 16 species red-listed in *Population Status of Birds in the UK* (Gregory et al 2002). Of these 16, nine declined significantly durning the ten year period, including Yellowhammer (*Emberiza citrinella*), Linnet (*Carduelis cannabina*) and Corn Bunting (*Miliaria calandra*). The good news however, is that four species, birds such as Tree Sparrow (*Passer montanus*) and Song Thrush (*Turdus philomelos*), increased their numbers significantly during the same period.

The Woodpigeon remained top of the heap as the most abundant species counted, with a staggering 52,502 recorded. The Starling was second with 33,557 records, the Blackbird (*Turdus merula*) third with 28,408 individuals recorded. The Chaffinch (*Fringilla* coelebs) earned the title of the most widespread species, being recorded on 2,306 survey squares.

Three species made their debut to BBS in 2004. A Glossy Ibis (*Plegadis falcinellus*) was seen in Oxfordshire, a Wryneck (*Jynx torquilla*) was spotted in Hampshire, while Bitterns (*Botaurus stellaris*) were recorded at sites in Suffolk, Norfolk, and North Lincolnshire.

A number of migrant species have increased markedly between 2003 and 2004. The Sand Martin (*Riparia riparia*) increased by a staggering 247%, with three times as many being recorded in 2004 as in 2003. The Cuckoo (*Cuculus canorus*), which has been in long-term decline, increased by 31%. Other migrants to increase included the Whitethroat (*Sylvia cantillans*), up by 19%, the Chiffchaff (*Phylloscopus collybita*), up 17% and the Willow Warbler (*P. trochilus*), which increased by 12%.

The BBS relies on volunteers to make it successful and ensure that the populations of our common and widespread species continue to be monitored. If you would like more information see www.bto.org/bbs

An army of volunteers visit their survey plots, which are based on the 1km squares of the national grid, three times during the year. The first visit is used to establish a survey route, record habitat and land use, while the remaining two visits are used to survey the birds. From this, data populations indices are calculated, allowing population changes to be measured for 100 species.

Sources:
Raven, Noble and Baillie (2005) The Breeding Bird Survey 2004. *BTO Research Report 403*. British Trust for Ornithology.

Now concern switches to woodland species

A REVIEW of Britain's woodland birds undertaken by the BTO, on behalf on the JNCC and published in the journal *British Birds,* revealed that its not just our farmland birds that are in dire straits. Woodland species such as the Lesser Spotted Woodpecker

(*D. minor*), Spotted Flycatcher (*Muscicapa striata*), and Lesser Whitethroat (*S. curruca*) have declined alarmingly in recent decades.

A wide range of factors were reviewed as potential causes for the declines, with seven emerging as the strongest contenders:
- Climate change affecting breeding grounds, which may change the timing in insect emergence, as well as drying out of woodlands
- Reduction in the numbers of insect prey
- Impacts of land use on woodland edges and other habitats
- Pressures on migrants during transit and in the wintering grounds
- Reduced lowland management in lowland woodland
- Habitat changes brought on by deer removing understorey vegetation and preventing regeneration
- Changing predation pressures

No single cause has come to the forefront, and the declines have occurred at a time of expanding area of woodland in Britain, and more sympathetic attitudes toward wildlife from woodland managers.

Source:
Fuller, Noble, Smith, Vanhinsbergh. *British Birds*, March 2005, pages 116-143

Habitat change benefits Nightjars

GOOD NEWS came from the results of the 2004 Nightjar Survey. Organised by the RSPB, BTO, English Nature and the Forestry Commission, the survey revealed that the Nightjar (*Caprimulgus europaeus*) has increased by over a third since 1992.

The Nightjar's population and range took a nose dive during the 1950s, and the species was placed on the Red List of birds of high conservation concern. However, the results from the effort of nearly 1,000 volunteers heading out at dusk to count churring males on heathland and forestry plantations, was that at least 4,500 males graced the UK.

One species to benefit from more sympathetic land management is the Nightjar.

The increase, which beats the UK Action Plan Target of 4,000 males by 2003, is believed to be the result of habitat restoration and the more sympathetic management of plantation forests.

Source: RSPB

LPO ACTS AS A SPUR TO CONSERVATION IN FRANCE

Ken Hall has been an enthusiastic supporter of the Ligue pour la Protection des Oiseaux for many years. Here the Bristol-based author summarises some of the achievements made by the organization and some of the avian attractions waiting on the other side of the Channel.

BIRD CONSERVATION in France doesn't date back quite as far as the Royal Society for the Protection of Birds' starting point but the centenary of the French equivalent, the Ligue pour la Protection des Oiseaux (LPO), founded in 1912, is not too far off.

Like the RSPB, it started with a single-issue campaign, trying to stop the indiscriminate shooting of auks breeding on the Brittany coast. This successful campaign resulted in the LPO's first reserve – Les Sept-Îles – and the adoption of the Puffin for its logo. The Puffins are still there today, though ironically it may be global warming that is now their biggest threat, here at the southernmost point of their range.

The LPO has since grown steadily, with membership currently around 35,000. It is the French representative of BirdLife International, and though its main focus, of course, is France, it inevitably has to take a pan-European outlook.

We recently celebrated the 25th anniversary of the EU's Birds Directive, a key piece of legislation that all conservationists in Europe should strive to maintain, if not improve on. In France, few months go by without some attempt being made by the shooting lobby to modify the Directive in order to change the periods when migrant birds can be hunted, either by extending the end of the season or by starting it earlier. Currently the season starts in August, a compromise between the July of the past, and the September preferred by the LPO, and ends in February.

The list of species that can legally be shot is still too long, and includes Sky Larks and Turtle Doves, which have suffered population crashes over so much of western Europe. However, the illegal shooting of the latter on spring migration is at last in

Gun implacement in Picardy

decline: it has taken quite some courage by LPO staff to face up physically to armed hunters around the Gironde estuary over several years and to get the local authorities, who have turned a blind eye for so long, to enforce the law.

In fact, legislation in France concerning hunting is quite strict, and the LPO works hard to ensure that contraventions are brought to court as often as possible. As everywhere, justice is expensive and moves slowly, but there have been many exemplary fines and even jail sentences handed down at the end of months of work by its legal department.

THE ROLE OF LPO IN FRENCH CONSERVATION

FORMING NATIONAL LINKS

A significant feature of conservation in France is that historically it has been, and still is, very fragmented. Across the country there are myriad organisations dedicated to the protection of wildlife in their immediate vicinity.

LPO relies on local volunteers

Local involvement is of course important, but global numbers also count, especially with politicians. The LPO has worked hard to bring many pre-existing groups 'into the fold', forming a coherent network under the LPO banner, thereby boosting the overall membership. In addition, the specialist group, FIR, dedicated to the protection of birds-of-prey, has also recently been integrated into the LPO family as the 'Mission-Rapaces'.

Raptors, of course, are a group that have always attracted special attention. Both Hen and Montagu's Harriers breed widely across France, often in cereal fields, where harvesting operations inadvertently destroy the hapless eggs and young. An army of volunteers spends every summer surveying the fields to locate the nests, so that the farmers can avoid them – a task of herculean proportions but which every year saves literally hundreds of young harriers from an otherwise certain death.

The Mission Rapaces is also the prime mover in the successful reintroduction of Griffon and Black Vultures to the Grands Causses of the southern Cevennes, following this up with similar reintroduction schemes in gorges elsewhere in south-east France. Other projects include nurturing the remnant population of Lesser Kestrels in Provence, by providing secure nest sites and active protection. Again the results have been satisfying, with more than 100 pairs now breeding and new centres of population springing up away from the original one on La Crau.

Other species that have been targeted by the LPO in recent years have been Corn Crakes, White Storks, Bitterns and Little Bustards, all of which have suffered one way or another from habitat loss. Habitat protection is another facet of the LPO's work but being that much smaller, it cannot match the RSPB in number of reserves. Nevertheless, it currently manages around 13,000 ha within the seven sites it does have.

LPO also works closely with farmers and other land managers right across the country to promote environmentally-friendly practices, and is very active in the EU's Natura 2000 programme to ensure that a representative sample of habitats across France is properly protected.

PROMOTING GARDEN FEEDING

With the intensification of agriculture it has become recognised that gardens can and do play an extremely important role in maintaining suitable habitat for what were once widespread and common birds. The British have had a long record of feeding and protecting birds in their gardens, but it is good to report that this practice is becoming more and more widespread in France too. One measure of this is the growth of the network of 'Refuges LPO', where owners formally protect the birds in their gardens, displaying the fact by means of a panel prominently displayed, partially to encourage others in the neighbourhood to do the same. The take-up has been remarkable, with currently more than 8,000 participants. Leaflets are available on how to make gardens more attractive to birds, and LPO's quarterly 'house magazine', *L'Oiseau*, carries a regular column on the same topic.

This is just one feature in what is an essential publication for anyone interested in bird conservation in France. The range of subjects covered is wide: in addition to reports on the activities of each of the LPO's groups, there is conservation news from France and elsewhere, identification articles aimed at 'the average birdwatcher', travelogues covering sites both in France and worldwide, articles on conservation projects and species of special interest, book and equipment reviews, and so on.

There is usually a biographical 'portrait' of someone active in the world of wildlife, often including artists, sculptors, writers, philosophers even – the sorts of people often overlooked in the birding press of the UK. For the 'more serious birder', the journal *Ornithos* carries annual reports on rare birds seen in France, rare breeding birds, wintering gull and wildfowl counts, specialist identification articles, detailed accounts of birding 'hot-spots' in France, etc. Either (or both) publication is well worth subscribing to, especially if you want to improve your grasp of the language and learn about birds at the same time.

CONTACT INFORMATION

The LPO is always glad of new members, and is very happy to welcome those from the UK. If you wish to join, or even just to subscribe to the publications, more information is available on their web site (The HQ of the LPO is La Corderie Royale, BP 90263, 17305 Rochefort, France (Tel: +33 (0)5 46 82 12 34).

I run a web site in English (www.kjhall.org.uk/lpo.htm) which includes news items with a French slant, and I can also arrange membership. Contact me at The Anchorage, The Chalks, Chew Magna BS40 8SN (Tel: 01275 332 980).

THE JOY OF FRENCH BIRDING

Griffon Vultures have been reintroduced in southern Cevennes

Protecting the environment and ensuring that there will still be something worthwhile for future generations is one thing, but we mustn't forget to get some enjoyment and satisfaction ourselves, in the here and now.

So how does France rate as a place to go birding? Well, I've been exploring, on and off, since the 1960s, and haven't tired of it yet. Back then, driving down through the Massif Central, I ticked off my first Osprey on its way north, maybe to a nest in recently colonised Speyside.

Cirl Buntings, Black Redstarts, Serins – rarities

THE ROLE OF LPO IN FRENCH CONSERVATION

in the UK, and all new to a schoolboy birdwatcher – were, and still are, common birds everywhere. My first Black Kites were at Entressen rubbish tip, on La Crau; I haven't been there for some time – can it still be as squalid as it was then? But the Little Bustards and Pin-tailed Sandgrouse I added to my list are still there, and even the Calandra Larks maintain a toe-hold.

Later, as a scruffy student, I was threatened with arrest for using a telescope close to the nearby military base, though my actual target was a pair of migrant Red-footed Falcons. Nowadays Entressen features in the rarities reports as a site for wintering Richard's Pipits, even the odd Blyth's, reminding us that winter in France can be just as rewarding as summer.

I've frozen on the shores of the Lac du Der in February, waiting for the thousands of Cranes to come in to roost, one of the great spectacles wild Europe has to offer, and scoped the 'grand chêne' for one of the White-tailed Eagles that now winter there so regularly. From the observation hide (provided, I have to say, by the much maligned chasseurs) at the Réserve de Barthes near Bayonne, I've watched Spotted Eagles circling over the alder woodland, dwarfing the accompanying Buzzards, and eyed warily by several Great White Egrets on the marsh below.

Rock Sparrows breed at the Abbey of Fontevraud

Winter is the best time to locate Wallcreepers – admittedly in June I've seen a speck crossing the Brèche de Roland at Gavarnie, but for close-up views there was nothing to beat the bird climbing the impossibly vertiginous Château de Penne in the Aveyron valley one March morning. I was even sipping a beer at a nearby café at the time.

Talking of historic buildings, where easier to combine culture and birds than at the Abbey of Fontevraud, in the Loire valley? It is the burial place of Richard the Lionheart, with the most northerly Rock Sparrows in Europe. And talking about drinking, what could be more pleasant after a day searching for Red-billed Leiothrixes, one of various naturalised species now tickable in France, among the rolling hills of Jurançon, than a chilled glass of the region's excellent white wine?

Or a rosé from Ensérune after searching for Lesser Grey Shrikes along the 'sentier pie-grièche à poitrine rose' managed by the vignerons of that part of Languedoc-Roussillon. I can't forget days at Le Teich, the wonderful west-coast bird reserve near Arcachon, with its Spoonbills, Bluethroats and White Storks.

Nor the high Pyrenees, which have kept me in trim while hunting for Lammergeiers, White-backed Woodpeckers and Snowfinches among rocky hillsides and quiet beech woods.

Nor, closer to home, the Normandy coast – Middle Spotted Woodpeckers and Fan-tailed Warblers just a short hop across the Channel. What are you waiting for?

ENGLISH LANGUAGE BIRD MAGAZINES

RECOGNISING that birdwatchers in Great Britain increasingly want information about birds and birding opportunities overseas, we have compiled this Directory of English language magazines (a mix of commercial and society-based titles). We welcome further reader recommendations for future Editions of *The Yearbook*.

GREAT BRITAIN

Birding World

A subscription-only title produced by the Bird Information Service team that runs the national Birdline telephone news service. The magazine caters for serious birders with a keen interest in UK rarities and overseas travel. Each monthly issue details the most significant bird sightings in Britain and the remainder of the Western Palearctic.

Other regulars include first-person accounts of the finding of rare birds in the UK and accounts of overseas birding trips.

Editor: Steve Gantlett

Contact details: Birding World, Sea Lawn, Coast Road, Cley-next-the-Sea, Holt, Norfolk NR25 7RZ. Tel: 01263 740 913.
E-mail: Steve@birdingworld.co.uk
Web-site: www.birdingworld.co.uk

Birds

The quarterly members-only magazine issued by the Royal Society for the Protection of Birds is a full-colour super-A4 magazine that promotes the organisation's work in conservation and education.

In addition to an extensive general news section, the magazine carries features on RSPB reserves, international initiatives with partner organisations, tips on developing birdwatching skills, members' letters and book reviews. *Birds* is available on tape for visually-impaired members.

Editor: Rob Hume

Contact details: Birds magazine, The Lodge, Sandy, Bedfordshire SG19 2DL. Tel: 01767 680 551. Web-site: www.rspb.org.uk

Birds Illustrated

A subscription only quarterly publication, *Birds Illustrated* was launched in August 2003 and now has subscribers in 32 countries.

Its focus is on the aesthetic appreciation of wild birds anywhere in the world and features profiles of leading bird artists and photographers as well as a broad range of ornithological topics. A regular column on secondhand books is a unique feature.

Editor: David Cromack.

Contact details: Buckingham Press, 55 Thorpe Park Road, Peterborough PE3 6LJ. 01733 561739.
E-mail: editor@ buckinghampress.com
www.birdsillustrated.com

Birdwatch

A monthly full-colour A4 magazine available on subscription and from main newsagents in Britain. Contains a range of features on identification, birding tips, UK and foreign birding areas and taxonomic issues.

Also contains news, events, readers' letters, product reviews and summaries of British, Irish and Western Palearctic bird sightings.

Editor: Dominic Mitchell

Contact details: Solo Publishing Ltd, The Chocolate Factory, 5 Clarendon Road, London N22 6XJ. Tel: 020 8881 0550. Fax: 020 8881 0990. Email: editorial@birdwatch.co.uk
Web-site: www.birdwatch.co.uk

ENGLISH LANGUAGE BIRD MAGAZINES

Bird Watching

Britain's best-selling monthly bird magazine available from all leading newsagents and on subscription. This A4 full-colour title caters for all active birdwatchers and includes a new eight-page section devoted entirely to beginners.

Every issue contains articles on garden birds and identification, plus news, readers' letters, leading columnists such as John Gooders and Ian Wallace, and the *Go Birding* pull-out guide to bird walks and reserves. The *UK Bird Sightings* section is the world's largest monthly round-up of bird news. Individual product reviews, plus surveys of leading optical products. The annual travel supplement *Destinations* appears with the November issue.

Editor: David Cromack
Contact details: Emap Active Ltd, Bretton Court, Peterborough PE3 8DZ. Tel: 01733 282 601. E-mail: david.cromack@emap.com www.birdwatching.co.uk

British Birds

A long-established subscription-only journal of record that aims to publish material on behaviour, conservation, distribution, ecology, identification, status and taxonomy for birders throughout the Western Palearctic. Organises the *BB Bird Photograph of the Year* competition. Publishes the annual report of the British Birds Rarities Committee.

Editor: Roger Riddington
Contact details: BB 2000 Ltd, Chapel Cottage, Dunrossness, Shetland ZE2 9JH.
Tel: 01950 460 080.
E-mail: editor@britishbirds.co.uk

BTO News

A bi-monthly A4 colour magazine sent to all members of the British Trust for Ornithology. Features include articles about the full range of BTO research projects, plus the status of various species, book reviews and an events guide.

The Bird Table is another BTO pubication sent to all participants in its garden bird survey.
Editor: Derek Toomer
Contact details: BTO, The Nunnery, Thetford, Norfolk IP24 2PU. Tel; 01842 750050.

E-mail: btonews@bto.org
www.bto.org

World Birdwatch

The long-established quarterly subscription-only magazine from BirdLife International promotes global bird conservation activities. An extensive round-up of world bird-related news regularly includes exciting reports of new bird species discoveries and rediscoveries, and is supported by features on birdwatching hotspots, Red Data species profiles, and articles on all aspects of conservation.

Editor: Richard Thomas
Contact details: BirdLife International, Wellbrook Court, Girton Road, Cambridge CB3 0NA. E-mail: birdlife@birdlife.org

FINLAND

Alula

Started in 1995, *Alula* is an independent journal for people interested in birds and bird identification. Regular topics include ID papers by field experts, articles about birding sites, tests of optical equipment, literature reviews, competitions and current birding issues. In order to widen its sales appeal, this high-quality quarterly A4 magazine is now available in an English-language edition as well as Finnish.

Editor: Antero Topp
Contact details: Alula Oy, P.O. Box 68, FI-02101 Espoo, Finland. E-mail: anttu1952@alula.fi
Web-site: www.alula.fi

HOLLAND

Dutch Birding

Many of the articles in this long-established bi-monthly journal that caters for serious birders and twitchers are published in English, and the Dutch pieces generally have an English summary. Regular topics include ID papers by field experts, extensive

coverage of scarcer birds in Holland and the Western Palearctic generally, literature reviews, competitions and current birding issues. Good range of quality rarity photographs (613 colour pictures in 2004).
Editor: Arnoud van den Berg,
E-mail: editors@dutchbirding.nl
www.dutchbirding.nl

SOUTH AFRICA

Africa Birds & Birding

This award-winning glossy A4 bi-monthly colour magazine enjoys support from BirdLife South Africa but remains totally independent. It strives to foster an awareness of the continent's birdlife and encourages birdwatching as a pastime and for its ecotourism potential. High quality photographic features are included along with articles on sites, book and product reviews, news and letters.
Editor-in-Chief/Publisher: Peter Borchert
Contact details: Black Eagle Publishing, PO Box 44223, Claremont 7735, Cape Town, South Africa. E-mail: wildmags@blackeaglemedia.co.za
Web-site: www.africa-geographic.com

USA

Birder's World

A popular title for American birdwatchers of all experience levels, this A4 all-colour monthly magazine carries features on garden birds, profiles of individual bird species, guides to top birding locations in the US, conservation issues, bird behaviour, book and optics reviews. In addition to readers' letters, there is usually a section devoted to answering reader questions and a photo-quiz. Copies can be found in larger branches of leading British newsagents.
Editor: Charles J Hagner.
Contact details: Klambach Publishing Co,

Birder's World editorial dept, PO Box 1612, Waukesha, WI 53187-1612, USA.
E-mail: mail@birder'sworld.com
www.birdersworld.com

Birding

Issued exclusively to members six times a year by the American Birders Association, a not-for-profit organisation that aims to help field birders develop their knowledge, skills and enjoyment of wild birds.

The organisation (membership open to all birdwatchers) also encourages the conservation of birds and their habitats. In practice this means the magazine carries full-colour features on bird-finding in the USA and Canada, in-depth ID articles, book and product reviews, photo quizzes, fieldcraft, taxonomy and conservation articles.
Editor: Tes Floyd
Contact details: American Birding Association, PO Box 7974, Boulder, CO 80306 - 7974, USA.
Web-site: www.americanbirding.org

Bird Watcher's Digest

This pocket-sized full-colour bi-monthly magazine

is unashamedly populist in its approach and features a high proportion of articles about backyard birding and readers' birding tales as well as helpful advice on ID and fieldcraft. Top birding areas in the USA and abroad are spotlighted and Book Notes covers the latest publications. Subscribers get free access to additional material in the magazine's Members Only section. BWD also offers an environmentally friendly electronic edition for readers who do not wish to wait for delivery of a printed copy from the USA.
Editor: William H Thompson, III
Contact details: Bird Watcher's Digest, PO Box 110, Marietta, Ohio 45750, USA.
E-mail: editor@birdwatchersdigest.com
Web-site: www.birdwatchersdigest.com

BRITISH BIRD-RELATED WEBSITES AND MAILING LISTS

Gordon Hamlett reviews the state of websites maintained by UK bird clubs and other related groups on a county-by-county basis to see what practical benefits they offer active birdwatchers.

IN COMPILING this survey of British bird club websites, I have concentrated on those sites that are trying to reach out beyond their core membership with material of interest to day visitors or those spending longer holidays in the area covered.

In particular, I've been looking for sites that recommend birdwatching locations and which provide up-to-date information on birds seen (not just rarities). My published opinions are based on the final visit to the website before the publisher's deadline and I apologise to clubs who may have subsequently updated their site.

Some of my views may be seen to be rather trenchant, though I hope most readers will view the comments as being constructive. I am a great enthusiast for the service these websites provide and hope that by celebrating the best it may inspire others to redouble their efforts.

Since my last club-orientated round-up (*Yearbook*, 2003 edition) there have been some noticeable changes, mainly fuelled by the boom in digiscoping. The ability of ordinary club members to combine digital camera and telescope to capture images means many sites now carry plenty of pictures, ranging from simple record shots to some that are aesthetically pleasing or which record interesting behaviour.

While these illustrations make for easy-on-the-eye viewing, site designers must think carefully about how long it takes for some pages to load. Are some forgetting that not everyone has broadband.

Page design has shown a noticeable turn for the better. Gone are the days of vivid hues of purple and orange and many sites look superb. The major gripe this year is that a lot of the on-screen printing is too small for easy reading or accurate mouse clicking. A site where you have to alter the default settings on your computer every time you visit is a site that quickly gets deleted from the favourites list.

Now that I have moved away from the area, I have no compunction at all at naming the Peterborough Bird Club's website as one of my sites of the year. It is not the prettiest site around, but it shows just what can be done with accepting, and then presenting, bird sightings. All the records get fed into one database, which can then be searched by date, site, species etc.

Imagine if there was a unified system such as this in place across the whole country. Think how much extra useful conservation data could be extracted, whether for planning applications or survey work.

At the moment every club seems to have a different system, none of which work together. Too many clubs are still making it too difficult for casual visitors to send in records. I feel I'm a fairly typical birdwatcher and while I am happy to send off an e-mail of my records, I am not prepared to download forms and learn the precise formatting details to use. Clubs that struggle to get enough local coverage should be working extra-hard to encourage people who can offer casual records.

Mind you, some clubs still insist on keeping their records solely for the benefit of their members. Presumably, they are happy to use the information on other websites while at the same time, denying the same courtesy to others. You can understand it if it is a pair of rare breeding birds, but what's the rationale of suppressing a passage Green Sandpiper? To these clubs I award my turkey of the year.

ENGLAND

Websites are listed alphabetically under their county headings.

AVON

http://www.avonwildlifetrust.org.uk/
The Avon Wildlife Trust's site is brash and dynamic and encourages you to get out and do things. These include taking part in their latest bird survey and visiting one of their 37 reserves – a page is devoted to each one.

http://www.cvlbirding.co.uk/index.html
An excellent site covering the birds and birding around Chew Valley Lake (see also Somerset). At the time of writing, recent sightings were right up to date and there is a full systematic list. An excellent map and access details help would-be visitors. Having to return to the home page all the time when navigating is the only downside.

http://www.geocities.com/ bristolornithologicalclub/
Despite a clean, uncluttered interface, the Bristol Ornithological Club's pages are ultimately frustrating. They tell you about the wonderful meetings, walks and field trips that they organise but don't bother to say what or when they are! Five local sites get a brief description.

BERKSHIRE

http://www.berksbirds.co.uk/
This is a massive resource telling you most of what you will want to know about birding in the county. As well as a huge online database of recent sightings, there is a gazetteer of local sites, though apart from a link to a map of the area, little information is given about each site. Submitting records, especially for the casual visitor is awkward but plenty of articles and news stories add to the site's value.

http://www.ndoc.org.uk/
The Newbury District Ornithological Club's site is a model of simplicity with details of all forthcoming meetings and trips, together with details of local birding hotspots. Details of recent sightings would have been nice but I suspect that most members will use Berksbirds (see above) instead.

http://www.theroc.org.uk/index.shtml
A nice, clean site detailing the activities of the Reading Ornithological Club. There are a few news items and articles, plus trip reports from previous club outings. Noticeably missing though are any site guides, recent sightings or reporting facilities. Again, you are referred to Berksbirds (see above).

Mailing lists: To join, send a blank email to **BERKSBIRDS-subscribe@yahoogroups.com**

BEDFORDSHIRE

http://www.bedsbirdclub.org.uk/
Though the pages are not easy to read, this site has everything about birding in Bedfordshire, apart from a list of recent sightings. Instead news is disseminated via the mailing list (see below). Copies of the club's newsletter – The Hobby – dating back to mid-2001 are all online.

http://www.fly.to/blows.downs
Blows Down is Bedfordshire's major migration hotspot, especially for Ring Ouzels in spring. This site details all the recent sightings as well as suggesting the best ways to work the site. Charts show when the migrants are likely to move through and the areas they particularly favour.

Mailing lists
To join, send a blank email to **bedsbirds-subscribe@yahoogroups.com**

ANNUAL WEBSITE SURVEY

FEATURES

BUCKINGHAMSHIRE

http://www.hawfinches.freeserve.co.uk/
The home pages of the Buckinghamshire Bird Club
seem determined to use as many colours as there are
in the palette and this gives the site a slightly old
fashioned feel. This is echoed when you look through
some of the sections and see that they have not been
updated for some time. Most of the local birders, you
feel, exclusively use the mailing list (see below).

Mailing lists: To join, send a blank email to
Bucksbirders-subscribe@yahoogroups.com

CAMBRIDGESHIRE

http://www.cambridgebirdclub.org.uk/
The Cambridgeshire Bird Club's site has improved
immeasurably during the last three years. The pages
look clean and there is an excellent selection of
photographs. Other sections include recent sightings
and a selection of the best places in the county to go
birding.

http://pbc.codehog.co.uk/
Peterborough Bird Club's site is superb. The mainstay
is a fully searchable database of recent sightings and
these have been cleverly linked to the top birding sites
so that you can see what has turned up over the
years at any particular site. All the club's annual
reports are about to go online in an enhanced version.
Despite all the extra work involved, sightings are
accepted in any format and a good selection of photos
rounds off the website nicely. It is no coincidence that
the mailing list Peterbirder (see below) is one of the
most dynamic and puts many major birding counties to
shame.

http://www.greatfen.org.uk
There are plans to restore 3,000 hectares of
Cambridgeshire back to the sort of fenland habitat
that once existed across the county. This well-
presented site allows you to follow the progress of the
project as well as looking at the wildlife and
archaeology of the area.

http://www.paxton-pits.org.uk/
Paxton Pits is one of the best reserves in the country
to see and hear Nightingales but with 229 species
recorded, there is far more to the area than just one
bird. This site covers everything you need to know
about birding and other activities.

Mailing lists: To join, send a blank email to
cambirds-subscribe@yahoogroups.com
and **peterbirder-subscribe@yahoogroups.com**

CHESHIRE

http://www.deeestuary.co.uk/
This is an excellent site covering both the English and

Welsh sides of the Dee. Recent sightings are bang up
to date and there is a first class site guide. Many
photographs illustrate the assorted articles and there
are links to tide tables. As a personal website, it is
hard to think of one better.

http://www.hilbrebirdobs.co.uk/
This is a fine, unfussy site devoted to the bird
observatory on Hilbre Island. There are some
interesting articles including one on Greenland
Wheatears. The sightings section hadn't been updated
for six months – they suggest that you use the Dee
Estuary site (above) instead.

http://www.ukbis.net/cawos/cawosportal.htm
Once you get past the shocking pink on the home
page, the best part of the Cheshire and Wirral
Ornithological Society's website is the ability to
download back issues of its magazine, though these
stop at 2001. This sums up the site really. The events
guide is up to date but the rest needs a lot more
attention.

http://www.10x50.com/
This site illustrates the transient, and sometimes
painful nature of the net. The Knutsford Ornithological
Society had a charming little site that was cut off in its
prime by an unexpected family bereavement. Our
condolences to the family and hopefully it won't be too
long before we see you online again.

Mailing lists: To join, send a blank email to
CheshireBirds-subscribe@yahoogroups.com

CLEVELAND

http://www25.brinkster.com/teesmouthbc/
The Teesmouth Bird Club's site is a typical club site,
with recent sightings, downloadable newsletters, site
guide etc. Text is very small, even on the largest
setting and though there are links to maps for the
various sites, there are no directions as such – a bit of
hand-holding wouldn't go amiss. Like a lot of sites, it
accepts adverts from Google. The first one I saw was
for bird pest control solutions, based in America. Not
ideal.

Mailing lists: To join, send a blank email to
**NorthEastBirding-
subscribe@yahoogroups.co.uk**

CORNWALL

http://www.cbwps.org.uk/
This is the site for the Cornwall Bird Watching and
Preservation Society. There is a good section on the
Cornish Choughs, together with details of the society's
three reserves, brief details of the best places to
watch in Cornwall and a summary of sightings for the
past three months.

http://www.cornwallwildlifetrust.org.uk
The local Wildlife Trust's site has details on 40 nature

27

reserves scattered around the county as well as some general articles on nature in Cornwall. As with several 'major' birding counties, there are plenty of websites and information around on other wildlife, but precious little on birds.

http://kernowbirds.250free.com/
Kernow (Cornish for Cornwall) Birds covers birds in the county and on the isles of Scilly. The site was still very much under construction when I last looked with recent sightings six weeks old.

http://www.scillybirding.co.uk/
Recent sightings were six weeks out of date when I last looked at the site of the Isles of Scilly Bird Group, which won't please the ardent twitchers. There is plenty of news supplied but little or no practical advice for anyone visiting the islands.

Mailing lists: To subscribe, send a blank email to **CornishWildlife-subscribe@yahoogroups.com** and **Kernowbirds-subscribe@yahoogroups.com**

CUMBRIA

http://www.cumbriabirdclub.freeserve.co.uk/
Three years ago, I described this as a modest little site and so it remains. A lot more of the county's old newsletters and reports are archived but you feel as if the site is always looking backwards, with no recent sightings or ways to submit your own records.

http://www.ospreywatch.co.uk
This is a far more dynamic site, detailing the ongoing success story of the Bassenthwaite Ospreys. There is a daily diary and webcam pictures, updated every ten minutes or so during the breeding season.

Mailing lists: Send a blank email to **BirdingCumbria-subscribe@yahoogroups.com**

DERBYSHIRE

http://www.carsingtonbirdclub.co.uk/
A simple site but one that has everything you need, including a detailed map, bird list, recent sightings and details of all the Carsington Bird Club's events. The overall feel is of a friendly and welcoming club.

http://www.derbyshireos.org.uk/
You feel that there is a lot going on at the pages of the Derbyshire Ornithological Society, too much in fact. You may baulk when you see what is crammed onto the home page, with loads of very small text, and drop-down menus, but persevere –there is plenty of good information here.

http://www.ogstonbirdclub.co.uk/
The Ogston Bird Club's site is another where you are presented with a bewildering number of choices –20 different subjects to choose from. Unfortunately, for every page you visit, you have to return to the home page before moving on. A good selection of

information though, presented in a clean, effective way.

Mailing lists: To subscribe, send a blank email to **derbybirder-subscribe@yahoogroups.com**

DEVON

http://www.dawlishwarren.co.uk/
As well as being an excellent nature reserve, Dawlish Warren has an excellent website to match. There is plenty of information about all sorts of wildlife, not just birds. There are tide tables and plenty of details on where to watch, though the map is a bit disappointing and needs to work better with the sites.

http://www.devonwildlifetrust.org/
In another county that is poorly served by birding websites, the Devon Wildlife Trust's site has details on 40 reserves, though these are not exactly birding hotspots. You are encouraged to send in your sightings but I couldn't find any details from other observers. Crazy. Just to emphasise the point, the site for Dartmoor National Park has no section on wildlife at all. Nor is there a mailing list for the county's birders.

DORSET

http://www.chog.org.uk/
The Christchurch Harbour Ornithological Group's website is an uncluttered site, complete with map of the area, recent sightings, systematic list, future meetings etc. The 'where to watch' section could usefully be expanded, including better access details than a mere grid reference.

http://www.naturalist.co.uk/nothe/index.htm
This is a first class site detailing one man's observations of his local patch over a period of over 20 years. The Nothe juts out into Weymouth Bay and attracts an excellent range of migrants. There is an excellent annotated systematic list and regularly updated diary.

http://www.portlandbirdobs.btinternet.co.uk
This has always been one of my favourite sites and a must for anyone visiting Portland and its Bird Observatory, south of Weymouth. There is an excellent summary of recent sightings, including other wildlife and links to local weather and tide table sites. And the tongue-in-cheek maps are sure to make you laugh.

Mailing lists: To subscribe, send a blank email to **dorsetbirding-subscribe@yahoogroups.com**

DURHAM

http://www.durhambirdclub.org/
The home page of the Durham Bird Club's site looks very inviting but when you click on the individual items

they are presented in a rather uninspiring way. The systematic list is just that – a list, whereas other clubs go that little bit further and have status details etc. Maps would be nice on the site guide.

http://www.whitburnbirding.co.uk/
This site deals with birding in and around Whitburn Observatory. Sightings are up to date and it is easy to add to them. There are pages on moths, cetaceans and other wildlife too. Site navigation takes a bit of getting used to and it would be helpful to link the map with the site guide but that apart, this is an impressive site.

http://www.wwt.org.uk/visit/washington/news.asp
I last looked at the Wildfowl and Wetland Trust's pages three years ago when they were awful. Have they improved? Not really. 'Latest news' is more than three months old and there isn't really anything on the site to make me want to visit. Come, on, enthuse about the place. A totally wasted opportunity.

Mailing lists: To subscribe, send a blank email to **NorthEastBirding-subscribe @yahoogroups.co.uk**

ESSEX

http://www.elbf.co.uk/index.htm
The East London Birders forum is set up to conduct and report on survey work in the area. If you live locally and want to get involved, this is a good place to start.

http://www.essexbirdwatchsoc.co.uk/
Given how good their annual report is, I was really disappointed by the website of the Essex Birdwatching Society which feels behind the times. Recent sightings are out of date, there is no systematic list or site guide and even the membership page fails to make you want to join. Considering this club is one of the country's most active groups, it merits a much better website.

http://www.southendrspb.co.uk/index.htm
Having vowed not to mention local RSPB groups, which tend to focus inwardly and cater only for existing members, the site of the South East Essex RSPB group deserves special mention, just for the sheer amount of information here. Layout and design is never going to win any prizes but the chances are, if you want to know about birding in the area, it's in there somewhere.

GLOUCESTERSHIRE

http://www.birder.pwp.blueyonder.co.uk/
This is another county with very poor website coverage and no mailing list. The entry for Slimbridge is another WWT missed opportunity. This personal site, more concerned with the author's personal travels, at least has some recent sightings and five site guides and is well done as far as it goes.

http://www.gloucestershirewildlifetrust.co.uk
The pages of the local Wildlife Trust have details about their reserves including some bird information. When I last looked, the recent sightings page was six weeks out of date.

GREATER LONDON

http://website.lineone.net/~andrewself/Londonsbirding.htm
London's Birding has a regularly updated sightings page, systematic list of birds seen in the area and a useful collection of links to other sites (some of them covered in other county sections here). The site needs a major update as some of the links were broken and all the events detailed had been, come and gone.

http://website.lineone.net/~edwardmayer/
Swifts. Nothing but Swifts. The author of this site is an obvious fanatic and has devoted pages of stuff to the current status of this species in our capital. There are sound and video clips to download.

Mailing lists: To subscribe, send a blank email to **LondonOrnithology-subscribe @yahoogroups.com**

GREATER MANCHESTER

http://www.gmbirds.freeserve.co.uk/
This is an excellent place to find out about all aspects of birding in the county. There's a decent site guide, systematic list, summary of the previous month's highlights, recent news and a comprehensive list of local clubs. The only downside was a couple of broken links.

http://www.leighos.org.uk/
Leigh Ornithological Society's site is a basic club site which suffers from being set up at the beginning of each September with details of indoor and outdoor meetings, and then not being updated again until the following year. A few site details but no trip reports nor recent sightings to whet the appetite.

http://www.northwestswanstudy.org.uk
This site, devoted to the study of Mute Swans in the region, was undergoing a major revamp at the time of writing so that there were only a few scant details available to read about the biology of the birds and how they are ringed.

HAMPSHIRE

http://www.hos.org.uk/
This is a major site from the Hampshire Ornithological Society. All the features you would expect are here but there no sense of passion igniting the pages. Not that there's anything wrong with that per se, but there was nothing to make me, an outsider, want to go straight back to one of my favourite areas, the New Forest.

ANNUAL WEBSITE SURVEY

http://mysite.wanadoo-members.co.uk/birdsofhants/index.html
If it's quirkiness that you want, then this is the place, with all sorts of facts, figures and statistics about Hampshire's birds. There is discussion of records not accepted and day-by-day analyses of what birds and how many have turned up where. Great stuff.

Mailing lists: To subscribe, send a blank email to **hoslist-subscribe@yahoogroups.com**

HEREFORDSHIRE

http://www.herefordshirebirding.net/
The site of the Herefordshire Ornithological Club was undergoing reconstruction at the time of writing. The home page has a friendly tone and there were small sections on recent sightings, including words and pictures of the history-making Bee-eaters, plus places to visit as well as the latest club news.

HERTFORDSHIRE

http://www.hertsbirdclub.org.uk/
Bang up to date with recent sightings, the pages of the Herts Bird Club are well designed and the overall site works very well with plenty of articles and site guides to keep the casual visitor interested. A county list was about all that was missing.

http://www.nhm.ac.uk/visit-us/galleries/tring/collections/collections.html
The town of Tring holds one of the largest collections of bird skins in the world, available to bona fide researchers. There are some details about the collection here, as well as information on the Walter Rothschild Zoological Museum though the site was being redesigned at the time of writing.

http://www.tringreservoirs.btinternet.co.uk/
The Friends of Tring Reservoirs is an organisation promoting conservation at the reserve. Though the design of site is fairly basic, there is plenty of useful information, most of which is freely available, though there is a separate exclusive section for members.

ISLE OF WIGHT

http://dbhale.members.beeb.net/IOW.htm
There are details of recent sightings on the island, a map showing the key sites and information about birding on the island, plus some photographs. Site navigation could be made a lot easier and the pulsing lights are distracting.

KENT

http://www.dungenessbirdobs.org.uk/
Beautifully designed, the website of the Dungeness Bird Observatory is a model of simplicity with pages detailing birds, flora and other wildlife. There are recent sightings, a calendar of what to look for through the year and information on the observatory itself. Whets the appetite perfectly.

http://www.kentos.org.uk/
A major birding county like Kent needs a major website and the newly-designed Kent Ornithological Society site provides it, though for the latest bird news there is no summary, so you have to click every site in turn. There is a lot of information here but it is presented in a rather formal way which may fail to inspire birders from outside the county.

http://www.sbbo.co.uk/
The site of the Sandwich Bay Bird Observatory is decidedly lacklustre. Recent bird sightings were ten days old and non-bird sightings more than six weeks out of date. There was no map of the area, nor details of the best places to watch, birding calendar etc. Disappointing.

Mailing lists: To subscribe, send a blank email to **kosnet-subscribe@yahoogroups.com**

LANCASHIRE

http://www.eastlancashirebirding.nstemp.net/
This is the new address for everything to do with birding in East Lancashire. The site was in the process of changing server at the time of writing so I can offer no information as to current content.

http://www.fyldebirdclub.org/
Another impressive site. As well as the usual features, the Fylde Bird Club's pages contain articles on breeding Black-tailed Godwits, identifying Water and Rock Pipits, and Pink-footed Geese – all local specialities.

http://www.lancasterbirdwatching.org.uk/
A very comprehensive systematic list is the highlight of the Lancaster and District Birdwatching Society's site. Recent sightings were up to date. The only downside was the local site guide, which consisted of nothing more than numbers on a map.

LEICESTERSHIRE

http://www.birdfair.org.uk/
The British Birdwatching Fair at Rutland Water every August is one of the major birding events in the world. This site details past conservation projects supported and is the best place to look for information on the next fair.

http://www.lros.org.uk/
The Leicestershire and Rutland Ornithological Society's site has good details of 15 top places to watch birds in the area. Recent sightings were a few days old when I last looked, with most of the sightings coming from the well-watched Rutland Water.

ANNUAL WEBSITE SURVEY

http://www.rutlandwater.org.uk/
This site, devoted entirely to Rutland Water tells you everything that you need to know about the two reserves, ranging from the best places to watch, to the Osprey re-introduction scheme and the Tree Sparrow project. A page devoted to the staff gives the site a friendly feel.

LINCOLNSHIRE

http://www.lincsbirdclub.co.uk/
There are more than 30 articles to peruse on the Lincolnshire Bird Club's site as well as the usual latest sightings page and systematic list. There's a good 'where to watch' section, though, given the size of the county, this is far from comprehensive. There is an email forum conducted through the website, rather than a general email list.

http://www.lincstrust.org.uk/reserves/gib/index.php
This is a fairly dry introduction to the bird observatory at Gibraltar Point. There is an overview of the wildlife – not just birds – and a history of the place but I feel it fails to sell those qualities that make the reserve such a special place.

http://www.wanderingbirders.com/
Two young birders based in South Lincs detail birding on their local patches and trips further afield. Both are heavily into digital photography and there is a good selection of images on the site.

NORFOLK

http://www.noa.org.uk/
For arguably Britain's best-loved birding county, Norfolk's presence on the Internet remains at best, extremely disappointing. There are mailing lists, but they are seldom used. This is the site of the Norfolk Ornithological Association. There is a brief description of the organisation's reserves and links to other observatories but not a lot else.

http://www.wildlifetrust.org.uk/norfolk/home/index.htm
There is basic information on more than 30 of the Norfolk Wildlife Trust's reserves and more information on their top five sites but the site is uninspiring in the extreme.

http://www.wvbs.co.uk
http://www.accessbs.com/narvos/index.htm
Two local sites dealing with birding in the Wensum Valley and Nar Valley. Both sites were down when I last looked but they do have some good local information when functioning properly.

Mailing lists: To subscribe, send blank emails to **Norfolkbirdnews-subscribe@yahoogroups.com** **norfolkbirds-subscribe@yahoogroups.com**

NORTHAMPTONSHIRE

http://atschool.eduweb.co.uk/jblincow/npton/nptonind.htm
This is a detailed systematic list of birds in the county. As it hasn't been updated for six years(!), it is now slightly out of date.

http://homepage.ntlworld.com/northantsbirds/
This Northampton Bird Club site offers only details of recent sightings in the county (up to date when I last looked) and a summary of the previous month.

Mailing lists: To subscribe, send a blank email to **birdingnorthants-subscribe @yahoogroups.co.uk**

NORTHUMBERLAND

http://www.farne-islands.com/boat-trips/
For anyone wanting to get across to the amazing Farne Islands in summer, this is the place to start. There are details on all the trips currently available.

http://www.northumberland.gov.uk/vg/wildlife.html
Basic information about some of the key habitats in the county. Part of a very large tourism page, it is useful as far it goes but local and visiting birders need a better resource than this.

http://www.ntbc.org.uk/
The site of the Northumberland and Tyneside Bird Club is a very simple affair, concerned totally with club activities. Much of the information is out of date and there is nothing at all to encourage a visiting birder in the way of recent sightings, site guide etc.

Mailing lists: To subscribe, send a blank email to **NorthEastBirding-subscribe @yahoogroups.co.uk**

NOTTINGHAMSHIRE

http://mysite.wanadoo-members.co.uk/eakringbirds/index.htm
It just goes to show what the enthusiastic amateur can achieve. Eakring is most likely not on your list of must-visit birding sites but this charming site is full of passion and beautifully designed. A great tribute to a local patch.

http://www.nottmbirds.org.uk/
There's a good friendly feeling about this Notts Birders site and you feel that you would be really welcomed as a visitor. There's a basic list of recent sightings and some good information on more than 40 prime birding sites as well as details of all the club's activities.

http://www.paulos.uk7.net/anrhome.html
This is an excellent site covering the wildlife of

Attenborough Nature Reserve. As well as plenty of information about birds, mammals, plants etc, there is a very clear map. A simple site, but one that does its job admirably.

Mailing lists: To subscribe, send a blank email to **Nottsbirdsnews-subscribe@yahoogroups.com**

OXFORDSHIRE

**http://
www.banburyornithologicalsociety.org.uk/**
The Banbury Ornithological Society site is very off-putting to non-members, not least because you feel as if you need a degree in computer science to enter sightings. Would I bother sending in my observations? No. Is that what the society wants? Who knows?

http://www.oos.org.uk/news.php
There is a lot of information on the Oxford Ornithological Society's site but a minimalist design means you have to scour the sub-menus carefully. This is another site that feels unfriendly to outsiders. Non-members can only gain access to the last two days' sightings. Why? What are they trying to hide?

Mailing lists: To subscribe, send a blank email to **OBIS-subscribe@yahoogroups.com** and **Oxonbirds-subscribe@yahoogroups.com**

SHROPSHIRE

http://www.shropshirebirds.com/
Just reading the trip reports on the pages of the Shropshire Ornithological Society' site suggests a really friendly club with the birders enjoying the birds for what they are, rather than just another tick on the list. There's plenty of other information here too, both on the club itself and birding in the county. There is a large section on Venus Pool, the county's premier birding spot.

Mailing lists: To subscribe, send a blank email to **Shropshirebirds-subscribe@yahoogroups.com**

SOMERSET & BRISTOL

**http://www.somersetbirds.uko2.co.uk/
index.htm**
This is a site that's developing nicely. It looks good and has some excellent sections on site guides and the birds of Somerset. There is a message board with access restricted to members of the Somerset Ornithological Society, rather than a mailing list with another discussion group.

SUFFOLK

http://www.goldenoriolegroup.org.uk/
The Golden Oriole Group is devoted to the

conservation of this spectacular bird in East Anglia. There are plenty of stunning pictures as well as details of the bird's biology etc. There's information on where to see the Orioles at RSPB Lakenheath.

http://home.clara.net/ammodytes/
The Lowestoft Bird Club certainly doesn't take itself too seriously. Calling themselves the Lounge Lizards, the members nevertheless provide a large amount of information about birds and birding along this stretch of the Suffolk coast. Plenty of information on other wildlife too.

http://www.lbo.org.uk/
Though it is a fairly simple site at the time of the writing, the pages of the Landguard Bird Observatory have plenty of potential to expand. Whether they will remains to be seen. At the present, there are details of recent sightings, an archive of old stuff and an article about finding the Trumpeter Finch.

SURREY

http://www.sbclub.ukonline.co.uk/
There is nothing about the Surrey Bird Club's site that really makes you want to join the club. Designed primarily for members, with the sightings being a week old, there is nothing at all to attract a visiting birder. Or, such is the paucity of information, there's not much to encourage members if it comes to that.

http://www.sdbws.ndo.co.uk/
The site of the Surbiton and District Bird Watching Society is marginally more dynamic as there are a couple of short trip reports at least. Looking at these two sites, you feel that either there is nothing to see in Surrey or that they don't want to encourage outsiders.

Mailing lists: To subscribe, send a blank email to **SurreyBirders-subscribe@yahoogroups.com**

SUSSEX

http://home.clara.net/yates/index.html
This is an excellent personal site, maintained by the warden at Rye Harbour and contains everything you need to know about birds and other wildlife in the area.

http://www.rxwildlife.org.uk/
Covering wildlife news along the coast from Hastings to Dungeness, this slick-looking site offers details of recent sightings searchable by location and category. Before a visit to the area, this would be the first place I would check.

http://www.sos.org.uk/
There's a lot of information on the Sussex Ornithological Society's site but it doesn't make for inspired reading. The gazetteer lists 2,800 places but doesn't offer any advice as to the best places to visit. The best bit is the fully annotated systematic list.

Sightings are up to date but record submission is unfriendly.

http://www.thebirdsofsussex.co.uk/
Part local patch survey (Ovingdean) and part survey of Sussex avifauna, complete with plenty of pictures, this is another good site to check out.

Mailing lists: To subscribe, send a blank email to
SussexBirds-subscribe@yahoogroups.com
hastings_wildlife-subscribe@yahoogroups.com
**Rye_Bay_Wildlife-subscribe
@yahoogroups.com**

WEST MIDLANDS

http://www.westmidlandbirdclub.com/
Covering the counties of Staffordshire, Warwickshire, Worcestershire and the West Midlands, the site of the West Midlands Bird Club should be your first stop for information about the area. There is plenty of information on where to watch, club news, recent publications and so on. Strangely, no recent sightings are detailed, so subscribe to the various mailing lists from the website.

Page design is very easy on the eye and many others would do well to take note.

There are six branches of the club, each with their own programme of events and organised trips. For more information about the area, there is an excellent selection of links too.

Mailing lists: To subscribe, send a blank email to
WMBC-Staffs-subscribe@yahoogroups.com
**WMBC-WestMidlands-subscribe
@yahoogroups.com**
WMBC-Worcs-subscribe@yahoogroups.com
warwickbirds-subscribe@yahoogroups.com

WILTSHIRE

http://www.greatbustard.com
The attempt to reintroduce Great Bustards to Salisbury Plain is an ongoing project and this site keeps you up to date with the latest developments. There are plenty of pictures and details about the bustards themselves as well as the project.

http://www.waterpark.org/wildlife.html
Like most country parks, the Cotswold Water Park supports a wide variety of activities. For information on wildlife, hides and nature trails in the park, start here and try some of the other links from this page. You might have to wander a fair bit to find what you want.

http://www.wiltshirewildlife.org/
There are details of 34 nature reserves on the Wiltshire Wildlife Trust's site though wildlife details per se are scanty at best. It makes you wonder why they think people go to nature reserves.

YORKSHIRE

http://www.blactoft.demon.co.uk/fog/
Considering the reputation of Flamborough Head as a major seawatching and migration point, I was hoping for rather more than this Flamborough Ornithological Group site offers, not least because several broken links meant that I couldn't explore it properly. There were no sightings highlights after May when checked in September.

http://www.bradfordbirding.org/
Sites, sightings, articles and news from the Bradford area. A section on visible migration was one I hadn't come across before.

http://eastyorksbirdwatchers.co.uk/
This is a fairly basic club website, probably of interest only if you live in the immediate area of East Yorkshire, centred around Spurn, and want to join your local club.

http://www.fbog.co.uk/
Plenty of information about Filey Brigg but the site's navigation system is as easy as seeing a skua in thick fog.

http://www.filey2000.co.uk/birding.html
If you are in the Filey area, you should check out the Filey Dams reserve site as well which has recent sightings for the whole Filey area.

http://www.sbsg.org/
I didn't like the site of the Sheffield Bird Study Group three years ago but it has come a long way since then, with a whole host of interesting articles and plenty of photos too.

**http://www.spurnpoint.com/
Spurn_bird_observatory.htm**
Spurn Observatory has a very basic site, detailing migrants month by month (one month in arrears) but very little else.

http://www.tka.co.uk/yoc/
The York Ornithological Club's site has a reasonable 'where to watch' section but the rest of the site needs some attention, in terms of better navigation and more regular updating.

Mailing lists: To subscribe, send a blank email to
airebirding-subscribe@yahoogroups.com
YorkshireBirds-subscribe@yahoogroups.com

ISLE OF MAN

http://www.iombirding.co.uk/
This is the only website I looked at that had a really garish colour scheme, but if you can get past that, there is a good site guide, searchable database, recent sightings page and information on other wildlife.

NORTHERN IRELAND

http://www.cbo.org.uk/
The Copeland Bird Observatory's site has a lot of good information, especially about shearwaters, but the recent sightings page was two months out of date. There is a good selection of articles taken from old copies of the annual report.

http://www.nioc.fsnet.co.uk/
The Northern Ireland Ornithologists' Club site has past issues of the club's newsletter online and some good photography but little else to appeal to the casual surfer.

SCOTLAND

ABERDEENSHIRE

http://www.abdn.ac.uk/nsbc/
The North Sea Bird Club suffers from a lack of up-to-date information, with the latest posting ending in 2004. Nevertheless, there are some fascinating pictures of birds that turn up on oil rigs.

http://www.abdn.ac.uk/~src074/main.htm
The Aberdeen University Bird Club is one of those sites to have suffered since the original, enthusiastic founders graduated and left. There is still some useful information to be gleaned from old newsletters but they are now four years old.

http://www.wildlifeweb.co.uk/
This is one of the oldest websites around. Navigation is still creaky – you have to click on 'local community' for details of latest sightings. There are still gaps in the 'where to watch' section that haven't been filled in 15 years.

Mailing lists: To subscribe, send a blank email to **ABZ-Rare-Birds-subscribe@yahoogroups.com**

ANGUS AND DUNDEE

http://angusbirding.homestead.com/
The Angus and Dundee Bird Club's site is still incredibly slow to load. There is a grapevine of recent sightings. That apart, there is little of interest for non-members – and little to encourage you to join either.

http://www.montrosebasin.org.uk/
Montrose Basin is probably the premier reserve in the region and this site does ample justice to the birds and other wildlife on the reserve.

ARGYLL AND BUTE

http://www.argyllbirdclub.org/
While the 'where to watch' section of the Argyll Bird Club's site is still being written, the rest of the pages look very smart, and, even better, load very quickly. Past copies of the club's newsletter can be downloaded. An impressive site.

ANNUAL WEBSITE SURVEY

http://www.mullbirds.com/
With Mull becoming an ever-increasingly popular holiday destination, you need a website that gives you all the information you need and this is it. Details of where to go, birds seen, tours to be taken and accommodation details make this an excellent pre-holiday planner.

AYRSHIRE

http://www.fssbirding.org.uk/index.htm
This personal site has a mini site guide to birding in Ayrshire among a range of other interesting material such as phone-scoped pictures and overseas trip reports.

http://www.south-ayrshire.gov.uk/ environment/bird_dg.htm
A brief introduction to birding in Southern Ayrshire, with a downloadable leaflet giving further information.

Mailing lists: To subscribe, send a blank email to **ayrshirebirding-subscribe@yahoogroups.com**

CENTRAL

http://www.argatyredkites.co.uk/
It's not just Wales that has Red Kite feeding stations. Following the reintroduction of birds into the area, you can come and get close up views of these magnificent raptors. Full details here.

DUMFRIES AND GALLOWAY

http://www.wwt.org.uk/visit/caerlaverock/
Given that the local tourist office publishes an excellent free booklet on birding in D&G, I was surprised to find little presence for the county on the web. These pages from The Wildfowl and Wetland Trust remain at their usual appalling standard.

Mailing lists: To subscribe, send a blank email to **DumfriesandGallowayBirding-subscribe @yahoogroups.com**

FIFE

http://www.fifebirdclub.org.uk/
Though you get some recent sightings, there is also a separate mailing list for club members only. Fife Bird Club is the only one I've come across so far that won't willingly disseminate its information. There's a site guide where you can read about the club's hides – again, members only.

HIGHLANDS AND ISLANDS

http://www.roydennis.org/
An organisation dedicated to conservation and research, The Highland Foundation for Wildlife's site includes details and maps of satellite tracking Honey Buzzards, Marsh Harriers and Ospreys (coming soon).

http://www.skye-birds.com/
Yet again, the islands do what the mainland can't, by creating a first class site to local birding – this time on Skye. There is an excellent site guide, species list, recent sightings and tips on where to find Golden and White-tailed Eagles.

http://www.wildcaithness.org/
There's a good selection of information about all sorts of wildlife in Caithness on this site. There's a good selection of guided walks to go on too, taking advantage of the special local knowledge of the leaders.

http://www.wildlifehebrides.com/
A cracking site and one that everyone should visit before their trip to the Outer Hebrides. This is one of ten linked websites that cover a spectrum of Hebridean activity including culture, cycling and music.

LOTHIAN

http://www.andrewsi.freeserve.co.uk/lothian-sites.htm
Though it was written some 15 years ago, this is still an excellent 'where to watch' guide for the region.

http://www.seabird.org/
The Scottish Seabird Centre's site looks and feels as if a lot of money has been thrown at it. They have certainly improved the wildlife coverage too with much more detail on what has been seen. You can check it out for yourself too with five webcams showing pictures of Gannets on the Bass Rock, Puffins on the Isle of May etc.

Mailing lists: To subscribe, send a blank email to **lothianbirdnews-subscribe@yahoogroups.com**

ORKNEY

http://www.nrbo.f2s.com/
While the site for the North Ronaldsay Bird Observatory has some interesting information, the fact that they are still advertising for staff needed in 2003 suggests that maybe they didn't get any applicants and there is no-one to keep the site up-to-date!

http://www.visitorkney.com/ bird_reports.html
Here are extremely detailed archived reports of birds recorded over the various months. There are other sections on birdwatching on the site too but the poor design means you will struggle to find them. Where are the links? Check under active Orkney and Mainland east and west though there may be others.

SHETLAND

http://www.nature.shetland.co.uk/
Another terrific site that brings together all Shetland's wildlife sites. There are latest sightings for birds, sea mammals, insects and other wildlife and I even noticed some astronomical sightings. Start here and then follow the links to all the other organisations.

STRATHCLYDE

http://www.clydebirds.com/
Another site – covering the area round the Clyde - still under development, especially the 'where to watch' section. There are recent sightings listed and plenty of photos.

http://wildlife.glasgow.gov.uk/sites.htm
This is a fairly basic site, detailing some of the places to watch wildlife in Glasgow. There are plenty of places listed, but information on the wildlife is limited.

Mailing lists: To subscribe, send a blank email to **cbirds-subscribe@yahoogroups.com**

WALES

CLWYD

http://www.birdingconwy.co.uk/
This is very much a personal site, dealing with birds in and around Conwy. There are memorable days written up and a warden's day detailed. There is an excellent site guide and plenty of photos.

http://www.deeestuary.co.uk/
See Cheshire for full write up.

DYFED

http://members.aol.com/skokholm/holm.htm
There is a lot of information about Skokholm here, but be warned. The site hasn't been updated since 1997!

http://www.rosemoor.com/IPFiles/Nationalpark/skomer.html
Another Pembrokeshire island, this time Skomer and another poor site, with limited information and the bird page 'not ready yet'. Don't hold your breath.

GLAMORGAN

http://www.glamorganbirds.org.uk/
This site is a joint venture between the Glamorgan Bird Club and Gower Ornithological Society. Chunks of it, such as the local site guide are still under development but it all looks promising. There is an active discussion forum on the site.

GWENT

http://www.gwentbirds.org.uk/
Design on the Gwent Ornithological Society's site is fairly simple but what there is works admirably, with plenty of online material to browse, including the latest newsletter, recent sightings and systematic list.

GWYNEDD

http://www.anglesey-history.co.uk/angnatur.html
This is a quick guide to some of the best natural history sites on Anglesey.

http://www.bbfo.org.uk/
Tiny point sizes makes navigating the site of Bardsey Bird Observatory difficult. Most of the information is geared towards booking a place at the observatory but there are details of recent sightings.

http://mysite.wanadoo-members.co.uk/cambrianos
The Cambrian Ornithological Society covers Gwynedd, Anglesey and half of Conwy. Though the site hadn't been updated for six months when I looked, there is a useful site guide to the area. There is also an English-Welsh bird name translator.

POWYS

http://www.gigrin.co.uk/
Gigrin Farm is famous for its Red Kite feeding station. This site tells you everything you need to know as well as having plenty of pages about the birds themselves. There are videos to download too. Page design is over-enthusiastically messy and could do with simplifying.

EVENTS DIARY 2006

JANUARY

28 – Sussex Ornithological Society annual conference Haywards Heath. e-mail: adrianrspb@btinternet.com

28-29 – Big Garden Birdwatch Contact RSPB on 01767 680551; www.rspb.org.uk

FEBRUARY

11-12 – Great West Bird Fair WWT Slimbridge, Glos, 01453 890333; www.wwt.org.uk

18-19 – Lea Valley Bird Fair Lee Valley Park Farms, Nr Fishers Green, Essex. 01992 702200. www.leevalleypark.org.uk

18 –West Midlands Ringers' Conference Cotswold Water Park. Contact Robin Ward, e-mail: robin.ward@wwt.org.uk

25 – Oxfordshire/BTO Birdwatchers' One Day Meeting. Didcot Civic Hall, Contact: Dawn Balmer at BTO email: dawn.balmer@bto.org.

MARCH

11 – North East Ringers' Conference, University of Durham Contact: Dawn Balmer at BTO email: dawn.balmer@bto.org

APRIL

1-3 – Woodland Birds: their ecology and conservation. BOU conference, University of Leicester.

7-9 – RSPB Members Weekend in York Contact RSPB on 01767 680551; www.rspb.org.uk

MAY

6-7 – Leighton Moss Bird Fair Contact Leighton Moss RSPB reserve, 01524 701601.

7- June 11 – Wildlife Photographer of the Year Exhibition Pensthorpe, Fakenham, Norfolk. 01328 851465; www.pensthorpe.com

JULY

NEWA Annual Exhibition date/venue TBC Contact: 11 Dibbins Hey, Poulton Lancelyn, Bebington, Wirral.CH63 9JU; e-mail: NEWA@mtuffrey.freeserve.co.uk www.newa.cwc.net

AUGUST

1-6 – Wildlife Photographer of the Year Exhibition Pensthorpe, Fakenham, Norfolk. 01328 851465; www.pensthorpe.com

3-4 – Family Day Minsmere RSPB Reserve, 01728 648281; www.rspb.org.uk

18-20 – The British Birdwatching Fair Rutland Water www.birdfair.org.uk

SEPTEMBER

27-October 8 – SWLA Annual Exhibition Mall Galleries, London. Contact SWLA.

OCTOBER

7 – RSPB AGM Queen Elizabeth II Conference Centre, London Contact RSPB on 01767 680551; www.rspb.org.uk

28 – Feed the Birds Day Contact RSPB on 01767 680551; www.rspb.org.uk

NOVEMBER

18-19 – North West Bird Fair (Date to be confirmed) WWT Martin Mere, 01704 895181; www.wwt.org.uk

DECEMBER

8-10 – BTO Annual Conference. Hayes Conference Centre, Swanwick, Derbyshire. Contact BTO.

Contact details for the organisations listed here can be found under their entries in the County and National Directories.

DIARY - JANUARY 2006

1	Sun	*New Year's Day*
2	Mon	*Holiday (Scotland)*
3	Tue	
4	Wed	
5	Thu	
6	Fri	
7	Sat	
8	Sun	
9	Mon	
10	Tue	
11	Wed	
12	Thu	
13	Fri	
14	Sat	
15	Sun	
16	Mon	
17	Tue	
18	Wed	
19	Thu	
20	Fri	
21	Sat	
22	Sun	
23	Mon	
24	Tue	
25	Wed	
26	Thu	
27	Fri	
28	Sat	
29	Sun	
30	Mon	
31	Tues	

BIRD NOTES - JANUARY 2006

DIARY - FEBRUARY 2006

1	Wed	
2	Thu	
3	Fri	
4	Sat	
5	Sun	
6	Mon	
7	Tue	
8	Wed	
9	Thu	
10	Fri	
11	Sat	
12	Sun	
13	Mon	
14	Tue	
15	Wed	
16	Thu	
17	Fri	
18	Sat	
19	Sun	
20	Mon	
21	Tue	
22	Wed	
23	Thu	
24	Fri	
25	Sat	
26	Sun	
27	Mon	
28	Tues	

DIARY 2006

DIARY - MARCH 2006

1	Wed	
2	Thu	
3	Fri	
4	Sat	
5	Sun	
6	Mon	
7	Tue	
8	Wed	
9	Thu	
10	Fri	
11	Sat	
12	Sun	
13	Mon	
14	Tue	
15	Wed	
16	Thu	
17	Fri	
18	Sat	
19	Sun	
20	Mon	
21	Tue	
22	Wed	
23	Thu	
24	Fri	
25	Sat	
26	Sun	*British Summertime begins* *Mothering Sunday*
27	Mon	
28	Tue	
29	Wed	
30	Thu	
31	Fri	

DIARY - APRIL 2006

1	Sat	
2	Sun	
3	Mon	
4	Tue	
5	Wed	
6	Thu	
7	Fri	
8	Sat	
9	Sun	
10	Mon	
11	Tue	
12	Wed	
13	Thu	
14	Fri	*Good Friday*
15	Sat	
16	Sun	*Easter Day*
17	Mon	*Easter Monday*
18	Tue	
19	Wed	
20	Thu	
21	Fri	
22	Sat	
23	Sun	
24	Mon	
25	Tue	
26	Wed	
27	Thu	
28	Fri	
29	Sat	
30	Sun	

DIARY - MAY 2006

1	Mon	*May Day*
2	Tue	
3	Wed	
4	Thu	
5	Fri	
6	Sat	
7	Sun	
8	Mon	
9	Tue	
10	Wed	
11	Thu	
12	Fri	
13	Sat	
14	Sun	
15	Mon	
16	Tue	
17	Wed	
18	Thu	
19	Fri	
20	Sat	
21	Sun	
22	Mon	
23	Tue	
24	Wed	
25	Thu	
26	Fri	
27	Sat	
28	Sun	
29	Mon	*Spring Bank Holiday*
30	Tue	
31	Wed	

BIRD NOTES - MAY 2006

DIARY - JUNE 2006

1	Thu
2	Fri
3	Sat
4	Sun
5	Mon
6	Tue
7	Wed
8	Thu
9	Fri
10	Sat
11	Sun
12	Mon
13	Tue
14	Wed
15	Thu
16	Fri
17	Sat
18	Sun
19	Mon
20	Tue
21	Wed
22	Thu
23	Fri
24	Sat
25	Sun
26	Mon
27	Tue
28	Wed
29	Thu
30	Fri

DIARY - JULY 2006

1	Sat
2	Sun
3	Mon
4	Tue
5	Wed
6	Thu
7	Fri
8	Sat
9	Sun
10	Mon
11	Tue
12	Wed
13	Thu
14	Fri
15	Sat
16	Sun
17	Mon
18	Tue
19	Wed
20	Thu
21	Fri
22	Sat
23	Sun
24	Mon
25	Tue
26	Wed
27	Thu
28	Fri
29	Sat
30	Sun
31	Mon

DIARY - AUGUST 2006

1	Tue
2	Wed
3	Thu
4	Fri
5	Sat
6	Sun
7	Mon
8	Tue
9	Wed
10	Thu
11	Fri
12	Sat
13	Sun
14	Mon
15	Tue
16	Wed
17	Thu
18	Fri
19	Sat
20	Sun
21	Mon
22	Tue
23	Wed
24	Thu
25	Fri
26	Sat
27	Sun
28	Mon	*Late Summer Holiday*
29	Tue
30	Wed
31	Thu

DIARY - SEPTEMBER 2006

1	Fri
2	Sat
3	Sun
4	Mon
5	Tue
6	Wed
7	Thu
8	Fri
9	Sat
10	Sun
11	Mon
12	Tue
13	Wed
14	Thu
15	Fri
16	Sat
17	Sun
18	Mon
19	Tue
20	Wed
21	Thu
22	Fri
23	Sat
24	Sun
25	Mon
26	Tue
27	Wed
28	Thu
29	Fri
30	Sat

BIRD NOTES -SEPTEMBER 2006

DIARY - OCTOBER 2006

1	Sun	
2	Mon	
3	Tue	
4	Wed	
5	Thu	
6	Fri	
7	Sat	
8	Sun	
9	Mon	
10	Tue	
11	Wed	
12	Thu	
13	Fri	
14	Sat	
15	Sun	
16	Mon	
17	Tue	
18	Wed	
19	Thu	
20	Fri	
21	Sat	
22	Sun	
23	Mon	
24	Tue	
25	Wed	
26	Thu	
27	Fri	
28	Sat	
29	Sun	*British Summertime ends*
30	Mon	
31	Tue	

DIARY 2006

DIARY - NOVEMBER 2006

1	Wed	
2	Thu	
3	Fri	
4	Sat	
5	Sun	
6	Mon	
7	Tue	
8	Wed	
9	Thu	
10	Fri	
11	Sat	
12	Sun	*Remembrance Sunday*
13	Mon	
14	Tue	
15	Wed	
16	Thu	
17	Fri	
18	Sat	
19	Sun	
20	Mon	
21	Tue	
22	Wed	
23	Thu	
24	Fri	
25	Sat	
26	Sun	
27	Mon	
28	Tue	
29	Wed	
30	Thu	

DIARY - DECEMBER 2006

1	Fri	
2	Sat	
3	Sun	
4	Mon	
5	Tue	
6	Wed	
7	Thu	
8	Fri	
9	Sat	
10	Sun	
11	Mon	
12	Tue	
13	Wed	
14	Thu	
15	Fri	
16	Sat	
17	Sun	
18	Mon	
19	Tue	
20	Wed	
21	Thu	
22	Fri	
23	Sat	
24	Sun	
25	Mon	*Christmas Day*
26	Tue	*Boxing Day*
27	Wed	*Bank Holiday*
28	Thu	
29	Fri	
30	Sat	
31	Sun	

BIRD NOTES -DECEMBER 2006

DIARY 2006

YEAR PLANNER 2007

January

February

March

April

May

June

July

August

September

October

November

December

LOG CHARTS

Corn Bunting by Nick Williams

LOG CHARTS

NEW ORDER OF THE BRITISH LIST
an explanation

NEWCOMERS to birdwatching are sometimes baffled when they examine their first fieldguide as it is not immediately clear why the birds are arranged the way they are. The simple answer is the order is meant to reflect the evolution of the included species. If one were to draw an evolutionary tree of birds, those families that branch off earliest (i.e are the most ancient) should be listed first.

Previously the British List was based on Voous Order (BOU 1977), the work of an eminent Dutch taxonomist. However, more than 26 phylogenetic studies, many using DNA analysis, have been published in recent years that together form a large body of evidence showing that the order of birds in the British List did not properly reflect their evolution. A change in order was required.

The British Ornithologists' Union's Records Committee (BOURC) is responsible for maintaining the British List and it relies on its Taxonomic Sub-Committee (BOURC-TSC) to advise on taxonomic issues relating to the species that form the British List. This advice usually takes the form of recommendations relating to the status of a species or sub-species which sometimes results in 'splitting' (creating two or more species from a single species) and 'lumping' (creating a single species from two or more).

At the end of 2002, BOURC-TSC recommended that the order of species on the British List be changed as it accepted the most likely hypotheses for bird evolution stemmed from the following key characters:

1. That the deepest branch point in the evolutionary tree of birds splits them into the Palaeognathae (tinamous and 'ratites') and the Neognathae (all other birds).

2. That within the Neognathae, the deepest branch-point splits them into Galloanserae (composed of two 'sister' groups – Anseriformes (waterfowl) and Galliformes (turkeys, guineafowl, megapodes, grouse, pheasants etc) and Neoaves (all remaining birds).

3. The World list would therefore start with Palaeognathae, but because only Neognathae occur in Britain, the new British List starts with the Galloanserae, as the deepest split from all other birds (Neoaves).

Within the Galloanserae there are fewer species of Anseriformes than Galliformes, therefore Anseriformes are listed first in accordance with normal custom. The orders of families within these groups remains unchanged, so the British List now starts with Anatidae (swans, ducks, geese), followed by Tetraonidae and Phasianidae (grouse, pheasants, quail and partridges), followed by all remaining families as in the old order (divers, grebes etc).

These recommendations have been accepted by the British Ornithologists' Union who have advised all book, magazine and bird report editors and publishers to begin using the new order as soon as possible and preferably no later than the publication of reports covering the year 2003.

Martin Collinson & Steve Dudley - British Ornithologists' Union

SPECIES, CATEGORIES, CODES AND GUIDE TO USE

Species list
The charts include all species from categories A, B and C on the British List and is based on the latest BOU listing. Selected species included in categories D and E are listed separately at the end of the log chart.

Vagrants which are not on the British List, but which may have occurred in other parts of the British Isles, are not included. Readers who wish to record such species may use the extra rows provided on the last page. In this connection it should be noted that separate lists exist for Northern Ireland (kept by the Northern Ireland Birdwatchers' Association) and the Isle of Man (kept by the Manx Ornithological Society), and that Irish records are assessed by the Irish Rare Birds Committee.

The commoner speces in the log charts are indicated by the • symbol to help make record-keeping easier.

Taxonomic changes introduced in 2002 mean there is a new order of species (as outlined above). The species names are those most widely used in the current fieldguides (with some proposed changes shown in parentheses); each is followed by its scientific name, printed in italics.

Species categories
The following categories are those assigned by the British Ornithologists' Union.

A Species which have been recorded in an apparently natural state at least once since January 1, 1950.

LOG CHARTS

B Species which would otherwise be in Category A but have not been recorded since December 31, 1949.

C Species that, although originally introduced by man, either deliberately or accidentally, have established breeding populations derived from introduced stock that maintain themselves without necessary recourse to further introduction. (This category has been subdivided to differentiate between various groups of naturalised species, but these subdivisions are outside the purpose of the log charts).

D Species that would otherwise appear in Categories A or B except that there is reasonable doubt that they have ever occurred in a natural state. (Species in this category are included in the log charts, though they do not qualify for inclusion in the British List, which comprises species in Categories A, B and C only. One of the objects of Category D is to note records of species which are not yet full additions, so that they are not overlooked if acceptable records subsequently occur. Bird report editors are encouraged to include records of species in Category D as appendices to their systematic lists).

E Species that have been recorded as introductions, transportees or escapees from captivity, and whose populations (if any) are thought not to be self-sustaining. They do not form part of the British List.

E U Species not on the British List, or in Category D, but which either breed or occur regularly elsewhere in Europe.

Life list
Ticks made in the 'Life List' column suffice for keeping a running personal total of species. However, added benefit can be obtained by replacing ticks with a note of the year of first occurrence. To take an example: one's first-ever Marsh Sandpiper, seen on April 14, 2006, would be logged with '06' in the Life List and '14' in the April column (as well as a tick in the 2006 column). As Life List entries are carried forward annually, in years to come it would be a simple matter to relocate this record.

First and last dates of migrants
Arrivals of migrants can be recorded by inserting dates instead of ticks in the relevant month columns. For example, a Common Sandpiper on March 11 would be recorded by inserting '11' against Common Sandpiper in the March column. The same applies to departures, though dates of last sightings can only be entered at the end of the year after checking one's field notebook.

Unheaded columns
The three unheaded columns at the right hand end of each chart are for special (personal) use. This may be, for example, a, second holiday, a particular county or a 'local patch'. Another use could be to indicate species on, for example, the Northern Ireland List or the Isle of Man List.

BTO species codes
British Trust for Ornithology two-letter species codes are shown in brackets in the fourth column from the right. They exist for many species, races and hybrids recorded in recent surveys. Readers should refer to the BTO if more codes are needed. In addition to those given in the charts, the following are available for some well-marked races or forms - Whistling Swan (WZ), European White-fronted Goose (EW), Greenland White-fronted Goose (NW), dark-bellied Brent Goose (DB), pale-bellied Brent Goose (PB), Black Brant (BB), domestic goose (ZL), Green-winged Teal (TA), domestic duck (ZF), Yellow-legged Gull (YG), Kumlien's Gull (KG), Feral Pigeon (FP), White Wagtail (WB), Black-bellied Dipper (DJ), Hooded Crow (HC), intermediate crow (HB).

Rarities
Rarities are indicated by a capital letter 'R' immediately preceding the 'Euring No.' column.

EURING species numbers
EURING species numbers are given in the last column. As they are taken from the full Holarctic bird list there are many apparent gaps. It is important that these are not filled arbitrarily by observers wishing to record species not listed in the charts, as this would compromise the integrity of the scheme. Similarly, the addition of a further digit to indicate sub-species is to be avoided, since EURING has already assigned numbers for this purpose. The numbering follows the Voous order of species.

Rare breeding birds
Species monitored by the Rare Breeding Birds Panel (see National Directory) comprise all those on Schedule 1 of the Wildlife and Countryside Act 1981 (see Quick Reference) together with all escaped or introduced species breeding in small numbers. The following annotations in the charts (third column from the right) reflect the RBBP's categories:

(b)^A Rare species. All breeding details requested.
(b)^B Less scarce species. Totals requested from counties with more than 10 pairs or localities; elsewhere all details requested.
(b)^C Less scarce species (specifically Barn Owl, Kingfisher, Crossbill). County summaries requested.
(b)^D Escaped or introduced species. Treated as less scarce species.

65

SWANS, GEESE, DUCKS

			Life list	2006 list	24 hr	Garden	Holiday	Jan	Feb	Mar	Apr	May	Jun	Jul	Aug	Sep	Oct	Nov	Dec		BTO	RBBP	Bou	EU No
• AC	Mute Swan	Cygnus olor																			MS			0152
• A	Bewick's (Tundra) Swan	C. columbianus																			BS			0153
• A	Whooper Swan	C. cygnus																			WS	bAD		0154
• A	Bean Goose	Anser fabalis																			BE	bD		0157
• A	Pink-footed Goose	A. brachyrhynchus																			PG	bAD		0158
• A	White-fronted Goose	A. albifrons																			WG	bD		0159
• A	Lesser White-fr Goose	A. erythropus																			LC	bB	R	0160
• AC	Greylag Goose	A. anser																			GJ			0161
• A	Snow Goose	A. caerulescens																			SJ	bD		0163
• AC	Canada Goose	Branta canadensis																			CG			0166
• A	Barnacle Goose	B. leucopsis																			BY	bD		0167
• A	Brent Goose	B. bernicla																			BG	bD		0168
• A	Red-breasted Goose	B. ruficollis																			EB	bB	R	0169
• C	Egyptian Goose	Alopochen aegyptiacus																			EG	bD		0170
• B	Ruddy Shelduck	Tadorna ferruginea																			UD	bD		0171
• A	Shelduck	T. tadorna																			SU			0173
• C	Mandarin Duck	Aix galericulata																			MN			0178
• A	Wigeon	Anas penelope																			WN	bB		0179
• A	American Wigeon	A. americana																			AW		R	0180
• AC	Gadwall	A. strepera																			GA	bB		0182
• A	Eurasian Teal	A. crecca																			T			0184
• A	Green-winged Teal	A. carolinensis																						
• AC	Mallard	A. platyrhynchos																			MA			0186
• A	American Black Duck	A. rubripes																			BD		R	0187

Sub-total

66

DUCKS continued

	Species	Scientific	Life list	2006 list	24 hr	Garden	Holiday	Jan	Feb	Mar	Apr	May	Jun	Jul	Aug	Sep	Oct	Nov	Dec	BTO	RBBP	Bou	EU No
• A	Pintail	A. acuta																		PT	bA		0189
• A	Garganey	A. querquedula																		GY	bA		0191
• A	Blue-winged Teal	A. discors																		TB	bB	R	0192
• A	Shoveler	A. clypeata																		SV			0194
• A	Red-crested Pochard	Netta rufina																		RQ	bD		0196
• A	Canvasback	Aythya valisineria																				R	0197
• A	Pochard	A. ferina																		PO	bB		0198
• A	Redhead	A. americana																		AZ		R	0199
• A	Ring-necked Duck	A. collaris																		NG			0200
• A	Ferruginous Duck	A. nyroca																		ED			0202
• A	Tufted Duck	A. fuligula																		TU			0203
• A	Scaup	A. marila																		SP	bA		0204
• A	Lesser Scaup	A. affinis																		AY		R	0205
• A	Eider	Somateria mollissima																		E			0206
• A	King Eider	S. spectabilis																		KE		R	0207
• A	Steller's Eider	Polysticta stelleri																		ES		R	0209
• A	Harlequin	Histrionicus histrionicus																		HQ		R	0211
• A	Long-tailed Duck	Clangula hyemalis																		LN	bA		0212
• A	Common Scoter	Melanitta nigra																		CX	bA		0213
• A	Surf Scoter	M. perspicillata																		FS			0214
• A	Velvet Scoter	M. fusca																		VS			0215
• A	Bufflehead	Bucephala albeola																		VH			0216
• A	Barrow's Goldeneye	B. islandica																				R	0217
• A	Goldeneye	B. clangula																		GN	bAD		0218

Sub-total

67

DUCKS, GAMEBIRDS, DIVERS, GREBES

	Species	Scientific	Life list	2006 list	24 hr	Garden	Holiday	Jan	Feb	Mar	Apr	May	Jun	Jul	Aug	Sep	Oct	Nov	Dec	BTO	RBBP	Bou	EU No
A	Smew	Mergellus albellus																		SY			0220
A	Red-breasted Merganser	Mergus serrator																		RM			0221
A	Goosander	M. merganser																		GD			0223
C	Ruddy Duck	Oxyura jamaicensis																		BY			0225
A	Red (Willow) Grouse	Lagopus lagopus																		RG			0329
A	Ptarmigan	L. mutus																		PM			0330
A	Black Grouse	Tetrao tetrix																		BK			0332
BC	Capercaillie	T. urogallus																		CP			0335
C	Red-legged Partridge	A. rufa																		RL			0358
AC	Grey Partridge	Perdix perdix																		P			0367
A	Quail	Coturnix coturnix																		Q	bB		0370
C	Pheasant	Phasianus colchicus																		PH			0394
C	Golden Pheasant	Chrysolophus pictus																		GF	bD		0396
C	Lady Amherst's Pheasant	C. amherstiae																		LM	bD		0397
A	Red-throated Diver	Gavia stellata																		RH	bB		0002
A	Black-throated Diver	G. arctica																		BV	bA		0003
A	Great Northern Diver	G. immer																		ND			0004
A	White-(Yellow)billed Diver	G. adamsii																		IW		R	0005
A	Pied-billed Grebe	Podilymbus podiceps																		PJ		R	0006
A	Little Grebe	Tachybaptus ruficollis																		LG			0007
A	Great Crested Grebe	Podiceps cristatus																		GG			0009
A	Red-necked Grebe	P. grisegena																		RX	bA		0010
A	Slavonian Grebe	P. auritus																		SZ	bA		0011
A	Black-necked Grebe	P. nigricollis																		BN	bA		0012
Sub-total																							

ALBATROSS, FULMAR, PETRELS, SHEARWATERS, CORMORANTS, BITTERN

	Species	Scientific name	Life list	2006 list	24 hr	Garden	Holiday	Jan	Feb	Mar	Apr	May	Jun	Jul	Aug	Sep	Oct	Nov	Dec	BTO	RBBP	Bou	EU No
A	Black-browed Albatross	Thalassarche melanophris																		AA		R	0014
•A	Fulmar	Fulmarus glacialis																		F			0020
A	*'Soft-plumaged Petrel'	Pterodroma mollis/madeira/feae																				R	0026
B	Capped Petrel	Pterodroma hasitata																				R	0029
B	Bulwer's Petrel	Bulweria bulwerii																				R	0034
A	Cory's Shearwater	Calonectris diomedea																		CQ			0036
A	Great Shearwater	Puffinus gravis																		GQ			0040
•A	Sooty Shearwater	P. griseus																		OT			0043
•A	Manx Shearwater	P. Puffinus																		MX			0046
A	Mediterranean Shearwater	P. mauretanicus																					0046
A	Little Shearwater	P. assimilis																				R	0048
•A	Wilson's Petrel	Oceanites oceanicus																				R	0050
B	White-faced Petrel	Pelagodroma marina																				R	0051
•A	Storm Petrel	Hydrobates pelagicus																		TM			0052
•A	Leach's Petrel	Oceanodroma leucorhoa																		TL	bB		0055
•A	Swinhoe's Petrel	O. monorhis																				R	0056
B	Madeiran Petrel	O. castro																				R	0058
A	Red-billed Tropicbird	Phaethon aethereus																					
•A	Gannet	Morus bassanus																		GX			0071
•A	Cormorant	Phalacrocorax carbo																		CA			0072
A	Double-crested Cormorant	P. auritus																				R	0078
•A	Shag	P. aristotelis																		SA			0080
A	Ascension Frigatebird	Fregata aquila																				R	
•A	Bittern	Botaurus stellaris																		BI	bA		0095
	Sub-total																						

*Alternative sub-species

69

HERONS, STORKS, SPOONBILL, RAPTORS

	Common Name	Scientific Name	Life list	2006 list	24 hr	Garden	Holiday	Jan	Feb	Mar	Apr	May	Jun	Jul	Aug	Sep	Oct	Nov	Dec	BTO	RBBP	Bou	EU No
A	American Bittern	B. lentiginosus																		AM		R	0096
A	Little Bittern	Ixobrychus minutus																		LL		R	0098
A	Night Heron	Nycticorax nycticorax																		NT	bAD	R	0104
A	Green Heron	Butorides virescens																		HR		R	0107
A	Squacco Heron	Ardeola ralloides																		QH		R	0108
A	Cattle Egret	Bubulcus ibis																		EC		R	0111
A	Little Egret	Egretta garzetta																		ET	bA		0119
• A	Great White Egret	Ardea alba																		HW		R	0121
• A	Grey Heron	A. cinerea																		H			0122
A	Purple Heron	A. purpurea																		UR			0124
A	Black Stork	Ciconia nigra																		OS		R	0131
A	White Stork	C. ciconia																		OR			0134
A	Glossy Ibis	Plegadis falcinellus																		IB			0136
• A	Spoonbill	Platalea leucorodia																		NB	bA		0144
• A	Honey Buzzard	Pernis apivorus																		HZ	bA		0231
• A	Black Kite	Milvus migrans																		KB		R	0238
• AC	Red Kite	M. milvus																		KT	bA		0239
• A	White-tailed Eagle	Haliaeetus albicilla																		WE	bA		0243
BD	Egyptian Vulture	Neophron percnopterus																				R	0247
A	Short-toed Eagle	Circaetus gallicus																					0256
• A	Marsh Harrier	Circus aeruginosus																		MR	bA		0260
• A	Hen Harrier	C. cyaneus																		HH	bB		0261
A	Pallid Harrier	C. macrourus																				R	0262
• A	Montagu's Harrier	C. pygargus																		MO	bA		0263
	Sub-total																						

RAPTORS, RAILS AND CRAKES

	English	Scientific	Life list	2006 list	24 hr	Garden	Holiday	Jan	Feb	Mar	Apr	May	Jun	Jul	Aug	Sep	Oct	Nov	Dec	BTO	RBBP	BOU	EU No
• AC	Goshawk	Accipiter gentilis																		GI	bB		0267
• A	Sparrowhawk	A. nisus																		SH			0269
• A	Buzzard	Buteo buteo																		BZ			0287
• A	Rough-legged Buzzard	B. lagopus																		RF		R	0290
B	Greater Spotted Eagle	A. clanga																					0293
• A	Golden Eagle	A. chrysaetos																		EA	bB		0296
• A	Osprey	Pandion haliaetus																		OP	bA		0301
A	Lesser Kestrel	Falco naumanni																				R	0303
• A	Kestrel	F. tinnunculus																		K			0304
A	American Kestrel	F. sparverius																				R	0305
• A	Red-footed Falcon	F. vespertinus																		FV		R	0307
• A	Merlin	F. columbarius																		ML	bB		0309
• A	Hobby	F. subbuteo																		HY	bB		0310
A	Eleonora's Falcon	F. eleonorae																				R	0311
• A	Gyrfalcon	F. rusticolus																		YF		R	0318
• A	Peregrine	F. peregrinus																		PE	bB		0320
• A	Water Rail	Rallus aquaticus																		WA			0407
• A	Spotted Crake	Porzana porzana																		AK	bA		0408
A	Sora	P. carolina																				R	0409
A	Little Crake	P. parva																		JC		R	0410
A	Baillon's Crake	P. pusilla																		VC		R	0411
• A	Corn Crake	Crex crex																		CE	bA		0421
• A	Moorhen	Gallinula chloropus																		MH			0424
B	Allen's Gallinule	Porphyrula alleni																				R	0425
Sub-total																							

71

GALLINULES AND WADERS

			Life list	2006 list	24 hr	Garden	Holiday	Jan	Feb	Mar	Apr	May	Jun	Jul	Aug	Sep	Oct	Nov	Dec	BTO	RBBP	Bou	EU No
A	American Purple Gallinule	P. martinica																				R	0426
• A	Coot	Fulica atra																		CO		R	0429
A	American Coot	F. americana																				R	0430
• A	Crane	Grus grus																		AN	bA		0433
A	Sandhill Crane	G. canadensis																				R	0436
A	Little Bustard	Tetrax tetrax																				R	0442
C	Houbara Bustard	Chlamydotis undulata																				R	0444
B	Macqueen's Bustard	Chlamydotis macqueenii																				R	0444
A	Great Bustard	Otis tarda																		OC		R	0446
• A	Oystercatcher	Haematopus ostralegus																		OC			0450
A	Black-winged Stilt	Himantopus himantopus																		IT		R	0455
• A	Avocet	Recurvirostra avosetta																		AV	bA		0456
• A	Stone Curlew	Burhinus oedicnemus																		TN	bA		0459
• A	Cream-coloured Courser	Cursorius cursor																				R	0464
A	Collared Pratincole	Glareola pratincola																				R	0465
A	Oriental Pratincole	G. maldivarum																		GM		R	0466
A	Black-winged Pratincole	G. nordmanni																		KW		R	0467
• A	Little Ringed Plover	Charadrius dubius																		LP	bB		0469
• A	Ringed Plover	C. hiaticula																		RP			0470
• A	Semipalmated Plover	C. semipalmatus																		TV		R	0471
A	Killdeer	C. vociferus																		KL		R	0474
A	Kentish Plover	C. alexandrinus																		KP			0477
A	Lesser Sand Plover	C. mongolus																				R	0478
A	Caspian Plover	C. asiaticus																				R	0480
Sub-total																							

WADERS continued

Tracking columns (Life list, 2006 list, 24 hr, Garden, Holiday, Jan, Feb, Mar, Apr, May, Jun, Jul, Aug, Sep, Oct, Nov, Dec) are blank for all rows.

	Species	Scientific name	BTO	RBBP	Bou	EU No
• A	Dotterel	C. morinellus	DO	bB		0482
A	American Golden Plover	Pluvialis dominica	ID		R	0484
A	Pacific Golden Plover	P. fulva	IF		R	0484
• A	Golden Plover	P. apricaria	GP			0485
• A	Grey Plover	P. squatarola	GV			0486
A	Sociable Lapwing	Vanellus gregarius	IP		R	0491
A	White-tailed Lapwing	V. leucurus			R	0492
• A	Lapwing	V. vanellus	L			0493
A	Great Knot	Calidris tenuirostris	KO		R	0495
• A	Knot	C. canutus	KN			0496
• A	Sanderling	C. alba	SS			0497
A	Semipalmated Sandpiper	C. pusilla	PZ		R	0498
A	Western Sandpiper	C. mauri	ER		R	0499
A	Red-necked Stint	C. ruficollis			R	0500
• A	Little Stint	C. minuta	LX			0501
• A	Temminck's Stint	C. temminckii	TK	bA		0502
A	Long-toed Stint	C. subminuta			R	0503
A	Least Sandpiper	C. minutilla	EP		R	0504
A	White-rumped Sandpiper	C. fuscicollis	WU		R	0505
A	Baird's Sandpiper	C. bairdii	BP		R	0506
A	Pectoral Sandpiper	C. melanotos	PP		R	0507
A	Sharp-tailed Sandpiper	C. acuminata	W			0508
• A	Curlew Sandpiper	C. ferruginea	CV		R	0509
• A	Purple Sandpiper	C. maritima	PS	bA		0510
Sub-total						

WADERS continued

	Species	Scientific name	Life list	2006 list	24 hr	Garden	Holiday	Jan	Feb	Mar	Apr	May	Jun	Jul	Aug	Sep	Oct	Nov	Dec			BTO	RBBP	Bou	EU No
•A	Dunlin	C. alpina																				DN			0512
A	Broad-billed Sandpiper	Limicola falcinellus																				OA		R	0514
A	Stilt Sandpiper	Micropalama himantopus																				MI		R	0515
A	Buff-breasted Sandpiper	Tryngites subruficollis																				BQ			0516
•A	Ruff	Philomachus pugnax																				RU	bA		0517
•A	Jack Snipe	Lymnocryptes minimus																				JS			0518
•A	Snipe	Gallinago gallinago																				SN			0519
A	Great Snipe	G. media																				DS		R	0520
A	Short-billed Dowitcher	Limnodromus griseus																				LD		R	0527
A	Long-billed Dowitcher	L. scolopaceus																				WK			0529
•A	Woodcock	Scolopax rusticola																				BW	bA		0532
•A	Black-tailed Godwit	Limosa limosa																				HU		R	0533
A	Hudsonian Godwit	L. haemastica																				BA			0534
A	Bar-tailed Godwit	L. lapponica																						R	0536
A	Little Whimbrel (Curlew)	N. minutus																						R	0537
B	Eskimo Curlew	N. borealis																				WM	bA		0538
•A	Whimbrel	N. phaeopus																				CU			0541
•A	Curlew	N. arquata																				UP		R	0544
A	Upland Sandpiper	Bartramia longicauda																				DR			0545
•A	Spotted Redshank	T. erythropus																				RK			0546
•A	Redshank	T. totanus																				MD		R	0547
A	Marsh Sandpiper	T. stagnatilis																				GK	bB		0548
•A	Greenshank	T. nebularia																				LZ		R	0550
A	Greater Yellowlegs	T. melanoleuca																							

Sub-total

74

WADERS continued, SKUAS, GULLS

	Species	Scientific name	Life list	2006 list	24 hr	Garden	Holiday	Jan	Feb	Mar	Apr	May	Jun	Jul	Aug	Sep	Oct	Nov	Dec		BTO	RBBP	Bou	EU No
A	Lesser Yellowlegs	T. flavipes																			LY		R	0551
A	Solitary Sandpiper	T. solitaria																			I		R	0552
•A	Green Sandpiper	T. ochropus																			GE			0553
•A	Wood Sandpiper	T. glareola																			OD	bA		0554
A	Terek Sandpiper	Xenus cinereus																			TR		R	0555
•A	Common Sandpiper	Actitis hypoleucos																			CS			0556
A	Spotted Sandpiper	A. macularia																			PQ		R	0557
A	Grey-tailed Tattler	Heteroscelus brevipes																			YT		R	0558
•A	Turnstone	Arenaria interpres																			TT			0561
•A	Wilson's Phalarope	Phalaropus tricolor																			WF		R	0563
•A	Red-necked Phalarope	P. lobatus																			NK	bA		0564
•A	Grey Phalarope	P. fulicarius																			PL			0565
•A	Pomarine Skua	Stercorarius pomarinus																			PK			0566
•A	Arctic Skua	S. parasiticus																			AC			0567
•A	Long-tailed Skua	S. longicaudus																			OG			0568
•A	Great Skua	Catharacta skua																			NX			0569
B	Great Black-headed (Pallas's) Gull	L. ichthyaetus																					R	0573
•A	Mediterranean Gull	L. melanocephalus																			MU	bA		0575
A	Laughing Gull	L. atricilla																			LF		R	0576
A	Franklin's Gull	L. pipixcan																			FG		R	0577
•A	Little Gull	L. minutus																			LU			0578
•A	Sabine's Gull	L. sabini																			AB			0579
A	Bonaparte's Gull	L. philadelphia																			ON		R	0581
•A	Black-headed Gull	L. ridibundus																			BH		R	0582
	Sub-total																							

75

GULLS AND TERNS

	Species	Scientific	Life list	2006 list	24 hr	Garden	Holiday	Jan	Feb	Mar	Apr	May	Jun	Jul	Aug	Sep	Oct	Nov	Dec	BTO	RBBP	Bou	EU No
A	Slender-billed Gull	L. genei																		EI			0585
• A	Ring-billed Gull	L. delawarensis																		IN			0589
A	Common (Mew) Gull	L. canus																		CM			0590
• A	Lesser Black-backed Gull	L. fuscus																		LB			0591
• A	Herring Gull	L. argentatus																		HG			0592
• A	Iceland Gull	L. glaucoides																		IG			0598
• A	Glaucous Gull	L. hyperboreus																		GZ			0599
• A	Great Black-backed Gull	L. marinus																		GB			0600
A	Ross's Gull	Rhodostethia rosea																		QG		R	0601
• A	Kittiwake	Rissa tridactyla																		KI			0602
A	Ivory Gull	Pagophila eburnea																		IV		R	0604
• A	Gull-billed Tern	S.nilotica																		TG		R	0605
• A	Caspian Tern	S.caspia																		CJ		R	0606
A	Royal Tern	S.maxima																		QT		R	0607
A	Lesser Crested Tern	S.bengalensis																		TF	bA	R	0609
• A	Sandwich Tern	S.sandvicensis																		TE			0611
• A	Roseate Tern	S.dougallii																		RS	bA		0614
• A	Common Tern	S.hirundo																		CN			0615
• A	Arctic Tern	S.paradisaea																		AE			0616
A	Aleutian Tern	S.aleutica																				R	0617
A	Forster's Tern	S.forsteri																		FO		R	0618
A	Bridled Tern	S.anaethetus																				R	0622
A	Sooty Tern	S.fuscata																				R	0623
• A	Little Tern	S.albifrons																		AF	bB		0624

Sub-total

LOG CHARTS

TERNS continued, AUKS, DOVES, CUCKOOS AND BARN OWL

			Life list	2006 list	24 hr	Garden	Holiday	Jan	Feb	Mar	Apr	May	Jun	Jul	Aug	Sep	Oct	Nov	Dec	BTO	RBBP	Bou	EU No
A	Whiskered Tern	Chlidonias hybrida																		WD		R	0626
A	Black Tern	C. niger																		BJ			0627
A	White-winged Black Tern	C. leucopterus																		WJ		R	0628
A	Guillemot	Uria aalge																		GU			0634
A	Brünnich's Guillemot	U. lomvia																		TZ		R	0635
A	Razorbill	Alca torda																		RA			0636
A	Black Guillemot	Cepphus grylle																		TY			0638
A	Ancient Murrelet	Synthliboramphus antiquus																				R	0645
A	Little Auk	Alle alle																		LK			0647
A	Puffin	Fratercula arctica																		PU			0654
A	Pallas's Sandgrouse	Syrrhaptes paradoxus																				R	0663
AC	Rock Dove	Columba livia																		DV			0665
A	Stock Dove	C. oenas																		SD			0668
A	Woodpigeon	C. palumbus																		WP			0670
A	Collared Dove	Streptopelia decaocto																		CD			0684
A	Turtle Dove	S. turtur																		TD			0687
A	Rufous (Oriental) Turtle Dove	S. orientalis																				R	0689
A	Mourning Dove	Zenaida macroura																				R	0695
C	Rose-ringed Parakeet	Psittacula krameri																		RI	bD		0712
A	Great Spotted Cuckoo	Clamator glandarius																		UK		R	0716
A	Cuckoo	Cuculus canorus																		CK			0724
A	Black-billed Cuckoo	Coccyzus erythrophthalmus																				R	0727
A	Yellow-billed Cuckoo	C. americanus																				R	0728
A	Barn Owl	Tyto alba																		BO	bC		0735

Sub-total

77

OWLS, NIGHTJARS, SWIFTS, KINGFISHERS AND BEE-EATERS

	Species	Scientific name	Life list	2006 list	24 hr	Garden	Holiday	Jan	Feb	Mar	Apr	May	Jun	Jul	Aug	Sep	Oct	Nov	Dec		BTO	RBBP	Bou	EU No
A	Scops Owl	Otus scops																			SO		R	0739
• A	Snowy Owl	Nyctea scandiaca																				bA	R	0749
A	Hawk Owl	Surnia ulula																					R	0750
• C	Little Owl	Athene noctua																			LO			0757
• A	Tawny Owl	Strix aluco																			TO			0761
• A	Long-eared Owl	Asio otus																			LE			0767
• A	Short-eared Owl	A. flammeus																			SE			0768
A	Tengmalm's Owl	Aegolius funereus																					R	0770
• A	Nightjar	Caprimulgus europaeus																			NJ			0778
B	Red-necked Nightjar	C. ruficollis																					R	0779
A	Egyptian Nightjar	C. aegyptius																					R	0781
A	Common Nighthawk	Chordeiles minor																					R	0786
A	Chimney Swift	Chaetura pelagica																					R	0790
A	White-throated Needletail	Hirundapus caudacutus																			NI		R	0792
• A	Swift	Apus apus																			SI			0795
A	Pallid Swift	A. pallidus																					R	0796
A	Pacific Swift	A. pacificus																					R	0797
A	Alpine Swift	A. melba																			AI		R	0798
A	Little Swift	A. affinis																					R	0800
• A	Kingfisher	Alcedo atthis																			KF	bC		0831
A	Belted Kingfisher	Ceryle alcyon																					R	0834
A	Blue-cheeked Bee-eater	Merops superciliosus																					R	0839
• A	Bee-eater	M. apiaster																			MZ			0840
A	Roller	Coracias garrulus																					R	0841
	Sub-total																							

WOODPECKERS, LARKS, SWALLOWS AND PIPITS

	Species	Scientific name	Life list	2006 list	24 hr	Garden	Holiday	Jan	Feb	Mar	Apr	May	Jun	Jul	Aug	Sep	Oct	Nov	Dec	BTO	RBBP	Bou	EU No
A	Hoopoe	Upupa epops																		HP			0846
• A	Wryneck	Jynx torquilla																		WY	bA		0848
• A	Green Woodpecker	P. viridis																		G			0856
A	Yellow-bellied Sapsucker	Sphyrapicus varius																				R	0872
• A	Great Spotted Woodpecker	Dendrocopos major																		GS			0876
• A	Lesser Spotted Woodpecker	D. minor																		LS			0887
A	Eastern Phoebe	Sayornis phoebe																				R	0909
A	Calandra Lark	Melanocorypha calandra																				R	0961
A	Bimaculated Lark	M. bimaculata																				R	0962
A	White-winged Lark	M. leucoptera																				R	0965
A	Short-toed Lark	Calandrella brachydactyla																		VL			0968
A	Lesser Short-toed Lark	C. rufescens																				R	0970
A	Crested Lark	Galerida cristata																				R	0972
• A	Wood Lark	Lullula arborea																		WL	bB		0974
• A	Sky Lark	Alauda arvensis																		S			0976
• A	Shore (Horned) Lark	Eremophila alpestris																		SX			0978
A	Sand Martin	Riparia riparia																		SM			0981
• A	Tree Swallow	Tachycineta bicolor																				R	0983
A	Crag Martin	Ptyonoprogne rupestris																				R	0991
• A	Swallow	Hirundo rustica																		SL			0992
A	Red-rumped Swallow	H. daurica																		VR		R	0995
A	Cliff Swallow	H. pyrrhonota																				R	0998
• A	House Martin	Delichon urbica																		HM			1001
A	Richard's Pipit	Anthus novaeseelandiae																		PR			1002

Sub-total

79

PIPITS, WAGTAILS, WAXWINGS AND CHATS

	English name	Scientific name	Life list	2006 list	24 hr	Garden	Holiday	Jan	Feb	Mar	Apr	May	Jun	Jul	Aug	Sep	Oct	Nov	Dec		BTO	RBBP	Bou	EU No
A	Blyth's Pipit	A. godlewskii																					R	1004
A	Tawny Pipit	A. campestris																			TI			1005
A	Olive-backed Pipit	A. hodgsoni																			OV		R	1008
• A	Tree Pipit	A. trivialis																			TP			1009
A	Pechora Pipit	A. gustavi																					R	1010
• A	Meadow Pipit	A. pratensis																			MP			1011
A	Red-throated Pipit	A. cervinus																			VP		R	1012
• A	Rock Pipit	A. petrosus																			RC			1014
• A	Water Pipit	A. spinoletta																			WI			1014
A	Buff-bellied Pipit	A. rubescens																					R	1014
• A	Yellow Wagtail	Motacilla flava																			YW			1017
A	Citrine Wagtail	M. citreola																					R	1018
• A	Grey Wagtail	M. cinerea																			GL			1019
• A	Pied (White) Wagtail	M. alba																			PW			1020
A	Cedar Waxing	Bombycilla cedrorum																					R	1046
• A	(Bohemian) Waxwing	Bombycilla garrulus																			WX			1048
• A	Dipper	Cinclus cinclus																			DI			1050
• A	Wren	Troglodytes troglodytes																			WR			1066
A	Northern Mockingbird	Mimus polyglottos																					R	1067
A	Brown Thrasher	Toxostoma rufum																					R	1069
A	Gray Catbird	Dumetella carolinensis																						
• A	Dunnock	Prunella modularis																			D			1084
A	Alpine Accentor	P. collaris																					R	1094
A	Rufous-tailed Scrub Robin	Cercotrichas galactotes																					R	1095

Sub-total

CHATS, WHEATEARS AND THRUSHES

		Species	Scientific name	Life list	2006 list	24 hr	Garden	Holiday	Jan	Feb	Mar	Apr	May	Jun	Jul	Aug	Sep	Oct	Nov	Dec	BTO	RBBP	Bou	EU No
•	A	Robin	Erithacus rubecula																		R			1099
	A	Thrush Nightingale	Luscinia luscinia																		FN		R	1103
•	A	Nightingale	L. megarhynchos																		N			1104
	A	Siberian Rubythroat	L. calliope																				R	1105
	A	Bluethroat	L. svecica																		BU			1106
	A	Siberian Blue Robin	L. cyane																					
	A	Red-flanked Bluetail	Tarsiger cyanurus																				R	1113
	A	White-throated Robin	Irania gutturalis																				R	1117
•	A	Black Redstart	Phoenicurus ochruros																		BX	bA		1121
•	A	Redstart	P. phoenicurus																		RT			1122
	A	Moussier's Redstart	P. moussieri																				R	1127
•	A	Whinchat	Saxicola rubetra																		WC			1137
	A	Stonechat	S. torquata																		SC			1139
	A	Isabelline Wheatear	Oenanthe isabellina																				R	1144
•	A	Wheatear (Northern)	O. oenanthe																		W			1146
	A	Pied Wheatear	O. pleschanka																		PI		R	1147
	A	Black-eared Wheatear	O. hispanica																				R	1148
	A	Desert Wheatear	O. deserti																				R	1149
	A	White-crowned(-tailed) Black Wheatear O. leucopyga																						1157
	A	Rock Thrush	Monticola saxatilis																		OH		R	1162
	A	Blue Rock Thrush	M. solitarius																				R	1166
	A	White's Thrush	Zoothera dauma																				R	1170
	A	Siberian Thrush	Z. sibirica																				R	1171
	A	Varied Thrush	Z. naevia																		VT		R	1172

Sub-total

81

THRUSHES AND WARBLERS

	Common Name	Scientific Name	Life list	2006 list	24 hr	Garden	Holiday	Jan	Feb	Mar	Apr	May	Jun	Jul	Aug	Sep	Oct	Nov	Dec	BTO	RBBP	Bou	EU No
A	Wood Thrush	Hylocichla mustelina																				R	1175
A	Hermit Thrush	Catharus guttatus																				R	1176
A	Swainson's Thrush	C. ustulatus																				R	1177
A	Grey-cheeked Thrush	C. minimus																				R	1178
A	Veery	C. fuscescens																				R	1179
• A	Ring Ouzel	Turdus torquatus																		RZ			1186
• A	Blackbird	T. merula																		B			1187
A	Dusky Thrush	T. naumanni																				R	1196
A	Dark-throated Thrush	T. ruficollis																		XC		R	1197
• A	Fieldfare	T. pilaris																		FF	bA		1198
• A	Song Thrush	T. philomelos																		ST			1200
• A	Redwing	T. iliacus																		RE	bA		1201
• A	Mistle Thrush	T. viscivorus																		M			1202
• A	American Robin	T. migratorius																		AR		R	1203
• A	Cetti's Warbler	Cettia cetti																		CW	bA		1220
A	Zitting Cisticola (Fan-tailed Warbler)	Cisticola juncidis																				R	1226
A	Pallas's Grasshopper Warbler	Locustella certhiola																				R	1233
A	Lanceolated Warbler	L.lanceolata																				R	1235
• A	Grasshopper Warbler	L.naevia																		GH			1236
A	River Warbler	L.fluviatilis																		VW		R	1237
A	Savi's Warbler	L.luscinioides																		VI	bA	R	1238
A	Moustached Warbler	Acrocephalus melanopogon																				R	1241
A	Aquatic Warbler	A. paludicola																		AQ			1242
• A	Sedge Warbler	A. schoenobaenus																		SW			1243
	Sub-total																						

WARBLERS continued

	Species	Scientific name	BTO	RBBP	BOU	EU No
A	Paddyfield Warbler	A. agricola	PY		R	1247
A	Blyth's Reed Warbler	A. dumetorum			R	1248
• A	Marsh Warbler	A. palustris	MW	bA		1250
• A	Reed Warbler	A. scirpaceus	RW			1251
A	Great Reed Warbler	A. arundinaceus	QW		R	1253
A	Thick-billed Warbler	A. aedon			R	1254
A	Eastern Olivaceous Warbler	Hippolais pallida			R	1255
A	Western Olivaceous Warbler	Hippolais opaca			R	
A	Booted Warbler	H. caligata				1256
A	Syke's Warbler	H. rama				
A	Icterine Warbler	H. icterina	IC			1259
A	Melodious Warbler	H. polyglotta	ME			1260
A	Marmora's Warbler	Sylvia sarda	MM			1261
• A	Dartford Warbler	S. undata	DW	bB		1262
A	Spectacled Warbler	S. conspicillata				1264
A	Subalpine Warbler	S. cantillans				1265
A	Sardinian Warbler	S. melanocephala				1267
A	Rüppell's Warbler	S. rueppelli				1269
A	Desert Warbler	S. nana				1270
A	Orphean Warbler	S. hortensis				1272
• A	Barred Warbler	S. nisoria	RR			1273
• A	Lesser Whitethroat	S. curruca	LW			1274
• A	Whitethroat	S. communis	WH			1275
• A	Garden Warbler	S. borin	GW			1276

Sub-total

(Columns: Life list, 2006 list, 24 hr, Garden, Holiday, Jan, Feb, Mar, Apr, May, Jun, Jul, Aug, Sep, Oct, Nov, Dec)

WARBLERS continued, FLYCATCHERS AND TITS

	Common name	Scientific name	Life list	2006 list	24 hr	Garden	Holiday	Jan	Feb	Mar	Apr	May	Jun	Jul	Aug	Sep	Oct	Nov	Dec	BTO	RBBP	Bou	EU No
• A	Blackcap	S. atricapilla																		BC			1277
A	Greenish Warbler	Phylloscopus trochiloides																		NP			1293
A	Arctic Warbler	P. borealis																		AP			1295
A	Pallas's Warbler	P. proregulus																		PA			1298
• A	Yellow-browed Warbler	P. inornatus																		YB			1300
A	Hume's Leaf Warbler	P. humei																					1300
A	Radde's Warbler	P. schwarzi																					1301
A	Dusky Warbler,	P. fuscatus																		UY		R	1303
A	Western Bonelli's Warbler	P. bonelli																		IW		R	1307
A	Eastern Bonelli's Warbler	P. orientalis																				R	1307
• A	Wood Warbler	P. sibilatrix																		WO			1308
• A	Common Chiffchaff	P. collybita																		CC			1311
A	Iberian Chiffchaff	P. ibericus																				R	1311
• A	Willow Warbler	P. trochilus																		WW			1312
• A	Goldcrest	Regulus regulus																		GC			1314
• A	Firecrest	R. ignicapilla																		FC	bA		1315
• A	Spotted Flycatcher	Muscicapa striata																		FY			1335
• A	Red-breasted Flycatcher	Ficedula parva																					1343
A	Collared Flycatcher	F. albicollis																				R	1348
• A	Pied Flycatcher	F. hypoleuca																		PF			1349
• A	Bearded Tit	Panurus biarmicus																		BR	bB		1364
• A	Long-tailed Tit	Aegithalos caudatus																		LT			1437
• A	Marsh Tit	Parus palustris																		MT			1440
• A	Willow Tit	P. montanus																		WT			1442

Sub-total

TITS, CREEPERS, SHRIKES AND CROWS

	Common name	Scientific name	Life list	2006 list	24 hr	Garden	Holiday	Jan	Feb	Mar	Apr	May	Jun	Jul	Aug	Sep	Oct	Nov	Dec	BTO	RBBP	Bou	EU No
A	Crested Tit	P.cristatus																		CI	bB		1454
A	Coal Tit	P. ater																		CT			1461
A	Blue Tit	P. caeruleus																		BT			1462
A	Great Tit	P. major																		GT			1464
A	Red-breasted Nuthatch	S. canadensis																				R	1472
A	Nuthatch	S. europaea																		NH			1479
A	Wallcreeper	Tichodroma muraria																				R	1482
A	Treecreeper	Certhia familiaris																		TC			1486
A	Short-toed Treecreeper	C. brachydactyla																		TH		R	1487
A	Penduline Tit	Remiz pendulinus																		DT		R	1490
A	Golden Oriole	Oriolus oriolus																		OL	bA		1508
A	Brown Shrike	Lanius cristatus																				R	1513
A	Isabelline Shrike	L. isabellinus																		IL		R	1514
A	Red-backed Shrike	L. collurio																		ED	bA		1515
A	Lesser Grey Shrike	L. minor																				R	1519
A	Great Grey Shrike	L. excubitor																		SR			1520
A	Southern Grey Shrike	L. meridionalis																				R	1520
A	Woodchat Shrike	L. senator																		OO			1523
A	Jay	Garrulus glandarius																		J			1539
A	Magpie	Pica pica																		MG			1549
A	Nutcracker	Nucifraga caryocatactes																		NC		R	1557
A	Chough	P. pyrrhocorax																		CF	bB		1559
A	Jackdaw	Corvus monedula																		JD			1560
A	Rook	C. frugilegus																		RO			1563
	Sub-total																						

CROWS continued, SPARROWS, VIREOS AND FINCHES

	English name	Scientific name	Life list	2006 list	24 hr	Garden	Holiday	Jan	Feb	Mar	Apr	May	Jun	Jul	Aug	Sep	Oct	Nov	Dec	BTO	RBBP	Bou	EU No
• A	Carrion (Hooded) Crow	C. corone																		C			1567
• A	Hooded Crow	C. cornix																					1572
• A	Raven	C. corax																		RN			1582
• A	Starling	S. vulgaris																		SG			1594
• A	Rose-coloured (Rosy) Starling	Sturnus roseus																		OE		R	1591
• A	House Sparrow	Passer domesticus																		HS			1592
• A	Spanish Sparrow	P. hispaniolensis																				R	1598
• A	Tree Sparrow	P.r montanus																		TS			1604
• A	Rock Sparrow	Petronia petronia																				R	1628
• A	Yellow-throated Vireo	Vireo flavifrons																				R	1631
• A	Philadelphia Vireo	V. philadelphicus																				R	1633
• A	Red-eyed Vireo	V.olivaceus																		EV		R	1636
• A	Chaffinch	Fringilla coelebs																		CH			1638
• A	Brambling	F. montifringilla																		BL	bA		1640
• A	Serin	Serinus serinus																		NS	bA		1649
• A	Greenfinch	Carduelis chloris																		GR			1653
• A	Goldfinch	C carduelis																		GO			1654
• A	Siskin	C. spinus																		SK			1660
• A	Linnet	C. cannabina																		LI			1662
• A	Twite	C. flavirostris																		TW			1663
• A	Lesser Redpoll	C. cabaret																		LR			1663
• A	Mealy Redpoll	C. flammea																					1664
• A	Arctic Redpoll	C. hornemanni																		AL		R	1665
• A	Two-barred Crossbill	Loxia leucoptera																		PD		R	

Sub-total

FINCHES continued, NEW WORLD WARBLERS

	Common name	Scientific name	Life list	2006 list	24 hr	Garden	Holiday	Jan	Feb	Mar	Apr	May	Jun	Jul	Aug	Sep	Oct	Nov	Dec		BTO	RBBP	Bou	EU No
• A	Crossbill	L. curvirostra																			CR	bC		1666
• A	Scottish Crossbill	L. scotica																			CY	bB		1667
• A	Parrot Crossbill	L. pytyopsittacus																			PC	bA	R	1668
A	Trumpeter Finch	Bucanetes githagineus																					R	1676
• A	Common Rosefinch	Carpodacus erythrinus																			SQ	bA		1679
A	Pine Grosbeak	Pinicola enucleator																					R	1699
• A	Bullfinch	Pyrrhula pyrrhula																			BF			1710
• A	Hawfinch	Coccothraustes coccothraustes																			HF			1717
A	Evening Grosbeak	Hesperiphona vespertina																					R	1718
A	Black-and-white Warbler	Mniotilta varia																					R	1720
A	Golden-winged Warbler	Vermivora chrysoptera																					R	1722
A	Tennessee Warbler	V. peregrina																					R	1724
A	Northern Parula	Parula americana																					R	1732
A	Yellow Warbler	Dendroica petechia																					R	1733
A	Chestnut-sided Warbler	D. pensylvanica																					R	1734
A	Blackburnian Warbler	D. fusca																					R	1747
A	Cape May Warbler	D. tigrina																					R	1749
A	Magnolia Warbler	D. magnolia																					R	1750
A	Yellow-rumped Warbler	D. coronata																					R	1751
A	Blackpoll Warbler	D. striata																					R	1753
A	Bay-breasted Warbler	D. castanea																					R	1754
A	American Redstart	Setophaga ruticilla																			AD		R	1755
A	Ovenbird	Seiurus aurocapilla																			AD		R	1756
A	Northern Waterthrush	S. noveboracensis																					R	1757

Sub-total

NEW WORLD WARBLERS continued AND BUNTINGS

	Species	Scientific name	Life list	2006 list	24 hr	Garden	Holiday	Jan	Feb	Mar	Apr	May	Jun	Jul	Aug	Sep	Oct	Nov	Dec		BTO	RBBP	Bou	EU No
A	Yellowthroat	Geothlypis trichas																					R	1762
A	Hooded Warbler	Wilsonia citrina																					R	1771
A	Wilson's Warbler	Wilsonia pusilla																					R	1772
A	Summer Tanager	Piranga rubra																					R	1786
A	Scarlet Tanager	P. olivacea																					R	1788
A	Eastern Towhee	Pipilo erythrophthalmus																					R	1798
A	Lark Sparrow	Chondestes grammacus																					R	1824
A	Savannah Sparrow	Passerculus sandwichensis																					R	1826
A	Song Sparrow	Melospiza melodia																					R	1835
A	White-crowned Sparrow	Zonotrichia leucophrys																					R	1839
A	White-throated Sparrow	Z. albicollis																					R	1840
A	Dark-eyed Junco	Junco hyemalis																			JU		R	1842
A	Lapland Bunting	Calcarius lapponicus																			LA			1847
•A	Snow Bunting	Plectrophenax nivalis																			SB	bA		1850
A	Black-faced Bunting	Emberiza spodocephala																					R	1853
A	Pine Bunting	E. leucocephalos																			EL		R	1856
•A	Yellowhammer	E. citrinella																			Y			1857
•A	Cirl Bunting	E. cirlus																			CL	bA		1958
A	Rock Bunting	E. cia																					R	1860
A	Ortolan Bunting	E. hortulana																			OB			1866
A	Cretzschmar's Bunting	E. caesia																					R	1868
A	Yellow-browed Bunting	E. chrysophrys																					R	1871
A	Rustic Bunting	E. rustica																					R	1873
A	Little Bunting	E. pusilla																			LJ			1874

Sub-total

BUNTINGS continued

	Name	Scientific name	Life list	2006 list	24 hr	Garden	Holiday	Jan	Feb	Mar	Apr	May	Jun	Jul	Aug	Sep	Oct	Nov	Dec		BTO	RBBP	Bou	EU No
A	Yellow-breasted Bunting	E. aureola																						1876
• A	Reed Bunting	E. schoeniclus																			RB		R	1877
A	Pallas's Bunting	E. pallasi																					R	1878
A	Black-headed Bunting	E. melanocephala																					R	1881
• A	Corn Bunting	Miliaria calandra																			CB			1882
A	Rose-breasted Grosbeak	Pheucticus ludovicianus																					R	1887
A	Indigo Bunting	Passerina cyanea																					R	1892
A	Bobolink	Dolichonyx oryzivorus																					R	1897
A	Brown-headed Cowbird	Molothrus ater																					R	1899
A	Baltimore Oriole	Icterus galbula																					R	1918

CATEGORY D & E SPECIES PLUS
EUROPEAN SPECIES

	Name	Scientific name	Life list	2006 list	24 hr	Garden	Holiday	Jan	Feb	Mar	Apr	May	Jun	Jul	Aug	Sep	Oct	Nov	Dec		BTO	RBBP	Bou	EU No
D	Falcated Duck	A. falcata																			FT	R		0181
D	Baikal Teal	A. formosa																			IK	R		0183
D	Marbled Duck	Marmaronetta angustirostris																				R		0195
EU	White-headed Duck	O. Leucocephala																			WQ			0226
EU	Hazel Grouse	Bonasa bonasia																						0326
EU	Rock Partridge	Alectoris graeca																						0357
EU	Barbary Partridge	A. barbara																						0359
EU	Pygmy Cormorant	P. pygmeus																						0082
D	Great White Pelican	Pelecanus onocrotalus																			YP	R		0088
EU	Dalmatian Pelican	P. crispus																						0089
D	Greater Flamingo	Phoenicopterus roseus																			FL	R		0147
EU	Black-winged Kite	Elanus caeruleus																						0235

Sub-total

CATEGORY D & E SPECIES PLUS EUROPEAN SPECIES

			Life list	2006 list	24 hr	Garden	Holiday	Jan	Feb	Mar	Apr	May	Jun	Jul	Aug	Sep	Oct	Nov	Dec			BTO	RBBP	Bou	EU No
D	Bald Eagle	H. leucocephalus																						R	0244
EU	Lammergeier	Gypaetus barbatus																							0246
D	Black (Monk) Vulture	Aegypius monachus																						R	0255
EU	Levant Sparrowhawk	A. brevipes																							0273
EU	Long-legged Buzzard	B. rufinus																							0288
EU	Lesser Spotted Eagle	Aquila pomarina																							0292
EU	Imperial Eagle	A. heliaca																							0295
EU	Booted Eagle	Hieraaetus pennatus																							0298
EU	Bonelli's Eagle	H. fasciatus																				FB			0299
EU	Lanner	F. biarmicus																							0314
D	Saker	F. cherrug																				JF		R	0316
EU	Andalusian Hemipode	Turnix sylvatica																							0400
EU	Purple (Swamp-hen) Gallinule	Porphyrio porphyrio																							0427
EL	Crested Coot	F. cristata																							0431
FA	Greater Sand Plover	C. leschenaultii																				DP		R	0479
EU	Spur-winged Plover	Hoplopterus spinosus																				UW			0487
EU	Black-bellied Sandgrouse	Pterocles orientalis																							0661
EU	Pin-tailed Sandgrouse	P. alchata																							0662
EU	Eagle Owl	Bubo bubo																				EO	bO		0744
EU	Pygmy Owl	Glaucidium passerinum																							0751
EU	Ural Owl	S. uralensis																							0765
EU	Great Grey Owl	S. nebulosa																							0766
EL	White-rumped Swift	A. melba																							0799
EU	Grey-headed Woodpecker	Picus canus																							0855
Sub-total																									

CATEGORY D & E SPECIES PLUS EUROPEAN SPECIES

LOG CHARTS

	Species	Scientific name	EU No	Bou	RBBP	BTO	Dec	Nov	Oct	Sep	Aug	Jul	Jun	May	Apr	Mar	Feb	Jan	Holiday	Garden	24 hr	2006 list	Life list
EU	Black Woodpecker	Dryocopus martius	0863																				
EU	Syrian Woodpecker	D. syriacus	0878																				
EU	Middle Spotted Woodpecker	D. medius	0883																				
EU	White-backed Woodpecker	D. leucotos	0884																				
EU	Three-toed Woodpecker	Picoides tridactylus	0898																				
EU	Dupont's Lark	Chersophilus duponti	0959																				
EU	Thekla Lark	G. theklae	0973																				
EU	Black Wheatear	O. Leucura	1158	R																			
IA	Eyebrowed Thrush	T. obscurus	1195	R																			
EU	Olive-tree Warbler	H. olivetorum	1258																				
EU	Cyprus Warbler	S. melanothorax	1268																				
D	Asian Brown Flycatcher	Muscicapa dauurica	1335																				
D	Mugimaki Flycatcher	F.mugimaki	1344	R																			
EU	Semi-collared Flycatcher	F. semitorquata	1347																				
EU	Sombre Tit	P. lugubris	1441																				
EU	Siberian Tit	P. cinctus	1448																				
EU	Krüper's Nuthatch	Sitta krueperi	1469																				
EU	Corsican Nuthatch	S. whiteheadi	1470																				
EU	Rock Nuthatch	S. neumayer	1481																				
EU	Masked Shrike	L. nubicus	1524																				
EU	Siberian Jay	Perisoreus infaustus	1543																				
EU	Azure-winged Magpie	Cyanopica cyana	1547																				
EU	Alpine Chough	Pyrrhocorax graculus	1558																				
D	Daurian Starling	Sturnus sturninus	1579	R																			
	Sub-total																						

91

CATEGORY D & E SPECIES PLUS EUROPEAN SPECIES

	Species	Scientific name	Life list	2006 list	24 hr	Garden	Holiday	Jan	Feb	Mar	Apr	May	Jun	Jul	Aug	Sep	Oct	Nov	Dec				BTO	RBBP	Bou	EU No
EU	Spotless Starling	S. unicolor																								1583
D	Snow Finch	Montifringilla nivalis																							R	1611
D	Palm Warbler	D. palmarum																							R	1752
EU	Cinereous Bunting	E. cineracea																								1865
D	Chestnut Bunting	E. rutila																							R	1875
D	Red-headed Bunting	E. bruniceps																								1880
D	Blue Grosbeak	Guiraca caerulea																							R	1891

Sub-total

DIRECTORY OF ART, PHOTOGRAPHY AND LECTURERS

Gannet by Nick Williams

DIRECTORY OF WILDLIFE ART GALLERIES

BIRDS BIRDS BIRDS

Paul and Sue Cumberland opened Birds; Birds; Birds in June 2001. Now in its fourth year it is becoming one of the nations leading bird galleries. The steady increase in sales, has encouraged professional wildlife artists onto the portfolio. Prints are now being produced and published in-house, using the giclee system.
Address: 4, Limes Place, Preston St, Faversham, Kent ME13 8PQ; 01795 532370; email:birdsbirdsbirds@birdsbirdsbirds.co.uk www.birdsbirdsbirds.co.uk

NATURE IN ART MUSEUM AND ART GALLERY

The world's first museum dedicated to art inspired by nature, with work from Picasso to David Shepherd, Flemish Masters to contemporary crafts. Permanent collection plus regular special exhibitions and 70 artists in residence each year. See website for details.
Opening times: Tues-Sun (10am-5pm) and Bank holidays. Closed Dec 24-26.
Address: Wallsworth Hall, A38, Twigworth, Gloucester, GL2 9PA; 01452 731422; (Fax)01452 730937. www.nature-in-art.org.uk e-mail: ninart@globalnet.co.uk

OLD BREWERY STUDIOS

Changing exhibitions of work the whole year through. Various painting and drawing courses available.
Opening times: Variable, best to telephone first.
Address; The Manor House, Kings Cliffe, Peterborough, PE8 6XB; 01780 470247; (Fax)01780 470334.
www.oldbrewerystudios.co.uk

THE WILDLIFE ART GALLERY

Opened in 1988 as a specialist in 20th Century and contemporary wildlife art. It exhibits work by many of the leading European wildlife artists, both painters and sculptors, and has published several wildlife books.
Opening times: Mon-Sat (10am-4.30pm) and Sun (2pm-4.30pm).
Address: 97 High Street, Lavenham, Suffolk CO10 9PZ; (Tel)+44 (0)1787 248562; (Fax)+44 (0) 1787 247356.
E-mail: wildlifeartgallery@btinternet.com www.wildlifeartgallery.com

DIRECTORY OF WILDLIFE ARTISTS

ALLEN, Richard

Watercolour paintings and illustrations of birds, wildlife and flowers, recently published work in *Sunbirds* (Helm). Also in RSPB *Birds magazine*, *Garden Bird Behaviour*, *Bird Migration* and *The Complete Back Garden Birdwatcher*, (all by New Holland). Sets of stamps designed for The Solomon Islands, Ascension Island, Navru and Kiribati. Also black and white scraperboard illustrations for a variety of publications.
Products for sale: Limited edition prints: Paphos Lighthouse, Cyprus, spring migrants as *Birding World* cover Vol 16, no 6.
Address: 34 Parkwood Avenue, Wivenhoe, Essex, CO7 9AN; 01206 826753.
e-mail: richardallen@romanriver.freeserve.co.uk www.birdingart.com

BURTON, Philip

Acrylics on canvas; current enthusiasm seabirds. Many book illustrations e.g. in recent *Raptors of the World*. Founder member of Society of Wildlife Artists.
Exhibitions for 2006: SWLA, London, annually.
Address: High Kelton, Doctors Commons Road, Berkhamsted, Herts, HP4 3DW; 01442 865020. e-mail: pjkburton@supanet.com

DIRECTORY OF WILDLIFE ARTISTS

CALE, Steve
Steve is a keen naturalist and specialises in painting in acrylics. His paintings have gone as far afield as Hong Kong and New Zealand. He undertakes work for The Mareeba Wetland Foundation in Australia and for Pensthorpe Waterfowl Park in Norfolk. Steve painted a mural for the RSPB at Titchwell.
Products for sale: Commissioned originals, plus cards, book marks and other printed goods.
Address: Bramble Cottage, Westwood Lane, Gt Rysburgh, Fakenham, Norfolk, NR21 7AP; 01328 829589.

CHEUNG, Mabel
Originally from Newton Abbot in Devon, now based in Cardiff working for RSPB Cymru as Assistant Conservation Officer. Amongst varied work history, highlights include working with sea turtle conservation charities in Costa Rica and BirdWatch Ireland. Works in pencil, ink pens, charcoal, pastels and acrylics, from life and/or photos and has contributed illustrations to the recently published 'Birds of Inishbofin Connemara' by Tim Gordon.
e-mail: chinita@talk21.com

HAMPTON, Michael SWLA
Watercolours of birds and mammals. John Aspinall, the late zoo owner, commissioned many paintings of his animals. Via the Federation of British Artists, he painted Arabian mammals for the Sultan of Oman in 2000. Michael's work has appeared on many magazine covers and calendars.
Exhibitions for 2006: SWLA, London (Sept).
Address: 13 Sandy Way, Shirley, Croydon, Surrey, CR0 8QT.
www.michael-hampton.com

JONES, Chris
Painter of all wildlife subjects (primarily in oils), especially birds and poultry. International Young Artist of the Year 1998. Gold Award Winner the Wildlife Art Society 2000. Featured in 'Birds In Art 2005', USA.
Exhibitions for 2006: Peter Hedley Gallery (Wareham, Dorset), Falconry Fair (Shropshire), Marwell Zoo (Hants).
Products for sale: Original paintings and drawings, prints, cards (by Medici) and postcards. Commission and illustration work undertaken. See website for examples.
Address: 47 Church Lane, North Bradley, Trowbridge, Wilts, BA14 0TE; 01225 769717.
e-mail: chrisjonesart@yahoo.co.uk
www.chrisjoneswildlifeart.com

LEAHY, Ernest
Original watercolours and drawings of Western Palearctic birds, wildlife and country scenes. Illustrations for many publications including Poysers. Wide range of framed and unframed originals available. Commissions accepted and enquiries welcome.
Exhibitions for 2006: E-mail for details.
Products for sale: Original watercolours,

ART/PHOTOGRAPHY/LECTURERS

Kittiwakes by Dan Powell

95

acrylics, pastels and line drawings £20-£500.
Address: 32 Ben Austins, Redbourn, Herts,
AL3 7DR; 01582 793 144
e-mail: ernest.leahy@ntlworld.com
www.wildlifewatercolours.co.uk

POWELL, Dan
Dan studied Wildlife Art at Carmarthen
College of Art and considers that the best part
of the course was meeting his wife Rosemary.
In 1995 he won an award from the Artists for
Nature Foundation which resulted in a trip to
the Pyrennees and an exhibition at the Gaudi-
designed La Pedrera in Barcelona. He has also
won British Birds *Bird Illustrator of the year.*
Dan is well known for his fieldguide *The
Dragonflies of Great Britain* which involved
many hours of research standing in bogs. His
work has appeared in numerous wildlife books
and publications.
Products for sale: Illustrations from various
publications, plus a varying number of
paintings.
Address: 4 Forth Close, Stubbington, Hants
PO14 3SZ. E-mail: dan.powell@care4free.net

MACKAY, Andrew
Colour and line artwork of birds and insects.
Illustrations in *Concise BWP, RSPB Birds of
Britain and Europe, Birds of South-east Asia*
etc. Commissions for paintings and
illustrations welcome.
Products for sale: Paintings in acrylic and
gouache, also limited edition prints.
Address: 50 Cordery Road, Evington,
Leicester, LE5 6DE:0116 210 3740.
e-mail: andy@ajm-wildlife-art.co.uk
www.ajm-wildlife-art.co.uk

ROSE, Chris
Originals in oils and acrylics of birds and
animals in landscapes. Particularly interested
in painting water and its myriad effects.
Limited edition prints available. Illustrated
many books including *Grebes of the World*
(publ. end 2002), *Handbook to The Birds of
The World* and *Robins and Chats of the World*
(in progress). A large format book, *In a
Natural Light - the Wildlife Art of Chris Rose*
was released in summer 2005 by Langford

Press. Copies available from the artist.
Exhibitions for 2006: SWLA, London,
September, British Birdwatching Fair, Rutland,
August.
Products for sale: Original drawings and
paintings, illustrations, limited edition
reproductions, postcards.
Address: Maple Cottage, Holydean, Bowden,
Melrose, Scotland TD6 9HT; (Tel/Fax)01835
822547. e-mail: chrisroseswla@onetel.com
www.chrisrose-artist.co.uk

PHILLIPS, Antonia
Antonia's work is about being near the sea,
from Purbeck to the coastal fringes of the far
North, inspired by the light and mood of the
sea; bird shapes, trails and patterns. On
location or from sketches and memory, she
works with freedom and expression, with a
love of seabirds and their habitat. Antonia also
designs educational and interpretation material
for several conservation groups, including
WWT and Dorset Wildlife Trust.
Exhibitions for 2006: WWT Caelaverock
(Wetlands for Life Bursary - travelling show);
Wellbeloved Gallery, Portland, Dorset - May,
Dorset art weeks - May, SWLA, Mall
Galleries, London - September.
Address: 116 Sandbanks Road, Poole, Dorset,
BH14 8DA.1202 748001.
e-mail: antonia@antoniaphillips.co.uk
www.antoniaphillips.co.uk

POMROY, Jonathan
Works in watercolour and oils, always from
sketches, made on trips across the British Isles,
most recently to Anglesey & Snowdonia,
North West Scotland, North Yorkshire Moors
and coast and Slimbridge as well as around
home in Wiltshire. Recent exhibitions at WWT
Slimbridge and Barnes, annually at The
Gallery, Cirencester.
Exhibitions for 2006: Slimbridge, Great West
Bird Fair-February 2006, Wykeham Gallery
Stockbridge- February, West Barn, Bradford
on Avon-July. See website for latest details.
Products for sale: Chiefly selling
original watercolours and oils at one
man exhibitions and from website.
Commissions accepted only for subjects I have

observed and sketched. Cards.
Address: 10, Bobbin Lane, Westwood,
Bradford on Avon, Wiltshire BA15 2DL;
01225 864726.
e-mail: jonathanpomroy@supanet.com
www.jonathanpomroy.co.uk

SNOW, Philip

Original & atmospheric paintings, sketches,
reproductions, illustrations & writings on
wildlife, mainly birds in landscape. Has
illustrated or contributed to/written, over 50
books and in many magazines, including *"The
Real Life of Birds"*, *2005*, DayOne publishers;
*Collins Guide: "Birds by
Behaviour"* 2003; *"Tall Tales from an
Estuary"*, & *"A Hebridean Wildlife and
Landscape Sketchbook"*, publication delayed.
Exhibitions for 2006: Biennial: Sept 11 - 25,
Tegfryn Gallery, Menai Bridge, Anglesey,
tel:01248 715128, open daily. Pensychnant
Conservation Centre, Conwy, N Wales,
summer, or by appointment, Tel: 01492
592595 .
Products for sale: A wide selection of limited
edition reproduction prints, sketches
& originals, from many countries. Website:
http://artofcreation.org.uk.
Address: 2 Beach Cottages, Malltraeth,
Anglesey, North Wales,LL62 5AT; (Tel/
Fax)01407 840512.
e-mail: philip@snow4083.freeserve.co.uk
http://artofcreation.org.uk

SYKES, Thelma K (SWLA)

Artist printmaker: (haunts coastal marsh and
estuaries). Original linocuts, woodcuts of
British birds. Published work includes BTO
Atlases, *Birdwatcher's Yearbooks* to 1996,
RSPB and Medici Society greetings cards.
Exhibits SWLA, Society of Wood Engravers.
Artist in residence with Nature in Art,
Gloucester.
Exhibitions for 2006: Solo exhibition (June 17
- Aug 28), Grosvenor Museum, Chester. SWLA
at WWT Slimbridge (Feb 19 - Mar 28). SWLA
London (September), NEWA Cheshire, (July).
Address: Blue Neb Studios, 18 Newcroft,
Saughall, Chester, CH1 6EL; 01244 880209.
e-mail: thelmasykes@tiscali.co.uk
www.thelmasykes.co.uk

WARREN, Michael

Original watercolour paintings of birds, all
based on field observations. Books, calendars,
cards and commissions.
Exhibitions for 2006: Wildlife Art Gallery,
Lavenham, Suffolk, (Autumn). British
Birdwatching Fair, Rutland, (August), SWLA,
London, (September).
Products for sale: Original watercolour
paintings of birds, all based on field
observations. Books, calendars, cards.
Commissions welcomed.
Address: The Laurels, The Green, Winthorpe,
Nottinghamshire, NG24 2NR; 01636 673554;
(Fax)01636 611569.
e-mail: mike.warren.birdart@care4free.net
www.mikewarren.co.uk

WOOLF, Colin

Beautiful original watercolour paintings. The
atmosphere of a landscape and the character of
his subject are his hallmark, also the pure
watercolour technique that imparts a softness
to the natural subjects he paints. Owls, birds
of prey and ducks are specialities. Wide range
of Limited Editions and greetings cards,
special commissions also accepted.
Exhibitions for 2006: British Birdwatching
Fair, Rutland Water, (August). Ring for
details.
Products for sale: Original paintings, limited
edition prints and greetings cards.
Address: Tremallt, Penmachno, Betws y Coed,
Conwy, LL24 0YL; +44 (0) 1690 760 308.
e-mail: colin@wildart.co.uk
www.wildart.co.uk

ART/PHOTOGRAPHY/LECTURERS

97

DIRECTORY OF WILDLIFE PHOTOGRAPHERS

BASTON, Bill
Photographer, lecturer.
Subjects: East Anglian rarities and common birds, Mediterranean birds and landscapes, UK wildlife and landscapes, Florida birds and landscapes, Northern Greece, Spain.
Products for sale: Prints, slides, digital, mounted/unmounted.
Address: 86 George Street, Hadleigh, Ipswich, IP7 5BU; 01473 827062. www.billbaston.com
e-mail: bill.baston@bt.com

BATES,Tony
Photographer and lecturer.
Subjects: Mainly British wildlife, landscapes and astro landscapes.
Products for sale: Prints (mounted or framed), original handmade photo greetings cards.
Address: 22 Fir Avenue, Bourne, Lincs, PE10 9RY; 01778 425137.
e-mail: mtr@masher.f9.co.uk

BORG, Les
Photographer, course leader.
Subjects: Mostly British wildlife, with some from Florida, Jamaica, Europe and elsewhere.
Products for sale: Formats, 6x6 and 35mm scans and CDs. Mounted, unmounted or framed, inkjet prints or Ilfochromes if required. Table mats and coasters, greetings cards.
Address: 17 Harwood Close, Tewin, Welwyn, Herts, AL6 0LF; 01438 717841, (Fax)01438 840459.
e-mail: les@les-borg-photography.co.uk
www.les-borg-photography.co.uk

BROADBENT, David
Professional photographer.
Subjects: UK birds and wild places.
Products for sale: Top quality photographic prints.
Address: Rose Cottage, Bream Road, Whitepool, St Briavels, Lydney, GL15 6TL; 07771 664973.

e-mail: info@davidbroadbent.com
www.davidbroadbent.com

BROOKS, Richard
Wildlife photographer, writer, lecturer.
Subjects: Owls (Barn especially), raptors, Kingfisher and a variety of European birds (Lesvos especially) and landscapes.
Products for sale: Mounted and unmounted computer prints (6x4 - A3+ size), framed pictures, A5 greetings cards, surplus slides for sale.
Address: 24 Croxton Hamlet, Fulmodeston, Fakenham, Norfolk, NR21 0NP; 01328 878632.
e-mail: email@richard-brooks.co.uk
www.richard-brooks.co.uk

CONWAY, Wendy (PSA4. AFIAP)
Award-winning wildlife photographer.
Subjects: Birds, mammals, landscapes: UK, USA, Lesvos and Africa.
Products for sale: 35mm and medium format. Prints matted and unmatted, greetings cards.
Address: The Oaks, Parkend Walk, Coalway, Coleford, Glos GL16 7JR; 01594 832956.
e-mail: wendy@terry-wall.com

DENNING, Paul
Wildlife photographer, lecturer.
Subjects: Birds, mammals, reptiles, butterflies and plants from UK, Europe, Canaries, North and Central America.
Products for sale: 35mm transparencies and prints.
Address: 17 Maes Maelwg, Beddau, Pontypridd, CF38 2LD; (H)01443 202607; (W)02920 673243.
e-mail: pgdenning.naturepics@virgin.net

FEATHERBE, David
Wildlife and landscape photographer.
Subjects: General selection of wildlife and landscape photography, the majority of which are of UK subjects.

Products for sale: Mounted and framed images can be purchased from an extensive website.
Address: 16 East Cliff Gardens, Folkestone, Kent CT19 6AP. 01303 244489; e-mail: david.featherbe@mac.com http://homepage.mac.com/david.featherbe

HARROP, Hugh
Professional wildlife guide, photographer and author.
Subjects: European birds, cetaceans, wild flowers, butterflies and dragonflies. I specialise in all Shetland subjects.
Products for sale: 35mm transparencies, digital images on CD or via modem. Commercial enquiries only please.
Address: Longhill, Maywick, Shetland, ZE2 9JF; 01950 422483; (Fax)01950422430. e-mail: hugh@hughharrop.com www.hughharrop.com

LANE, Mike
Wildlife photographer, workshops, photo shoots.
Subjects: Birds and wildlife from around the world, also landscapes and the environment .
Products for sale: 35mm, medium format and digital. Work for private buyers and publishing.
Address: 36 Berkeley Road, Shirley, Solihull, West Midlands, B90 2HS; 0121 744 7988,. e-mail: mikelane@nature-photography.co.uk www.nature-photography.co.uk

LANGSBURY, Gordon FRPS
Professional wildlife photographer, lecturer, author.
Subjects: Birds and mammals from UK, Europe, Scandinavia, N America, Gambia, Kenya, Tanzania, Morocco and Falklands.
Products for sale: 35mm transparencies for publication, lectures and prints.
Address: Sanderlings, 80 Shepherds Close, Hurley, Maidenhead, Berkshire, SL6 5LZ; (Tel/fax)01628 824252. e-mail: gordonlangsbury@birdphoto.org.uk www.birdphoto.org.uk

McKAVETT, Mike
Wildlife photographer and lecturer.
Subjects: Birds and mammals from India,

Kenya, The Gambia, Lesvos, N.America and UK.
Products for sale: 35mm transparencies for publication and commercial use, prints and lectures.
Address: 34 Rectory Road, Churchtown, Southport, PR9 7PU; 01704 231358; e-mail: mike.mckavett@btinternet.com.

OFFORD, Keith
Photographer, writer, tour leader, conservationist.
Subjects: Raptors, UK wildlife and scenery, birds and other wildlife of USA, Africa, Spain, Australia, India.
Products for sale: Conventional prints, greetings cards, framed pictures.
Address: Yew Tree Farmhouse, Craignant, Selattyn, Nr Oswestry, Shropshire, SY10 7NP; 01691 718740. e-mail: keith-offord@virgin.net www.keithofford.co.uk

PARKER, Susan and Allan (ARPS)
Professional photographers (ASPphoto - Images of Nature) lecturers and tutors.
Subjects: Birds plus other flora and fauna from the UK, Spain, Lesvos, Cyprus, Florida and Texas.
Products for sale: 35mm 645 medium format, slides, mounted digital prints, greetings cards and digital images on CD for reproduction (high quality scans up to A3+).
Address: Ashtree House, 51 Kiveton Lane, Todwick, Sheffield, South Yorkshire, S26 1HJ; 01909 770238.
e-mail: aspaspphoto@clara.co.uk

PIKE, David
Photographer, presenter and writer
Subjects: Wildlife, including birds from Japan, N America and Africa.
Products for sale: 35mm mounted. Conventional prints and digital.
Address: The Old Rectory, Water Newton, Cambs, PE8 6LU; 01733 238171; (Fax) 01733 237383.
e-mail: david.pike@infocado.co.uk

READ, Mike
Photographer (wildlife and landscapes), tour leader, writer.

ART/PHOTOGRAPHY/LECTURERS

Subjects: Birds, mammals, plants, landscapes, and some insects. UK, France, USA, Ecuador (including Galapagos).
Products for sale: Prints, greetings cards, books, including *Red Kite Country* (Wildguides).
Address: Claremont, Redwood Close, Ringwood, Hampshire, BH24 1PR; 01425 475008. e-mail: mike@mikeread.co.uk
www.mikeread.co.uk

SWASH, Andy
Photographer, author, tour leader.
Subjects: Birds, habitats/landscapes and general wildlife from all continents; photographic library currently 1,700 bird species.
Products for sale: Slides for publication and duplicates for lectures. High resolution scans on CD-Rom. Conventional and digital prints, unmounted, mounted or framed.
Address: Stretton Lodge, 9 Birch Grove, West Hill, Ottery St Mary, Devon, EX11 1XP; (H&fax)01404 815383, (W)01392 822901.
e-mail: andy_swash@wildguides.co.uk
www.wildguides.co.uk

SIMPSON, Geoff
Professional natural history and landscape photographer.
Subjects: Specialises in evocative images of Britain's wildlife and landscape.
Products for sale: 35mm and panoramic. Slides and CD.
Address: Camberwell, 1 Buxton Road, New Mills, High Peak, Derbyshire, SK22 3JS; 01633 743089.
e-mail: info@geoffsimpson.co.uk
www.wildphoto.demon.co.uk

TIPLING, David
Wildlife and landscape photographer, photographic tour leader, author, with a passion for birds.

Subjects: Worldwide wildlife and landscapes, with an emphasis on birds.
Products for sale: Greetings cards, book and limited edition prints for sale. Prints are truly archival, using the finest printing techniques and finest papers.
Address: 84 Dolphin Quays, Clive Street, North Shields, NE29 6HJ.
www.davidtipling.com
e-mail: dt@windrushphotos.demon.co.uk

WILKES, Mike FRPS
Professional wildlife photographer, tour leader.
Subjects: African, European and British birds.
Address: 43 Feckenham Road, Headless Cross, Redditch, Worcestershire, B97 5AS; 01527 550686.
e-mail: wilkes@photoshot.com

WALL, Terry (ARPS EFIAP PPSA)
Wildlife photographer.
Subjects: Birds, mammals, landscapes from UK, USA, Lesvos, Africa and Galapagos.
Products for sale: 35mm/medium. Prints matted/unmatted, greetings cards. 35mm scanning service and restoration and retouching service. Quality printing service. *One-to-One Photoshop Tuition* book and individual lessons.
Address: The Oaks, Parkend Walk, Coalway, Coleford, Glos GL16 7JR; 01594 832956. e-mail: wildimages@terry-wall.com
www.terry-wall.com

WILLIAMS, Nick
Photographer, lecturer, author, tour leader.
Subjects: W.Palearctic including Cape Verde Islands and Falkland Islands.
Products for sale: Duplicate slides, some originals, prints also available.
Address: Owl Cottage, Station Road, Rippingale, Lincs, PE10 0TA; (Tel/Fax)01778 440500.
e-mail: birdmanandbird@hotmail.com

DIRECTORY OF LECTURERS

Lecturers who have indicated that they are willing to travel to all parts of Britain are listed first. For the remainder we have grouped them geographically in the following regions: England, Eastern; North-eastern; North-western; South-eastern; South-western; West Midlands and Wales; Scotland.

To ensure this valuable section continues strongly in the future, we would be grateful if you would mention the *Yearbook* when contacting any of the listed lecturers.

If your group has enjoyed a talk from anyone not listed here, we would appreciate receiving contact details so they might be included in the 2007 edition.

NO LIMITS

BATES,Tony
Photographer and lecturer.
Subjects: Mainly British wildlife, landscapes and astro landscapes, folklore..
Fees: £75 plus travel. **Limits:** None. **Times:** To suit.
Address: 22 Fir Avenue, Bourne, Lincs, PE10 9RY; 01778 425137.
e-mail: mtr@masher.f9.co.uk

BELL, Graham
Cruise lecturer worldwide, photographer, author.
Subjects: Arctic, Antarctic, America, Siberia, Australia, Canada, Iceland, Seychelles, UK - identification, behaviour, seabirds, garden birds, entertaining bird sound imitations, birds in myth and fact, bird names etc.
Fees: £35 plus travel. **Limits:** None.
Times: Any.
Address: Ros View, South Yearle, Wooler, Northumberland, NE71 6RB; (Tel/fax)01668 281310. e-mail: seabirdsdgb@hotmail.com

BOND, Terry
International consultant, ex-bank director, photographer, group field leader, conference speaker worldwide.
Subjects: 8 talks (including Scilly Isles, Southern Europe, USA - shorebirds and inland birds, Birdwatching Identification - a new approach).
Fees: By arrangement (usually only expenses).
Limits: Most of UK. **Times:** Evenings.
Address: 3 Lapwing Crescent, Chippenham, Wiltshire, SN14 6YF; 01249 462674.
e-mail: terryebond@btopenworld.com

BRIGGS, Kevin
Freelance ecologist.
Subjects: General wildlife in NW and S England; specialist topics - Raptors, Oystercatcher, Ringed Plover, Goosander, Yellow Wagtail, Ring Ouzel.
Fees: £60 + petrol. **Limits:** None.
Times: Any.
Address: 2 Osborne Road, Farnborough, Hampshire, GU14 6PT; 01252 519881.
e-mail: kbbriggs@yahoo,com

BROOKS, Richard
Wildlife photographer, writer, lecturer.
Subjects: 12 talks (including Lesvos, Evros Delta, Israel, Canaries, E.Anglia, Scotland, Wales, Oman).

DIRECTORY OF LECTURERS

Fees: £75 plus petrol. **Limits:** None if accom provided. **Times:** Any.
Address: 24 Croxton Hamlet, Fulmodeston, Fakenham, Norfolk, NR21 0NP; 01328 878632.
e-mail: email@richard-brooks.co.uk
www.richard-brooks.co.uk

BUCKINGHAM, John
Lecturer, photographer, tour leader.
Subjects: 60+ titles covering birds, wildlife, botany, ecology and habitats in UK, Europe, Africa, Australia, India and the Americas.
Fees: £60 plus expenses. **Limits:** None.
Times: Any.
Address: 3 Cardinal Close, Tonbridge, Kent, TN9 2EN; (Tel/fax) 01732 354970.

BURROWS, Ian
Tour leader.
Subjects: Papua New Guinea, Cape Clear Island and 'Food from the Wild'.
Fees: £70 plus mileage over 100. **Limits:** Anything considered. **Times:** Evenings preferable but other times considered.
Address: Trefor, Creake Road, Sculthorpe, Fakenham, Norfolk, NR21 9NQ;01328 856 925. e-mail: Ian@sicklebill.demon.co.uk
www.sicklebill.com

CARRIER, Michael
Lifelong interest in natural history.
Subjects: 1) 'Birds in Cumbria', 2) 'The Solway and its Birds' and 3)'The Isle of May', 4)'A look at Bird Migration'.
Fees: £20. **Limits:** None but rail connection essential. **Times:** Sept-March inc. afternoons or evenings.
Address: Lismore Cottage, 1 Front Street, Armathwaite, Carlisle, Cumbria, CA4 9PB; 01697 472218.

CHARTERS, Roger
Experienced wildlife sound recordist and one-time professional photographer.
Subjects: 'Aural and visual Images from Around the World', 'The Importance of Sound', 'Scandinavia with emphasis on the Arctic', 'Ukraine – a birdwatcher's paradise' 'Spain', including the Coto Donàna, 'The Australian Outback'. Each talk lasts about one hour with extensive use of sound recordings and visual sequences.
Fees: £35 made payable to Warwickshire WT, + 28p per mile. **Limits:** None.
Address: 11 Eastnor Grove, Leamington Spa, Warwickshire CV31 1LD: 01926 882583.
e-mail: Roger.Charters@btinternet.com

CROUCHER, Roy
Wildlife tour leader, former local authority ecologist, wildlife consultant.
Subjects: Three talks (Northern France, Montenegro, Managing Britain's Habitats).
Fees: £50 plus petrol from Leicester. **Limits:** Mainland Britain. **Times:** December tour only.
Address: Place de L'Eglise, 53700, Averton, France; 0033 2430 06969.
e-mail: nfwt@online.fr

DUGGAN, Glenn
Ex-Commander Royal Navy, tour leader, researcher.
Subjects: Ten talks including, birds of paradise and bower birds, history of bird art (caveman to present day), modern day bird art, famous Victorian bird artists (John Gould, the Birdman and John James Andubohon.
Fees: £50 plus expenses. **Limits:** none with o.n accom. **Times:** Any.
Address: 25 Hampton Grove, Fareham, Hampshire, PO15 5NL; 01329 845976, (M)07771 605320.
e-mail: glenn.m.duggan@ntlworld.com
www.birdlectures.com

EYRE, John
Author, photographer, conservationist and chairman Hampshire Ornithological Society
Subjects: World birding (Europe, Africa, Australasia and the Americas), plus special Hampshire subjects (eg. Gilbert White's birds and heathland birds).
Fees: £60 plus travel. **Limits:** Any location negotiable. **Times:** Any.
Address: 3 Dunmow Hill, Fleet, Hampshire, GU51 3AN; 01252 677850.
e-mail: John.Eyre@ntlworld.com

GALLOP, Brian
Speaker, photographer, tour leader.
Subjects: 30 talks covering UK, Africa, India,

DIRECTORY OF LECTURERS

Galapagos and Europe - all natural history subjects.
Fees: £50 plus 20p per ml. **Limits:** None - o.n acc. if over 100 mls. **Times:** Any.
Address: 13 Orchard Drive, Tonbridge, Kent, TN10 4LT; 01732 361892.

GARCIA, Ernest
Writer/editor *Gibraltar Bird Report*.
Subjects: Raptor and seabird migration at Gibraltar, birding in southern and western Spain (Andalucia/Extremadura)..
Fees: £50 plus expenses. **Limits:** None..
Times: Any.
Address: Woodpecker House, 2 Pine View Close, Chilworth, Surrey GU4 8RS; 01483 539053. e-mail: EFJGarcia@aol.com

GARNER, David
Wildlife photographer.
Subjects: 17 live talks and audio-visual shows on all aspects of wildlife in UK and some parts of Europe - list available.
Fees: £35 plus 20p per ml. **Limits:** None.
Times: Any.
Address: 73 Needingworth Road, St Ives, Cambridgeshire, PE27 5JY; (H)01480 463194; (W)01480 463194.
e-mail: davidgarner@hushwing.freeserve.co.uk
http://hushwing.mysite.wanadoo-members.co.uk

GUNTON, Trevor
Ex.RSPB Staff member, recruitment advisor, lecturer and consultant.
Subjects: 'In Search of Vikings' (Norse culture and wildlife), 'I Know an Island' (UK Bird Islands), 'A Norwegian Coastal Adventure', 'Birds and Pits', 'Birds of The Broadacres' (Yorkshire), 'Nature in Trust' (NT), 'Look Again at Garden Birds', 'Shetland – Isles of the Simmer Dim' also membership recruitment workshops for wildlife organisations.
Fees: Variable (basic £60 plus expenses).
Limits: None. **Times:** Anytime, anywhere.
Address: 15 St James Road, Little Paxton, St Neots, Cambs, PE19 6QW; (tel/fax)01480 473562. e-mail: trevor.gunton@tesco.net

HARROP, Hugh
Professional wildlife guide, photographer and author.

Subjects: 15 talks including Shetland wildlife, Shetland birds, polar bears, whales and dolphins, Galapagos, seals and sea lions, Alaska, Iceland, general wildlife photography.
Fees: £125 plus accom at cost and return flight from Shetland. **Limits:** UK only. **Times:** November to March.
Address: Longhill, Maywick, Shetland, ZE2 9JF; 01950 422483; (Fax)01950 422430.
e-mail: hugh@hughharrop.com
www.hughharrop.com

HASSELL, David
Birdwatcher and photographer.
Subjects: Six talks (including British Seabirds, Shetland Birds, British Birds, USA Birds, including Texas, California, Florida etc).
Fees: £45 plus petrol. **Limits:** None.
Times: Any.
Address: 15 Grafton Road, Enfield, Middlesex, EN2 7EY; 020 8367 0308.
e-mail: dave@davehassell.com
www.davehassell.com

KNYSTAUTAS, Algirdas
Ornithologist, photographer, writer, tour leader.
Subjects: Birds and natural history of Russia, Baltic States, S America, Indonesia, Birding the Great Silk Route (seven talks).
Fees: £1 per person. £70 minimum plus £25 travelling. **Limits:** None - in UK, o.n accom needed. **Times:** Oct and Nov.
Address: 7 Holders Hill Gardens, London, NW4 1NP; 020 8203 4317.
e-mail ibisbill@talk21.com

LANE, Mike
Wildlife photographer.
Subjects: Seven talks from the UK and worldwide, mostly on birds.
Fees: Varies. **Limits:** None. **Times:** Any.
Address: 36 Berkeley Road, Shirley, Solihull, West Midlands, B90 2HS; 0121 744 7988,.
e-mail: mikelane@nature-photography.co.uk
www.nature-photography.co.uk

LANGSBURY, Gordon FRPS
Professional wildlife photographer, lecturer, author.
Subjects: 20 talks - Africa, Europe, USA,

ART/PHOTOGRAPHY/LECTURERS

103

DIRECTORY OF LECTURERS

Falklands and UK. Full list provided.
Fees: £80 plus travel expenses. **Limits:** None.
Times: Any.
Address: Sanderlings, 80 Shepherds Close,
Hurley, Maidenhead, Berkshire, SL6 5LZ;
(Tel/fax)01628 824252.
e-mail: gordonlangsbury@birdphoto.org.uk
www.birdphoto.org.uk

McKAVETT, Mike
Photographer.
Subjects: Five talks, Birds and Wildlife of
India, North and Western Kenya and the
Gambia, Bird Migration in North America.
Fees: £40 plus expenses. **Limits:** None.
Times: Any.
Address: 34 Rectory Road, Churchtown,
Southport, PR9 7PU; 01704 231358;
e-mail: mike.mckavett@btinternet.com.

MOIR, Geoffrey DFC, FRGS, FRPSL
Ret. Schoolmaster, lecturer, writer, philatelist,
lived in Falkland Islands.
Subjects: Fully illustrated talks on subjects
including: 'Falklands 2000', 'The Island of
South Georgia and its Wildlife', 'The Flora of
the Falkland Islands', 'Falkland's Wildlife'.
Fees: £20. **Limits:** None. **Times:** Any.
Address: 37 Kingscote Road, Croydon,
Surrey, CR0 7DP; Tel/fax 020 8654 9463.

MORRIS, Rosemary and YATES, Bas
Both members of Cookhill and Studley Camera
Club (Wildlife photographers / Travel).
Subjects: England Naturally, Ecuador-
Rainforests and Galapagos Islands, Tanzania
up Close, A Wild Special Place (Lesvos),
Trinidad and Tobago, Southern Ireland to the
Highlands, Scotland (in the pipeline).
Fees: Worcs/Warwicks £35 + mileage if over
40 miles. Distant talks negotiable. **Limits:**
Please ask, will travel. **Times:** Any.
Address: Wrens Nest, Droitwich Rd, New End,
Astwoodbank, Redditch Worcs B96 6NE.
e-mail: gilbert.morris1@btopenworld,com.uk

OFFORD, Keith
Photographer, writer, tour leader,
conservationist.

Subjects: 14 talks covering raptors, uplands,
gardens, migration, woodland wildlife,
Australia, Southern USA, Tanzania, Gambia,
Spain, SW.Africa.
Fees: £80 plus petrol. **Limits:** None.
Times: Sept - April.
Address: Yew Tree Farmhouse, Craignant,
Selattyn, Nr Oswestry, Shropshire, SY10 7NP;
01691 718740. e-mail: keith-offord@virgin.net
www.keithofford.co.uk

PALMER, Phil
Tour leader for Bird Holidays.
Subjects: Mostley birds, but includes
mammals, insects, reptiles, whale watching etc.
Many foreign trips including Alaska, Midway
Atoll, Antarctica and India. British birds -
'First for Britain' from Phil's book, The Secret
Life of the Nightjar, twitching in the UK and
bird photography
Fees: To suit all club budgets **Limits:** None.
Times: Any.
Address: 72 Grove Road, Retford,
Nottinghamshire DN22 7JN; (Tel/fax)0113
391 0510.
e-mail: Phil@birdholidays.fsnet.co.uk

ROBINSON, Peter
Consultant ornithologist and former Scilly
resident, author of Birds of the Isles of Scilly.
Subjects: Sea and landbirds of Scilly and life
in an island environment, 'Going Wild in
Devon' making of the 'Bill Oddie Goes Wild'
BBC Programmes..
Fees: £45 plus Petrol. **Limits:** None.
Times: Any.
Address: 19 Pine Park Road, Honiton, Devon,
EX14 2HR; 01404 549873 (M) 07768 538132.
e-mail: pjrobinson2@aol.com

RUMLEY-DAWSON, Ian
Photographer, course leader, cruise lecturer.
Subjects: 96 talks using twin dissolving
projectors. Birds, mammals, insects, plants,
habitats, ethology. Arctic, Antarctic,
Falklands, N and S America, N.Z, Seychelles,
North Pacific islands. Albatrosses, penguins,
Snowy Owls, polar bears etc.
Fees: £50 plus expenses. **Limits:** None.
Times: Any.

Address: Oakhurst, Whatlington Road, Battle, East Sussex, TN33 0JN; 01424 772673.

SCOTT, Ann and Bob

Ex-RSPB staff, tour leaders, writers, lecturers, tutors, trainers.

Subjects: 16+ talks (including nature reserves, RSPB, tours, gardening, Europe, Africa, S America, after-dinner talks etc).

Fees: £60 plus travel over 50 mls. **Limits:** None (by arrangement). **Times:** Any.

Address: 8 Woodlands, St Neots, Cambridgeshire, PE19 1UE; 01480 214904; (fax)01480 473009.

e-mail: abscott@tiscali.co.uk

SWASH, Andy

Photographer, author, tour leader.

Subjects: Birds, general wildlife, scenery and tales from travels in: Antarctica, Argentina, Australia, Brazil, Chile, China, Costa Rica, Cuba, Galápagos, Kenya, Namibia, South Africa, USA or Venezuela .

Fees: £85 plus petrol. **Limits:** None. **Times:** Evenings.

Address: Stretton Lodge, 9 Birch Grove, West Hill, Ottery St Mary, Devon, EX11 1XP; (H&fax)01404 815383, (W)01392 822901.

e-mail: andy_swash@wildguides.co.uk

www.wildguides.co.uk

TAYLOR, Mick

Co-ordinator South Peak Raptor Group, photographer, ornithologist, writer.

Subjects: Several talks including (Merlins, Peak District birds, Peak District raptors, Alaskan wildlife).

Fees: £50 plus petrol. **Limits:** Negotiable. **Times:** Evenings preferred.

Address: 76 Hawksley Avenue, Chesterfield, Derbyshire, S40 4TL;01246 277749.

TODD, Ralph

Tour leader, lecturer & photographer.

Subjects: 13 talks incl. 'Galapagos Wildlife', 'On the Trail of the Crane', 'Polar Odyssey', 'Operation Osprey', 'Iceland & Pyrenees', 'Man & Birds-Travels through time', 'A summer in Northern Landscapes', 'Where Yeehaa meets Ole'.

Fees: £60 plus expenses. **Limits:** None, neg over 120 miles. **Times:** Any - also short notice.

Address: 9 Horsham Road, Bexleyheath, Kent, DA6 7HU; (Tel/fax)01322 528335.

e-mail: rbtodd@todds9.fsnet.co.uk

WILKES, Mike FRPS

Professional wildlife photographer, tour leader.

Subjects: 13 talks - natural history - Africa, America, South America, Europe, Gt Britain.

Fees: According to distance on request.

Limits: None. **Times:** Any.

Address: 43 Feckenham Road, Headless Cross, Redditch, Worcestershire, B97 5AS; 01527 550686.

e-mail: wilkes@photoshot.com

WILLIAMS, Nick

Photographer, lecturer, author, tour leader.

Subjects: Several audio visual shows (including Spain, Camargue, Turkey, Canaries and Cape Verdi Islands, Falklands) and birds of prey.

Fee: £90-£110 depending on group size and distance. **Limits:** None. **Times:** Any

Address: Owl Cottage, Station Road, Rippingale, Lincs, PE10 0TA; (Tel/Fax)01778 440500. e-mail: birdmanandbird@hotmail.com

WREN, Graham ARPS

Wildlife photographer, lecturer, tour guide.

Subjects: 22 talks, birds - UK and Scandinavia, the environment - recent habitat changes and effect on bird populations, wildlife - Ohio and Kenya. Detailed information package supplied on request.

Fees: £50-70 plus petrol. **Limits:** None. **Times:** Any.

Address: The Kiln House, Great Doward, Whitchurch, Ross-on-Wye, Herefordshire, HR9 6DU; 01600 890488, (Fax)01600 890294. e-mail: grahamjwren@aol.com

ART/PHOTOGRAPHY/LECTURERS

105

DIRECTORY OF LECTURERS

EASTERN ENGLAND

APPLETON, Tim
Reserve Manager, Rutland Water
Subjects: Rutland Water, British Birdwatching Fair, Return of Ospreys to England, Trips and birds of Spain, Australia, Papua New Guinea, various African countries, Argentina and more.
Fees: Negotiable. **Limits:** Preferably within 2hrs of Rutland. **Times:** Winter preferred but can be flexible.
Address: Fishponds Cottage, Stamford Road, Oakham, Rutland, LE15 8AB; (H)01572 724101, (W)01572 770651: (Fax) 01572 755931.

BROOKS, David
Freelance naturalist.
Subjects: Various talks on wildlife, principally birds, in UK and overseas.
Fees: £50 plus petrol. **Limits:** 50 mls without o.n. accom. **Times:** Any.
Address: 2 Malthouse Court, Green Lane, Thornham, Norfolk, PE36 6NW; 01485 512548.
e-mail: david.g.brooks@tesco.net

COOK, Tony MBE
35 years employed by WWT. Travelled in Europe, Africa and N. America.
Subjects: 22 talks from Birds of The Wash, garden birds to travelogues of Kenya, E and W North America, Europe (Med to North Cape).
Fees: £35 plus 20p per ml. **Limits:** 100 mls.
Times: Any.
Address: 11 Carnoustie Court, Sutton Bridge, Spalding, Lincs, PE12; 01406 350069.

CROMACK, David
Editor of *Bird Watching* and *Birds Illustrated* magazine, bird tour leader.
Subjects: 1) 'Bird Magazines and the Art of Bird Photography', 2) 'Birds of Arizona and California', 3) 'World Class Bird Images (International Wildbird Photographic Competition), 4) 'Birdwatching For Beginners.
Fees: 1) No fee - expenses only, 2,3 and 4) £60 plus expenses. **Limits:** 175 mls. **Times:** All requests considered from Jan 2006 onwards.

Address: c/o Bird Watching Magazine, Bretton Court, Peterborough, PE3 8DZ.
e-mail: david.cromack@.emap.com

PIKE, David
Photographer, presenter and writer
Subjects: Winter Birds of Japan', 'Wildlife Photography'.
Fees: £100.
Address: The Old Rectory, Water Newton, Cambs, PE8 6LU; 01733 238171; (Fax) 01733 237383.
e-mail: david.pike@infocado.co.uk

NORTH EASTERN ENGLAND

MATHER, John Robert
Ornithologist, writer, tour guide, lecturer.
Subjects: Birds and wildlife of: Kenya, Tanzania, Uganda, Ethiopia, Namibia, South Africa, Costa Rica, Romania/Bulgaria, India, Nepal, Algonquin to Niagara - a tour around the Great Lakes'. 'Landscapes, Flowers and Wildlife of the American West'. 'Bird on the Bench' - a fascinating account of bird biology
Fees: £65 plus 20p per ml. **Limits:** 100 mls.
Times: Evenings.
Address: Eagle Lodge, 44 Aspin Lane, Knaresborough, North Yorkshire, HG5 8EP; 01423 862775.

PARKER, Susan and Allan (ARPS)
Professional photographers, (ASPphoto – Images of Nature), lecturers and tutors.
Subjects: Talks on birds and natural history, natural history photography - countries include UK, USA (Texas, Florida), Spain, Greece, Cyprus.
Fees: On application. **Limits:** Up to 120 mls without o.n accom. **Times:** Any.
Address: Ashtree House, 51 Kiveton Lane, Todwick, Sheffield, South Yorkshire, S26 1HJ; 01909 770238.
e-mail: aspaspphoto@clara.co.uk

DIRECTORY OF LECTURERS

SOUTH EASTERN ENGLAND

BORG, Les
Photographer, course leader.
Subjects: Florida (mostly birds), nature photography (parts I, II and III), cetaceans of the Azores, 'A Year of Nature Photography', 'A Hint of Finland and Norway', 'Hot & Cold (Jamaica and Finland Birds)', 'The Pembrokeshire island of Stokholm'.
Fees: Negotiable. **Limits:** 150 mls without o.n accom. **Times:** Any.
Address: 17 Harwood Close, Tewin, Welwyn, Herts, AL6 0LF; 01438 717841.
e-mail: les@les-borg-photography.co.uk
www.les-borg-photography.co.uk

CLEAVE, Andrew MBE
Head of environmental education centre, author, tour leader.
Subjects: 30 talks (including Galapagos, Arctic, Mediterranean and Indian birds, dormice, woodlands). List available.
Fees: £60 plus petrol. **Limits:** 60 mls without o.n accom. **Times:** Evenings, not school holidays.
Address: 31 Petersfield Close, Chineham, Basingstoke, Hampshire, RG24 8WP; (H)01256 320050, (W)01256 882094, (Fax)01256 880174.
e-mail: andrew@bramleyfirth.co.uk

COOMBER, Richard
Tour leader, photographer, writer.
Subjects: Alaska, Australia, Botswana, Namibia, Seychelles, Falklands, Galapagos, S America, USA, seabirds.
Fees: £65 plus petrol. **Limits:** 50 mls without o.n. accom, 150 mls otherwise. **Times:** Afternoons or evenings.
Address: 1 Haglane Copse, Lymington, Hampshire, SO41 8DT; 01590 674471.

FEATHERBE, David
Wildlife and landscape photographer.
Subjects: Wrecks and Wildlife of the Dover Straits, Kalahari Gemsbok and South West Cape Province, Zimbabwe, Western Matabeleland, wildlife photography.
Fees: £60. **Limits:** Primarily S.E England, further a field by negotiation.
Times: Evenings.
Address: 16 East Cliff Gardens, Folkestone, Kent CT19 6AP. 01303 244489;
e-mail: david.featherbe@mac.com
http://homepage.mac.com/david.featherbe

FURNELL, Dennis
Natural history writer, radio and television broadcaster, artist and wildlife sound recordist.
Subjects: British and European wildlife, France, (*The Nature of France*). Wildlife sound recording, wildlife and disability access issues.
Fees: £100. **Limits:** 50 miles, further with o.n. accom. **Times:** Afternoons or evenings according to commitments.
Address: 19 Manscroft Road, Gadebridge, Hemel Hempstead, Hertfordshire, HP1 3HU; 01442 242915, (Fax)01442 242032.
e-mail; dennis.furnell@btinternet.com
www.natureman.co.uk

NOBBS, Brian
Amateur birdwatcher and photographer.
Subjects: Wildlife of the Wild West, Israel, Mediterranean, Florida, Wildlife Gardening, Reserved for Birds.
Fees: £30 plus 25p per ml. **Limits:** Kent, Surrey, Sussex. **Times:** Evenings.
Address: The Grebes, 36 Main Road, Sundridge, Sevenoaks, Kent, TN14 6EP; 01959 563530.
e-mail: Brian.nobbs@tiscali.co.uk

WARD, Chris
Photographer.
Subjects: 20 talks on UK and worldwide topics (W.Palearctic, Americas, Africa, Australasia) - primarily birds, some other wildlife.
Fees: £45 plus petrol. **Limits:** 100 mls.
Times: Evenings, afternoons poss.
Address: 22 Whitsun Pasture, Willen Park, Milton Keynes, MK15 9DQ; 01908 669448.
e-mail: cwphotography@hotmail.com
www.cwardphotography.co.uk

ART/PHOTOGRAPHY/LECTURERS

DIRECTORY OF LECTURERS

COOMBER, Richard
Tour leader, photographer, writer.
Subjects: Alaska, Australia, Botswana, Namibia, Seychelles, Falklands, Galapagos, S America, USA, seabirds.
Fees: £65 plus petrol. **Limits:** 50 mls without o.n. accom, 150 mls otherwise. **Times:** Afternoons or evenings.
Address: 1 Haglane Copse, Lymington, Hampshire, SO41 8DT; 01590 674471.

COUZENS, Dominic
Full-time birdwatcher, tour leader (UK and overseas), writer and lecturer.
Subjects: 'The Secret Habits of Garden Birds', 'Birds Behaving Badly – the trials and tribulations of birds through the year', 'Bird Sounds – As You've Never Heard Them Before', 'Have Wings Will Travel' – the marvel of bird migration, 'Vive la Difference' – a look at the lives of some European birds.
Fees: £70 plus travel. **Limits:** London and south. **Times:** Any.
Address: 3 Clifton Gardens, Ferndown, Dorset, BH22 9BE; (Tel/fax) 01202 874330.
e-mail: Dominic@birdwords.freeserve.co.uk
www.birdwords.co.uk

READ, Mike
Photographer, tour leader, writer.
Subjects: 12 talks featuring British and foreign subjects (list available on receipt of sae).
Fees: £70 plus travel. **Limits:** 175 mls.
Times: Any.
Address: Claremont, Redwood Close, Ringwood, Hampshire, BH24 1PR; 01425 475008. e-mail: mike@mikeread.co.uk
www.mikeread.co.uk

BROADBENT, David
Photographer.
Subjects: UK birds and wild places. In praise of natural places.
Fees: £70 plus travel. **Limits:** 50mls without o.n accom. Anywhere otherwise.
Times: Any.
Address: Rose Cottage, Bream Road, Whitepool, St Briavels, Lydney, GL15 6TL.
e-mail: info@davidbroadbent.com
www.davidbroadbent.com

CONWAY, Wendy (PSA4. AFIAP)
Award-winning wildlife photographer.
Subjects: Several talks, birds, mammals, landscapes from UK, USA, Lesvos and Africa.
Fees: £45 plus petrol. **Limits:** Over 50mls please contact. **Times:** Any.
Address: The Oaks, Parkend Walk, Coalway, Coleford, Glos GL16 7JR; 01594 832956.
e-mail: wendy@terry-wall.com

DENNING, Paul
Wildlife photographer, lecturer.
Subjects: 15 talks, (birds, mammals, reptiles, butterflies etc, UK, western and eastern Europe, north and central America, Canaries).
Fees: £40 plus petrol. **Limits:** 100 mls.
Times: Evenings, weekends.
Address: 17 Maes Maelwg, Beddau, Pontypridd, CF38 2LD; (H)01443 202607; (W)02920 673243.
e-mail: pgdenning.naturepics@virgin.net

LINN, Hugh (ARPS)
Experienced lecturer, photographer.
Subjects: 12 talks, covering Uk, Europe, Africa and bird-related subjects. List available
Fees: £40 plus petrol. **Limits:** 100 mls without o.n. accom 150 mls otherwise.
Times: Flexible.
Address: 4 Stonewalls, Rosemary Lane, Burton, Rossett, Wrexham, LL12 0LG; 01244 571942.

ROGERS, Paul
Tour leader, lecturer.
Subjects: Tunnicliffe's Anglesey, 20+ talks on USA, Africa, Antarctica, Ecology, 'Birds as Flying Machines', 'The World of Wildfowl'
Fees: £35 + petrol. **Limits:** 3-4 hrs travelled, (150-200 mls). **Times:** Not weekends.
Address: Shorelands, Malltraeth, Bodorgan, Anglesey, LL62 5AT; 01407 840396.

WALL, Terry (ARPS EFIAP PPSA)
Wildlife photographer.
Subjects: Several (birds, mammals, landscapes – USA, UK, Lesvos, Africa and Galapagos).
Fees: £45 plus petrol. **Limits:** Over 50 mls please contact. **Times:** Any.
Address: The Oaks, Parkend Walk, Coalway, Coleford, Glos GL16 7JR; 01594 832956.
e-mail: wildimages@terry-wall.com
www.terry-wall.com

Directory of BTO Speakers

This directory has been compiled to assist bird clubs and similar organisations in locating speakers for their indoor meetings.

Each entry within the directory consists of a named individual along with their e-mail address, a list of talks/lectures undertaken and details of fees, expenses and travelling distances.

If you wish to contact anyone listed in this directory, you may do so directly by letter or telephone or by e-mail as indicated by the individual speaker's entry.

Clubs that are part of the BTO/Bird Clubs Partnership, and are more than 120 miles from Thetford should contact Derek Toomer about special rates.

APPLETON, Graham
Head of Fundraising and Publicity.
Subjects: The Work of the BTO, The BTO Migration Atlas, *Time to Fly* – bird Migration. **Fee:** BTO fee £35–£40. **Expenses:** Negotiable. **Distance:** Dependant on expenses.
E-mail: graham.appleton@bto.org

BAILLIE, Dr Stephen
Director of Populations Research.
Subjects: Migration Watch, Population Monitoring.
Fee: BTO fee £35–£40. **Expenses:** Travel.
Distance: By agreement. E-mail: stephen.baillie@bto.org

BAKER, Jeff
Head of Membership
Subjects: The work of the BTO, 'Little brown jobs' – warblers and how to identify them.
Fee: £40. **Expenses:** Travel expenses.
Distance: Dependent on expenses.
E-mail: jeff.baker@bto.org

BALMER, Dawn
Demography Unit Population Biologist.
Subjects: Bird Ringing, Ringing in Lesvos, BTO work in general, Golden Pheasants.
Fee: BTO fee £35–£40. **Expenses:** Travel.
Distance: Anywhere.
e-mail: dawn.balmer@bto.org

BEAVEN, Peter
Nest Records Officer.
Subjects: Barn Owls in Britain, Nests and nest recording , Insearch of British Sand Martins - ringing in West Africa, The wonders of bird migration, The work of the BTO. **Fee:** BTO fee £35–£40, £50 for private talks. **Expenses:** Travel. **Distance:** Anywhere.
e-mail: peter.beaven@bto.org

BLACKBURN, Jez
Ringing Unit Recoveries and Licencing Team Leader.
Subjects: Bird moult (suitable for ringers), Sule Skerry.
Fee: BTO fee £35–£40, £50 for private talks.
Expenses: Travel. **Distance:** East Anglia.
e-mail: jez.blackburn@bto.org

CHAMBERLAIN, Dr Dan
Senior Research Ecologist, BTO Scotland.
Subjects: Garden BirdWatch, Breeding Bird Survey.

DIRECTORY OF LECTURERS

Fee: BTO fee £35–£40. **Expenses:** Travel.
Distance: By agreement.
e-mail: dan.chamberlain@bto.org

CLARK, Jacquie
Head of Ringing Unit.
Subjects: Waders and Severe Weather, Why
Ring Birds? Ringing for Conservation, Birds
and Weather, The Migration Atlas. **Fee:** BTO
fee £35–40. **Expenses:** Petrol. **Distance:** 100
mile radius of Thetford.
e-mail: jacquie.clark@bto.org

CLARK, Dr Nigel
Head of Projects Development Unit.
Subjects: Waders, Man and Estuaries,
Horseshoe Crabs and Waders, Migration
through Delaware in Spring.
Fee: BTO fee £35–40. **Expenses:** Petrol.
Distance: 100 mile radius of Thetford.
E-mail: nigel.clark@bto.org

Crick, Dr Humphrey
Head of Demography Unit.
Subjects: Climate Change and Birds, One
Million Nests.
Fee: BTO fee £35–£40 £50 for private.
Expenses: Mileage @ 25p per mile. **Distance:**
Willing to travel, prefer less than 100 miles
from Cambridge.
e-mail: humphrey.crick@bto.org

FULLER, Dr Rob
Director of Habitats Research.
Subjects: Monitoring Woods for Biodiversity,
Woodland Management and Birds, The
Importance of BTO Surveys to Bird
Conservation. **Fee:** BTO fee £35–£40.
Expenses: Travel. **Distance:** Anywhere.
E-mail: rob.fuller@bto.org

GILLINGS, Dr Simon
Terrestrial Ecology Unit Research Ecologist.
Subjects: Winter Golden Plovers and Lapwings,
Winter Farmland Birds. **Fee:** BTO fee £35–£40.
Expenses: Travel. **Distance:** Anywhere.
e-mail: simon.gillings@bto.org

GOUGH, Su
Terrestrial Ecology Unit Research Ecologist.

Subjects: Bird Biology, London Bird Project,
Wildlife of Canada (non-BTO talk),
Wildlife of European Mountains (non-BTO
talk).
Fee: BTO fee £35–£40, expenses for private
talks. **Expenses:** Travel. **Distance:** Negotiable.
e-mail: su.gough@bto.org

GRANTHAM, Mark
Ringing Unit Recoveries Officer.
Subjects: A range of general talks on ringing,
migration plus Bird Observatories,
Oiled sea-birds. **Fee:** BTO fee £35–£40.
Expenses: Travel. **Distance:** 100 miles.
e-mail: mark.grantham@bto.org

**GREENWOOD,
Professor J J D**
BTO Director.
Subjects: How to Change Government Policy
by Counting Birds, Why Ring Birds? The
Future for Birds.. and People, GM and Birds:
What's the Problem?
Fee: BTO fee £35–£40. £50 for private talks.
Expenses: Public transport or 35p/mile.
Distance: 100 miles from Thetford - further by
arrangement.
e-mail: jeremy.greenwood@bto.org

HENDERSON, Dr Ian
Terrestrial Ecology Unit Research Manager.
Subjects: Arable Farming and Birds.
Fee: BTO fee £35–£40. **Expenses:** Travel.
Distance: By agreement.
e-mail: ian.henderson@bto.org

HOLLOWAY, Steve
Wetland & Coastal Ecology Unit Research
Ecologist.
Subjects: Wetland Bird Survey. **Fee:** BTO fee
£35–£40. **Expenses:** Travel. **Distance:** By
agreement.
E-mail: steve.holloway@bto.org

LACK, Dr Peter
Head of Information Systems Unit.
Subjects: Bird Atlassing, Palearctic Migrants in
Africa, On Foot in Rwanda and Zambia, Bird
Ecology in East African Savannahs, General
Natural History of Eastern Africa. All are given

as non-BTO talks **Fee:** Negotiable. **Expenses:** Travel. **Distance:** 60 miles from Bury St Edmunds.
e-mail: peter.lack@bto.org

MARCHANT, John
Census Unit Team Leader.
Subjects: Heronries, Waterways Bird Survey/ Waterways Breeding Bird Survey, Breeding Bird Trends in the UK. **Fee:** BTO fee £35–£40. **Expenses:** Travel. **Distance:** By agreement.
e-mail: john.marchant@bto.org

MUSGROVE, Dr Andy
Wetland & Coastal Ecology Unit Research Manager.
Subjects: The Wetland Bird Survey, Little Egrets in the UK, Recording Moths in Your Garden (non-BTO).
Fee: BTO fee £35–£40, £30 for private talk. **Expenses:** Travel. **Distance:** By agreement.
e-mail: andy.musgrove@bto.org

NOBLE, Dr David
Head of Census Unit.
Subjects: The Farmland Bird Indicator, Population Trends.
Fee: BTO fee £35–£40. **Expenses:** Travel. **Distance:** By agreement.
e-mail: david.noble@bto.org

RAVEN, Mike
Breeding Bird Survey Organiser.
Subjects: Latest Findings from the Breeding Bird Survey. **Fee:** BTO fee £35–£40.
Expenses: Travel. **Distance:** By agreement.
E-mail: mike.raven@bto.org

REHFISCH, Dr Mark
Head of Wetland & Coastal Ecology Unit.
Subjects: Introduced Species Waterbird Alerts, Golden Pheasants, Wetland Work at the BTO, Climate Change, Water Quality and Waterbirds, Monitoring Waterbirds, Sea Level Rise and Climate Change.
Fee: BTO fee £35–£40 up to £40 for private talk. **Expenses:** Travel. **Distance:** By agreement.
e-mail: mark.rehfisch@bto.org

ROBINSON, Dr Rob
Demography Unit Senior Population Biologist.
Subjects: Farming and Birds, House Sparrows, Starling Population Declines.
Fee: BTO fee £35–£40. **Expenses:** Travel.
Distance: By agreement.
e-mail: rob.robinson@bto.org

SIRIWARDENA, Dr Gavin
Terrestrial Ecology Unit Research Manager.
Subjects: Marsh and Willow Tits – Analysis of BTO Data, Evidence of Impacts of Nest Predation and Competition, Quantifying Migratory Strategies. Currently all short talks.
Fee: BTO fee £35–£40. **Expenses:** Travel.
Distance: 50 miles further with accommodation.
e-mail: gavin.siriwardena@bto.org

TOMS, Mike
Garden BirdWatch Organiser.
Subjects: The BTO Garden BirdWatch. **Fee:** £40.00. **Expenses:** Petrol. **Distance:** 50 miles, further by Arrangement.
e-mail: michael.toms@bto.org

TOOMER, Dr Derek
Membership Development Officer.
Subjects: Making Your Birding Count – the Work of the BTO. **Fee:** BTO fee £35–£40.
Expenses: Travel. **Distance:** Negotiable.
e-mail: derek.toomer@bto.org

VICKERY, Dr Juliet
Head of Terrestrial Ecology Unit.
Subjects: Farmland Birds.
Fee: BTO fee £35–£40. **Expenses:** Travel.
Distance: 100 miles.
e-mail: juliet.vickery@bto.org

WERNHAM, Dr Chris
Senior Research Ecologist, BTO Scotland.
Subjects: The BTO's Migration Research including the Migration Atlas and later developments, The work of BTO Scotland.
Fee: £40. **Expenses:** Petrol. **Distance:** Scotland and northeast England.
e-mail: chris.wernham@bto.org

ART/PHOTOGRAPHY/LECTURERS

Wildlife Books and Resources

WILDGuides

Wildguides Ltd is a publishing company that produces definitive yet simple-to use field guides to wildlife and site guides to top wildlife destinations at home or abroad.

Full details of our current titles and our exciting planned titles for 2006 can be found on our website or by calling the Order Line and requesting a brochure.

Why not check out our current online field identification and distribution resources to Britain's orchids at

www.orchids.fieldguide.co.uk

and British arable plants at

www.arableplants.fieldguide.co.uk

Sales of many of our titles generate funds that are channelled back into conservation

ORDER LINE 01628 529297

Wildguides Ltd, Parr House, 63 Hatch Lane, Old Basing, Hampshire RG24 7EB

www.wildguides.co.uk

TRADE DIRECTORY

Yellow Wagtail by Nick Williams

BIRD GARDEN SUPPLIERS

BIRD GARDEN SUPPLIERS

BAMFORDS TOP FLIGHT

Company ethos: Family-owned manufacturing company providing good quality bird foods via a network of UK stockists or mail order. RSPB Corporate Member, BTO Business Ally, Petcare Trust Member.
Key product lines: A range of wild bird mixtures containing the revolutionary new 'Pro-tec Health Aid', developed by Bamfords, to protect and promote the welfare of wild birds. Vast array of other foods and seeds for birds.
Other services: Trade suppliers of bulk and pre-packed bird and petfoods. Custom packing/own label if required.
Opening times: Mon - Fri 8.00-5.30pm; Sat 8.00-12.00noon; (Sunday 10.00-12.00noon, Mill Shop only).
Address; Globe Mill, Midge Hall, Leyland, Lancashire , PR26 6TN:01772 456300;(Fax) 01772 456302; email: sales@bamfords.co.uk web: www.bamfords.co.uk

CJ WILDBIRD FOODS LTD

Company ethos: High quality products, no-quibble guarantee, friendly, professional service.
Key product lines: Complete range of CJ Wildlife feeders, food, nest boxes, bird tables and accessories, alongside a collection of other wildlife-related products.
Other services: Mail order company, online ordering, 24hr delivery service, free delivery over £50 (only £1.99 below). Free handbook.
Opening times: Mon-Fri (9am-5pm) Sat (9am-12pm).
Address; The Rea, Upton Magna, Shrewsbury, Shropshire, SY4 4UR; 0800 731 2820; Fax: 0870 010 9699.
e-mail: enquiries@birdfood.co.uk
www.birdfood.co.uk

ERNEST CHARLES

Company ethos: Member of Birdcare Standards Assoc. ISO 9001 registered. Offering quality birdfoods/wildlife products through a friendly mail-order service.

Key productlines: Bird foods,feeders, nest boxes and other wildlife products.
Other services: Own label work for other companies considered and trade enquiries.
Opening times: Mon to Fri(8am-5pm).
Contact: Stuart Christophers, Copplestone Mills, Crediton, Devon EX17 5NF; 01363 84842;(Fax)01363 84147.
e-mail: stuart@ernest-charles.com
www.ernest-charles.com

FIELD AND GARDEN LTD

Company ethos: To provide an inspirational range of quality garden wildlife products to attract and care for creatures great and small.
Key product lines: Wild bird food, feeders and accessories, wildlife habitats and attractants.
Opening times: Telephone and internet mail order 24/7.
Address; Neil Spinks. 01553 844055; (Fax)01553 842162.
email: info@fieldandgarden.co.uk
www.fieldandgarden.co.uk

foodforbirds.co.uk

Company ethos: Specialist mail order company supplying high quality wild bird foods via a fast and friendly next day service. Supporter of RSPB and BTO through parent company.
Key product lines: A great range of tried and tested, freshly made wild bird mixtures, together with a whole host of straight foods - peanuts, sunflowers, niger seed, fat foods etc.
Other services: Vast array of bird feeders for peanuts and seed, plus other wildlife foods, all of which can be ordered via a secure on-line website. Send for free catalogue.
Opening times: Telesales (freephone) 8.00 - 5.30pm (order before midday for next-day delivery). Answer phone outside these hours. On-line ordering and fax, 24 hours.
Contact: Foodforbirds, PO Box 247, Leyland, Lancashire PR26 6TN; (Freephone)0800 043 9022; (Fax)01772 456 302.
e-mail: sales@foodforbirds.co.uk
www.foodforbirds.co.uk

JACOBI JAYNE & CO

Company Ethos: Supplying market-leading products of highest quality and proven

BOOK PUBLISHERS

conservation worth for almost 20 years. Offering expertise and special prices to wildlife groups, schools and colleges.
Key products: Birdfeeders, birdfoods, nest boxes & accessories. UK distributor of Schwegler WoodcretePLUS™ nest boxes, Droll Yankees birdfeeders & accessories and Jacobi Jayne wildlife foods.
Other Services: *Living with Birds* mail-order catalogue.
Opening Times: 24hrs (use websites or answering service when office is closed)
Contact: Graham Evans/Sally Haynes, Jacobi Jayne & Co, Wealden Forest Park, Herne Bay, Kent, CT6 7LQ; 0800 072 0130; Fax 01227 719235; e-mail: enquiries@jacobijayne.com
www.jacobijayne.com
www.livingwithbirds.com

JAMIE WOOD LTD
Company ethos: A comprehensive range of quality hand-maide products at competitive prices as supplied to the RSPB, universities, film units, householders. Thirty years' experience.
Key products: Owl boxes, nest boxes, bird tables. High quality bird food supplier, feeders, hides, photographic electronics.
Other services: Mail order, delivery ex-stock, within seven days. Hides and screens. Individual items can be made to order.
New for 2006: New showroom and sales area. Call for opening times.
Opening times: Mon to Fri (9am-6pm), Sat (9am-12.30pm).
Contact: John Miller, Unit 17, Oaks Farm Workshops, Blackboys Road, Framfield, East Sussex TN22 5PN. 01825 890990.
www.birdbox.co.uk
e-mail: sales@birdbox.co.uk

VINE HOUSE FARM BIRD FOODS
Company ethos: Growing and selling black sunflowers, red and white millet, canary seed, naked oats and now niger direct from the farm to the public.
Key product lines: Full range of bird food plus feeders and other accessories.
Other services: Open days in the winter. to view all the finches and buntings feeding at our farm. Farm walks in the summer.

Opening times: Mon to Sat (8am-5pm).
Contact: Nicholas Watts, Vine House Farm, Deeping St Nicholas, Spalding, PE11 3DG;01775 630208; (Fax) 01775 630244.
e-mail p.n.watts@farming.co.uk
www.vinehousefarmbirdfoods.co.uk

BOOK PUBLISHERS

BRITISH ORNITHOLOGISTS' UNION
Expected during 2006: *The Birds of Sâo Tome, Principé and Annobon; The Birds of Borneo.*
Address: BOU, Dept of Zoology, University of Oxford,South Parks Road, Oxford OX1 3PS. (Tel/fax) 01865 281842.
e-mail: sales@bou.org.uk
www.bou.org.uk

BUCKINGHAM PRESS
Imprints: Single imprint company - publishers of *The Birdwatcher's Yearbook* since 1980, *Who's Who in Ornithology* (1997), *Best Birdwatching Sites* series covering Norfolk, Sussex and the Highlands of Scotland. Also publishes *Birds Illustrated*, a quarterly, subscription-only magazine (see survey of English language bird magazines page 22).
New for 2006: *Best Birdwatching Sites in Norfolk* (up-dated second edition) *Best Birdwatching Sites in North Wales*, *Who's Who in Ornithology*.
Address: 55 Thorpe Park Road, Peterborough, PE3 6LJ. Tel/fax: 01733 561739.
e-mail: admin@buckinghampress.com

CHRISTOPHER HELM PUBLISHERS
Imprints: Christopher Helm – the leading publisher of ornithology books in the world; includes many fieldguides, identification guides, family guides, county and country avifaunas, and a Where to Watch Birds series. T & AD Poyser – an acclaimed series of respected ornithology monographs. Birds of Africa – the standard series of handbooks on African birds. A & C Black – publisher of definitive natural history books.
New for 2006: *Birds of Northern South America, Volumes 1 & 2; Where to Watch Birds in World Cities; Birds of the Thai-Malay*

TRADE DIRECTORY

115

Peninsula Volume 2; Pocket Guide to Albatrosses, Petrels and Shearwaters; The Barn Swallow; Bewick's Swan; A Complete Guide to Arctic Wildlife; Digital Wildlife Photography; Orchids of Europe, North Africa and the Middle East; The Star Guide; Pocket Guide to the Reptiles and Amphibians of East Africa; Where to Watch Mammals in Britain and Ireland; Guide to the Mammals of Madagascar.
Address: 38 Soho Square, London, W1D 3HB; 020 7758 0200; (Fax)020 7758 0222.
e-mail: nredman@acblack.com
www.acblack.com/christopherhelm

HARPER COLLINS PUBLISHERS

Imprints: Collins Natural History — the leading publisher of fieldguides to the natural world.
Collins New Naturalist Series, the encyclopaedic reference for all areas of British natural history.
HarperCollins, publisher of the best illustrated books.
New for 2006: *New Naturalist: Bumblebees* by Ted Benton, *Galloway and the Borders* by Derek Ratcliffe; Collins: *How to Identify Birds* by Nicholas Hammond; *Wild Guide Whales & Dolphins* by Mark Carwardine; *Field Guide British Wildlife Sounds* by Geoff Sample; Habitat Explorers: *Seashore, Rivers, Parks & Gardens* and *Woodlands & Forest* by Nick Baker; *Illustrated Checklist Birds of Mexico* by Ber Van Perlo.
Address: 77-85 Fulham Palace Rd, Hammersmith, London, W6 8JB; 020 8307 4998; (Fax)020 8307 4037.
e-mail: Myles.Archibald@harpercollins.co.uk
www.collins.co.uk

NEW HOLLAND PUBLISHERS (UK) LTD

Imprints; New Holland, illustrated bird books, general wildlife and personality-led natural history.
New for 2006: *Nick Baker's British Wildlife* by Nick Baker (Feb), *Identification Guide to Birds of South East Asia* by Craig Robson (March), *Bill Oddies's Birding Map of Britain & Ireland* (5th edn) by Bill Oddie (March), *Complete Garden Wildlife Book* by Mark Golley (April),

The Private Life of Birds by Stephen Moss (October).
Address: Garfield House, 86-88 Edgware Road, London, W2 2EA; 020 7724 7773; (Fax)020 7258 1293.
e-mail: postmaster@nhpub.co.uk
www.newhollandpublishers.com

WILDGuides LTD

Imprints; WILD*Guides* — definitive natural history fieldguides. Hardback and flexicover.
OCEAN*Guides* — definitive identification guides to marine wildlife. Hardback and flexicover.
WILD eARTh — lavishly illustrated celebrations of wildlife and natural places. Hardback.
Your Countryside Guides — regional heritage guides for walkers. Hardback. Sales of all books benefit conservation.
New for 2006: *Birds and Mammals of the Falkland islands, Walking on the Wildside - Thames Valley and the Chilterns, Whales and Dolphins of the North American Pacific, Wildlife of the Seychelles, A Visitors Guide to South Georgia, A Photographic Guide To The Nightjars of The World.*
Address: PO Box 680, Maidenhead, Berkshire, SL6 9ST; 01628 529 297; (Fax)01628 525314.
e-mail: info@wildguides.co.uk
www.wildguides.co.uk

BOOK SELLERS

ATROPOS

Company ethos: Lively magazine for butterfly, moth and dragonfly enthusiasts. Mail order book service providing key titles swiftly at competitive prices.
Key subjects: Butterflies, moths, dragonflies and other insects.
Address: 36 Tinker Lane, Meltham, Holmfirth, West Yorkshire HD9 4EX.
e-mail: atropos@atroposed.freeserve.co.uk
www.atropos.info

NHBS ENVIRONMENT BOOKSTORE

Company ethos: A unique natural history, conservation and environmental bookstore.
Key subjects: Natural history, conservation,

environmental science, zoology, habitats and ecosystems, botany, marine biology.
Other services: NHBS.com offers a searchable and browseable web catalogue with more than 95,000 titles.
Opening times: Mon-Fri (9am-5pm). for mail-order service.
Address: 2-3 Wills Road, Totnes, Devon, TQ9 5XN; 01803 865913; (Fax) 01803 865280. e-mail: nhbs@nhbs.co.uk
www.nhbs.com

ORNITHOLIDAYS BOOK STOP

Company ethos: Friendly and helpful staff on hand to assist in the purchase of ornithological and natural history books.
Key subjects: Ornithology and natural history books by mail order.
Other services; We are primarily a tour operator sending birdwatching and natural history tours worldwide. Since 2000 we have complemented our business by successfully supplying a wide range of books published by the leading companies at a 10% discount with free postage and packing within the UK.
Opening times: Mail order/internet only. Telephone lines are open Mon-Fri (9am-5pm).
Address: 29 Straight Mile, Romsey, Hampshire SO51 9BB;01794 523500; (Fax)01794 523544.
e-mail: ruth@ornitholidays.co.uk
www.ornitholidays.co.uk

PICTURE BOOK

Company ethos: Knowledgeable staff, natural history books new and secondhand available during shop hours to browse.
Key subjects: Birdwatching, natural history, travel guides.
Other services: Mail order, catalogue available.
Opening times: Mon-Sat (9.15am-5pm).
Address: Picture Book, 6 Stanley Street, Leek ST13 5HG;01538 384337; (Fax)01538 399696. e-mail: info@leekbooks.co.uk
www.leekbooks.co.uk

PORTLAND OBSERVATORY BOOK SHOP

Company ethos: To meet the needs of amateur and professional naturalists.

Key subjects: Ornithology, general natural history, topography, art and local history. New and secondhand.
Other services: Mail order, discount on new books, increased discount for observatory members.
Opening times: Wednesday, Friday, Saturday and Sunday; (10am to 4pm) throughout the year. Other times on request.
Address; Bird Observatory, Old Lower Light, Portland Bill, Dorset, DT5 2JT; 01305 820553, (shop) 01305 826625, (home) 01225 700728.
e-mail: petermowday@tiscali.co.uk
www.portlandbirdobs.btinternet.co.uk

SECOND NATURE

Company ethos: Buying and selling out-of-print/secondhand/antiquarian books on natural history, topography and travel.
Key subjects: Birds, mammals and travel with a natural history interest. Very large specialist stock.
Other services; Occasional catalogues issued. Often exhibiting at bird/natural history fairs.
Opening times: Mail order only.
Address; Knapton Book Barn, Back Lane, Knapton, York, YO26 6QJ; (Tel/fax) 01904 339493.
e-mail: SecondnatureYork@aol.com

SUBBUTEO BOOKS

Company ethos: Specialist knowledge on all aspects of wildlife, travel and natural history, friendly service.
Key subjects: Wildlife, natural history and travel books.
Other services: Source any natural history book from around the world. Online ordering, free delivery over £50 (UK orders), free catalogue.
Opening times: Mon-Fri (9am-5pm) Sat 9am-12pm.
Address: The Rea, Upton Magna, Shrewsbury, Shropshire, SY4 4UR; 0870 010 9700; (Fax) 0870 010 9699.
e-mail: info@wildlifebooks.com
www.wildlifebooks.com

CLOTHING AND EQUIPMENT SUPPLIERS

WILDSIDE BOOKS AND GALLERY

Company ethos: Suppliers of quality antiquarian and out-of-print books, efficient and personal customer service by Chris and Christine Johnson.

Key subjects: Fine and rare antiquarian, out-of-print and new, bird and general natural history books, wildlife art (paintings, sculpture, drawings, prints).

Other services: Catalogues issued quarterly, UK and international mail order, secure on-line purchasing from website.

Opening times: By appointment

Address: Rectory House, 26 Priory Road, Great Malvern, Worcestershire WR14 3DR; 01684 562 818; (Fax)01684 566 491.

e-mail: enquiries@wildsidebooks.co.uk
www.wildsidebooks.co.uk

CLOTHING SUPPLIERS

COUNTRY INNOVATION

Company ethos: Friendly advice by well-trained staff.

Key product lines: Full range of outdoor wear: Jackets, smock, fleeces, trousers, walking boots, poles, lightweight clothing, hats, gloves, bags and pouches. Ladies fit available.

Other services: Mail order.

Opening times; Mon-Fri (9am-5pm).

Address: 1 Broad Street, Congresbury, North Somerset, BS49 5DG; 01934 877333; (Fax)01934 77999.

e-mail: sales@countryinnovation.com
www.countryinnovation.com

EQUIPMENT AND SERVICES

ALWYCH BOOKS

Company ethos: The Bird Watcher's All-weather Flexible Pocket Book.

Key product lines: Alwych all-weather notebooks.

Address: Jennifer Mallory, JR Reid Print and Media Group, 79-109 Glasgow Road, Blantyre, Glasgow G72 0LY;01698 307415.

www.alwych.co.uk

ART DECOY

Company ethos: Beautiful hand finished wooden birds.

Key product lines: Carved ducks, waders etc.

New for 2006: New birds always being introduced.

Address: Rod Dennis, Beck Cottage, Wash Lane, Forncett St Peter, Norfolk NR16 1LS; 01508 530 977.

e-mail: info@artdecoy.co.uk
www.artdecoy.co.uk

BIRDGUIDES LTD

Company ethos: Top quality products and services, especially using new technologies such as CD-ROM, DVD, websites, plus one-stop on-line shop for books, bird food etc.

Key product lines: DVD-ROM, CD-ROM, DVD, video guides to British, European and American birds. Rare bird news services via e-mail, website and SMS.

Address: Dave Gosney, Jack House, Ewden, Sheffield, S36 4ZA; 0114 2831002; order line (freephone) 0800 919391.

e-mail: birdguides@birdfood.co.uk
www.birdguides.com

BIRD IMAGES

Company ethos: High quality products at affordable prices.

Key product lines: Bird videos and DVDs.

New for 2006: Double DVD on waders.

Opening times: Telephone first.

Address: Paul Doherty, 28 Carousel Walk, Sherburn in Elmet, North Yorkshire LS25 6LP; 01977 684666.

e-mail: paul@birdvideodvd.com
www.birdvideodvd.com

EagleEye OpticZooms

Company ethos: Innovative design, quality manufacturing, custom products, comprehensive and expert advice on all aspects of digital photography and digiscoping.

Key product lines: Telephoto lenses for fixed lens digital cameras/camcorders, digiscoping adapters and accessories, custom digital camera accessories.

Opening times: Mon-Fri (9am-5.30pm).

Address: Carlo Bonacci, Wentshaw Lodge,

Fairseat, Sevenoaks, Kent, TN15 7LR, UK;
Tel/(Fax) 01474 871219.
e-mail: info@eagleeyeuk.com
www.eagleeyeuk.com

EVERETT ASSOCIATES LTD

Company ethos: Friendly support for all customers, especially for those new to computing.
Key product lines: CDs of Western Palearctic, world birds, butterflies and moths databases.
New for 2006: Fishing and house contents databases.
Address: Peter Everett, Longnor House, Gunthorpe, Norfolk NR24 2NS; Tel/fax 01263 860035.
e-mail: everett:birdsoftware.co.uk
www.birdsoftware.co.uk

GOLDEN VALLEY INSURANCE SERVICES

Company ethos: Knowledgeable, friendly insurance services. Free information pack on request. Freephone telephone number for all enquiries.
Key product lines: Insurance for optical/photographic/video/computer equipment for birdwatchers. Also, public liability for ornithological clubs and societies.
Opening times: Mon-Fri (9am-5pm), answering machine at other times
Address: Sharron or Marion, Golden Valley Insurance Services, The Olde Shoppe, Ewyas Harold, Herefordshire HR2 0ES;0800 015 4484; (Fax)01981 240451.
e-mail: gvinsurance@aol.com

HARVEY MAPS

Company ethos: Over 25 years, HARVEY has gained a reputation for high quality maps for outdoor recreation. These award winning maps are compiled from original aerial surveys and field checked by our experienced surveyors, themselves hill-walkers.
Key Product lines: Detailed maps for walking
New for 2006: Dartmoor National Park Atlas, SW Coast Path maps.
Opening times: 8.30-17.30 Monday to Friday
Contact: Susan Harvey, 01786 841202; (fax)01786 841098; www.harveymaps.co.uk
e-mail: sales@harveymaps.co.uk

WILDSOUNDS

Company ethos: Donates a significant portion of profit to bird conservation, committed to sound environmental practices, official bookseller to African Bird Club (ABC) and Ornithological Society of the Middle East (OSME). Operates a Commission for Conservation Programme.
Key product lines: Mail order, post-free books, multi-media guides and eGuides for PDAs - mobile versions of popular field guides complete with bird sounds and listing software. i.e. the award winning *Collins Bird eGuide* and a mobile version of the best-selling *Sasol Birds of Southern Africa* fieldguide. Field recording equipment.
Opening times: Weekdays (9.30am-5pm).
Address: Cross Street, Salthouse, Norfolk, NR25 7XH; (Tel/fax) +44(UK) (0) 1263 741100. e-mail: duncan@wildsounds.com
www.wildsounds.com

WILDLIFE WATCHING SUPPLIES

Company ethos: To bring together a comprehensive range of materials, clothing and equipment to make it easier and more comfortable for you to blend in with the environment. Quick and friendly service.
Key product lines: Hides, camouflage, bean bags, lens and camera covers, clothing etc.
Opening times: Mon to Fri (9am-5pm), Mail order. Visitors by appointment.
Address: Town Living Farmhouse, Puddington, Tiverton, Devon, EX16 8LW; 01884 860692(24hr); (Fax) 01884 860994.
e-mail: enquiries@wildlifewatchingsupplies.co.uk
www.wildlifewatchingsupplies.co.uk

HOLIDAY COMPANIES

AVIAN ADVENTURES

Company ethos: Top quality, value for money tours, escorted by friendly, experienced leaders at a fairly relaxed pace. ATOL 3367.
Types of tours: Birdwatching, birds and wildlife photography and wildlife safaris, all suitable for both the first-time and the more experienced traveller.

HOLIDAY COMPANIES

Destinations: More than 70 tours worldwide.
Brochure from: 49 Sandy Road, Norton, Stourbridge, DY8 3AJ; 01384 372013; (Fax) 01384 441340.
e-mail: aviantours@argonet.co.uk
www.avianadventures.co.uk

BRITISH-BULGARIAN FRIENDSHIP SOCIETY
Company ethos: To introduce people to the beauty of Bulgarian wildlife at exceptional value prices with expert leaders.
Types of tours: Birdwatching tours in Spring and Autumn, also butterfly, wild flower and natural history tours. Group size 12-14 persons.
Destinations: Specialists to Bulgaria, over 30 years experience.
New for 2006: Wildlife photography tour.
Brochure from: Our ATOL-bonded agents, Balkania Travel Ltd, Room 40, Third Floor, Morley House, 320 Regent Street, London W1B 3BE; (Tel) 020 7538 8654 or 020 7636 8338. General enquiries: Dr Annie Kay, 020 7237 7616. e-mail: annie.kay@btinternet.com
www.bbfs.org.uk

BIRDFINDERS
Company ethos: Top-value birding tours to see all specialities/endemics of a country/area, using top UK and local guides. ATOL 5406.
Types of tours: Birdwatching tours for all abilities.
Destinations: 47 tours in UK, Europe, Africa, Asia, Australasia, North and South America and Antarctica.
New for 2006: Georgia & Armenia, Ethiopia and Montana, USA.
Brochure from: Vaughan Ashby, Westbank, Cheselbourne, Dorset, DT2 7NW. 01258 839066.
e-mail: info@birdfinders.co.uk
www.birdfinders.co.uk
Our office is open seven days a week.

BIRD HOLIDAYS
Company ethos: Relaxed pace, professional leaders, small groups, exciting itineraries.
Types of tours: Birdwatching for all levels, beginners to advanced.

Destinations: Worldwide (40 tours, 5 continents).
New for 2006: Japan, Kenya, Slovakia, Antarctica.
Brochure from: 10 Ivegate, Yeadon, Leeds, LS19 7RE; (Tel/fax)0113 3910 510.
e-mail: info@birdholidays.co.uk
www.birdholidays.co.uk

CARPATHIAN WILDLIFE SOCIETY
Company ethos: A non-profit organisation bringing together people with a shared interest in conservation of large mammals and birds.
Types of tours: Wildlife tours for everyone contributing to research. Tracking of wolves, bears and lynx. Enjoyable birdwatching holidays.
Destinations: Slovakia. National parks, primeval forests and wetlands.
New for 2006: Top birding sites, Senne wetland and Polana Wildlife Reserve.
Brochure from: Driftwood, The Marrams, Sea Palling, Norfolk NR12 0UN; 01692 598135; (Fax) 01692 598141.
e-mail: cws@szm.sk
www.cws.szm.sk

CLASSIC JOURNEYS
Company ethos: Professional and friendly company, providing well organised and enjoyable birdwatching holidays.
Types of tours: General birdwatching and wildlife holidays on the Indian sub-continent.
Destinations: Nepal, India, Bhutan, Sri Lanka.
New for 2006: South India.
Brochure from: 33 High Street, Tibshelf, Alfreton, Derbyshire, DE55 5NX; 01773 873497; (Fax)01773 590243.
e-mail: birds@classicjourneys.co.uk
www.classicjourneys.co.uk

EXPEDITION CRUISES & CRUISES FOR NATURE
Company ethos: We are specialists in cruises on expedition ships, with expertise for birdwatchers, photographers and naturalists. We also have excorted cruises, using professional guides, to amazing destinations. ATOL licensed.
Types of cruises: Both escorted and

120

unescorted on expedition ships.
Destinations: Worldwide
New for 2006: Bridging Beringia (Kamchatka & Alaska), Galapagos Islands, Spitsbergen, Antarctica, Humboldt Current - and much more!
Brochure from: 29 Straight Mile, Romsey, Hampshire, SO51 9BB; 01794 523 500; (Fax)01794 523544.
e-mail: info@expeditioncruising.co.uk
www.expeditioncruising.co.uk

HEATHERLEA

Company ethos: Exciting holidays to see all the birds of Scotland and selected overseas destinations. Experienced guides and comfortable award winning hotel to give great customer service.
Types of tours: Birdwatching and other wildlife watching tours.
Destinations: Scottish Highlands, including holidays from our base in Nethybridge, plus Outer Hebrides, Orkney, Shetlands and more. Selected destinations include Pyrenees, Lesvos, Mallorca and Trinidad.
New for 2006: Scottish weeks including Ardnamurchan, Lewis, Harris and Skye, new 'go further' destinations including St Kilda and Shetland, and Brand New overseas holidays to Hungary, Extremadura, Finland and more.
Brochure from: The Mountview Hotel, Nethybridge, Inverness-shire, PH25 3EB; 01479 821248; (Fax)01479 821515.
e-mail: hleabirds@aol.com
www.heatherlea.co.uk

HONEYGUIDE WILDLIFE HOLIDAYS

Company ethos: Relaxed natural history holidays with wildlife close to home. Quality accommodation, expert leaders, beginners welcome.
Types of holidays: Birds, flowers and butterflies, with a varied mix depending on the location.
Destinations: Europe, including Algarve, Extremadura, Spanish Pyrenees, Crete, Lesvos, Menorca, Dordogne, Danube Delta and Hungary.
New for 2006: From the Alps to the Adriatic - E Italy and Slovenia.

Brochure from: 36 Thunder Lane, Thorpe St Andrew, Norwich, Norfolk, NR7 0PX; 01603 300552(Evenings).
e-mail: honeyguide@tesco.net
www.honeyguide.co.uk

HOSKING TOURS LTD

Company ethos: The best in wildlife photographic holidays.
Types of tours: Wildlife photography for all levels of experience.
Destinations: Africa, America, Europe.
New for 2006: Australia, Galapagos.
Brochure from: Pages Green House, Wetheringsett, Stowmarket, Suffolk, IP14 5QA; 01728 861113; (Fax)01728 860222.
www.hosking-tours.co.uk

LIMOSA HOLIDAYS

Company ethos: The very best in birdwatching and wildlife holidays - expertly led, fun, friendly and packed with great birding and wildlife. ATOL 2950. AITO member. AITO Trust 1049.
Types of tours: Birdwatching and wildlife tours and cruises.
Destinations: Celebrating our 21st year of operation, our ALL NEW 2006 brochure features more than 100 departures worldwide.
New for 2006: Twenty-three new tours including - seven in Europe, two in South Africa (one for owls), and two in Canada. New cruises include Scottish islands, Cape Verde, Gambia and Senegal and the Gulf, Finland. Also Fair Isle and Shetland and Islay and the Scottish Highlands.
Brochure from: Limosa Holidays, Suffield House, Northrepps, Norfolk, NR27 0LZ; 01263 578143; (Fax)01263 579251.
e-mail: info@limosaholidays.co.uk
www.limosaholidays.co.uk

NATURETREK

Company ethos: Friendly, gentle-paced, birdwatching holidays with broad-brush approach. Sympathetic to other wildlife interests, history and local culture. ATOL no 2962.
Types of tours: Escorted birdwatching, botanical and natural history holidays worldwide.

HOLIDAY COMPANIES

Destinations: Worldwide - see brochure.
New for 2006: Carmargue, Catalonia, China, Northern Cyprus, Dominica, India's vultures, Northern Madagascar, Tasmania, West Texas and Mexico, Yellowstone Park.
Brochure from: Cheriton Mill, Cheriton, Alresford, Hampshire, SO24 0NG; 01962 733051; (Fax)01962 736426.
e-mail: info@naturetrek.co.uk
www.naturetrek.co.uk

NORTHERN FRANCE WILDLIFE TOURS
Company ethos: Friendly personal attention. Normally a maximum of five in a group. Totally flexible.
Types of tours: Mini-bus trips catering for all from beginners to experienced birders. Local birds include Bluethroat, Black Woodpecker, Melodious Warbler.
Destinations: Brittany, Normandy and Pays de la Loire.
Brochure from: Place de L'Eglise, 53700, Averton, Mayenne, France; 0033 243 006 969. e-mail: nfwt@online.fr
www.northernfrancewildlifetours.com

NORTH WEST BIRDS
Company ethos: Friendly, relaxed and unhurried, but targetted to scarce local birds.
Types of tours: Very small groups (up to four) based on large family home in South Lakes with good home cooking. Short breaks with birding in local area. Butterflies in season.
Destinations: Local to Northwest England. Lancashire, Morecambe Bay and Lake District.
Brochure from: Mike Robinson, Barn Close, Beetham, Cumbria, LA7 7AL; (Tel/fax) 015395 63191. www.nwbirds.co.uk
e-mail: mike@nwbirds.co.uk

RAY NOWICKI
Company ethos: To provide a professional guiding service, ensuring that everyone, from beginner to expert, enjoys a unique wildlife experience.
Types of tours: Individuals or small groups requiring help in finding specific birds, plants, mammals and more!
Destinations: Scottish Highlands and islands.

New for 2006: Spain - birds.
Brochure from: Linwood, Nethybridge, Inverness-shire PH25 3DR;01479 821 215.
e-mail: nowicki@freeuk.com

ORKNEY ISLAND WILDLIFE
Company ethos: Provide high quality holidays for discerning travellers to learn about Orkney and Shetland from knowledgeable local experts, with an emphasis on fun and enjoyment, in a relaxed informal atmosphere.
Types of tours: Multi-interest holidays, exploring all aspects of Orkney and Shetland, thieir birds, flowers, archaeology, history and culture.
Destinations: Specialising in Orkney and Shetland.
New for 2006: Celebrating 20 years of holidays in Orkney.
Brochure from: Paul and Louise Hollinrake, Orkney Island Holidays, Furrowend, Balfour, Shapinsay 19, Orkney, KW17 2DY; 01856 711 373. e-mail: holidays@orkney.com
www.orkneyislandholidays.com

ORNITHOLIDAYS
Company ethos: Full-time tour leaders and a company with more than 40 years experience. ABTA member. ATOL no 0743.
Types of tours: Escorted birdwatching and natural history tours.
Destinations: More than 70 worldwide.
New for 2006: Ghana, Taiwan, China & Tibet, Gabon, India - Gujarat, Southern India at Christmas.
Brochure from: 29 Straight Mile, Romsey, Hampshire, SO51 9BB; 01794 519445; (Fax)01794 523544.
e-mail: info@ornitholidays.co.uk
www.ornitholidays.co.uk

SHETLAND WILDLIFE
Company ethos: Award-winning small group travel with the very best naturalist guides.
Types of tours: A unique blend of itineraries to bring you the very best of Shetland. Week-long or three-day holidays dedicated to wildlife, photography, walking and archaeology.
Destinations: All corners of Shetland including Fair Isle.
Brochure from: Longhill, Maywick, Shetland, ZE2 9JF; 01950 422483; (Fax)01950 422430.

HOLIDAY COMPANIES

e-mail; info@shetlandwildlife.co.uk
www.shetlandwildlife.co.uk

SICKLEBILL SAFARIS LTD
Company ethos: Qualified and very experienced, genial leader, to show you the real natural world. Under ATOL 4002.
Types of tours: Birdwatching and general natural history tours including mammals, insects, higher plants and macrofungi.
Destinations: East Anglia, Papua New Guinea Irian Jaya and Myanmar.
New for 2006: Solomon Islands.
Brochure from: Trefor, Creake Road, Sculthorpe, Fakenham, Norfolk, NR21 9NQ;01328 856925.
e-mail: Ian@sicklebill.demon.co.uk
www.sicklebill.com

SPEYSIDE WILDLIFE
Company ethos: Expert leaders, personal attention and a sense of fun - it's your holiday. ATOL no 4259.
Types of tours: Experts in Scotland and leaders worldwide - birdwatching, mammals and whale watching.
Destinations: Speyside and the Scottish Islands, Scandinavia, the Arctic, Europe, N America, Africa and Australia.
New for 2006: Gambia, Australia, California and Arctic Canada.
Brochure from: Garden Office, Inverdruie House, Inverdruie, Aviemore, Inverness-shire, PH22 1QH; (Tel/fax)01479 812498.
e-mail: enquiries@speysidewildlife.co.uk
www.speysidewildlife.co.uk

SUNBIRD
Company ethos: Enjoyable birdwatching tours led by full-time professional leaders. ATOL no 3003
Types of tours: Birdwatching, Birds & Music, Birds & History, Birds & Butterflies, Sunbirder events.
Destinations: Worldwide.
New for 2006: Sicuan; China – Crested Ibis and terracotta warriors; Northern China; Taiwan; Karelia (NE Finland & NW Russia); N Spain - Birds and Butterflies; Transylvania; New England; Zambia.

Brochure from: PO Box 76, Sandy, Bedfordshire, SG19 1DF; 01767 682969; (Fax) 01767 692481.
e-mail: sunbird@sunbirdtours.co.uk
www.sunbirdtours.co.uk

THE BIRD ID COMPANY - PAUL LAURIE
Company ethos: Expert tour guides teaching bird watchers of all levels bird identification and field craft skills.
Types of tours: Daily guided tours £30. Weekend breaks, five day migration tours, rare breeding bird tours to see Golden Oriole, Montagu's Harrier and Honey Buzzard. Personalised and customised tours UK and abroad.
Destinations: Norfolk, Britain, Europe, America, Middle East.
New for 2006: Ireland, Shetland, Dorset
Brochure from: Paul Laurie, 9B Chapel Yard, Albert Street, Holt, Norfolk, NR25 6HG; 1263 710203. e-mail: paul.seethebird@fsmail.net
www.birdtour.co.uk

THE TRAVELLING NATURALIST
Company ethos: Friendly, easy-going, expertly-led birdwatching and wildlife tours. Atol no.3435. AITO 1124.
Types of tours: Tours include birds and history, birds and bears, whale-watching, birds and flowers.
Destinations: Worldwide.
New for 2006: North & West Australia, Cévennes, North-west Greece, Southern India, Peninsular Malaysia, Manitoba in Spring, Norway for Orcas, Provence, Sikkim & Assam, Sutherland, Central Sweden, Turkey (Birds and Total Eclipse).
Brochure from: PO Box 3141, Dorchester, Dorset, DT1 2XD; 01305 267994; (Fax) 01305 265506.
e-mail: jamie@naturalist.co.uk
www.naturalist.co.uk

THE ULTIMATE TRAVEL COMPANY
Company ethos: Quality wildlife experiences and shared enjoyment of the natural world.
Types of tours: Relaxed wildlife and birdwatching holidays with friendly groups and Britain's most experienced leaders.

OPTICAL MANUFACTURERS AND IMPORTERS

Destinations: Various locations in Africa, South America, Indian Ocean and Europe
Brochure from: The Ultimate Travel Company, 25-27 Vanston Place, London, SW6 1AZ; 020 7386 4676; (Fax)020 7381 0836.
email:
enquiry@theultimatetravelcompany.co.uk

THINKGALAPAGOS
Company ethos: Specialists in the Galapagos Islands, with expert guides and personal attention to ensure a once-in-a-lifetime adventure travel experience.
Types of tours: Friendly and relaxing holidays that are educationally orientated for people with a keen interest in wildlife and photography. Suitable for both the first-time and more experienced traveller.
Destinations: Galapagos and mainland Ecuador.
Brochure from: Rachel Dex, 25 Trinity Lane, Beverley, East Yorkshire HU17 0DY; 01482 872716. www.thinkgalapagos.com
e-mail: info@thinkgalapagos.karoo.co.uk

WILD INSIGHTS
Company ethos: Our goal is to enable clients to savour, understand and enjoy birds and wildlife fully, rather than simply build large tick lists. ATOL no 5429 (in association with Wildwings)
Types of tour: Relaxed UK workshops, Reader Breaks for *Bird Watching* magazine and skills-building UK courses, plus selected overseas tours.
Destinations: Various UK locations, plus USA, Africa, Northern India and Europe.
Calender brochure from: Yew Tree Farmhouse, Craignant, Selattyn, Oswestry, Salop SY10 7NP. (Tel/fax) 01691 718 7401; e-mail: keith.offord@virgin.net
www.wildinsights.co.uk

WILDWINGS
Company ethos: Superb value holidays led by expert guides.
Types of tours: Birdwatching holidays, whale and dolphin watching holidays, wildlife cruises, ecovolunteers.
Destinations: Europe, Arctic, Asia, The

Americas, Antarctica, Africa, Trinidad and Tobago, Jamaica.
New for 2006: Emperor Penguins, Christmas Island, Slovakia, Ukraine, The Western Pacific Odyssey, Ecovolunteers Brazil Toucans Project.
Brochure from: 577-579 Fishponds Road, Fishponds, Bristol, BS16 3AF;
e-mail: wildinfo@wildwings.co.uk

WINGSPAN BIRD TOURS
Company ethos: To seek out and enjoy the rich bird life of Spain whilst taking into account its delicate and sensitive nature. Professionally led tours, years of experience, fun, relaxed and easy going, a truely memorable holiday experience.
Types of Tours: Bird-watching at its best with Relaxed Day trips, Tailor-made tours, fully inclusive 7-day tours and 2 week Grand Tours of Spain. Great accommodation, Self catering, B & B or Full Board, superb food, magnificent scenery, English guide - Spanish delights.
Destinations: Spain: Andalucia - Coto Donana, Extremadura, Pyrenees and Picos de Europa. Greece -Lesvos.
Brochure from: Bob Buckler, 34 Southway Drive, Yeovil, Somerset BA21 3ED. 01935 422590; e-mail: bobbuckler49@hotmail.com
www.wingspanbirdtours.com

OPTICAL MANUFACTURERS AND IMPORTERS

ACE OPTICS
Company ethos: To be the best - service, price and stock.
Product lines: Importers of Optolyth, Ace Avian, Questar and other products. All the best tripods and an array of optical related accessories.
Address: 16 Green Street, Bath, BA1 2JZ; 01225 466364; (fax) 01225 469761.
e-mail: aceoptics@ba12jz.freeserve.co.uk
www.acecameras.co.uk

CARL ZEISS LTD
Company ethos: World renowned, high

quality performance and innovative optical products.

Product lines: Product ranges of stabilised, Victory, Dialyt, compacts and binoculars, now enhanced by the introduction of the FL family of binoculars, and Diascope telescopes.
Address: PO Box 78, Woodfield Road, Welwyn Garden City, Hertfordshire, AL7 1LU; 01707 871350; (Fax) 01707 871287.
e-mail: binos@zeiss.co.uk
www.zeiss.co.uk

INTRO 2020 LTD

Company ethos: Experienced importer of photo and optical products.
Product lines: Summit (binoculars and scopes), Steiner (binoculars), Velbon (tripods), Kenko (range of scopes), Slik (tripods), Hoya and Cokin (filters), Oyster and Crumpler (bags).
Address: Unit 1, Priors Way, Maidenhead, Berkshire, SL6 2HR; 01628 674411; (Fax) 01628 771055.
e-mail: sales@intro2020.co.uk
www.intro2020.co.uk

LEICA CAMERA LTD

Company ethos: Professional advice from Leica factory-trained staff.
Product lines: Duovid the world's first dual magnification binocular. Trinovid and new Ultravid full-size binoculars, Trinovid compacts. Televid 62 and 77 spotting scopes with angled or straight view with a choice of five eyepieces, a photo-adapter and a digiscope adaptor for the Leica Digilux 1 digital camera.
Address: Leica Camera Limited, Davy Avenue, Knowlhill, Milton Keynes, MK5 8LB; 01908 256400; (Fax) 01908 671316.
e-mail: info@leica-camera.co.uk
www.leica-camera.com

MARCHWOOD

Company ethos: Quality European optics offering outstanding value for money.
Product lines: Binoculars and Meopta telescopes from the Czech Republic and Eschenbach Optik binoculars from Germany. Dowling and Rowe optics for the discerning observer.
Address: Unit 308, Cannock Chase Enterprise

Park, Hednesford, Staffordshire, WS15 5QU; 01543 424255; (Fax) 01543 422082.
e-mail: john@lancashirej.freeserve.co.uk

MINOX UK LIMITED

Company ethos: A very old brand name with new company technology and attitude.
Product lines: Minox binoculars, telescopes and cameras.
Address: Old Sawmills Road, Faringdon, Oxon SN7 7DS; 01367 243535; (Fax) 01367 241124. www.minox.com
e-mail: sales@minoxuk.co.uk

OPTICRON

Company ethos: To provide the highest quality, value-for-money optics for today's birdwatcher.
Product lines: Official importers of Opticron binoculars and telescopes, plus mounting systems and accessories.
Address: PO Box 370, Unit 21, Titan Court, Laporte Way, Luton, LU4 8YR; 01582 726522: (Fax) 01582 273559.
e-mail: info@opticron.co.uk

PYSER-SGI LTD

Company ethos: Our primary aim is customer care and satisfaction, achieved through technological leadership, quality technical assistance and post-sales service support.
Product lines: Pyser-SGI (13 full-size, two compact binoculars/accessories), Swift Optics (16 full-size, four compact binoculars, one telescope and one spotting scope), Kowa (five full-size, four compact binoculars/one digital spotting scope, 13 spotting scopes/cases/eyepieces/adaptors), Skua (cases), Niggeloh (straps and cushions).
Address: Fircroft Way, Edenbridge, Kent, TN8 6HA. 01732 864111; (Fax)01732 865544.
e-mail: sales@pyser-sgi.com
www.pyser-sgi.com

SWAROVSKI UK

Company ethos: Constantly improving on what is good in terms of products and committed to conservation world-wide.
Product lines: ATS/STS 80 spotting scope and EL 8x32 and 10x32 binoculars, the latest

OPTICAL DEALERS

additions to a market-leading range of telescopes and SLC binoculars. Swarovski tripods also available, plus a range of branded rucksacks and other travel bags.
Address: Perrywood Business Park, Salfords, Surrey, RH1 5JQ; 01737 856812; (Fax) 01737 856885.
e-mail: christine.percy@swarovskioptik.co.uk
www.swarovskioptik.com

VICKERS SPORTS OPTICS
Company ethos: Importers of world renowned products from American-based company Bushnell.
Product lines: High performance binoculars (including compacts) and telescopes, includes market-leading Natureview range and Legacy compacts.
Address: Unit 9, 35 Revenge Road, Lordswood, Kent, ME5 8DW; 01634 201284; (Fax) 01634 201286.
e-mail: info@jjvickers.co.uk
www.jjvickers.co.uk

OPTICAL DEALERS

EAST MIDLANDS AND EAST ANGLIA

BIRDNET OPTICS LTD
Company ethos: To provide the birdwatcher with the best value for money on optics, books and outdoor clothing.
Viewing facilities: Clear views to distant hills for comparison of optics at long range and wide variety of textures and edges for clarity and resolution comparison.
Optical stock: Most leading binocular and telescope ranges stocked. If we do not have it we will endeavour to get it for you.
Non-optical stock: Books incl. New Naturalist Series and Poysers, videos, CDs, audio tapes, tripods, hide clamps, accessories and clothing.

SWAROVSKI OPTIK: COMMITTED TO CONSTANTLY RAISING STANDARDS

IN LESS than two decades Swarovski Optik has established itself as one of the world's leading manufacturers of high quality binoculars and telescopes for the birdwatching market throughout the world.

Skilled technicians and innovative designers working at the family-owned factory in Austria's picturesque Tyrol region create new products that constantly set new standards for optical performance.

The launch of the Swarovski EL binoculars and ATS spotting scopes revolutionised the design of modern birding optics and the craftsmanship of their hand-made products means that each and every product will give years of trouble-free performance.

Swarovski Optik recognises the close relationship between industry, society and the environment and continues to support more than 20 conservation projects around the globe.

To discover more about the company, its ethics and the full product range of binoculars and telescopes contact Swarovski UK at Perrywood Business Park, Salfords, Surrey RH1 5JQ or call 01737 856 812 for a brochure. Website: www.swarovskioptik.com

EL 8.5 x 42

Opening times: Mon-Sat (9:30am-5:30pm). Sundays (9:30am-5pm).
Address: 5 London Road, Buxton, Derbyshire, SK17 9PA;01298 71844; (Fax)01298 27727.
e-mail: paulflint@birdnet.co.uk
www.birdnet.co.uk

In-focus

Company ethos: The binocular and telescope specialists, offering customers informed advice at birdwatching venues throughout the country. Main sponsor of the British Birdwatchng Fair.
Viewing facilities: Available at all shops (contact your local outlet), or at field events (10am-4pm) at bird reserves (see *Bird Watching* magazine or website www.at-infocus.co.uk for calendar)
Optical stock: Many leading makes of binoculars and telescopes, plus own-brand Delta range of binoculars and tripods.
Non-optical stock: Wide range of tripods, clamps and other accessories. Repair service available.
Opening times: Vary - please contact local shop or website before travelling.
NORFOLK; Main Street, Titchwell, Nr King's Lynn, Norfolk, PE31 8BB; 01485 210101.
RUTLAND; Anglian Water Birdwatching Centre, Egleton Reserve, Rutland Water, Rutland, LE15 8BT; 01572 770656.

LONDON CAMERA EXCHANGE

Company ethos: To supply good quality optical equipment at a competitive price, helped by knowlegeable staff.
Viewing facilities: In shop and at local shows. Contact local branch.
Optical stock: All leading makes of binoculars and scopes.
Non-optical stock: All main brands of photo, digital and video equipment.
Opening times: Mon-Sat (9am-5.30pm).
CHESTERFIELD: 1A South Street, Chesterfield, Derbyshire, S40 1QZ; 01246 211891; (Fax) 01246 211563;
e-mail: chesterfield@lcegroup.co.uk
DERBY: 17 Sadler Gate, Derby, Derbyshire, DE1 3NH; 01332 348644; (Fax) 01332 369136; e-mail: derby@lcegroup.co.uk

LINCOLN; 6 Silver Street, Lincoln, LN2 1DY; 01522 514131; (Fax) 01522 537480; e-mail: lincoln@lcegroup.co.uk
NOTTINGHAM: 7 Pelham Street, Nottingham, NG1 2EH; 0115 941 7486; (Fax) 0115 952 0547;
e-mail: nottingham@lcegroup.co.uk

WAREHOUSE EXPRESS

Company ethos: Mail order and website.
Viewing facilities: By appointment only.
Optical stock: All major brands including, Leica, Swarvoski, Opticron, Kowa, Zeiss, Nikon, Bushnell, Canon, Minolta etc.
Non-optical stock: All related accessories including hides, tripods and window mounts etc.
Opening times: Mon-Fri (9am-5.30pm).
Address: PO Box 659, Norwich, Norfolk, NR2 1UJ; 01603 626222; (Fax) 01603 626446.
www.warehouseexpress.com

NORTHERN ENGLAND

FOCALPOINT

Company ethos: Friendly advice by well-trained staff, competitive prices, no 'grey imports'.
Viewing facilities: Fantastic open countryside for superb viewing from the shop, plenty of wildlife. Parking for up to 20 cars.
Optical stock: All leading brands of binoculars and telescopes from stock, plus many pre-owned binoculars and telescopes available.
Non-optical stock: Bird books, outdoor clothing, boots, tripods plus full range of Skua products etc. available from stock.
Opening times: Mon-Sat (9:30am-5pm).
Address: Marbury House Farm, Bentleys Farm Lane, Higher Whitley, Warrington, Cheshire, WA4 4QW; 01925 730399; (Fax)01925 730368. e-mail: focalpoint@dial.pipex.com
www.fpoint.co.uk

In-focus

(see entry in Eastern England).
LANCASHIRE: WWT Martin Mere, Burscough, Ormskirk, Lancs, L40 0TA: 01704 897020.

OPTICAL DEALERS

WEST YORKSHIRE: Westleigh House Office Est. Wakefield Road, Denby Dale, West Yorks, HD8 8QJ: 01484 864729.

LONDON CAMERA EXCHANGE
(See entry in Eastern England).
CHESTER: 9 Bridge Street Row, CH1 1NW; 01244 326531.
MANCHESTER: 37 Parker Street, Picadilly, M1 4AJ; 0161 236 5819.

PHOTO EXPRESS (LAKELAND) LTD
Company ethos: Friendly and independent advice, UK dealers for all top brands, comptitive prices.
Viewing facilities: Dedicated optical viewing room with full showroom facilities at Ulverston branch.
Optical stock: Leica, Swarovski, Zeiss, Opticron, Kowa and selection of other brands.
Non-optical stock: Digital cameras, camcorders and accessories, tripods,m clamps and cases.
Opening times: Mon-Sat (9am-6pm).
Address: 39 Market Street, Ulverston, Cumbria LA12 7LR; 01229 583050; (Fax) 01229 480135.
e-mail: dave@photo-express.net

SOUTH EAST ENGLAND

BINOCULARS-ONLINE LTD
Company ethos: Small retail and mail order binocular specialist. Friendly advice.
Viewing facilities: Shop has views across the Thames Estuary.
Optical stock: Wide range of binoculars including Swarovski, Leica, Zeiss, Opticron, Nikon and Steiner.
Non-optical stock: Manfrotto and Velbon tripods.
Opening times: Tue-Sat (10am-4pm).
Address: 144 Eastern Esplanade (seafront), Southend-on-Sea, Essex SS1 2YH; 01702 601603. www.binoculars-online.co.uk

In-focus
(see entry in Eastern England).
ST ALBANS; Bowmans Farm, London Colney, St Albans, Herts, AL2 1BB: 01727 827799:

(Fax) 01727 827766.
SOUTH WEST LONDON: WWT The Wetland Centre, Queen Elizabeth Walk, Barnes, London, SW13 9WT: 020 8409 4433.

KAY OPTICAL (1962)
Company ethos: Unrivalled expertise, experience and service, since 1962.
Viewing facilities: At Morden. Also field-days every weekend at reserves in South.
Optical stock: All leading makes of binoculars and telescopes stocked. Also giant binoculars and astronomical.
Non-optical stock: Tripods, clamps etc.
Opening times: Mon-Sat (9am-5pm) closed (1-2pm).
Address: 89(B) London Road, Morden, Surrey, SM4 5HP; 020 8648 8822; (Fax)020 8687 2021.
e-mail: info@kayoptical.co.uk
www.kayoptical.co.uk and
www.bigbinoculars.co.uk

LONDON CAMERA EXCHANGE
(See entry in Eastern England).
FAREHAM: 135 West Street, Fareham, Hampshire, PO16 0DU; 01329 236441; (Fax) 01329 823294;
e-mail: fareham@lcegroup.co.uk
PORTSMOUTH: 40 Kingswell Path, Cascados, Portsmouth, PO1 4RR; 023 9283 9933; (Fax) 023 9283 9955;
e-mail: portsmouth@lcegroup.co.uk
GUILDFORD: 8/9 Tunsgate, Guildford, Surrey, GU1 2DH; 01483 504040; (Fax) 01483 538216;
e-mail: guildford@lcegroup.co.uk
READING: 7 Station Road, Reading, Berkshire, RG1 1LG; 0118 959 2149; (Fax) 0118 959 2197;
e-mail: reading@lcegroup.co.uk
SOUTHAMPTON: 10 High Street, Southampton, Hampshire, SO14 2DH; 023 8022 1597; (Fax) 023 8023 3838;
e-mail: southampton@lcegroup.co.uk
STRAND, LONDON: 98 The Strand, London, WC2R 0AG; 020 7379 0200; (Fax) 020 7379 6991;
e-mail: strand@lcegroup.co.uk
WINCHESTER: 15 The Square, Winchester, Hampshire, SO23 9ES; 01962

866203; (Fax) 01962 840978;
e-mail: winchester@lcegroup.co.uk

SOUTH WEST ENGLAND

ACE OPTICS
Company ethos: To be the best for service, price and stock.
Viewing facilities: Bird of prey at 100 yds, Leica test card to check quality.
Optical stock: All the top brands, including Questar. Official importer for Optolyth products.
Non-optical stock: All the best tripods and an array of optical related accessories.
Opening times: Mon-Sat (8:45am-6pm).
Address: 16 Green Street, Bath, BA1 2JZ; 01225 466364; (fax) 01225 469761.
e-mail: aceoptics@balzjz.freeserve.co.uk

LONDON CAMERA EXCHANGE
(See entry in Eastern England).
BATH: 13 Cheap Street, Bath, Avon, BA1 1NB; 01225 462234; (Fax) 01225 480334. e-mail: bath@lcegroup,co.uk
BOURNEMOUTH: 95 Old Christchurch Road, Bournemouth, Dorset, BH1 1EP; 01202 556549; (Fax) 01202 293288;
e-mail: bournemouth@lcegroup.co.uk
BRISTOL: 53 The Horsefair, Bristol, BS1 3JP; 0117 927 6185; (Fax) 0117 925 8716;
e-mail: bristol.horsefair@lcegroup.co.uk
EXETER: 174 Fore Street, Exeter, Devon, EX4 3AX;01392 279024/438167; (Fax) 01392 426988. e-mail: exeter@lcegroup.co.uk
PAIGNTON: 71 Hyde Road, Paignton, Devon, TQ4 5BP;01803 553077; (Fax) 01803664081. e-mail: paignton@lcegroup.co.uk
PLYMOUTH: 10 Frankfort Gate, Plymouth, Devon, PL1 1QD; 01752 668894; (Fax) 01752 604248.
e-mail: plymouth@lcegroup.co.uk
SALISBURY: 6 Queen Street, Salisbury, Wiltshire, SP1 1EY; 01722 335436; (Fax) 01722 411670;
e-mail: salisbury@lcegroup.co.uk

TAUNTON: 6 North Street, Taunton, Somerset, TA1 1LH; 01823 259955; (Fax) 01823 338001.
e-mail: taunton@lcegroup.co.uk

WESTERN ENGLAND

FOCUS OPTICS
Company ethos: Friendly, expert service. Top quality instruments. No 'grey imports'.
Viewing facilities: Our own pool and nature reserve with feeding stations.
Optical stock: Full range of leading makes of binoculars and telescopes.
Non-optical stock: Waterproof clothing, fleeces, walking boots and shoes, bird food and feeders. Books, videos, walking poles.
Opening times: Mon-Sat (9am-5pm). Some bank holidays.
Address: Church Lane, Corley, Coventry, CV7 8BA; 01676 540501/542476; (Fax) 01676 540930. e-mail: focopt1@aol.com
www.focusoptics.co.uk

In-focus
(see entry in Eastern England).
GLOUCESTERSHIRE:
WWT Slimbridge, Gloucestershire, GL2 7BT: 01453 890978.

LONDON CAMERA EXCHANGE
(see entry in Eastern England).
CHELTENHAM: 10-12 The Promenade, Cheltenham, Gloucestershire, GL50 1LR; 01242 519851; (Fax) 01242 576771;
e-mail: cheltenham@lcegroup.co.uk
GLOUCESTER: 12 Southgate Street, Gloucester, GL1 2DH; 01452 304513; (Fax) 01452 387309;
e-mail: gloucester@lcegroup.co.uk
LEAMINGTON: Clarendon Avenue, Leamington, Warwickshire, CV32 5PP; 01926 886166; (Fax) 01926 887611;
e-mail: leamington@lcegroup.co.uk
WORCESTER: 8 Pump Street, Worcester, WR1 2QT; 01905 22314; (Fax) 01905 724585;
e-mail: worcester@lcegroup.co.uk

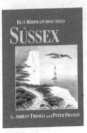

BIRD RESERVES
AND
OBSERVATORIES

Kingfisher by Nick Williams

PLEASE NOTE:
It is our intention to update this section annually, adding new reserves where we can. This means that space becomes limited and some reserves where deails have not changed for some time may have been removed. These sites have been shown in full in previous editions of *The Yearbook* and will so again in future ones. However, if you require more information on a particular site that is not shown in full, please contact Buckingham Press and we will do our best to provide you with contact details for that site.

Bedfordshire

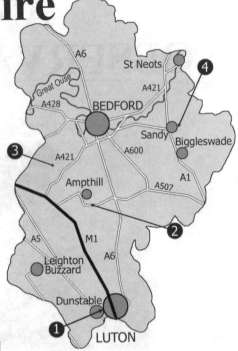

1. BLOW'S DOWNS

The Wildlife Trust for Beds, Cambs, Northants and Peterborough.
Location: TL 033 216. On the outskirts of Dunstable, W of Luton. Parking is at Skimpot roundabout, off Hatters Way on A505 and in Half Moon Lane, Dunstable.
Access: Open all year.
Facilities: None.
Public transport: None.
Habitat: Chalk downland, scrub and grassland that is a traditional resting place for incoming spring migrants.
Key birds: *Spring/summer*: Hobby, Turtle Dove, Grasshopper Warbler, Lesser Whitethroat, Cuckoo, Spotted Flycatcher, Golden Plover. *Winter*: Buzzard, winter thrushes, possible Brambling. *Passage*: Ring Ouzel, Wheatear, Redstart. *All year*: Marsh and Willow Tits, Bullfinch, Sparrowhawk.
Contact: Trust HQ, The Manor House, Broad Street, Great Cambourne, Cambridgeshire CB3 6DH. 01954 713500; fax 01954 710051.
e-mail: cambridgeshire@wildlifebcnp.org
www.wildlifebcnp.org

2. FLITWICK MOOR

The Wildlife Trust for Beds, Cambs, Northants and Peterborough.
Location: TL 046 354. SE of Ampthill. From Flitwick town centre, take the road towards Greenfield. After approx 0.8km turn L into Maulden Road. Head N to Folly Farm (approx 0.8km), opposite an industrial estate. Turn R at the farm and follow road to car park.
Access: Open all year.
Facilities: Car park. Please stick to public paths.
Public transport: None.
Habitat: SSSI, valley fen, woodland, sedge, reed.
Key birds: *Spring/summer*: Turtle Dove, possible Grasshopper Warbler, Chiffchaff, Cuckoo, warblers. *Winter*: Teal, Lapwing, Woodcock, Siskin. *All year*: Water Rail, Little Owl, Great Spotted and Lesser Spotted Woodpeckers, possible Willow and Marsh Tits, Jay.
Contact: Trust HQ, 01954 713500; fax 01954 710051. e-mail:cambridgeshire@wildlifebcnp.org
www.wildlifebcnp.org

3. MARSTON VALE MILLENNIUM CP

Marston Vale Trust (Regd Charity No 1069229).
Location: SW of Bedford off A421 at Marston Moretaine. Only five mins from J13 of M1.
Access: Park and forest centre open seven days. Summer 10am-6pm, winter 10am-4pm. No dogs in reserve, reserve path unsurfaced. Coaches fine.
Facilities: Cafe bar, gift shop, art gallery, exhibition. Free parking.
Public transport: Marston Vale Line – trains direct from Millbrook and Stewartby station.
Habitat: Lake – 284 acres/freshwater marsh (man-made), reedbed, hawthorn scrub and grassland.
Key birds: *Winter*: Iceland and Glaucous Gulls (regular), gull roost, wildfowl, Great Crested Grebe. *Spring*: Passage waders and terns (Black Tern, Arctic Tern, Garganey. *Summer*: Nine species of breeding warblers, Hobby, Turtle Dove, Nightingale. *Autumn*: Passage waders and terns. *Rarities*: White-winged Black Tern, Laughing Gull, divers, Manx Shearwater, Bittern.

132

Contact: Forest Centre, Station Road, Marston Moretaine, Beds MK43 0PR. 01234 767037. e-mail: info@marstonvale.org www.marstonvale.org

4. THE LODGE

RSPB Central England Office
Location: TL 190 479. Reserve lies one mile E of Sandy on B1402 to Potton.
Access: Reserve open daily 9am-9pm (or sunset when earlier); shop 9am to 5pm weekdays, 10am-5pm weekends.
Facilities: Shop, toilets, nature trails, up to 3.5 miles (5.6 km), one path (0.5 miles/0.8km) and gardens wheelchair/pushchair accessible. Hides, including one that offers wheelchair accessible-parking 50 yards(45m).

Public transport: Buses, frequently from Bedford to Sandy but infrequently past reserve. Walk from Sandy town centre (1.25 miles), along High Street, over railway bridge, up B1042 (Potton Road) with pavement. Train, Sandy (one mile), up B1042 (Potton Road). Bicycle, view the National Cycle Network map at The Lodge.
Habitat: This reserve is a mixture of woodland, heathland and includes the formal gardens of the RSPB.
Key birds: *Spring/summer*: Woodpeckers, Dartford warbler, and many more.
Contact: Peter Bradley, RSPB Central England Office, 46 The Green, South Bar, Banbury, Oxfordshire, OX16 9AB. 01295 253330. www.rspb.org.uk/reserves/

Berkshire

1. DINTON PASTURES

Wokingham District Council.
Location: SU 784 718. Country Park, E of Reading off B3030 between Hurst and Winnersh.
Access: Open all year, dawn to dusk.
Facilities: Hides, information centre, car park, café, toilets. Suitable for wheelchairs.
Public transport: Information not available.
Habitat: Mature gravel pits and banks of River Loddon.
Key birds: Kingfisher, Water Rail, Little Ringed Plover, Common Tern, Nightingale. *Winter*: Wildfowl (inc. Goldeneye, Wigeon, Teal, Gadwall).
Contact: The Ranger, Dinton Pastures Country Park, Davis Street, Hurst, Berks. 0118 934 2016. e-mail: countryside@wokingham.gov.uk

2. LAVELL'S LAKE

Wokingham District Council.
Location: SU 781 729. Via Sandford Lane off B3030 between Hurst and Winnersh, E of Reading.
Access: Dawn to dusk. No permit required.
Facilities: Hides.
Public transport: Information not available.
Habitat: Gravel pits, two wader scapes, although one congested with *crassula helmsii*, rough grassland, marshy area, between River Loddon and Emm Brook. To N of Lavell's Lake, gravel pits are being restored and attact birds. A lake is viewable walking N along River Lodden from Lavell's Lake over small green bridge. The lake is in a field immediately on R but is only viewable through hedge. No access is permitted.

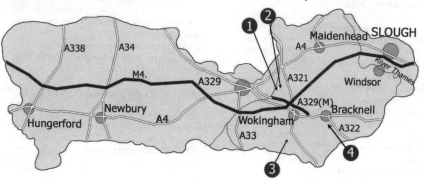

Key birds: Sparrowhawk. *Summer*: Garganey, Common Tern, Redshank, Lapwing, Hobby, Red Kite, Peregrine, Buzzard. Passage waders. *Winter*: Green Sandpiper, ducks (inc. Smew), Bittern.
Contact: See Dinton Pastures Country Park above.

3. MOOR GREEN LAKES

Blackwater Valley Countryside Partnership.
Location: SU 805 628. Main access and parking off Lower Sandhurst Road, Finchampstead. Alternatively, Rambler's car park, Mill Lane, Sandhurst (SU 820 619).
Access: Car parks open dawn-dusk. Two bird hides open to members of the Moor Green Lakes Group (contact BVCP for details). Dogs on leads. Site can be used by people in wheelchairs, though surface not particularly suitable.
Facilities: Two bird hides, footpaths around site, Blackwater Valley Long Distance Path passes through site.
Public transport: Nearest bus stop, Finchampstead (approx 1.5 miles from main entrance). Local bus companies – Stagecoach Hants & Surrey, tel 01256 464501, First Beeline & Londonlink, tel 01344 424938.
Habitat: 36 hectares (90 acres) in total. Three lakes with gravel islands, beaches and scrapes. River Blackwater, grassland, surrounded by willow, ash, hazel and thorn hedgerows.
Key birds: *Spring/summer*: Redshank, Little Ringed

Plover, Sand Martin, Willow Warbler and of particular interest, a flock of Goosander. Also Whitethroat, Sedge Warbler, Common Sandpiper, Common Tern and Dunlin. Black Terns on passage. Lapwings breed on site and several sightings of Red Kite. *Winter*: Ruddy Duck, Wigeon, Teal, Gadwall.
Contact: Blackwater VCP, Ash Lock Cottage, Government Road, Aldershot, Hants GU11 2PS. 01252 331353. www.blackwater-valley.org.uk e-mail: blackwater.valley@hants.gov.uk

4. WILDMOOR HEATH

Berks, Bucks & Oxon Wildlife Trust.
Location: SU 842 627. Between Bracknell and Sandhurst. From Sandhurst shopping area, take the A321 NW towards Wokingham. Turn E at the mini-roundabout on to Crowthorne Road. Continue for about one mile through one set of traffic lights. Car park is on the R at the bottom of the hill.
Access: Open all year. No access to woodland N of Rackstraw Road at Broadmoor Bottom. Please keep dogs on a lead.
Facilities: Car park. **Public transport:** None.
Habitat: Wet and dry lowland heath, bog, mixed woodland and mature Scots pine plantation.
Key birds: *Spring/summer*: Wood Lark, Nightjar, Dartford Warbler, Stonechat. Good for dragonflies.
Contact: Trust HQ, 01865 775476. e-mail: bbowt@cix.co.uk
www.wildlifetrust.org.uk/berksbucksoxon

Buckinghamshire

1. BURNHAM BEECHES NNR

Corporation of London.
Location: SU 950 850. N of Slough and on W side of A355, running between J2 of the M40 and J6 of M4. There are several entrances from A355. Also entrances from Hawthorn Lane and Pumpkin Hill to the S and Park Lane to the W. Small network of metalled roads, several meet at Victory Cross.
Access: Open all year. Main Lord Mayor's Drive open from 8am-dusk.
Facilities: Car parks, toilets, café, seasonal refreshment book.
Public transport: Train: nearest station is Slough on the main line from Paddington.
Habitat: Ancient woodland, streams, pools, heathland, grassland, scrub.

Key birds: *Spring/summer*: Whitethroat, Cuckoo, possible Turtle Dove. *Winter*: Siskin, Redpoll, Crossbill, regular large flocks c100 Brambling. Possible Woodcock. *All year*: Mandarin (good population), all three woodpeckers, Sparrowhawk, Marsh Tit, possible Willow Tit.
Contact: Corporation of London, Open Spaces Department, PO Box 270, Guildhall, London EC2P 2EJ. 020 7332 3514.

2. CHURCH WOOD RSPB RESERVE

RSPB Central England Office
Location: SU 972 872. Reserve lies three miles from J2 of M40 in Hedgerley. Park in village, walk down small track beside pond for approx 200m. Reserve entrance is on L.

Access: Open all year.
Facilities: Two marked paths.
Public transport: None.
Habitat: Mixed woodland.
Key birds: *Spring/summer*: Blackcap, Garden Warbler, Spotted Flycatcher, Swallow. *Winter*: Redpoll, Siskin. *All year*: Marsh Tit, Willow Tit, Nuthatch, all three woodpeckers.
Contact: RSPB Central England Office, 01295 253330. www.rspb.org.uk/wildlife/reserves

3. COLLEGE LAKE WILDLIFE CENTRE

Berks, Bucks & Oxon Wildlife Trust.
Location: SU 934 140. 2 miles N of Tring. Go N on B488. 1/4 mile N of canal bridge at Bulbourne turn L into unsigned entrance.
Access: Open daily 10am-5pm. Permits available on site or from Trust HQ.
Facilities: Hides, nature trails, visitor centre, toilets, wheelchair access.
Public transport: Tring railway station is two miles away.
Habitat: Marsh area, lake, islands, shingle.
Key birds: *Spring/summer*: Breeding Lapwing, Redshank, Little Ringed Plover. Hobby. Passage waders inc. Green Sandpiper.
Contact: The Warden, College Lake Wildlife Centre, Upper Icknield Way, Bulbourne, Tring, Herts HP23 5QG. 01296 662890.

4. LITTLE MARLOW GRAVEL PITS

Lefarge Redland Aggregates.
Location: SU 880 880. NE of Marlow from J4 of M40. Use permissive path from Coldmoorholm Lane to Little Marlow village. Follow path over a wooden bridge to N end of lake. Permissive path ends just

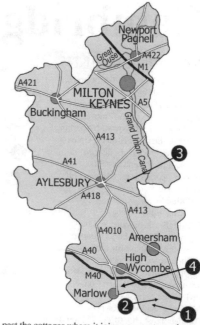

past the cottages where it joins a concrete road to sewage treatment works.
Access: Open all year. Please do not enter the gravel works.
Facilities: Paths. **Public transport:** None.
Habitat: Gravel pit, lake, scrub.
Key birds: *Spring*: Passage migrants, Sand Martin, Garganey, Hobby. *Summer*: Reed warblers, Kingfisher, wildfowl. *Autumn*: Passage migrants. *Winter*: Wildfowl, possible Smew, Goldeneye, Yellow-legged Gull, Lapwing, Snipe.

Cambridgeshire

1. BRAMPTON WOOD

The Wildlife Trust for Beds, Cambs, Northants and Peterborough.
Location: TL 185 698. Two miles E of Grafham village on N side of road to Brampton. From A14 take main road S from Ellington.
Access: Open daily. Coaches able to drop passengers off but unfortunately not sufficient space available to park.
Facilities: Car park, interpretative shelter.
Public transport: None.
Habitat: SSSI. Primarily ash and field maple with hazel coppice.
Key birds: *Summer*: Breeding Grasshopper Warbler, Nightingale, Spotted Flycatcher, Woodcock; all three woodpeckers. *Winter*: Thrushes.
Contact: Trust HQ, 01954 713500; fax 01954 710051. e-mail:cambridgeshire@wildlifebcnp.org www.wildlifebcnp.org

Cambridgeshire

2. DOGSTHORPE STAR PIT

The Wildlife Trust for Beds, Cambs,
Northants and Peterborough.
Location: TF 213 025. NE edge of
Peterborough. From A47 turn into White
Post Road on Eye by-pass. No parking
at reserve. Entrance is via the Green
Wheel cycle track in the disused
Welland Road behind Star Pit.
Access: Open all year. Dropping off
point for coaches only.
Facilities: None.
Public transport: None.
Habitat: Brackish water, scrub.
Key birds: *Winter*: Large numbers of
gulls, attracted to nearby rubbish tip inc.
Mediterranean, Glaucous and Yellow-
legged. *Passage*: Occasional waders, inc
Green Sandpiper. *All year*: Common water
fowl, Kingfisher, Green Woodpecker.
Summer: Common warblers.
Contact: Trust HQ, 01954 713500; fax 01954
710051. e-mail:cambridgeshire@wildlifebcnp.org

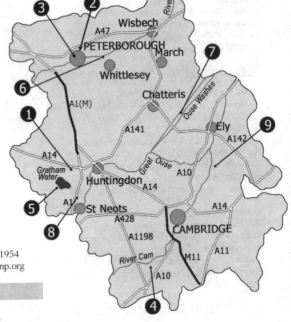

3. FERRY MEADOWS

Nene Park Trust.
Location: TL 145 975. Three miles W of
Peterborough town centre, signed off A605.
Access: Open all year.
Facilities: Car park (fee at weekends), visitor centre,
toilets, café, two hides.
Public transport: Tel. Traveline 0870 6082608 or
www.traveline.org.uk
Habitat: Lakes, meadows, scrub, woodland and
small nature reserve.
Key birds: *Spring*: Grey Heron, Common and
Arctic Terns, waders, Yellow Wagtail. *Winter*:
Siskin, Redpoll, Water Rail, occasional Hawfinch. *All
year*: Good selection of woodland and water birds,
Kingfisher.
Contact: Nene Park Trust, Ham Farm House,
Orton, Peterborough, PE2 5UU. 01733 234443.

4. FOWLMERE

RSPB (East Anglia Office).
Location: TL 407 461. Turn off A10 Cambridge to
Royston road by Shepreth and follow sign.
Access: Access at all times along marked trail.

Facilities: One and a half miles of trails. Three
hides, toilets. Space for one coach, prior booking
essential. Wheelchair access to one hide, toilet and
some of the trails.
Public transport: Shepreth railway station 2 miles.
Habitat: Reedbeds, meres, woodland, scrub.
Key birds: *Summer*: Nine breeding warblers. *All
year:* Water Rail, Kingfisher. *Winter*: Snipe, raptors.
Corn Bunting roost.
Contact: The Warden, The RSPB, Manor Farm,
High Street, Fowlmere, Royston, Herts SG8 7SH.
Tel/fax 01763 208978.

5. GRAFHAM WATER

The Wildlife Trust for Beds, Cambs, Northants and
Peterborough.
Location: TL 143 671. Follow signs for Grafham
Water from A1 at Buckden or A14 at Ellington.
Nature Reserve entrance is from Mander car park, W
of Perry village.
Access: Open all year. Dogs barred in wildlife

garden only, on leads elsewhere.

Facilities: Five bird hides in reserve. Two in wildlife garden accessible to wheelchairs. Cycle track through reserve also accessible to wheelchairs. Visitor centre with restaurant, shop and toilets. Disabled parking.

Public transport: None.

Habitat: Open water, ancient and plantation woodland, grassland.

Key birds: *Winter*: Wildfowl, gulls. *Spring/summer*: Breeding Nightingale, Reed, Willow and Sedge Warblers, Common and Black Terns. *Autumn*: Passage waders.

Contact: The Warden, Grafham Water Nature Reserve, c/o The Lodge, West Perry, Huntingdon, Cambs PE28 0BX. 01480 811075.
e-mail: grafham@cix.co.uk
www.wildlifetrust.org.uk/bcnp

6. NENE WASHES

RSPB (East of England Office).

Location: TL 300 995. N of Whittlesey and six miles E of Peterborough.

Access: Open at all times along South Barrier Bank, accessed at Eldernell, one mile NE of Coates, off A605. Group visits by arrangement. No access to fields. No access for wheelchairs along bank.

Facilities: Small car park.

Public transport: Bus and trains to Whittlesey, bus to Coates.

Habitat: Wet grassland with ditches. Frequently flooded.

Key birds: *Spring/early summer*: Breeding waders, including Black-tailed Godwit, duck, including Garganey, Marsh Harrier and Hobby. *Winter*: Waterfowl including Bewick's Swan and Pintail, Barn Owl, Hen Harrier.

Contact: Charlie Kitchin, RSPB Nene Washes, 21a East Delph, Whittlesey, Cambs PE7 1RH. 01733 205140.

7. OUSE WASHES

RSPB (East Anglia Office).

Location: TL 471 861. Between Chatteris and March on A141, take B1093 to Manea. Reserve signposted from Manea. Reserve office and visitor centre located off Welches Dam. Approximately ten miles from March or Chatteris.

Access: Access at all times from visitor centre (open every day except Christmas Day and Boxing Day). Welches Dam to public hides approached by marked paths behind boundary bank. No charge. Dogs to be kept on leads at all times. Disabled access to Welches Dam hide, 350 yards from car park.

Facilities: Car park and toilets. Visitor centre – unmanned but next to reserve office. Ten hides overlooking the reserve: nearest 350 yards from visitor centre (with disabled access) and furthest one mile from visitor centre. Boardwalk over pond – good for dragonflies in summer.

Public transport: None to reserve entrance. Train station at Manea – three miles from reserve.

Habitat: Lowland wet grassland – seasonally flooded. Open pool systems in front of some hides, particularly Stockdale's hide.

Key birds: *Summer:* Around 70 species breed including Black-tailed Godwit, Lapwing, Redshank, Snipe, Shoveler, Gadwall, Garganey and Spotted Crake. Also Hobby and Marsh Harrier. *Autumn*: Passage waders including Wood and Green Sandpipers, Spotted Redshank, Greenshank, Little Stint, plus terns and Marsh and Hen Harrier. *Winter:* Large numbers of Bewick's and Whooper Swans, Wigeon, Teal, Shoveler, Pintail, Pochard.

Contact: Site Manager, Ouse Washes Reserve, Welches Dam, Manea, March, Cambs PE15 0NF. 01354 680212. e-mail: cliff.carson@rspb.org.uk
www.rspb.org.uk

8. PAXTON PITS

Huntingdonshire District Council.

Location: TL 197 629. Access from A1 at Little Paxton, two miles N of St Neots.

Access: Free Entry. Open 24 hours. Visitor centre manned at weekends. Dogs allowed under control. Heron trail suitable for wheelchairs during summer.

Facilities: Toilets available most days 9am-5pm (including disabled), two bird hides (always open), marked nature trails.

Public transport: 565/566 run between Huntingdon and St Neots Mon-Sat. Tel: 0870 608 2608.

Habitat: Grassland, scrub, lakes.

Key birds: *Spring/summer*: Nightingale, Kingfisher, Common Tern, Sparrowhawk, Hobby, Grasshopper, Sedge and Reed Warblers, Lesser Whitethroat. *Winter*: Smew, Goldeneye, Goosander, Gadwall, Pochard.

Contact: The Ranger, The Visitor Centre, High Street, Little Paxton, St Neots, Cambs PE19 6ET. 01480 406795. e-mail: mail@paxtonpits.uklinux.net
www.paxton-pits.org.uk

9. WICKEN FEN

The National Trust.

Location: TL 563 705. Lies 17 miles NE of Cambridge and ten miles S of Ely. From A10 drive E along A1123.

Access: Daily except Mon (9am-5pm). Permit required from visitor centre. National Trust members free. Disabled access along approx 0.75 miles boardwalk. Dogs must be on leads.
Facilities: Toilets in car park. Visitor centre with hot/cold drinks, sandwiches at weekends, three nature trails, seven hides.
Public transport: Nearest rail link either Cambridge or Ely. Buses only on Thu and Sun (very restricted service).

Habitat: Open fen – cut hay fields, sedge beds, grazing marsh – partially flooded wet grassland, reedbed, scrub, woodland.
Key birds: *Spring*: Passage waders and passerines. *Summer*: Marsh Harrier, Long-eared Owl, breeding waders and warblers, good for butterflies. *Winter*: Wigeon, Hen Harrier, Merlin.
Contact: Martin Lester, Lode Lane, Wicken, Cambs CB7 5XP. 01353 720274. www.wicken.org.uk e-mail: martin.lester@nationaltrust.org.uk

Cheshire

1. FIDDLERS FERRY

SSE (Scottish and Southern Energy)
Location: SJ 552 853. Off A562 between Warrington and Widnes.
Access: Parking at main gate of power station. Summer (8am-8pm); winter (8am-5pm). For free permit; apply in advance with sae to Manager, Fiddlers Ferry Power Station, Warrington WA5 2UT.
Facilities: Hide, nature trail.
Public transport: Arriva bus 110 every 20 minutes.
Habitat: Ash and water lagoons, tidal and non-tidal marshes with phragmites and great reedmace, meadow grassland with small woods.
Key birds: *Summer*: Breeding Gadwall, Pochard, Buzzard, Peregrine and Raven. *Winter:* Glaucous and Iceland Gulls, Short-eared Owl, Peregrine, Jack Snipe and Twite. *Recent rarities:* Great White Egret, Black Kite, Baltic Gull, Marsh Harrier, Hobby, Caspian Gull, Mediterranean Gull, Kumlien's Gull and Chiffchaff (*tristus*).
Contact: Keith Massey, 4 Hall Terrace, Great Sankey, Warrington WA5 3EZ. 01925 721382.

2. GAYTON SANDS

RSPB Dee Estuary Office.
Location: SJ 275 785. On W side of Wirral, S of Birkenhead. View high tide activity from Old Baths car park near Boathouse pub, Parkgate.
Access: Open at all times. Viewing from public footpaths and car parks. Please do not walk on the saltmarsh, the tides are dangerous.
Facilities: Car park, picnic area, group bookings, guided walks, special events, wheelchair access. Toilets at Parkgate village opposite the Square.
Public transport: Bus – Parkgate every hour. Rail – Neston, two miles.
Habitat: Estuary – saltmarsh, pools, mud, sand.
Key birds: *Spring/summer/autumn*: Greenshank, Spotted Redshank, Curlew Sandpiper. *Winter*: Shelduck, Teal, Wigeon, Pintail, Oystercatcher, Black-tailed Godwit, Curlew, Redshank, Merlin, Peregrine, Water Rail, Short-eared Owl, Hen Harrier.
Contact: Colin E Wells, Burton Point Farm, Station Road, Burton, Nr Neston, Cheshire CH64 5SB. 0151 3367681. e-mail: colin.wells@rspb.org.uk

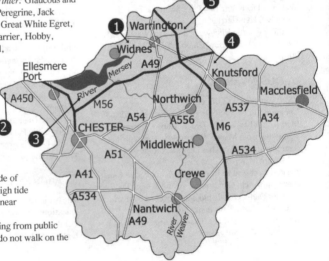

3. GOWY MEADOWS

Cheshire Wildlife Trust.
Location: SJ 435 740. Thornton Moors, Ellesmere Port.
Access: Through public footpath gate on Thornton Green Lane. Open all year.
Facilities: None.
Public transport: The Arriva bus service stops on the Thornton Green Lane opposite the church.
Habitat: Lowland grazing marsh.
Key birds: *Spring/summer*: Wildfowl, Green Sandpiper, Lapwing, Jack Snipe, Snipe, warblers, Whinchat. *Winter*: Reed Bunting. *Passage*: Stonechat, Wheatear.
Contact: Trust HQ, 01270 610180.
e-mail: cheshirewt@cix.co.uk
www.wildlifetrust.org.uk/cheshire/

4. ROSTHERNE MERE

English Nature (Cheshire to Lancashire team).
Location: SJ 744 843. Lies N of Knutsford and S of M56 (junction 8).
Access: View from Rostherne churchyard and lanes; no public access, except to A W Boyd Observatory (permits from D A Clarke, 1 Hart Avenue, Sale M33 2JY, 0161 973 7122). Not suitable for coach parties but can accommodate smaller group visits by prior arrangement
Facilities: None.
Public transport: None.

Habitat: Deep lake, woodland, willow bed, pasture.
Key birds: *Winter*: Good range of duck (inc Ruddy Duck and Pintail), gull roost (inc. occasional Iceland and Glaucous). Passage Black Terns.
Contact: Tim Coleshaw, Site Manager, English Nature, Attingham Park, Shrewsbury SY4 4TW. 01743 282000; fax 01743 709303; e-mail tim.coleshaw@english-nature.org.uk.

5. WOOLSTON EYES

Woolston Eyes Conservation Group.
Location: SJ 654 888. E of Warrington between the River Mersey and Manchester Ship Canal. Off Manchester Road down Weir Lane or from Latchford to end of Thelwall Lane.
Access: Open all year. Permits required from Chairman, £7 each, £14 per family (see address below).
Facilities: No toilets or visitor centre. Good hides, some elevated.
Public transport: Buses along A57 nearest stop to Weir Lane, or Thelwell Lane, Latchford.
Habitat: Wetland, marsh, scrubland, wildflower meadow areas.
Key birds: Breeding Black-necked Grebe, warblers (including Grasshopper Warbler), all raptors (Merlin, Peregrine, Marsh Harrier). SSSI for wintering wildfowl, many duck species breed.
Contact: B R Ankers, Chairman, 9 Lynton Gardens, Appleton, Cheshire WA4 5ED. 01925 267355. www.woolstoneyes.co.uk

Cornwall

1. BRENEY COMMON

Cornwall Wildlife Trust.
Location: SX 054 610. Three miles S of Bodmin. Take minor road off A390 one mile W of Lostwithiel to Lowertown.
Access: Open at all times but please keep to paths. Disabled access.
Facilities: Wilderness trail.
Public transport: None.
Habitat: Wetland, heath and scrub.
Key birds: Willow Tit, Nightjar, Tree Pipit, Sparrowhawk, Lesser Whitethroat, Curlew.
Contact: Sean O'Slea, 5 Acres, Allet, Cornwall TR4 9DJ. 01872 273939. e-mail: cornwt@cix.co.uk
www.cornwallwildlifetrust.org.uk

2. GOLITHA NNR

English Nature (Cornwall & Isles of Scilly Team).
Location: SX 227 690. Golitha is three miles NW of Liskeard in E Cornwall. Take minor roads N for 2.5 miles from Dobwalls on A38.
Access: Various paths from 0.5 mile to four miles. Can be muddy after rain. Limited disabled access.
Facilities: Toilets. **Public transport:** None.
Habitat: Ancient woodland, deep granite gorge.
Key birds: *All year*: Sparrowhawk, Buzzard, Kingfisher, all three woodpeckers, Jay, Grey Wagtail, Dipper, Marsh Tit, Treecreeper, Nuthatch. *Summer:* Redstart, Wood Warbler, Pied Flycatcher.
Contact: Cornwall & Isles of Scilly Team, 01872 265710. e-mail: cornwall@english-nature.org.uk

Cornwall

3. HAYLE ESTUARY

RSPB (South West England Office).
Location: SW 550 370. In town of Hayle. Follow
signs to Hayle from A30.
Access: Open at all times. No permits required. No
admission charges. Dogs on leads please. Sorry -
no coaches.
Facilities: Eric Grace Memorial Hide at
Ryan's Field has disabled parking and
viewing. Nearest disabled toilets at
Wyevale garden centre, Lelant, 600
yards W just off the roundabout.
No visitor centre but
information board at hide.
Public transport: Buses and
trains at Hayle.
Habitat: Intertidal mudflats,
saltmarsh, lagoon and islands,
sandy beaches and sand dunes.
Key birds: *Winter*: Wildfowl, gulls,
Kingfisher, Ring-billed Gull, Great Northern
Diver. *Spring/summer*: Migrant waders, breeding
Shelduck. *Autumn*: Rare waders, often from N
America, terns, gulls.
Contact: Dave Flumm, RSPB, The Manor Office,
Marazion, Cornwall TR17 0EF. Tel/fax; 01736
711682.

4. MARAZION MARSH

RSPB (South West England Office).
Location: SW 510 315. Reserve is one mile E of
Penzance, 500 yards W of Marazion. Entrance off
seafront road near Marazion.
Access: Open at all times. No permits required. No
admission charges. Dogs on leads please. Sorry - no
coaches.
Facilities: One hide. No toilets. No visitor centre.
Nearest toilets in Marazion and seafront car park.
Public transport: Bus from Penzance.
Habitat: Wet reedbed, willow carr.
Key birds: *Winter*: Wildfowl, Snipe, occasional
Bittern. *Spring/summer*: Breeding Reed, Sedge and
Cetti's Warblers, herons, swans. *Autumn*: Occasional
Aquatic Warbler, Spotted Crake. Large roost of
swallows and martins in reedbeds, migrant warblers
and waders.
Contact: Dave Flumm, RSPB, The Manor Office,
Marazion, Cornwall TR17 0EF. Tel/fax; 01736
711682.

5. STITHIANS RESERVOIR

Cornwall Birdwatching & Preservation Society.
Location: SS 715 365. From B3297 S of Redruth.
Access: Good viewing from causeway. Hides
accessible to members only.
Facilities: None.
Public transport: None.
Habitat: Open water, marshland.
Key birds: Wildfowl and waders (inc. rarities, eg.
Pectoral and Semipalmated Sandpipers, Lesser
Yellowlegs).
Contact: Stuart Hutchings, 5 Acres, Allet, Cornwall
TR4 9DJ. 01872 273939. e-mail: cornwt@cix.co.uk
www.cornwallwildlifetrust.org.uk

6. TAMAR ESTUARY

Cornwall Wildlife Trust.
Location: SX 434 631. (Northern Boundary). SX
421 604 (Southern Boundary). From Plymouth, head
W on A38. Access parking at Cargreen and
Landulph from minor roads off A388.
Access: Open at all times.
Facilities: Information boards at Cargreen and
Landulph.
Public transport: None.
Habitat: Tidal mudflat with some saltmarsh.
Key birds: *Winter*: Avocet, Snipe, Black-tailed

Godwit, Redshank, Dunlin, Curlew, Whimbrel, Spotted Redshank, Green Sandpiper, Golden Plover, Kingfisher.
Contact: Stuart Hutchings, 5 Acres, Allet, Cornwall TR4 9DJ. 01872 273939. e-mail: cornwt@cix.co.uk www.cornwallwildlifetrust.org.uk

7. TAMAR LAKES

Tamar Lakes Country Park.
Location: SS 295 115. Leave A39 at Kilkhampton. Take minor road E to Thardon, car park off minor road running between upper and lower lakes.
Access: Open all year.
Facilities: Hide open all year round. New birdwatching centre on lower Tamar. Cafe, toilets (Apr-Sep).
Public transport: None.
Habitat: Two large bodies of water.
Key birds: Migrant waders (inc. North American vagrants). *Spring*: Black Tern. *Winter*: Wildfowl (inc. Goldeneye, Wigeon, Pochard).
Contact: Ranger, Tamar Lakes Water Park, Kilkhampton, N Cornwall, 01409 211514.

Cumbria

1. CAMPFIELD MARSH

RSPB (North of England Office).
Location: NY 207 620. On S shore of Solway estuary, W of Bowness-on-Solway. Follow signs from B5307 from Carlisle.
Access: Open at all times, no charge. Views high-tide roosts from roadside (suitable for disabled).
Facilities: Hide overlooking wetland areas, along nature trail (1.5 miles). No toilets or visitor centre.
Public transport: Nearest railway station – Carlisle (13 miles). Infrequent bus service to reserve.
Habitat: Saltmarsh/intertidal areas, open water, peat bog, wet grassland.
Key birds: *Winter*: Waders and wildfowl include Barnacle Goose, Shoveler, Scaup, Grey Plover. *Spring/summer*: Breeding Lapwing, Redshank, Snipe, Tree Sparrow and warblers. *Autumn*: Passage waders.
Contact: Dave Blackledge, North Plain Farm, Bowness-on-Solway, Wigton, Cumbria CA7 5AG. e-mail: dave.blackledge@rspb.org.uk www.rspb.org.uk

2. FOULNEY ISLAND

Cumbria Wildlife Trust.
Location: SD 246 640. Three miles SE of Barrow town centre on the A5087 from Barrow or Ulverston. At a roundabout 2.5 miles S of Barrow take a minor road through Rampside to Roa Island. Turn L into reserve car park. Walk to main island along stone causeway.
Access: Open all year. Access restricted to designated paths during bird breeding season. Slitch Ridge is closed at this time. No dogs allowed during bird breeding season.
Facilities: None.
Public transport: Bus: regular service from Barrow to Roa Island.
Habitat: Shingle, sand, grassland.
Key birds: *Summer*: Arctic and Little Terns, Oystercatcher, Ringed Plover, Eider Duck. *Winter*: Brent Goose, Redshank, Dunlin, Sanderling.
Contact: Trust HQ, Plumgarths, Crook Road, Kendal LA8 8LX. 01539 816300; (fax) 01539 816301. e-mail: mail@cumbriawildlifetrust.org.uk www.cumbriawildlifetrust.org.uk

3. HARRINGTON RESERVOIR LNR

Allerdale Borough Council
Location: NX 994 257. 2 miles S of Workington. Take A597 S to Harrington, turn L into Moorclose Road.
Access: Site open to the public from several points along Moorclose Road.
Facilities: None
Public transport: Bus service from the centre of Workington to Harrington and Moorclose Road
Habitat: Small steep-sided wooded valley with streamside habitat and reservoir.
Key birds: A variety of woodland species and wildfowl including Willow Warbler, Chiffchaff, Redshank, Sedge Warbler, Woodcock and Kingfisher.
Contact: Patrick Joyce, Allerdale BC, Allerdale House, Workington, Cumbria CA14 3YJ. 01900 326324; fax 01900 326346.

Cumbria

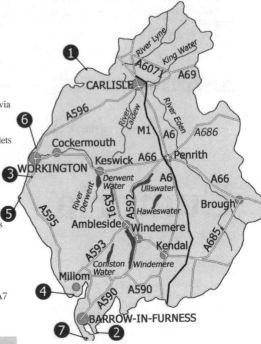

4. HODBARROW

RSPB (North of England Office).
Location: SD 174 791. Lying beside Duddon
Estuary on the outskirts of Millom. Follow signs via
Mainsgate Road.
Access: Open at all times, no charge.
Facilities: One hide overlooking island. Public toilets
in Millom (two miles). Nature trail around the
lagoon.
Public transport: Nearest trains at Millom (two
miles).
Habitat: Brackish coastal lagoon bordered by
limestone scrub and grassland.
Key birds: *Winter*: Waders and wildfowl includes
Redshank, Dunlin, Goldeneye, Red-breasted
Merganser. *Spring/summer*: Breeding gulls and
terns, Eider, grebes, Lapwing. *Autumn*: Passage
waders.
Contact: Dave Blackledge, Warden, North Plain
Farm, Bowness-on-Solway, Wigton, Cumbria CA7
5AG. e-mail: dave.blackledge@rspb.org.uk
www.rspb.org.uk

5. ST BEES HEAD

RSPB (North of England Office).
Location: NX 962 118. S of Whitehaven via the
B5345 road to St Bees village.
Access: Open at all times, no charge. Access via
coast-to-coast footpath. The walk to the viewpoints is
long and steep in parts.
Facilities: Three viewpoints overlooking seabird
colony. Public toilets in St Bees beach car park at
entrance to reserve.
Public transport: Nearest trains at St Bees (0.5
mile).
Habitat: Three miles of sandstone cliffs up to 300 ft
high.
Key birds: *Summer*: Largest seabird colony on W
coast of England: Guillemot, Razorbill, Puffin,
Kittiwake, Fulmar and England's only breeding Black
Guillemot.
Contact: Dave Blackledge, Warden, North Plain
Farm, Bowness-on-Solway, Wigton, Cumbria CA7
5AG. 01697 351330
e-mail: dave.blackledge@rspb.org.uk
www.rspb.org.uk

6. SIDDICK POND

Allerdale Borough Council/EN/Cumbria WT.
Location: NY 001 305. One mile N of
Workington, adjacent to A596.
Access: Access to hide by arrangement with
Iggesund Paperboard (Workington) Ltd on site.
Facilities: Site is crossed lengthways by a cycleway
which makes for easy movement and observation
of all points within the reserve.
Public transport: Bus service to Northside (on the
edge of the reserve) from the centre of Workington.
Habitat: Shallow pond with reedbeds.
Key birds: *Winter*: Wildfowl (inc. Goldeneye,
Shoveler, Whooper Swan). *Spring/summer*: 35
nesting species. Occasional visitors inc. Black-
necked Grebe, Black-tailed Godwit, Black Tern,
Ruff, Osprey.
Contact: Patrick Joyce, Technical Officer Leisure
Services, Parks Development Officer, Allerdale
BC, Allerdale House, Workington, Cumbria CA14
3YJ. 01900 326324; fax 01900 326346.

7. SOUTH WALNEY

Cumbria Wildlife Trust.
Location: SD 215 620. Six miles S of Barrow-in-Furness. From Barrow, cross Jubilee Bridge onto Walney Island, turn L at lights. Continue through Biggar village to South End Caravan Park. Follow unsurfaced road for one mile to reserve.
Access: Open daily (10am-5pm) plus Bank Holidays. No dogs except assistance dogs. Day permits £2 adults, 80p children. Cumbria Wildlife Trust members free.
Facilities: Toilets, nature trails, eight hides (two with wheelchair accessible), 200m boardwalk, cottage available to rent.
Public transport: Bus service as far as Biggar.
Habitat: Shingle, lagoon, sand dune, saltmarsh.
Key birds: *Spring/autumn*: Passage migrants. *Summer*: Breeding Eider, Herring, Greater and Lesser Black-backed Gulls, Shelduck. *Winter*: Teal, Wigeon, Goldeneye, Redshank, Greenshank, Curlew, Oystercatcher, Knot, Dunlin, Twite.
Contact: The Warden, No 1 Coastguard Cottages, South Walney Nature Reserve, Walney Island, Barrow-in-Furness, Cumbria LA14 3YQ. 01229 471066.
e-mail: mail@cumbriawildlifetrust.org.uk
www.cumbriawildlifetrust.org.uk

Derbyshire

1. CARR VALE NATURE RESERVE

Derbyshire Wildlife Trust.
Location: SK 45 70. 1km W of Bolsover on A632 from Chesterfield. Turn R at roundabout into Riverside Way. Car park at end of road. Reserve is reached via footpath to the R around Peter Fidler Reserve (reclaimed colliery tip.)
Access: Open all year.
Facilities: Car park, good disabled access, paths, viewing platform. Coach parking on approach to car park.
Public transport: Various Stagecoach services from Chesterfield (Stephenson Place) all pass close to the reserve: Mon to Sat - 83 serves Villas Road, 81, 82, 82A and 83 serve the roundabout on the A632. Sun - 81A, 82A serve the roundabout on the A632.
Habitat: Lakes, wader flashes, reed bed, sewage farm, scrub, arable fields.
Key birds: *Spring/summer*: Warblers, waders, farmland birds, wildfowl. *Winter*: Wildfowl.
Contact: Trust HQ. 01773 881188.
e-mail: enquiries@derbyshirewt.co.uk
www.derbyshirewildlifetrust.org.uk

2. CARSINGTON RESERVOIR

Severn Trent Water.
Location: SK 24 51. Follow B5035 from either Ashbourne or B5036, then B5035 from Cromford.
Access: Open all year except Dec 25 (7am to sunset). There are various access points. Track is very steep in places and can be slippery in winter.
Facilities: Car parks (charge made) visitor centre, toilets, restaurant, hides.
Public transport: D&G Coach & Bus 111 (Sun and BH Mon) from Derby. D&G Coach & Bus (daily) Matlock-Ashbourne via Wirksworth. D&G Coach & Bus (Wed only) from Derby. Some D&G Coach & Bus 109 journeys from Derby are extended to Carsington (Water Sun & BH Mon).
Habitat: Reservoir, woodland.
Key birds: *Spring/summer*: Gulls and wildfowl species. *Winter*: Winter thrushes, Usual woodland and farmland species, 200-plus species recorded and planned reedbeds and scrapes should increase bird diversity, gulls.
Contact: Severn Trent Water, Sherbourne House, 87 St Martin's Road, Finham, Coventry CV3 6SD.
e-mail: customer.relations@severntrent.co.uk

3. DRAKELOW WILDFOWL RESERVE

Powergen PLC.
Location: SK 22 72 07. Drakelow Power Station, one mile NE of Walton-on-Trent.
Access: Permit holders only for the time being. Reserve is subject to closure at short notice during demolition of power station. Scheduled to last to end of 2005. Any problems, please ring warden during evenings.
Facilities: Seven hides, no other facilities.
Public transport: None.
Habitat: Disused flooded gravel pits with wooded islands and reedbeds.
Key birds: *Summer*: Breeding Reed and Sedge

Derbyshire

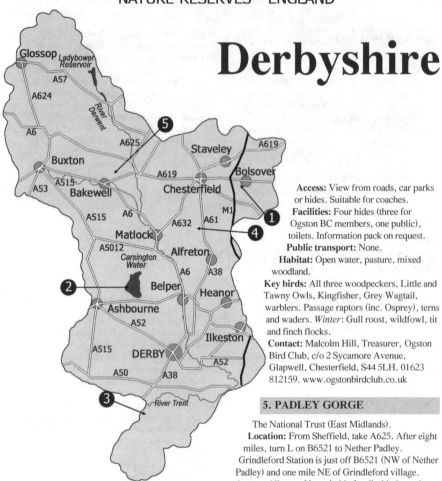

Access: View from roads, car parks or hides. Suitable for coaches.
Facilities: Four hides (three for Ogston BC members, one public), toilets. Information pack on request.
Public transport: None.
Habitat: Open water, pasture, mixed woodland.
Key birds: All three woodpeckers, Little and Tawny Owls, Kingfisher, Grey Wagtail, warblers. Passage raptors (inc. Osprey), terns and waders. *Winter*: Gull roost, wildfowl, tit and finch flocks.
Contact: Malcolm Hill, Treasurer, Ogston Bird Club, c/o 2 Sycamore Avenue, Glapwell, Chesterfield, S44 5LH. 01623 812159. www.ogstonbirdclub.co.uk

Warblers. Water Rail, Hobby. *Winter*: Wildfowl (Goldeneye, Gadwall, Smew), Merlin. Regular sightings of Peregrine in station area. Recent rarities include Great White and Little Egret, Cetti's Warbler and Golden Oriole, Bittern and Spotted Crake. Excellent for dragonflies and butterflies.
Contact: Tom Cockburn, Hon, Warden 1 Dickens Drive, Swadlincote, Derbys DE11 0DX. 01283 217146.

4. OGSTON RESERVOIR

Severn Trent Water Plc.
Location: From Matlock, take A615 E to B6014. From Chesterfield take A61 S of Clay Cross onto B6014.

5. PADLEY GORGE

The National Trust (East Midlands).
Location: From Sheffield, take A625. After eight miles, turn L on B6521 to Nether Padley. Grindleford Station is just off B6521 (NW of Nether Padley) and one mile NE of Grindleford village.
Access: All year. Not suitable for disabled people or those unused to steep climbs. Some of the paths are rocky. No dogs allowed.
Facilities: Café and toilets at Longshaw lodge.
Public transport: Bus: from Sheffield to Bakewell stops at Grindleford/Nether Padley. Tel: 01709 566 000. Train: from Sheffield to Manchester Piccadilly stops at Grindleford Station. Tel: 0161 228 2141.
Habitat: Steep-sided valley containing largest area of sessile oak woodland in south Pennines.
Key birds: *Summer*: Pied Flycatcher, Spotted Flycatcher, Redstart, Wheatear, Whinchat, Wood Warbler, Tree Pipit.
Contact: , High Peak Estate Office, Edale End, Edale Road, Hope S33 2RF. 01433 670368. www.nationaltrust.org.uk

Devon

1. AYLESBEARE COMMON

RSPB (South West England Office).
Location: SY 058 897. Five miles E of J30 of M5
at Exeter, 0.5 miles past Halfway Inn on B3052.
Turn R to Hawkerland, car park is on L. The
reserve is on opposite side of main road.
Access: Open all year. One track suitable for
wheelchairs and pushchairs.
Facilities: Car park, picnic area, group bookings,
guided walks and special events. Disabled access via
metalled track to private farm
Public transport: Buses (Exeter to Sidmouth, 53,
52a, 52b). Request stop at Joynes Grass (reserve
entrance).
Habitat: Heathland, wood fringes, streams and
ponds.
Key birds: *Spring/summer*: Nightjar, Stonechat. *All
year*: Dartford Warbler, Buzzard. *Winter*: Possible
Hen Harrier.
Contact: Toby Taylor, Hawkerland Brake Barn,
Exmouth Road, Aylesbeare, Nr Exeter, Devon, Nr
Exeter, Devon EX5 2JS. 01395 233655..

2. BOVEY HEATHFIELD

Devon Wildlife Trust
Location: SX 824 765. NW of Newton Abbott.
Take A832 towards Bovey Tracey, turn R on Battle
Road into Heathfield Industrial Estate. Turn L into
Cavalier Road and into Dragoon Close, reserve is
along gravel track.
Access: Open at all times. Please keep dogs on a
short lead and keep to the paths. Not suitable for
coaches, some paths are rough. Unlikely to be
suitable for most wheelchair users.
Facilities: An information hut is open when a warden
is on site.
Public transport: Nearest bus route is to Battle
Road, Heathfield.
Habitat: Heath
Key birds: Stonechat, Dartford Warbler, Nightjar.
Contact: Trust HQ, 01392 279244
e-mail: devonwt@cix.co.uk

3. BOWLING GREEN MARSH

RSPB (South West England Office).
Location: SX 972 876. On the E side of River Exe,
four miles SE of Exeter, 0.5 mile SE of Topsham.

Access: Open at all times. Please park at the public
car parks in Topsham, not in the lane by the reserve.
Facilities: One hide suitable for wheelchair access.
Viewing platform overlooking estuary reached by
steps from track. No toilets or visitor centre.
Public transport: Exeter to Exmouth railway has
regular (every 30 mins) service to Topsham station
(0.5 mile from reserve). Stagecoach Devon – 57 bus
has frequent service (Mon-Sat every 12 mins, Sun
every half-hour) from Exeter to Topsham.
Habitat: Coastal grassland, open water/marsh,
hedgerows.
Key birds: *Winter*: Wigeon, Shoveler, Teal, Black-
tailed Godwit, Curlew, Golden Plover. *Spring:*
Shelduck, passage waders, Whimbrel, passage
Garganey and Yellow Wagtail. *Summer*: Gull/tern
roosts, high tide wader roosts contain many passage
birds. *Autumn*: Wildfowl, Peregrine, wader roosts.
Contact: RSPB, Unit 3, Lions Rest Estate, Station
Road, Exminster, Exeter EX6 8DZ. 01392 824614.
www.rspb.org.uk

4. DART VALLEY

Devon Wildlife Trust.
Location: SX 680 727. On Dartmoor nine miles NW
from Ashburton. From A38 'Peartree Cross' near
Ashburton, follow signs towards Princetown. Access
from National Park car parks at New Bridge (S) or
Dartmeet (N).
Access: Designated 'access land' but terrain is rough
with few paths. It is possible to walk the length of
the river (eight miles). A level, well-made track runs
for a mile from Newbridge to give easy access to
some interesting areas. Not suitable for large coaches
(narrow bridges). Probably too rough for
wheelchairs.
Facilities: Dartmoor National Park toilets in car parks
at New Bridge and Dartmeet.
Public transport: Enquiry line 01392 382800.
Summer service only from Newton Abbot/Totnes to
Dartmeet.
Habitat: Upland moor, wooded valley and river.
Key birds: *All Year*: Raven, Buzzard. *Spring/
summer*: Wood Warbler, Pied Flycatcher, Redstart in
woodland, Stonechat and Whinchat on moorland,
Dipper, Grey Wagtail, Goosander on river.
Contact: Trust HQ, 01392 279244.
e-mail: devonwt@cix.co.uk
www.devonwildlifetrust.org

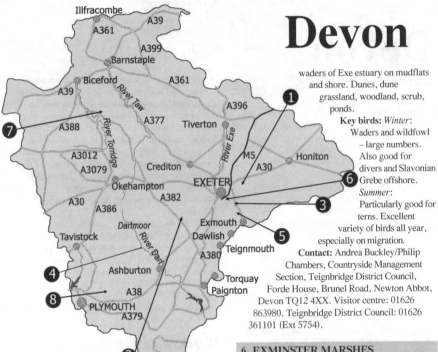

Devon

waders of Exe estuary on mudflats and shore. Dunes, dune grassland, woodland, scrub, ponds.

Key birds: *Winter*: Waders and wildfowl – large numbers. Also good for divers and Slavonian Grebe offshore. *Summer*: Particularly good for terns. Excellent variety of birds all year, especially on migration.

Contact: Andrea Buckley/Philip Chambers, Countryside Management Section, Teignbridge District Council, Forde House, Brunel Road, Newton Abbot, Devon TQ12 4XX. Visitor centre: 01626 863980. Teignbridge District Council: 01626 361101 (Ext 5754).

5. DAWLISH WARREN NNR

Teignbridge District Council.
Location: SX 983 788. At Dawlish Warren on S side of Exe estuary mouth. Turn off A379 at sign to Warren Golf Club, between Cockwood and Dawlish. Turn into car park adjacent to Lea Cliff Holiday Park. Pass under tunnel and turn L away from amusements. Park at far end of car park and pass through two pedestrian gates.
Access: Open public access, but avoid mudflats. Also avoid beach beyond groyne nine around high tide due to roosting birds. Parking charges apply. Restricted access for dogs, so please contact the wardens for more information.
Facilities: Visitor centre (01626 863980) open most weekends all year (10.30am-1pm and 2pm-5pm). Summer also open most weekdays as before, can be closed if warden on site. Toilets at entrance tunnel and in resort area only. Hide open at all times – best around high tide.
Public transport: Train station at site, also regular bus service operated by Stagecoach.
Habitat: High tide roost site for wildfowl and

6. EXMINSTER MARSHES

RSPB (South West England Office).
Location: SX 954 872. Five miles S of Exeter on W bank of River Exe. Marshes lie between Exminster and the estuary.
Access: Open at all times.
Facilities: No toilets or visitor centre. Information in RSPB car park and marked footpaths across reserve.
Public transport: Stagecoach Devon (01392 427711). Exeter to Newton Abbot/Torquay buses – stops are 400 yds from car park.
Habitat: Coastal grazing marsh with freshwater ditches and pools, reed and scrub-covered canal banks.
Key birds: *Winter*: Brent Goose, Wigeon, Water Rail, Short-eared Owl. *Spring*: Lapwing, Redshank and wildfowl breed, Cetti's Warbler on canal banks. *Summer*: Gull roosts, passage waders. *Autumn*: Peregrine, winter wildfowl increase.
Contact: RSPB, Unit 3, Lions Rest Estate, Station Road, Exminster, Exeter, Devon EX6 8DZ. 01392 824614. www.rspb.org.uk

7. HALSON NATURE RESERVE

Devon Wildlife Trust
Location: SS 554 131 and SS 560 117. From the

village of Dolton (Nr Great Torrington) turn up Fore Street. Continue out of village down West Lane until crossroads. Turn R, Quarry car park and entrance is on L or continue for a mile and follow track on L to Ashwell car park.
Access: Open all year. Please keep dogs on a lead and keep to the paths. Not suitable for coaches. The path from Quarry car park is level for 900m to the river although it is unsurfaced it is mostly suitable for wheelchairs.
Facilities: There are four waymarked trails on the reserve.
Public transport: None.
Habitat: Mixed deciduous river valley woodland.
Key birds: Kingfisher, Sand Martin, Heron, Dipper, Grey Wagtail, Buzzard, Pied Flycatcher, Great Spotted Woodpecker.
Contact: Trust HQ, 01392 279244.
e-mail: devonwt@cix.co.uk

8. WARLEIGH POINT

Devon Wildlife Trust
Location: SX 450 608. From village of Tamerton Foliot (just N of Plymouth) take road signposted to reserve (Station Road). Reserve is at end of this dead-end road about two miles from village.
Access: Open all year. Please keep dogs on a lead and keep to the paths. Not suitable for coaches. The path to the point is suitable for wheelchairs.
Facilities: There is a circular unsurfaced path around the reserve.
Habitat: Estuary and woodland.
Key birds: Tawny Owl, Great Spotted and Green Woodpecker, a large active rookery, Redshank, Shelduck, Great Crested Grebe, Pochard, Goldeneye, Little Egret.
Contact: Trust HQ, 01392 279244.
e-mail: devonwt@cix.co.uk

Dorset

1. ARNE

RSPB (South West England Office).
Location: SY 973 882. Four miles SE of Wareham, turn off A351 at Stoborough.
Access: Shipstal Point and Coombe birdwatchers' trails open all year. Bird hides available on both trails. Accessed from car park. Coaches and escorted parties by prior arrangement.
Facilities: Toilets in car park. Car park charge applies to non-members. Various footpaths. Reception hut (open end-May-early Sept).
Public transport: None.
Habitat: Lowland heath, woodland reedbed and saltmarsh, extensive mudflats of Poole Harbour.
Key birds: *All year*: Dartford Warbler, Little Egret, Stonechat. *Winter*: Hen Harrier, Red-breasted Merganser, Black-tailed Godwit. *Summer*: Nightjar, warblers. *Passage*: Spotted Redshank, Whimbrel, Greenshank, Osprey.
Contact: RSPB, Syldata, Arne, Wareham, Dorset BH20 5BJ. 01929 553360. www.rspb.org.uk

2. BROWNSEA ISLAND

Dorset Wildlife Trust.
Location: SZ 026 883. Half hour boat ride from

Poole Quay. Ten minutes from Sandbanks Quay (next to Studland chain-ferry).
Access: Apr, May, Jun, Sept and Oct. Access by self-guided nature trail. Costs £2 adults, £1 children. Jul, Aug access by afternoon guided tour (2pm daily, duration 105 minutes). Costs £2 adults, £1 children.
Facilities: Toilets, information centre, five hides, nature trail.
Public transport: Poole Rail/bus station for access to Poole Quay and boats.
Habitat: Saline lagoon, reedbed, lakes, coniferous and mixed woodland.
Key birds: *Spring*: Avocet, Black-tailed Godwit, waders, gulls and wildfowl. *Summer*: Common and Sandwich Terns, Yellow-legged Gull, Little Egret, Little Grebe, Golden Pheasant. *Autumn*: Curlew Sandpiper, Little Stint.
Contact: Chris Thain, The Villa, Brownsea Island, Poole, Dorset BH13 7EE. 01202 709445.
e-mail: dorsetwtisland@cix.co.uk
www.wildlifetrust.org.uk/dorset

3. DURLSTON COUNTRY PARK

Dorset County Council.
Location: SZ 032 774. One mile S of Swanage (signposted).

Dorset

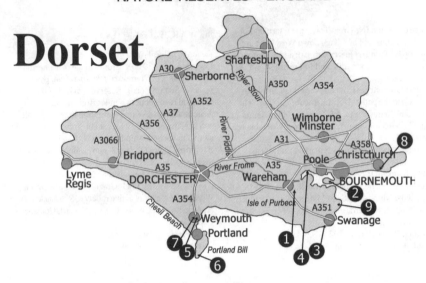

Access: Visitor centre in car park open weekends and holidays during winter and daily in other seasons (phone for times).
Facilities: Guided walks, toilets, bookshop.
Public transport: Two buses per day except Sundays and Bank Holidays.
Habitat: Grassland, hedges, cliff, meadows, downland.
Key birds: Cliff-nesting seabird colonies; good variety of scrub and woodland breeding species; spring and autumn migrants; seawatching esp. Apr/May & Aug/Nov.
Contact: The Ranger, Durlston Country Park, Swanage, Dorset BH19 2JL. 01929 424443. www.durlston.co.uk

4. HAM COMMON LNR

Poole Borough Council.
Location: SY 99. W of Poole. In Hamworthy, take the Blandford Road S along Lake Road, W along Lake Drive and Napier Road, leading to Rockley Park. Park in the beach car park by Hamworthy Pier or Rockley Viewpoint car park, off Napier Road, opposite the entrance to Gorse Hill Central Park.
Access: Open all year. Not suitable for coaches.
Facilities: None.
Public transport: None.
Habitat: Heathland, scrub, reedbeds, lake. Views over Wareham Channel and Poole Harbour.
Key birds: *Spring/summer:* Stonechat, Dartford Warbler. *Winter:* Brent Goose, Red-breasted

Merganser, occasional divers, rarer grebes, Scaup. Waders inc Whimbrel, Greenshank and Common Sandpiper. *All year:* Little Egret.
Contact: Poole Borough Council, Civic Centre, Poole BH15 2RU. 01202 633633.
e-mail: information@poole.gov.uk

5. LODMOOR

RSPB (South West England Office).
Location: SY 686 807. Adjacent Lodmoor Country Park, in Weymouth, off A353 to Wareham.
Access: Open all times.
Facilities: One viewing shelter, network of paths.
Public transport: Local bus service.
Habitat: Marsh, shallow pools, reeds and scrub, remnant saltmarsh.
Key birds: *Spring/summer:* Breeding Common Tern, warblers (including Reed, Sedge, Grasshopper and Cetti's), Bearded Tit. *Winter:* Wildfowl, waders. *Passage:* Waders and other migrants.
Contact: Nick Tomlinson, RSPB Visitor Centre, Swannery Car Park, Weymouth DT4 7TZ. 01305 778313. www.rspb.org.uk

6. PORTLAND BIRD OBSERVATORY

Portland Bird Observatory (registered charity).
Location: SY 681 690. Six miles S of Weymouth beside the road to Portland Bill.
Access: Open at all times. Parking only for members of Portland Bird Observatory. Self-catering

accommodation for up to 20. Take own towels, sheets, sleeping bags.
Facilities: Displays and information, toilets, natural history bookshop, equipped kitchen, laboratory.
Public transport: Bus service from Weymouth (First Dorset Transit Route 1).
Habitat: Scrub and ponds.
Key birds: *Spring/autumn*: Migrants including many rarities. *Summer*: Breeding auks, Fulmar, Kittiwake.
Contact: Martin Cade, Old Lower Light, Portland Bill, Dorset DT5 2JT. e-mail: obs@btinternet.com www.portlandbirdobs.btinternet.co.uk

7. RADIPOLE LAKE

RSPB (South West England Office).
Location: SY 677 796. In Weymouth. Enter from Swannery car park on footpaths.
Access: Visitor centre and nature trail open every day, summer (9am-5pm), winter (9am-4pm). Hide open (8.30am-4.30pm). Permit available from visitor centre required by non-RSPB members.
Facilities: Network of paths, one hide, one viewing shelter.
Public transport: Close to train station serving London and Bristol.
Habitat: Lake, reedbeds.
Key birds: *Winter*: Wildfowl. *Summer*: Breeding reedbed warblers (including Cetti's), Bearded Tit, passage waders and other migrants. Garganey regular in Spring. Good for rarer gulls.
Contact: Nick Tomlinson, RSPB Visitor Centre, Swannery Car Park, Weymouth DT4 7TZ. 01305 778313. www.rspb.org.uk

8. STANPIT MARSH LNR

Community Services, Christchurch Borough Council.
(Stanpit Marsh Advisory Panel)

Location: SZ 167 924. In Christchurch.
Access: Public open space. Limited disabled access across marshy terrain. Nearest coach parking is at Two Riversmeet Leisure Centre, Stony Lane.
Facilities: Information centre.
Public transport: Wilts & Dorset bus no 123 (tel 01202 673555) Stanpit recreation ground stop. Bournemouth Yellow Buses no 20 (tel 01202 636000) Purewell Cross roundabout stop.
Habitat: SSSI. Salt, fresh, brackish marsh, sand dune and scrub.
Key birds: *Estuarine*: Waders, winter wildfowl, migrants. Reedbed: Bearded Tit, Cetti's Warbler. *Scrub*: Sedge Warbler, Reed Warbler. *River/streams/bankside*: Kingfisher. Feeding and roosting site.
Contact: Peter Holloway, Christchurch Countryside Service, Steamer Point Nature Reserve, Highcliffe, Christchurch, Dorset BH23 4XX. 01425 272479. e-mail: countrysideservice@christchurch.gov.uk

9. STUDLAND & GODLINGSTON HEATHS

National Trust.
Location: SZ 030 846. N of Swanage. From Ferry Road N of Studland village.
Access: Open all year.
Facilities: Hides, nature trails.
Public transport: No150 bus hourly to and from Bournemouth. 142/3 from Swanage and Wareham.
Habitat: Woodland, heath, dunes, inter-tidal mudflats, saltings, freshwater lake, reedbeds, carr.
Key birds: Water Rail, Reed and Dartford Warblers, Nightjar, Stonechat. *Winter*: Wildfowl. Studland Bay, outside the reserve, has winter Black-necked and Slavonian Grebes, Scoter, Eider.
Contact: The National Trust, Countryside Office, Middle Beach Car Park, Studland, Swanage BH19 3AX.

Durham

1. CASTLE EDEN DENE

English Nature (Northumbria Team).
Location: NZ 435 397. Adjacent to Peterlee, signposted from A19 and Peterlee town centre.
Access: Open from 8am-8pm or sunset if earlier. Car park. Dogs under tight control please. Pre-booked coach parties welcome. Parking available for one coach only.

Facilities: Car parking with toilet block at Oakerside Dene Lodge. 12 miles of footpath, two waymarked trails.
Public transport: Bus service to Peterlee centre.
Habitat: Yew/oak/sycamore woodland, paramaritime, limestone grassland.
Key birds: More than 170 recorded, 50 regular breeding species, typical woodland species.

Durham

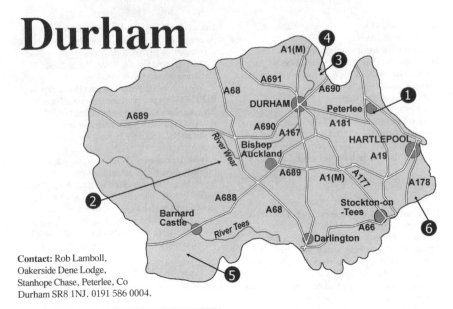

Contact: Rob Lamboll,
Oakerside Dene Lodge,
Stanhope Chase, Peterlee, Co
Durham SR8 1NJ. 0191 586 0004.

2. HAMSTERLEY FOREST

Forest Enterprise.
Location: NZ 093 315. Eight miles W of Bishop
Auckland. Main entrance is five miles from A68, S
of Witton-le-Wear and signposted through Hamsterley
village and Bedburn.
Access: Open all year. Toll charge. Vehicles should
not be left unattended after dark.
Facilities: Visitor centre, toilets, shop, access for
disabled. Visitors should not enter fenced farmland.
Public transport: None.
Habitat: Commercial woodland, mixed and
broadleaved trees.
Key birds: *Spring/summer*: Willow Warbler,
Chiffchaff, Wood Warbler, Redstart, Pied Flycatcher.
Winter: Crossbill, Redwing, Fieldfare. *All year*: Jay,
Dipper, Green Woodpecker.
Contact: Forest Enterprise, Eals Burn, Bellingham,
Hexham, Northumberland, NE48 2HP, 01434
220242. e-mail: richard.gilchrist@forestry.gsi.gov.uk

3. JOE'S POND NATURE RESERVE

Durham Wildlife Trust.
Location: NZ 32 48. Between Durham and
Sunderland on A690. N from Durham, leave A690 S
of Houghton-le-Spring on B21284 to Fence Houses
and Hetton-le-Hole. Head W towards Fence Houses
and turn L at 1st roundabout, after 800 metres, into
an opencast colliery site, signed Rye Hill Site.

Access: Open all year.
Facilities: Car park, bird hide.
Public transport: None.
Habitat: Scrub, pond, grassland.
Key birds: *Spring/summer*: Ruddy Duck,
hirundines, Whinchat, Lesser Whitethroat,
Whitethroat, Blackcap. Possible Yellow Wagtail,
Redstart, Grasshopper Warbler. *Passage*: Waders,
Wheatear. *Winter*: Teal, Pochard, Water Rail,
Woodcock, Short-eared Owl, Kingfisher, thrushes.
Chance of Merlin, Jack Snipe.
Contact: Rainton Meadows, Chilton Moor,
Houghton-le-Spring, Tyne & Wear, DH4 6PU.
01388 488 728. e-mail: durhamwt@cix.co.uk
www.wildlifetrust.org.uk/durham/

4. RAINTON MEADOWS

Durham Wildlife Trust, the City of Sunderland and
RJB Mining (UK) Ltd.
Location: NZ 326 486. Located W of A690
between Durham and Sunderland. Just S of
Houghton-le-Spring turn onto B1284, signposted to
Fence Houses and Hetton-le-Hole. Head W towards
Fence Houses and turn L at the first roundabout after
0.5 mile into Rye Hill Site.
Access: Park at Visitor Centre (entrance gate locked
at 4.30pm) or Mallard Way. Paths generally
wheelchair-accessible but there are some muddy
areas. Main circular walk. Dogs on lead.

Facilities: Visitor Centre, toilets, café, log book, shop, wildlife display.
Public transport: Buses from Sunderland and Durham (222 and 220) stop at Mill Inn. Reserve is reached via B1284 passing under A690. Bus from Chester-le-Street (231) stops at Fencehouses Station. Walk E along B1284. Tel: Traveline 0870 608 2608.
Habitat: Reedbed, ponds, grassland, young tree plantation.
Key birds: *Spring/summer*: Great Crested Grebe, Ruddy Duck, Whinchat, Reed Warbler. *Winter*: Water Rail, Kingfisher, Peregrine, Merlin, Long and Short-eared Owls. *Passage*: Waders.
Contact: Trust HQ, 0191 5843112.
e-mail: durhamwt@cix.co.uk

5. STANG FOREST AND HOPE MOOR

Forest Enterprise.
Location: NZ 022 075. The wood is six miles S of Barnard Castle. On A66 follow signs for Reeth after the turn-of to Barnard Castle on the W-bound carriageway. Stang is then about 3.5 miles from the A66 (car park for Hope Edge Walk).
Access: Open all year. Road dangerous in frost.
Facilities: Number of parking lay-bys and forest trails. For best birdwatching, head E and then N to Hope Edge.
Public transport: None.
Habitat: Woodland, moorland.
Key birds: *Spring/summer*: Whinchat, Wheatear, Cuckoo. *All year*: Red Grouse, Crossbill.
Contact: Forest Enterprise, Eals Burn, Bellingham, Hexham, Northumberland, NE48 2HP, 01434 220242.

6. TEESMOUTH

English Nature (Northumbria).
Location: Two components, centred on NZ 535 276 and NZ 530 260, three and five miles S of Hartlepool, E of A178. Access to northern component from car park at NZ 534 282, 0.5 miles E of A178. Access to southern compartment from A178 bridge over Greatham Creek at NZ 510 254. Car park adjacent to A178 at NZ 508 251 is currently closed.
Access: Open at all times. In northern component, no restrictions over most of dunes and North Gare Sands (avoid golf course, dogs must be kept under close control). In southern component, disabled access path to public hides at NZ 516 255 and NZ 516 252 (no other access).
Facilities: Nearest toilets at Seaton Carew, one mile to the N. Disabled access path and hides (see above), interpretive panels and leaflet. Teesmouth Field Centre (Tel: 01429 264912).
Public transport: Half-hourly bus service operates Mon-Sat between Middlesbrough and Hartlepool, along A178, Stagecoach Hartlepool, Tel: 01429 267082).
Habitat: Grazing marsh, dunes, intertidal flats.
Key birds: Passage and winter wildfowl and waders. Passage terns and skuas in late summer. Scarce passerine migrants and rarities. *Winter*: Merlin, Peregrine, Snow Bunting, Twite, divers, grebes.
Contact: Mike Leakey, English Nature, c/o British Energy, Tees Road, Hartlepool TS25 2BZ. 01429 853325. e-mail: northumbria@english-nature.org.uk
www.english-nature.org.uk

Essex

1. ABBERTON RESERVOIR

Essex Wildlife Trust.
Location: TL 963 185. Six miles SW of Colchester on B1026. Follow signs from Layer-de-la-Haye.
Access: Open Tue-Sun and Bank Holiday Mondays (9am-5pm). Closed Christmas Day and Boxing Day.
Facilities: Visitor centre, toilets, nature trail, five hides (disabled access). Also good viewing where roads cross reservoir.
Public transport: Phone Trust for advice.
Habitat: Nine acres on edge of 1200a reservoir.
Key birds: Nationally important for Mallard, Teal, Wigeon, Shoveler, Gadwall, Pochard, Tufted Duck, Goldeneye (most important inland site in Britain). Smew regular. Passage waders, terns, birds of prey. Tree-nesting Cormorants (largest colony in Britain); raft-nesting Common Tern. *Summer*: Yellow Wagtail, warblers, Nightingale, Corn Bunting; *Autumn*: Red-crested Pochard, Water Rail; *Winter*: Goosander.
Contact: Centre Manager, Essex Wildlife Trust, Abberton Reservoir Visitor Centre, Layer-de-la-Haye, Colchester CO2 0EU. 01206 738172.
e-mail: abberton@essexwt.org.uk

NATURE RESERVES - ENGLAND

2. ABBOTTS HALL FARM

Essex Wildlife Trust.
Location: TL 963 145. Seven miles SW from
Colchester. Turn E off B1026 (Colchester–Maldon
road) towards Peldon. Entrance about 0.5 mile on R.
Access: Weekdays (9am-5pm). No dogs please.
Working farm, so please take care.
Facilities: Toilets, hides, guided walks, fact-sheets.
Public transport: None.
Habitat: Saltmarsh, saline lagoons, grazing marsh,
farmland.
Key birds: *Winter*: Waders and wildfowl.
Contact: Trust HQ, 01621 862960.
e-mail: admin@essexwt.org.uk

3. BRADWELL BIRD OBSERVATORY

Essex Birdwatching Society
Location: 100 yards S of St Peter's Chapel,
Bradwell-on-Sea. Mouth of Blackwater estuary,
between Maldon and Foulness.
Access: Open all year.
Facilities: Accommodation for eight in hut; two
rooms each with four bunks; blankets, cutlery, etc.
supplied.
Public transport: None.
Habitat: Mudflats, saltmarsh.
Key birds: *Winter*: Wildfowl (inc. Brent Geese,
Red-throated Diver, Red-breasted Merganser), large
numbers of waders; small numbers of Twite, Snow
Bunting and occasional Shore Lark on beaches, also
Hen Harrier, Merlin and Peregrine. Good passage of
migrants usual in spring and autumn. *Summer*: Small
breeding population of terns and other estuarine
species.
Contact: Graham Smith, 48 The Meads,
Ingatestone, Essex CM4 0AE. 01277 354034.

4. CHIGBOROUGH LAKES

Essex Wildlife Trust.
Location: GR 877 086. Lies NE of Maldon, about
one mile from Heybridge on the B1026 towards
Tolleshunt d'Arcy, turn N into Chigborough Road.
Continue past fishery entrance and Chigborough
Farm buildings until you see an entrance gate to
Chigborough Quarry. The reserve entrance is just
beyond this point on the L.
Access: Open all year. do not obstruct the gravel pit
entrance.
Facilities: Car park with height restriction barrier of
5ft 9in.
Public transport: Bus: Colchester to Maldon
Leisure Centre along the B1026.

Habitat: Flooded gravel pits, small ponds, willow
carr, grassland, scrub.
Key birds: *Spring/summer*: Sedge and Reed
Warblers, Whitethroat, Reed Bunting, Willow
Warbler, Great Crested and Little Grebes, Kingfisher,
Water Rail. *Passage*: Waders including Greenshank.
Grass snakes and common lizards, good numbers of
common blue, small copper and ringlet butterflies and
dragonflies.
Contact: Trust HQ, 01621 862960.
e-mail: admin@essexwt.org.uk www.essexwt.org.uk

5. FINGRINGHOE WICK

Essex Wildlife Trust.
Location: TM 046 197. Colchester five miles. The
reserve is signposted from B1025 to Mersea Island,
S of Colchester.
Access: Open six days per week (not Mon or
Christmas or Boxing Day). No permits needed.
Donations invited. Centre/reserve open (9am-5pm).
Dogs must be on a lead.
Facilities: Visitor centre – toilets, shop, light
refreshments, car park, displays. Reserve – seven
bird hides, two nature trails, plus one that wheelchair
users could use with assistance.
Public transport: None.
Habitat: Old gravel pit, large lake, many ponds,
sallow/birch thickets, young scrub, reedbeds,
saltmarsh, gorse heathland.
Key birds: *Autumn/winter*: Brent Goose, waders,
Hen Harrier, Little Egret. *Spring*: 30 male
Nightingales. Good variety of warblers in scrub,
thickets, reedbeds and Turtle Dove, Green/Great
Spotted Woodpeckers. *Winter*: Little Grebe, Mute
Swan, Teal, Wigeon, Shoveler, Gadwall on lake.
Contact: Laurie Forsyth, Wick Farm, South Green
Road, Fingringhoe, Colchester, Essex CO5 7DN.
01206 729678. e-mail: admin@essexwt.org.uk

6. HANNINGFIELD RESERVOIR

Essex Wildlife Trust.
Location: TQ 725 972. Three miles N of Wickford.
Exit off Southend Road (Old A130) at Rettendon
onto South Hanningfield Road. Follow this for two
miles until reaching T-junction with Hawkswood
Road. Turn R, entrance to Visitor Centre and reserve
is one mile on R.
Access: Open Mon-Sun (9am-5pm) plus Bank
Holiday Mon from Easter to October. Disabled
parking, toilets, and adapted birdwatching hide. No
dogs. No cycling.
Facilities: Visitor centre, gift shop, optics,
refreshments, toilets, four bird hides, nature trails,

Essex

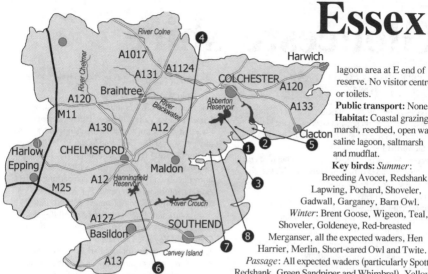

lagoon area at E end of reserve. No visitor centre or toilets.
Public transport: None.
Habitat: Coastal grazing marsh, reedbed, open water saline lagoon, saltmarsh and mudflat.
Key birds: *Summer*: Breeding Avocet, Redshank, Lapwing, Pochard, Shoveler, Gadwall, Garganey, Barn Owl. *Winter*: Brent Goose, Wigeon, Teal, Shoveler, Goldeneye, Red-breasted Merganser, all the expected waders, Hen Harrier, Merlin, Short-eared Owl and Twite. *Passage*: All expected waders (particularly Spotted Redshank, Green Sandpiper and Whimbrel), Yellow Wagtail, Whinchat and Wheatear.
Contact: Paul Charlton, Site Manager c/o 1 Old Hall Lane, Tolleshunt D'Arcy, Maldon, Essex CM9 8TP. 01621 869015.
e-mail: paul.charlton@rspb.org.uk

8. TOLLESBURY WICK

Essex Wildlife Trust.
Location: GR 970 104. On Blackwater Estuary eight miles E of Maldon. Follow B1023 to Tollesbury via Tiptree, leaving A12 at Kelvedon. Then follow Woodrolfe Road S towards the marina. Use car park at Woodrolfe Green. Small public car park near reserve suitable for cars and mini-buses only.
Access: Open all times along public footpath on top of sea wall.
Facilities: Public toilets at Woodrolfe Green car park.
Public transport: Bus services run to Tollesbury from Maldon, Colchester and Witham.
Habitat: Estuary with fringing saltmarsh and mudflats with some shingle. Extensive freshwater grazing marsh, brackish borrowdyke and small reedbeds.
Key birds: *Winter*: Wildfowl and waders, Short-eared Owl, Hen Harrier. *Summer*: Breeding Avocet, Redshank, Lapwing, occasional Little Tern, Reed and Sedge Warblers, Barn Owl. *Passage*: Whimbrel, Spotted Redshank.
Contact: Jonathan Smith, Tollesbury, Maldon, Essex CM9 8RJ. 01621 868628.
e-mail: jonathans@essexwt.org.uk

picnic area, coach parking, education room.
Public transport: Chelmsford to Wickford bus no 14 to Downham village and walk half mile down Crowsheath Lane.
Habitat: One hundred acre mixed woodland with grassy glades and rides, adjoining the 870 acre Hanningfield Reservoir, designated an SSSI due to its high numbers of wildfowl.
Key birds: *Spring*: Good numbers and mix of woodland warblers. *Summer*: Vast numbers of Swifts, Swallows and martins feeding over the water. Hobby and Osprey. *Winter*: Good numbers and mix of waterfowl. Large gull roost.
Contact: Bill Godsafe, Hanningfield Reservoir Visitor Centre, Hawkswood Road, Downham, Billericay CM11 1WT. 01268 711001.

7. OLD HALL MARSHES

RSPB (East Anglia Office).
Location: TL 97 51 25. Approx eight miles S of Colchester. From A12 take B1023, via Tiptree, to Tolleshunt D'Arcy. Then take Chapel Road (back road to Tollesbury), after one mile turn left into Old Hall Lane. Continue up Old Hall Lane, over speed ramp and through iron gates to cattle grid, then follow signs to car park.
Access: By permit only in advance from Warden, write to address below. Open 9am-9pm or dusk, closed Tues. No coaches.
Facilities: Two trails – one of three miles and one of 6.5 miles. Two viewing screens overlooking saline

153

Gloucestershire

1. ASHLEWORTH HAM AND MEEREND THICKET

Gloucestershire Wildlife Trust.
Location: SO 830 265. Leave Gloucester N on A417; R at Hartpury and follow minor road through Ashleworth towards Hasfield.
Access: Access prohibited at all times but birds may be viewed from new hide in Meerend Thicket.
Facilities: Bird viewing hide and screen, interpretation panels.
Public transport: None.
Habitat: Low-lying grassland flood plain.
Key birds: *Winter*: Wildfowl (inc. 4,000 Wigeon, 1,500 Teal, Pintail, Goldeneye, Bewick's Swan); passage waders; Peregrine, Hobby.
Contact: Trust HQ, 01452 383333.
e-mail: info@gloucestershirewildlifetrust.co.uk
www.wildlifetrust.org.uk/gloucswt/

2. COKES PIT LOCAL NATURE RESERVE (LAKE 34)

Cotswold Water Park Society.
Location: SU 026 957. Lake 34 is located adjacent to Keynes Country Park. From the A419, take the B4696 towards Ashton Keynes. At the staggered crossroads, go straight over, heading towards Somerford Keynes. Take the next R turn to Circencester. The entrance to Keynes Country Park is the second entrance on the R.
Access: Open at all times. Car parking charge applies.
Facilities: Paths are flat with footbridges. Some wheelchair access. Toilets, refreshments, car parking and information available from Keynes Country Park adjacent. A hide with log book is located on the E shore.
Public transport: Buses from Kemble, Cheltenham, Cirencester and Swindon. Tel: 08457 090 899. Nearest station is four miles away at Kemble. Tel: 08457 484 950.
Habitat: Small lake with wooded and reedbed margins with feeding station in winter.
Key birds: *Winter*: Common wildfowl, Red-crested Pochard. *Summer*: Breeding ducks, warblers, Nightingale, Hobby, Common Tern, Reed Bunting.
Contact: Cotswold Water Park Society, See details below.

3. COTSWOLD WATER PARK

Cotswold Water Park Society.
Location: The CWP comprises 140 lakes in the Upper Thames Valley, between Cirencester and Swindon. Many are accessible by using public rights of way. Start at The Cotswold Water Park Gateway Visitor Centre. SU 072 971. From A419, take B4696 towards Ashton Keynes. Visitor Centre is immediately on L after A419. Coach parking, toilets, cafe and information available. For Millennium Visitor Centre at Keynes Country Park SU 026 957, from A419, take B4696 towards Ashton Keynes. At staggered crossroads, go straight over, towards Somerford Keynes. Take next R turn to Cirencester. Entrance to Keynes Country Park is second entrance on R. Car parking charge applies. Coach parking on request - please call ahead. Toilets, refreshments and information available. Plus bathing beach, high ropes course and other activities.
Access: Open all year. The visitor centres are open every day except Christmas Day.
Facilities: Paths are flat with stiles and footbridges. Many are wheelchair accessible. Toilets, refreshments, car parking and information available from visitor centres. Hides at Cleveland Lakes/ Waterhay (lakes 68a and 68c), Shorncote Reed Bed (lakes 84/85), Cokes Pit (Lake 34) and Whelford Pools (Lake 111). Free copies of CWP Leisure Guide available from visitor centres. These have maps showing the lake numbering.
Public transport: Bus: from Kemble, Cheltenham, Cirencester and Swindon. Tel: 08457 090899. Train: nearest station is four miles away at Kemble. Tel: 08457 484950.
Habitat: Gravel extraction has created more than 1,000ha of standing open water or 140 lakes plus other associated wetland habitats, creating one of the largest man-made wetlands in Europe.
Key birds: *Winter*: Common wildfowl, Smew, Red-crested Pochard, Merlin, Peregrine. *Summer*: Breeding ducks, warblers, Nightingale, Hobby, Common Tern, Black-headed Gull colony, Reed Bunting, hirundines.
Contact: Cotswold Water Park Society, Keynes Country Park, Spratsgate Lane, Shorncote, Cirencester, Glos GL7 6DF.
01285 861459; (Fax)01285 860186.
e-mail: info@waterpark.org

4. HIGHNAM WOODS

RSPB (Central England Office).
Location: SO 778 190. Signed on A40 three miles W of Gloucester.
Access: Open at all times, no permit required. The nature trails can be very muddy. Dogs allowed on leads.
Facilities: One nature trail (approx 1.5 miles).
Public transport: Contact Glos. County Council public transport information line. Tel: 01452 425543.
Habitat: Ancient woodland in the Severn Vale with areas of coppice and scrub.
Key birds: *Spring/summer:* The reserve has about 12 pairs of breeding Nightingales. Resident birds include all three woodpeckers, Buzzard and Sparrowhawk. Ravens are frequently seen. *Winter*: feeding site near car park for woodland birds.
Contact: Ivan Proctor, The Puffins, Parkend, Lydney, Glos GL15 4JA. 01594 562852.
e-mail: ivan.proctor@rspb.org.uk

5. NAGSHEAD

RSPB (Central England Office).
Location: SO 097 085. In Forest of Dean, N of Lydney. Signed immediately W of Parkend village on the road to Coleford.
Access: Open at all times, no permit required. The reserve is hilly and there are some stiles on nature trails. Dogs must be kept under close control.
Facilities: There are two nature trails (one mile and 2.25 miles). Information centre open at weekends mid-Apr to end Aug. Schools education programme available.
Public transport: Contact Glos. County Council public transport information line, 01452 425543.
Habitat: Much of the reserve is 200-year-old oak plantations, grazed in some areas by sheep. The rest of the reserve is a mixture of open areas and conifer/mixed woodland.
Key birds: *Spring*: Pied Flycatcher, Wood Warbler, Redstart, warblers. *Winter*: Siskin, Crossbill in some years. *All year*: Buzzard, Raven, all three woodpeckers.
Contact: Ivan Proctor, The Puffins, Parkend, Lydney, Glos GL15 4JA. 01594 562852.
e-mail: ivan.proctor@rspb.org.uk
www.rspb.org.uk

6. SLIMBRIDGE

The Wildfowl & Wetlands Trust.
Location: SO 723 048. S of Gloucester. Signposted from M5 (exit 13 or 14).
Access: Open daily except Christmas Day, (9am-5.30pm, 5pm in winter). Group visits a speciality. Contact Bookings Officer 01453 891900.
Facilities: Hides, observatory, observation tower, Hanson Discovery Centre, wildlife art gallery, tropical house, facilities for disabled, worldwide collection of wildfowl species.
Public transport: None.
Habitat: Reedbed, saltmarsh, freshwater pools, mudflats.
Key birds: Kingfisher, waders, raptors. *Winter*; Wildfowl esp. Bewick's Swans, White-fronted Geese, Wigeon, Teal.
Contact: Jane Allen, Marketing Manager, The Wildfowl & Wetlands Trust, Slimbridge, Gloucester GL2 7BT. 01453 891900; (Fax)01453 890927.
e-mail: info.slimbridge@wwt.org.uk

7. SYMOND'S YAT

RSPB/Forest Enterprise.
Location: Hill-top site on the edge of Forest of Dean, three miles N of Coleford on the B4432 signposted from the Forest Enterprise car park. Also signposted from A40 S of Ross-on-Wye.

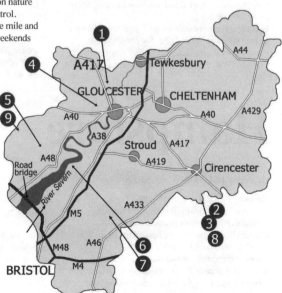

Access: Daily Apr-Aug only.
Facilities: Car park, toilets with adapted facilities for disabled visitors, picnic area, drinks and light snacks. Environmental education programmes available.
Public transport: None.
Habitat: Cliff above the River Wye and woodland.
Key birds: *Summer*: Peregrine, Buzzard, Raven and woodland species. Telescope is set up daily to watch the Peregrines.
Contact: The Puffins, Parkend, Lydney, Gloucestershire, GL15 4JA, 01594 562852.

8. WATERHAY AND CLEVELAND LAKES RESERVE

Cotswold Water Park Society.
Location: SU 060 933. The Waterhay is located to the S of Ashton Keynes, accessed from Waterhay Car Park (free of charge).
Access: Cleveland Lakes are a series of lakes located adjacent to the Waterhay. No public access exists at present, but access will be created in the future. In the intervening period, hides are being installed which overlook the lakes, accessed from the Thames National Trail. Until the appropriate rights of way are created with the necessary screening, please stay on the footpath. The site is still part of an active sand and gravel quarry and thus public access is not permitted. The Thames National Trail and Waterhay Car Park are open all year round. Note that the footpath and car park flood in winter and may not be easily accessible.
Facilities: Paths are even with footbridges. Hides are located at Lakes 68a and 68c. See the *Cotswold Water Park Leisure Guide* (available from the Cotswold Water Park Society or download from

www.waterpark.org).
Public transport: Bus: From Kemble, Cheltenham, Cirencester and Swindon. Tel: 08457 090 899.
Habitat: Wide variety of wetland habitats: open water, water channels, reedbeds, willow beds, silt lagoons, flooded fields in winter, open grassland. Series of restored and unrestored gravel workings.
Key birds: *Winter*: Large numbers of wildfowl, Teal, Goldeneye, Pintail, Bittern, Peregrine, Merlin, Cetti's Warbler, Hen and Marsh Harriers, Little Egret, Water Rail. Gull roost, Kumlien's Gull (occasional), Lapwing and Golden Plover flocks, Curlew. *Summer*: Breeding ducks, warblers, Hobby, Reed Bunting, Sand Martin, heronry. *Passage*: Whinchat, Stonechat, Wheatear, warblers, 30 species of waders in good years, inc Spotted Redshank, Whimbrel, Little Stint, plus large numbers of hirundines.
Contact: Cotswold Water Park Society, see above.

9. WOORGREENS LAKE AND MARSH

Gloucestershire Wildlife Trust.
Location: SO 630 127. Forest of Dean, W of Cinderford, N of B4226 Cannop road.
Access: Open at all times.
Facilities: None.
Public transport: None.
Habitat: Marsh, lake, heath on reclaimed opencast coalmine.
Key birds: Stonechat, Tree Pipit, Nightjar; birds of prey (inc, Buzzard, Goshawk, occasional Hobby). Passage waders (inc. Greenshank, Spotted Redshank, Green Sandpiper).
Contact: Trust HQ, 01452 383333.
e-mail: info@gloucestershirewildlifetrust.co.uk
www.wildlifetrust.org.uk/gloucswt/

Hampshire

1. FARLINGTON MARSHES

Hampshire & Isle of Wight Wildlife Trust.
Location: SU 685 045. E side of Portsmouth. Entrance of roundabout at junction of A2030 (Eastern Road) and A27, or from Harts Farm Way, Broadmarsh, Havant (S side of A27).
Access: Open at all times, no charge or permits, but donations welcome. Dogs on leads only. Wheelchair access via cycleway from E&W to building and

stream viewpoint. Groups please book to avoid clash of dates.
Facilities: Information at entrance and in shelter area of building. No toilets. **Public transport:** None.
Habitat: Coastal grazing marsh with pools and reedbed within reserve. Views over intertidal mudflats/saltmarshes of Langstone Harbour.
Key birds: *Autumn to spring*: Waders and wildfowl. *Winter*: Brent Goose, Wigeon, Pintail etc and waders (Dunlin, Grey Plover etc). On migration wide range

of waders including rarities. Reedbeds with Bearded Tit, Water Rail etc, scrub areas attract small migrants (Redstart, Wryneck, warblers etc).
Contact: Mark Kilby, Hampshire and Isle of Wight Wildlife Trust, Beechcroft House, Vicarage Lane, Curdridge, Hants SO32 2DP. 01489 774400. www.hwt.org.uk - go to 'Reserves' and then 'news' for sightings, etc

2. FLEET POND LNR

Hart District Council.
Location: SY 85. Located in Fleet. From the B3013, head to Fleet Station. Park in the long-stay car park at Fleet Station. Parking also available in Chestnut Grove and Westover Road.
Access: Open all year.
Facilities: None.
Public transport: None.
Habitat: Lake, reedbed, willow scrub.
Key birds: *Spring/autumn*: Migrant waders inc Little Ringed Plover, Dunlin, Greenshank, Little Gull, occasional Kittiwake, terns, Wood Lark, Sky Lark, occasional Ring Ouzel, Firecrest, Pied Flycatcher. *Summer*: Hobby, Common Tern, Tree Pipit. *Winter*: Bittern, wildfowl, occasional Smew, Snipe, occasional Jack Snipe, Siskin, Redpoll.
Contact: Hart District Council, Civic Office, Harlington Way, Fleet, Hampshire GU51 4AE. 01252 622122.

3. HOOK-WITH-WARSASH LNR

Hampshire County Council.
Location: SU 490 050. W of Fareham. Car parks by foreshore at Warsash. Reserve includes Hook Lake.
Access: Open all year.
Facilities: Public footpaths.
Public transport: None.
Habitat: Shingle beach, saltings, marsh, reedbed, scrape.
Key birds: *Winter*: Brent Geese on Hamble estuary. Waders. Stonechat, Cetti's Warbler.
Contact: Barry Duffin, Titchfield Haven Visitor Centre, Cliff Road, Hill Head, Fareham, Hants PO14 3JT. 01329 662145; fax 01329 667113.

4. LANGSTONE HARBOUR

RSPB (South East England Office).
Location: SU 695 035. Harbour lies E of Portsmouth, one mile S of Havant. Car parks at Broadmarsh (SE of A27/A3(M) junction) and West Hayling LNR (first right on A2030 after Esso garage).

Access: Restricted access. Good views from West Hayling LNR, Broadmarsh and Farlington Marshes LNR (qv). Winter boat trips may be booked from the nearby Portsmouth Outdoor Centre.
Facilities: None.
Public transport: Mainline trains all stop at Havant. Local bus service to W Hayling LNR.
Habitat: Intertidal mud, saltmarsh, shingle islands.
Key birds: *Summer*: Breeding waders and seabirds inc. Mediterranean Gull and Little Tern. *Passage/ winter*: Waterfowl, inc. Black-necked Grebes, c5000 dark-bellied Brent Geese, Shelduck, Shoveler, Goldeneye and Red-breasted Merganser. Waders inc. Oystercatcher, Ringed and Grey Plover, Dunlin, Black and Bar-tailed Godwit and Greenshank. Peregrine, Merlin and Short-eared Owl.
Contact: Chris Cockburn (Warden), RSPB Langstone Harbour, Unit B3, Wren Centre, Emsworth, Hants PO10 7SU. 01243 378784. e-mail: chris.cockburn@rspb.org.uk

5. LOWER TEST

Hampshire & Isle of Wight Wildlife Trust.
Location: SU 364 150. M271 S to Redbridge, three miles from Southampton city centre.
Access: Open at all times, guide dogs only. No coach parking facilities.
Facilities: One hide and two screens. Wheelchair access to bird hide planned for Autumn 2005. Hide open 9.30-4.30pm every day, screens open at all times.
Public transport: Totton train station and bus stops within easy walking distance.
Habitat: Saltmarsh, brackish grassland, wet meadows, reedbed, scrapes, meres, estuary.
Key birds: *Summer*: Breeding Little, Sandwich and Common Terns, Black-headed and Mediterranean Gulls, waders. *Passage/winter*: Waders. *Autumn/ winter*: Waders (inc. Black-tailed and Bar-tailed Godwits, Oystercatcher, Ringed and Grey Plover, Dunlin). Wildfowl (inc. Shelduck, Shoveler, Goldeneye, Red-breasted Merganser and c7000 dark-bellied Brent Geese). Black-necked Grebe, Short-eared Owl, Peregrine.
Contact: Clare Bishop, Hampshire and Isle of Wight Wildlife Trust, Beechcroft House, Vicarage Lane, Curdridge, Hants SO32 2DP. 023 8042 4206. e-mail: clareb@hwt.org.uk   www.hwt.org.uk

6. LYMINGTON REEDBEDS

Hampshire & Isle of Wight Wildlife Trust.
Location: SZ 324 965. From Lyndhurst in New Forest take A337 to Lymington. Turn L after railway

Hampshire

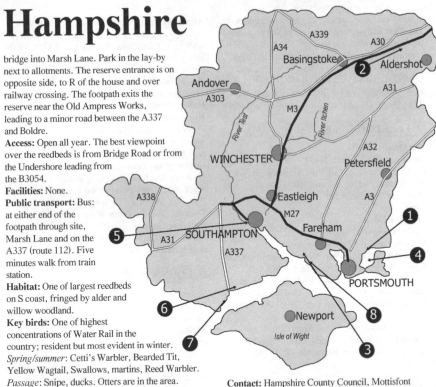

bridge into Marsh Lane. Park in the lay-by
next to allotments. The reserve entrance is on
opposite side, to R of the house and over
railway crossing. The footpath exits the
reserve near the Old Ampress Works,
leading to a minor road between the A337
and Boldre.

Access: Open all year. The best viewpoint
over the reedbeds is from Bridge Road or from
the Undershore leading from
the B3054.

Facilities: None.

Public transport: Bus:
at either end of the
footpath through site,
Marsh Lane and on the
A337 (route 112). Five
minutes walk from train
station.

Habitat: One of largest reedbeds
on S coast, fringed by alder and
willow woodland.

Key birds: One of highest
concentrations of Water Rail in the
country; resident but most evident in winter.
Spring/summer: Cetti's Warbler, Bearded Tit,
Yellow Wagtail, Swallows, martins, Reed Warbler.
Passage: Snipe, ducks. Otters are in the area.

Contact: Michael Boxall, Hampshire and Isle of
Wight Wildlife Trust, Beechcroft House, Vicarage
Lane, Curdridge Hants SO32 2DP. 01489 774 400.
e-mail: feedback@hwt.org.uk www.hwt.org.uk

7. LYMINGTON-KEYHAVEN NNR

Hampshire County Council.

Location: SZ 315 920. S of Lymington along
seawall footpath; car parks at Bath Road, Lymington
and at Keyhaven Harbour.

Access: Open all year

Facilities: None.

Public transport: None.

Habitat: Coastal marshland and lagoons.

Key birds: *Spring*: Passage waders (inc. Knot,
Sanderling, Bar-tailed and Black-tailed Godwits,
Whimbrel, Spotted Redshank), Pomarine and Great
Skuas. Breeding Oystercatcher, Ringed Plover and
Sandwich, Common and Little Terns. *Autumn*:
Passage raptors, waders and passerines. *Winter*:
Wildfowl (inc. Brent Goose, Wigeon, Pintail, Red-
breasted Merganser), waders (inc. Golden Plover),
Little Egret, gulls.

Contact: Hampshire County Council, Mottisfont
Court, High Street, Winchester, Hants SO23 8ZF.

8. TITCHFIELD HAVEN

Hampshire County Council.

Location: SU 535 025. From A27 W of Fareham;
public footpath follows derelict canal along W of
reserve and road skirts S edge.

Access: Open Wed-Sun all year, plus Bank Hols,
except Christmas and Boxing days.

Facilities: Centre has information desk, toilets, tea
room and shop. Guided tours (book in advance).
Hides.

Public transport: None.

Habitat: Reedbeds, freshwater scrapes, wet grazing
meadows.

Key birds: *Spring/summer*: Bearded Tit, waders
(inc. Black-tailed Godwit, Ruff), wildfowl, Common
Tern, breeding Cetti's Warbler, Water Rail. *Winter*:
Bittern.

Contact: Barry Duffin, Titchfield Haven Visitor
Centre, Cliff Road, Hill Head, Fareham, Hants PO14
3JT. 01329 662145; (fax) 01329 667113.

Herefordshire

HOLYWELL DINGLE

Hereford Nature Trust.
Location: SO 313 510, NE of Hay-on-Wye. Take A438 N then L on A4111. Park in lay-by on R, 1 mile N of Eardisley and take footpath to reserve.
Access: Open at all times. Some parts near the stream can be wet and muddy, so waterproof footwear is advised. The northern part of the reserve is very steep-sided in places, and there are precipitous drops into the stream-bed.
Facilities: Good network of marked paths and two footbridges crossing the stream.
Habitat: Narrow, steep-sided, wooded valley, mainly oak and ash. Fast-flowing freshwater stream.
Key birds: Good variety of woodland birds, including breeding Nuthatch, Pied Flycatcher, Marsh Tit, Treecreeper, Great Spotted Woodpecker, Chiffchaff, Rook and Blackcap.
Contact: Reserves Department, Herefordshire Nature Trust, Lower House Farm, Ledbury Road, Hereford HR1 1UT. 01432 356872.
www.wildlifetrust.org.uk/hereford/reserves

LEA AND PAGETS WOOD

Hereford Nature Trust.
Location: SO 598 343, SE of Hereford. Take B4224 S towards Ross-on-Wye. Turn L on small road 0.25 mile S of Fownhope towards Woolhope. Footpath is on R after turning for Common Hill. Parking for two cars at top of the hill.
Access: Open at all times. Upper paths generally dry in spring and summer, but the low-lying main track can be very muddy and treacherous in places in winter. Take extra care in the vicinity of the old quarry in Church Wood. This is partly fenced off, but there are still unguarded near-vertical drops.
Facilities: Footpaths.
Habitat: Ancient broad-leaved woodland.

Key birds: Small breeding population of Pied Flycatchers in nest-boxes, good range of woodland species, including all three woodpeckers (although Lesser Spotted Woodpeckers have not been recorded for some years) and warblers including Blackcap, Willow Warbler, Chiffchaff and the occasional Wood Warbler. Also Nuthatch, Treecreeper, Marsh Tit, Jay. Buzzards, Tawny Owls and Sparrowhawks also seen.
Contact: Reserves Department, Herefordshire Nature Trust, 01432 356872.
www.wildlifetrust.org.uk/hereford/reserves

LUGG MEADOW SSSI

Hereford Nature Trust.
Location: SO 539405, NE of Hereford. Take A438 from Hereford towards Ledbury. Near Lugwardine, park in lane on L adjacent to Lower House Farm, before crossing Lungwardine bridge.
Access: Access over Upper Lugg Meadow, unrestricted but do not walk in the growing hay between late April and July. In winter, the whole area may be flooded to a depth of more than 1m. for long periods, and access becomes impossible or distinctly dangerous. Take care when walking near the river as there are vertical cliffs along its banks.
Facilities: Permissive path and footpath. Not suitable for wheelchairs.
Public transport: None.
Habitat: Ancient hay meadows, flooded in winter, river.
Key birds: *Spring*: Curlew, Sky Lark. *Passage*: Greenshank, Redshank, Black-tailed Godwit, Snipe, Lapwing and Common Sandpiper *Winter*: Roosting gulls,wildfowl, swans and geese, Peregrine, Merlin.
Contact: Reserves Department, Herefordshire Nature Trust, 01432 356872.
www.wildlifetrust.org.uk/hereford/reserves

Hertfordshire

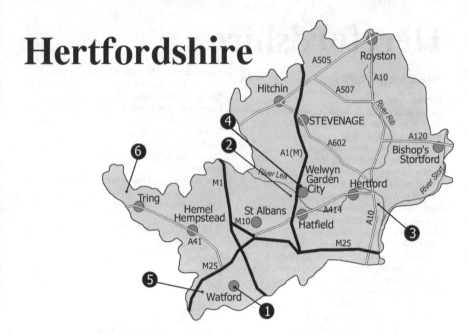

1. CASSIOBURY PARK

Welwyn & Hatfield Council.
Location: TL 090 970. Close to Watford town centre.
Access: Open all year.
Facilities: Car park, footpaths.
Public transport: Watford Metropolitan Underground station.
Habitat: Municipal park, wetland, river, alder/willow wood.
Key birds: *Spring/summer*: Kingfisher, Grey Wagtail. *Winter*: Snipe, Water Rail, occasional Bearded Tit.
Contact: Welwyn & Hatfield Council, 01707 357000. e-mail: council.services@welhat.gov.uk

2. LEMSFORD SPRINGS

Herts & Middlesex Wildlife Trust.
Location: TL 223 123. Lies 1.5 miles W of Welwyn Garden City town centre, off roundabout leading to Lemsford village on B197, W of A1(M).
Access: Access, via key, by arrangement with warden. Open at all times, unless work parties or group visits in progress. Keep to paths. Dogs on leads. Not ideal at present for disabled due to steps up to hides, but wheelchair access provision to hide should have been completed in summer 2005 (contact warden for up-to-date information). Coaches welcome and room to park on road, but limit of 30 persons.
Facilities: Two hides, chemical toilet, paths.
Public transport: Bus service to Valley Road, WGC & Lemsford Village No 366 (Centrebus-telephone Intalink 0870 608 2608). Nearest railway station Welwyn Garden City.
Habitat: Former water-cress beds, open shallow lagoons. Stretch of the River Lea, marsh, hedgerows. Nine acres.
Key birds: *Spring/summer*: Breeding warblers, Grey Wagtail, Kestrel. *Autumn/winter*: Green Sandpiper, Water Rail, Snipe, Siskin, occasional Jack Snipe. *All Year*: Kingfisher, Grey Heron, Sparrowhawk.
Contact: Barry Trevis, 11 Lemsford Village, Welwyn Garden City, Herts AL8 7TN. 01707 335517. e-mail: info@hmwt.org

3. RYE MEADS

RSPB (Central England Region)/Hertfordshire & Middlesex Wildlife Trust.
Location: TL 387 099. E of Hoddesdon, signed from A10, near Rye House railway station.
Access: Open every day 10am-5pm (or dusk if

earlier), except Christmas Day and Boxing Day.
Facilities: Disabled access and toilets. Drinks
machine, staffed reception, classrooms, picnic area,
car park, bird feeding area. Nature trails, hides.
RSPB reserve has close circuit TV on Kingfisher and
Common Terns in summer.
Public transport: Rail (Rye House) 55 metres, bus
(310) stops 600 metres from entrance.
Habitat: Marsh, willow scrub, pools, scrapes,
lagoons and reedbed.
Key birds: *Summer*: Breeding Tufted Duck,
Gadwall, Common Tern, Kestrel, Kingfisher, nine
species of warblers. *Winter*: Bittern, Shoveler, Water
Rail, Teal, Snipe, Jack Snipe, Redpoll and Siskin.
Contact: The Site Manager, RSPB Rye Meads
Visitor Centre, Rye Road, Stanstead Abbotts, Herts
SG12 8JS. 01992 708383; (Fax) 01992 708389.

4. STANBOROUGH REED MARSH

Herts & Middlesex Wildlife Trust.
Location: TL 230 105. Leave the A1M at J4 on
A6129 Stanborough Road. At next small roundabout,
turn R to Welwyn Garden City town centre. Continue
past lakes to next roundabout. Take a U turn and
then turn L into reserve car park. Follow path
between river and lake into reserve.
Access: Open all year. No access into reedbed.
Circular walk.
Facilities: None.
Public transport: Bus: stops on Stanborough Road.
Train: nearest station Welwyn Garden City.
Habitat: Willow woodland, river, reed marsh.
Key birds: *Summer*: Good numbers of Reed and
Sedge Warblers. *Winter*: Water Rail and Corn
Bunting roost.
Contact: Trust HQ, 01727 858901.
e-mail: info@hmwt.org
www.wildlifetrust.org.uk/herts

5. STOCKER'S LAKE

Herts & Middlesex Wildlife Trust.
Location: TQ 044 931. Rickmansworth, off A412
into Springwell Lane (TQ043932) L after bridge, or
via Bury Lake Aquadrome (parking).
Access: Open all year.
Facilities: None.
Public transport: Within 20 min walk of
Rickmansworth tube station.
Habitat: Mature flooded gravel pit with islands.
Key birds: 50 species breed; over 200 recorded.

Heronry. *Summer*: Breeding Pochard, Gadwall,
Common Tern. Large numbers of migrants. *Winter*:
Duck (inc. Goldeneye and nationally significant
numbers of Shoveler).
Contact: Trust HQ, 01727 858901.
e-mail: hertswt@cix.org.uk
www.wildlifetrust.org.uk/herts

6. TRING RESERVOIRS

Wilstone Reservoir – Herts & Middlesex Wildlife
Trust/British Waterways; other reservoirs – British
Waterways/Friends of Tring Res.
Location: Wilstone Reservoir, SP 90 51 34. Other
reservoirs SP 92 01 35. WTW Lagoon SP 92 31 34
adjacent to Marsworth Reservoir. Reservoirs 1.5
miles due N of Tring, all accessible from B489
which crosses A41 Aston Clinton By-pass, NB exit
from by-pass only Southbound, entry only
Northbound.
Access: Reservoirs – open at all times. All group
visits need to be cleared with British Waterways.
WTW Lagoon: open at all times by permit from
FOTR. Coaches can only drop off and pick up, for
advice contact FOTR. Wilston Reservoir has
restricted height access of 2.1 metres.
Facilities: Café and public house adjacent to Startops
Reservoir car park, safe parking for cycles. Also
disabled trail from here. Wilstone Reservoir: Public
house about 0.5 mile away in village and cafe about
0.25 mile from car park at Farm Shop. Hides with
disabled access at Startops/Marsworth Reservoir &
WTW Lagoon. Also other hides.
Public transport: Buses from Aylesbury & Tring
including a weekend service, tel. 0870 6082608.
Tring Station is 2.5 miles away via canal towpath.
Habitat: Four reservoirs with surrounding
woodland, scrub and meadows. Two of the
reservoirs with extensive reedbeds. WTW Lagoon
with islands, surrounding hedgerows and scrub.
Key birds: *Spring/summer*: Breeding warblers,
regular Hobby, occasional Black Tern, Marsh
Harrier, Osprey. *Autumn*: Passage waders and
wildfowl. *Winter*: Gull roost, large wildfowl flocks,
bunting roosts, Bittern.
Contact: Herts & Middsx Wildlife Trust: see
Directory entry, FOTR: see Peter Hearn in Bucks
BTO entry, or visit www.fotr.org.uk,
British Waterways,Ground Floor, Witangate House,
500-600 Witan Gate, Milton Keynes MK9 1BW

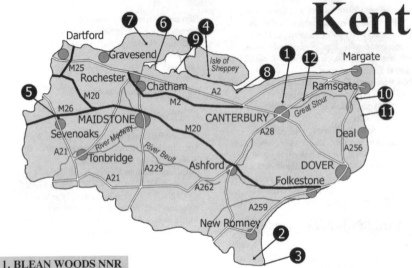

Kent

1. BLEAN WOODS NNR

RSPB (South East England Office).
Location: TR 126 592. From Rough Common (off A290, 1.5 miles NW of Canterbury).
Access: Open 8am-9pm. No parking for coaches-please drop passengers off in Rough Common village. Green Trail suitable for wheelchair users.
Facilities: Public footpaths and five waymarked trails.
Public transport: 24 and 24a buses from Canterbury to Rough Common. Local bus company Stagecoach 0870 243 3711.
Habitat: Woodland (mainly oak and sweet chestnut), relics of heath.
Key birds: Nightingale, Nightjar in summer, three species of woodpecker.
Contact: Michael Walter, 11 Garden Close, Rough Common, Canterbury, Kent CT2 9BP. 01227 455972.

2. DUNGENESS

RSPB (South East England Office).
Location: TR 063 196. SE of Lydd.
Access: Open daily 9am-9pm or sunset when earlier. Visitor centre open (10am-5pm, 4pm Nov-Feb). Parties over 20 by prior arrangement. Closed Dec 25th & 26th.
Facilities: Visitor centre, toilets (including disabled access), six hides, nature trail, wheelchair access to visitor centre and four hides.
Public transport: Service 12 from Lydd or Folkestone (not Sun) stops at reserve entrance on request – one mile walk to visitor centre.
Habitat: Shingle, flooded gravel pits, sallow scrub, reedbed, wet grassland.
Key birds: *Resident*: Bearded Tit. *Winter*: Bittern, Wildfowl (including Wigeon, Goldeneye, Goosander, Smew), divers and grebes. Migrant waders, landfall for passerines. *Summer*: Breeding Lapwing, Redshank, wildfowl, gulls, Cetti's Warbler.
Contact: Christine Hawkins/Bob Gomes, Boulderwall Farm, Dungeness Road, Lydd, Romney Marsh, Kent TN29 9PN. 01797 320588; (fax) 01797 321962. e-mail: dungeness@rspb.org.uk www.rspb.org.uk

3. DUNGENESS BIRD OBSERVATORY

Bird Observatory Trust.
Location: TR 085 173. Three miles SE of Lydd. Turn south off Dungeness Road at TR 087 185 and continue to end of road.
Access: Observatory open throughout the year.
Facilities: Accommodation available. Bring own sleeping bag/sheets and toiletries. Shared facilities including fully-equipped kitchen.
Public transport: Bus service between Rye and Folkestone, numbers 11, 12, 711, 712. Alight at the Pilot Inn, Lydd-on-Sea. Tel 01227 472082.
Habitat: Shingle promontory with scrub and gravel pits. RSPB reserve nearby.
Key birds: Breeding birds include Wheatear and Black Redstart and seabirds on RSPB Reserve. Important migration site.

Contact: David Walker, Dungeness Bird
Observatory, 11 RNSSS, Dungeness, Kent TN29
9NA. 01797 321309.
e-mail dungeness.obs@tinyonline.co.uk
www.dungenessbirdobs.org.uk

4. ELMLEY MARSHES

RSPB (South East England Office).
Location: TQ 93 86 80. Isle of Sheppey signposted
from A249, one mile beyond Kingsferry Bridge.
Reserve car park is two miles from the main road.
Access: Open every day except Tue, Christmas and
Boxing days. (9am-9pm or dusk if earlier). Free to
RSPB members. Dogs not allowed on reserve. Less
able may drive closer to the hides.
Facilities: Five hides. Disabled access to Wellmarsh
hide. No visitor centre. Toilets located in car park
1.25 miles from hides.
Public transport: Swale Halt, a request stop is
nearest railway station on Sittingbourne to Sheerness
line. From there it is a three mile walk to reserve
Habitat: Coastal grazing marsh, ditches and pools
alongside the Swale Estuary with extensive intertidal
mudflats and saltmarsh
Key birds: *Spring/summer*: Breeding waders –
Redshank, Lapwing, Avocet, Yellow Wagtail,
passage waders, Hobby. *Autumn*: Passage waders.
Winter: Spectacular numbers of wildfowl, especially
Wigeon and White-fronted Goose. Waders. Hunting
raptors – Peregrine, Merlin, Hen Harrier and Short-
eared Owl.
Contact: Barry O'Dowd, Elmley RSPB Reserve,
Kingshill Farm, Elmley, Sheerness, Kent ME12
3RW. 01795 665969.

5. JEFFERY HARRISON RESERVE, SEVENOAKS

Jeffery Harrison Memorial Trust.
Location: TQ 519 568. From J5 of M26, head S to
A25. Reserve is immediately N of Sevenoaks.
Access: Wed, Sat, Sun and bank holidays 10am-5pm
(or dusk). Closed Xmas to New Year. Coach
parking available. Access by car for disabled visitors.
Facilities: Visitor centre, nature trail, hides.
Public transport: 15 minutes walk from Bat & Ball
station, 20 minutes from Sevenoaks station.
Habitat: Flooded gravel pits.
Key birds: Wintering wildfowl, waders, woodland
birds.
Contact: John Tyler, Tadorna, Bradbourne Vale
Road, Sevenoaks, Kent TN13 3DH. 01732 456407.
e-mail: sevenoakswildfowl@kentwildlife.org.uk

6. NOR MARSH

RSPB (South East England Office).
Location: TQ 810 700. One mile NE of Gillingham
in the Medway Estuary.
Access: No access, to island. It is viewable from
Riverside Country Park (B2004) at the Horrid Hill
Peninsula, giving overviews of the Medway Estuary
saltmarsh and mudflats.
Facilities: None.
Public transport: Buses can be caught to Riverside
Country Park. Phone Medway Council for the bus
numbers, 01634 727777.
Habitat: Saltmarsh and mudflats.
Key birds: *Spring/summer*: Breeding Redshank,
Little Egret roost, Shelduck and Black-headed Gull.
Winter: Dunlin (2-3,000), Knot, Curlew, Redshank,
Oystercatcher, Turnstone, Brent Goose, Teal,
Wigeon, Shelduck.
Contact: Gordon Allison, Bromhey Farm.
Eastborough, Cooling, Rochester, Kent ME3 8DS.
01634 222480.

7. NORTHWARD HILL

RSPB (South East England Office).
Location: TQ 780 765. Adjacent to High Halstow,
off A228, approx six miles N of Rochester.
Access: Open all year, free access, trails in public
area of wood joining Saxon Shoreway link to grazing
marsh. Dogs allowed in public area on leads. Trails
often steep and not suitable for wheelchair users.
Facilities: Three nature trails in the wood and one
joining with long distance footpath. Toilets at village
hall, small car park adjacent to wood.
Public transport: Buses to village of High Halstow.
Contact Arriva buses for timetable details.
Habitat: Ancient and scrub woodland (approximately
130 acres), grazing marsh (approximately 350 acres).
Key birds: *Spring/summer*: Wood holds UK's
largest heronry (155 pairs in 2004), inc growing
colony of Little Egrets (48 pairs in 2004), breeding
Nightingale, Turtle Dove, scrub warblers and
woodpeckers. Marshes – breeding Lapwing,
Redshank, Avocet, Marsh Harrier, Shoveler,
Pochard. *Winter*: Wigeon, Teal, Shoveler. Passage
waders (ie Black-tailed Godwit), raptors, Tree
Sparrow, Corn Bunting. Long-eared Owl roost.
Contact: Gordon Allison, 01634 222480. (Address
above).

8. OARE MARSHES LNR

Kent Wildlife Trust.
Location: TR 01 36 48 (car park). Two miles N of

Faversham. From A2 follow signs to Oare and Harty Ferry.
Access: Open at all times. Access along marked paths only. Dogs under strict control to avoid disturbance to birds and livestock.
Facilities: Information centre, open weekends, Bank Holidays. Two hides.
Public transport: Bus to Oare Village one mile from reserve. Train: Faversham (two miles)
Habitat: Grazing marsh, mudflats/estuary.
Key birds: *All year*: Waders and wildfowl. *Winter*: Hen Harrier, Merlin, Peregrine. Divers, grebes and sea ducks on Swale. *Spring/summer*: Avocet, Garganey, Green and Wood Sandpipers, Little Stint, Black-tailed Godwit etc. Black Tern.
Contact: Trust HQ, 01622 662012. e-mail: info@kentwildlife.org.uk www.kentwildlife.org.uk

9. RIVERSIDE COUNTRY PARK

Medway County Council.
Location: TQ 808 683. From Rochester, take the A2 E into Gillingham and turn L onto the A289. After one mile, turn R onto the B2004 at Grange. After one mile, the visitor centre is on the L.
Access: Open all year, free access from 8.30am-4.30pm (winter) or 8.30am-8.30pm or dusk (summer).
Facilities: Visitor Centre with restaurant and toilets open every day except Christmas, Boxing and New Year's days (no parking on these days either) from 10am-5pm summer (4pm winter). Large car park and smaller one at Rainham Dock. Car park locked at dusk. Check times on arrival.
Public transport: Bus: contact Arriva tel: 08706 082 608. Train: nearest stations at Rainham and Gillingham. Cycle racks at visitor centre and a Sustrans cycle route.
Habitat: Mudflats, saltmarsh, ponds, reedbeds, grassland and scrub.
Key birds: *Winter*: Dunlin, Redshank, Grey Plover, Avocet, Brent Goose, Teal, Pintail, Goldeneye, Wigeon, Shelduck, Peregrine, Hen Harrier, thrushes, Short-eared Owl, Mediterranean Gull, Water Rail, Rock Pipit, Little Egret and Red-breasted Merganser, Whitethroat, Cetti's Warbler, Nightingale, Turtle Dove.
Contact: Riverside Country Park, Lower Rainham Road, Gillingham, Kent, ME7 2XH. 01634 378987. e-mail: riversidecp@medway.gov.uk

10. SANDWICH AND PEGWELL BAY

Kent Wildlife Trust.
Location: TR 34 26 35. Main carpark is off A256

Sandwich – Ramsgate road at Pegwell Bay.
Access: Open 8am-8pm or dusk.
Facilities: Toilets, hide, car parking and trails
Public transport: Bus stop within 400m (Stagecoach). Sustrans National Bike Route passes along the edge of the reserve.
Habitat: Saltmarsh, mudflats, sand dunes and coastal scrub.
Key birds: Range of wetland birds all year.
Contact: Trust HQ, 01622 662012. e-mail: info@kentwildlife.org.uk www.kentwildlife.org.uk

11. SANDWICH BAY BIRD OBSERVATORY

Sandwich Bay Bird Observatory Trust.
Location: TR 355 575. 2.5 miles from Sandwich, five miles from Deal, 15 miles from Canterbury. A256 to Sandwich from Dover or Ramsgate. Follow signs to Sandwich Station and then Sandwich Bay.
Access: Open daily. Disabled access.
Facilities: New Field Study Centre. Visitor centre, toilets, refreshments, hostel-type accommodation, plus self-contained flat.
Public transport: Sandwich train station two miles from Observatory. No buses but within walking distance.
Habitat: Coastal, dune land, farmland, marsh, two small scrapes.
Key birds: *Spring/autumn passage*: Good variety of migrants and waders, specially Corn Bunting. Annual Golden Oriole. *Winter*: Golden Plover.
Contact: Kevin Thornton, Sandwich Bay Bird Observatory, Guildford Road, Sandwich Bay, Sandwich, Kent CT13 9PF. 01304 617341. e-mail: sbbot@talk21.com www.sbbo.co.uk

12. STODMARSH NNR

English Nature (Kent Team).
Location: TR 222 618. Lies alongside River Stour and A28, five miles NE of Canterbury.
Access: Open at all times. Keep to reserve paths and keep dogs under control.
Facilities: Fully accessible toilets are available at the Stodmarsh entrance car park. Four hides (one fully accessible), easy access nature trail, footpaths and information panels. Car park, picnic area and toilets adjoining the Grove Ferry entrance with easily accessible path, viewing mound and two hides.
Public transport: There is a regular bus service from Canterbury to Margate/Ramsgate. Alight at Upstreet for Grove Ferry. Hourly on Sun.

Habitat: Open water, reedbeds, wet meadows, dry meadows, woodland.
Key birds: *Spring/summer*: Breeding Bearded Tit, Cetti's Warbler, Garganey, Reed, Sedge and Willow Warblers, Nightingale. Migrant Black Tern, Hobby, Osprey, Little Egret. *Winter*: Wildfowl. Hen Harrier, Bittern.
Contact: David Feast, English Nature, Coldharbour Farm, Wye, Ashford, Kent TN25 5DB. 01233 812525 or 07767 321058 (mobile).

Lancashire

1. CUERDEN VALLEY PARK

Cuerden Valley Park Trust.
Location: SD 565 238. S of Preston on the A6. Easy access from J28 and J29 of the M6 and J8 and J9 on the M62.
Access: Open all year.
Facilities: Visitor centre, toilets.
Public transport: None.
Habitat: Woodland, river, pond, agricultural grassland.
Key birds: *All year*: Kingfisher, Dipper, Great Spotted Woodpecker, Goldcrest and usual woodland birds.
Contact: Cuerden Valley Park Trust, The Barn, Berkeley Drive, Bamber, Preston PR5 6BY. 01772 324436. e-mail: rangers@cuerdenvalleypark.org

2. HEYSHAM NATURE RESERVE & BIRD OBSERVATORY

The Wildlife Trust for Lancashire, Manchester and North Merseyside, in conjunction with British Energy Estates.
Location: Main reserve is at SD 404 596 W of Lancaster. Take A683 to Heysham port. Turn L at traffic lights by Duke of Rothesay pub, then first right after 300m.
Access: Gate to reserve car park usually open 9.30am-6pm (longer in summer and shorter in winter). Pedestrian access at all times. Dogs on lead. Limited disabled access.
Facilities: Hide overlooking Power Station outfalls. Map giving access details at the reserve car park. No manned visitor centre or toilet access, but someone usually in reserve office, next to the main car park, in the morning. Latest sightings board can be viewed through the window if office closed.
Public transport: Train services connect with nearby Isle of Man ferry. Plenty of buses to Lancaster from various Heysham sites within walking distance (ask for nearest stop to the harbour).
Habitat: Varied: wetland, acid grassland, alkaline grassland, foreshore.
Key birds: Passerine migrants in the correct conditions. Good passage of seabirds in Spring, especially Arctic Tern. Storm Petrel and Leach's Petrel during strong onshore (SW-WWNW) winds in midsummer and autumn respectively. Good variety of breeding birds (e.g. eight species of warbler on the reserve itself). Two/three scarce land-birds each year, most frequent being Yellow-browed Warbler. Notable area for dragonflies.
Contact: Rueben Neville, Reserve Warden, The Barn, Berkeley Drive, Bamber Bridge, Preston, PR5 6BY. 07979 652138. Annual report from Leighton Moss RSPB reserve shop.

3. LEIGHTON MOSS

RSPB (North West England Office).
Location: SD 478 750. Four miles NW of Carnforth. Signposted from A6 N of Carnforth.
Access: Reserve open daily 9am-dusk. Visitor centre open daily 10am-5pm (except Christmas Day). No dogs. No charge to RSPB members.
Facilities: Visitor centre, shop, tea-room and toilets. Nature trails and five hides (four have wheelchair access).
Public transport: Silverdale train station 150 metres from reserve. Tel: 08457 484950.
Habitat: Reedbed, shallow meres and woodland.
Key birds: *All year*: Bittern, Bearded Tit, Water Rail, Pochard and Shoveler. *Summer*: Marsh Harrier, Reed and Sedge Warblers.
Contact: Robin Horner, Leighton Moss RSPB Nature Reserve, Myers Farm, Silverdale, Carnforth, Lancashire LA5 0SW. 01524 701601. www.rspb.org.uk

Lancashire

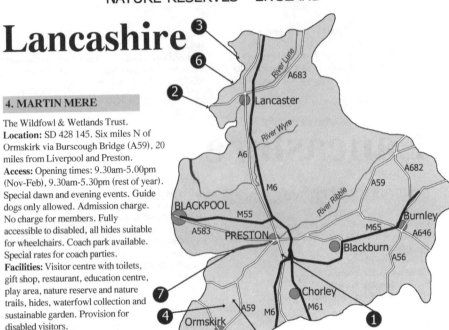

4. MARTIN MERE

The Wildfowl & Wetlands Trust.
Location: SD 428 145. Six miles N of
Ormskirk via Burscough Bridge (A59), 20
miles from Liverpool and Preston.
Access: Opening times: 9.30am-5.00pm
(Nov-Feb), 9.30am-5.30pm (rest of year).
Special dawn and evening events. Guide
dogs only allowed. Admission charge.
No charge for members. Fully
accessible to disabled, all hides suitable
for wheelchairs. Coach park available.
Special rates for coach parties.
Facilities: Visitor centre with toilets,
gift shop, restaurant, education centre,
play area, nature reserve and nature
trails, hides, waterfowl collection and
sustainable garden. Provision for
disabled visitors.
Public transport: Bus service to WWT
Martin Mere from Ormskirk. Train to
Burscough Bridge or New Lane Stations
(both 1.5 miles from reserve). For bus times
contact Traveline 0870 608 2608.
Habitat: Open water, wet grassland, moss, copses,
reedbed, parkland.
Key birds: *Winter*: Whooper and Bewick's Swans,
Pink-footed Goose, various duck, Ruff, Black-tailed
Godwit, Peregrine, Hen Harrier, Tree Sparrow.
Spring: Ruff, Little Ringed and Ringed Plover.
Summer: Marsh Harrier, Garganey, hirundines, Tree
Sparrow. Breeding Avocets, Lapwing, Redshank,
Shelduck. *Autumn*: Pink-footed Goose, waders on
passage.
Contact: Patrick Wisniewski, WWT Martin Mere,
Fish Lane, Burscough, Lancs L40 0TA. 01704
895181. e-mail: info.martinmere@wwt.org.uk
www.wwt.org.uk

5. MERE SANDS WOOD

The Wildlife Trust for Lancashire, Manchester and
North Merseyside.
Location: SD 44 71 57. Four miles inland of
Southport, 0.5 miles off A59 Preston – Liverpool
road, in Rufford along B5246 (Holmeswood Road).
Access: Visitor centre open 9am-5pm daily except
Christmas Day. Car park open until 8pm in summer.

Half mile of wheelchair-accessible path, leading to
two hides and viewpoint.
Facilities: Visitor centre with toilets (disabled), seven
hides, two trails, exhibition room, latest sightings
board. Feeding stations
Public transport: Bus: Southport-Chorley 347 stops
in Rufford, 0.5 mile walk. Train: Preston-Ormskirk
train stops at Rufford station, one mile walk.
Habitat: Freshwater lakes, mixed woodland, sandy
grassland/heath. 105h.
Key birds: *Winter*: Nationally important for Teal and
Gadwall, good range of waterfowl, Kingfisher.
Feeding stations attract Tree Sparrow, Bullfinch,
Reed Bunting. Woodland: Treecreeper. *Summer*:
Little Ringed Plover, Kingfisher, Lesser Spotted
Woodpecker. *Passage*: In most years, Osprey,
Crossbill, Green Sandpiper, Greenshank, Wood
Warbler and Turtle Dove.
Contact: Dominic Rigby, Warden, Mere Sands
Wood Nature Reserve, Holmeswood Road, Rufford,
Ormskirk, Lancs L40 1TG. 01704 821809.
e-mail: lancswtmsw@cix.co.uk
www.wildlifetrust.org/lancashire

6. MORECAMBE BAY

RSPB (North West England Office).
Location: SD 468 667. Two miles N of
Morecambe at Hest Bank.
Access: Open at all times. Do not venture onto
saltmarsh or intertidal area, there are dangerous
channels and quicksands.
Facilities: Viewpoint at car park.
Public transport: No 5 bus runs between
Carnforth and Morecambe. Tel: 0870 608 2608.
Habitat: Saltmarsh, estuary.
Key birds: *Winter*: Wildfowl (Pintail, Shelduck,
Wigeon) and waders – important high tide roost for
Oystercatcher, Curlew, Redshank, Dunlin, Bar-
tailed Godwit.
Contact: Robin Horner, Leighton Moss &
Morecambe Bay RSPB Reserves, Myers Farm,
Silverdale, Carnforth, Lancashire LA5 0SW. 01524
701601. www.rspb.org.uk

7. RIBBLE ESTUARY

English Nature (Cheshire to Lancashire team).
Location: SD 380 240.
Access: Open at all times.
Facilities: No formal visiting facilities.
Public transport: None.
Habitat: Saltmarsh, mudflats.
Key birds: High water wader roosts (of Knot,
Dunlin, Black-tailed Godwit, Oystercatcher and Grey
Plover) are best viewed from Southport, Marshside,
Lytham and St Annes. Pink-footed Geese and
wintering swans are present in large numbers from
Oct-Feb on Banks Marsh and along River Douglas
respectively. The large flocks of Wigeon, for which
the site is renowned, can be seen on high tides from
Marshside but feed on saltmarsh areas at night. Good
numbers of raptors also present in winter.
Contact: Site Manager, Old Hollow, Marsh Road,
Banks, Southport PR9 8EA. 01704 225624.

Leicestershire and Rutland

1. BEACON HILL COUNTRY PARK

Leicestershire County Council.
Location: SK 522 149. From Loughborough, take
A512 SW for 2.5 miles. Turn L onto Breakback
Road and follow it for 2.5 miles through Nanpantan.
Park in the car park on the left in Woodhouse Lane.
Access: Open all year from 8am-dusk. If opening
times are different, these will be clearly displayed at
the park. A permissive path from Deans Lane to
Woodhouse Lane is occasionally closed during the
year. Please check first.
Facilities: Two pay and display car parks, easy-to-
follow, well-waymarked tracks and woodland paths.
Several climbs to hill tops. Rocky outcrops slippy
after rain. Information boards. Toilets at lower car
park, The Outwoods car park and Woodhouse Eves.
Wheelchair access along park paths but no access to
summit. Refreshments at Bull's Head, Woodhouse
Eaves.
Public transport: Bus: No 123 Leicester to
Shepshed calls at Woodhouse Eaves. Tel: 0870 608
2608. Train: from Loughborough and Leicester.
Habitat: Forest, one of the oldest geological
outcrops in England and the second highest point in
Leicestershire.
Key birds: *All year*: Treecreeper, Nuthatch, Lesser

Spotted and Green Woodpeckers, Great and Coal
Tits, Little Owl, wagtails. *Summer*: Pied Flycatcher,
Whitethroat, Blackcap, Whinchat, Garden Warbler,
Stonechat.
Contact: Beacon Hill Country Park, Beacon Hill
Estate Office, Broombriggs Farm, Beacon Road,
Woodhouse Eaves, Loughborough, Leics LE12
8SR. 01509 890048.

2. EYEBROOK RESERVOIR

Corby & District Water Co.
Location: SP 853 964. Reservoir built 1940. S of
Uppingham, from unclassified road W of A6003 at
Stoke Dry.
Access: Access to 150 acres private grounds granted
to members of Leics and Rutland Ornithological
Society and Rutland Nat Hist Soc. Organised groups
with written permission (from Corby Water Co).
Facilities: SSSI since 1955. Good viewing from
public roads. Trout fishery season Apr-Oct.
Public transport: None.
Habitat: Open water, plantations and pasture.
Key birds: *Summer*: Good populations of breeding
birds, sightings of Ospreys and Red Kite. Passage
waders and Black Tern. *Winter*: Wildfowl (inc.
Goldeneye, Goosander, Bewick's Swan) and waders.

NATURE RESERVES - ENGLAND

Contact: Corby Water Co. PO Box 101, Weldon Road, Corby NN17 5UA, Fishing lodge. 01536 770264. www.eyebrook.com or www.eyebrook.org.uk

770651; fax 01572 755931; www.birdfair.org.uk e-mail: awbc@rutlandwater.org.uk www.rutlandwater.org.uk   www.ospreys.org.uk

3. RUTLAND WATER

Leics and Rutland Wildlife Trust.
Location: SK 866 6760 72. 1. Egleton Reserve: from Egleton village off A6003 S of Oakham. 2. Lyndon Reserve: south shore E of Manton village off A6003 S of Oakham.
Access: 1. Open daily 9am-5pm, (4pm Nov to Jan). 2. Open winter (Sat, Sun 10am-5pm), summer daily (10am-5pm). Day permits available for both.
Facilities: 1: Anglian Water Birdwatching Centre, now enlarged. Toilets and disabled access to 11 hides, electric buggies, conference facilities. 2: Interpretive centre. Marked nature trail leaflet.
Public transport: None.
Habitat: Reservoir, lagoons, scrapes, woods, meadows, plantations.
Key birds: *Spring/autumn*: Outstanding wader passage. Also harriers, owls, passerine flocks, terns (Black, Arctic, breeding Common, occasional Little and Sandwich). *Winter*: Wildfowl (inc Goldeneye, Smew, Goosander, rare grebes, all divers), Ruff. *Summer:* Breeding Ospreys.
Contact: Tim Appleton, Fishponds Cottage, Stamford Road, Oakham, Rutland LE15 8AB. 01572

4. SENCE VALLEY FOREST PARK

Forest Enterprise.
Location: SK 400 115. Ten miles NW of Leicester and two miles SW of Coalville, between Ibstock and Ravenstone. The car park is signed from the A447 N of Ibstock. Do not leave valuables in cars as there have been some break-ins.
Access: Open all year.
Facilities: Car park, information and recent sightings boards, hide, paths.
Public transport: None.
Habitat: Forest, rough grassland, pools, wader scrape.
Key birds: *Spring/summer*: Wheatear, Whinchat, Redstart, Common and Green Sandpiper, Ringed and Little Ringed Plovers, Redshank. Dunlin and Greenshank frequent, possible Wood Sandpiper. Reed Bunting, Meadow Pipit, Sky Lark, Linnet, Yellow Wagtail. Possible Quail. *Winter*: Stonechat, Redpoll, Short-eared Owl. Merlin, Peregrine, Buzzard occasionally seen. Goosander and Wigeon possible.
Contact: Forest Enterprise, 340 Bristol Business Park, Coldharbour Lane, Bristol BS16 1EJ. 0117 906 6000.

Lincolnshire

1. DONNA NOOK

Lincolnshire Wildlife Trust.
Location: TF 422 998. Near North Somercotes, off A1031 coast road, S of Grimsby.
Access: Donna Nook beach is closed on weekdays as this is an active bombing range, but dunes remain open. Dogs on leads. Some disabled access.
Facilities: No toilets or visitor centre.
Public transport: None.
Habitat: Dunes, slacks and intertidal areas, seashore, mudflats, sandflats.
Key birds: *Summer*: Little Tern, Ringed Plover, Oystercatcher. *Winter*: Brent Goose, Shelduck, Twite, Lapland Bunting, Shore Lark, Linnet.
Contact: Trust HQ, 01507 526667. e-mail: lincstrust@cix.co.uk www.lincstrust.co.uk

2. FAR INGS

Lincolnshire Wildlife Trust.
Location: TA 011 229 and TA 023 230. Off Far Ings Lane, W of Barton-on-Humber, the last turn off before the Humber Bridge, heading N.
Access: Open all year. No dogs. Limited disabled access.
Facilities: Toilets, visitor centre open some weekends and weekdays – not all week. Hides and paths.
Public transport: None.
Habitat: Chain of flooded clay pits and reedbeds.
Key birds: *Summer*: Marsh Harrier, Bittern, Bearded Tit, Water Rail. *Winter*: Wildfowl (Mallard, Teal, Gadwall, Pochard, Tufted and Ruddy Ducks).
Contact: Far Ings Visitor Centre, Far Ings Road, Barton-on-Humber DN18 5RG. 01652 634507. e-mail: farings@lincstrust.co.uk www.lincstrust.co.uk

NATURE RESERVES - ENGLAND

3. FRAMPTON MARSH

RSPB (East Anglia Office).
Location: TR 36 43 85. Four miles SE of
Boston. From A16 follow signs to Frampton
then Frampton Marsh.
Access: Open at all times. Free. Coaches by
prior arrangement.
Facilities: Footpaths, bench, car park, bicycle
rack. Free information leaflets available (please
contact the office), guided walks programme.
Public transport: None.
Habitat: Saltmarsh.
Key birds: *Summer*: Breeding Redshank,
passage waders (inc Greenshank, Ruff and
Black-tailed Godwit) and Hobby. *Winter*: Hen
Harrier, Short-eared Owl, Merlin, dark-bellied
Brent Goose, Twite, Golden Plover.
Contact: John Badley, RSPB Lincolnshire Wash
Office, 61a Horseshoe Lane, Kirton, Kirton,
Boston, Lincs PE20 1LW. 01205 724678. e-mail:
john.badley@rspb.org.uk www.rspb.org.uk

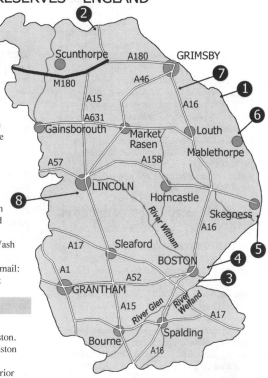

4. FREISTON SHORE

RSPB (Eastern England Office).
Location: TF 39 74 24. Four miles E of Boston.
From A52 at Haltoft End follow signs to Freiston
Shore.
Access: Open at all times, free. Coaches by prior
arrangement.
Facilities: Footpaths, two car parks, bird hide. Free
information leaflets available on site, guided walks
programme. Bicycle rack.
Public transport: None.
Habitat: Saltmarsh, saline lagoon, mudflats.
Key birds: *Summer:* Breeding waders including
Avocets, Ringed Plovers and Oystercatchers, Corn
Bunting and Tree Sparrow. *Winter:* Twite, dark-
bellied Brent Goose, wildfowl, waders, birds of prey
including Short-eared Owl and Hen Harrier. *Passage:*
Waders, including Curlew Sandpiper and Little Stint.
Autumn: Occasional seabirds including Arctic and
Great Skuas.
Contact: John Badley, RSPB Lincolnshire Wash,
(see Frampton Marsh).

5. GIBRALTAR POINT NNR & BIRD OBSERVATORY

Lincolnshire Wildlife Trust.
Location: TF 556 580. Three miles S of Skegness
on the N edge of The Wash. Signposted from
Skegness town centre.
Access: Reserve is open dawn-dusk all year.

Seasonal charges for car parking. Free admission to
reserve, visitor centre and toilets. Some access
restrictions to sensitive sites at S end, open access to
N. Dogs on leads at all times – no dogs on beach
during summer. Visitor centre and toilets suitable for
wheelchairs, as well as network of surfaced foot
paths. Bird observatory and four hides suitable for
wheelchairs. Day visit groups must be booked in
advance. Access for coaches. Contact Gibraltar Point
Field Station for residential or day visits.
Facilities: Site also location of Wash Study Centre
and Bird Observatory. Field centre is an ideal base
for birdwatching/natural history groups in spring,
summer and autumn. Visitor centre and gift shop
open daily (May-Oct) and weekends for remainder of
the year. Toilets open daily. Network of foot paths
bisect all major habitats. Public hides overlook
freshwater and brackish lagoons. Wash viewpoint
overlooks saltmarsh and mudflats.
Public transport: Bus service from Skegness runs in
occasional years but summer service only. Otherwise
taxi/car from Skegness. Cycle route from Skegness.
Habitat: Sand dune grassland and scrub, saltmarshes
and mudflats, freshwater marsh and lagoons.

169

Key birds: Large migration visible during spring and autumn passage – hirundines, chats, pipits, larks, thrushes and occasional rarities. Large numbers of waterfowl including internationally important populations of non-breeding waders. Sept-May impressive wader roosts on high tides. *Summer*: Little Tern and good assemblage of breeding warblers. *Winter*: Shore Lark, raptors, waders and wildfowl.
Contact: Kev Wilson, (Site Manager), Gibraltar Point Field Centre, Gibraltar Road, Skegness, Lincs PE24 4SU. 01754 762677.
e-mail: lincstrust@gibpoint.freeserve.co.uk
www.lincstrust.co.uk

6. SALTFLEETBY-THEDDLETHORPE DUNES

English Nature (East Midlands Team).
Location: TF 46 59 24-TF 49 08 83. Approx two miles N of Mablethorpe. All the following car parks may be accessed from the A1031: Crook Bank, Brickyard Lane, Churchill Lane, Rimac, Sea View
Access: Open all year at all times. Dogs on leads. A purpose-built easy access trail suitable for wheelchair users starts adjacent to Rimac car park, just over 0.5 miles long meanders past ponds. Includes pond-viewing platform and saltmarsh-viewing platform.
Facilities: Toilets, including wheelchair suitability at Rimac car park (next to trail) May to end of Sept, events programme.
Public transport: Grayscroft coaches (01507 473236) and Lincolnshire Roadcar (01522 532424). Both run services past Rimac entrance (Louth to Mablethorpe service). Lincs Roadcar can connect with trains at Lincoln. Applebys Coaches (01507 357900). Grimsby to Saltfleet bus connects with Grimsby train service.
Habitat: 13th Century dunes, freshwater marsh, new dune ridge with large areas of sea buckthorn, saltmarsh, shingle ridge and foreshore.
Key birds: *Summer*: Breeding birds in scrub include Nightingale, Grasshopper Warbler, Whitethroat, Lesser Whitethroat, Redpoll. *Winter*: Large flocks of Brent Goose, Shelduck, Teal and Wigeon. Wintering Short-eared Owl, Hen Harrier. Migrant birds in scrub and waders on Paradise scrape.
Contact: Simon Cooter, English Nature, 78 High Street, Boston, Lincs PE21 8SX. 01205 311674.
e-mail: simon.cooter@english-nature.org.uk
www.english-nature.org.uk

7. TETNEY MARSHES

RSPB (North of England Office).
Location: TA 345 025. Near Cleethorpes. Via gate or river bank E of Tetney Lock, which is two miles E of A1031 at Tetney, or through Humberston Fithes.
Access: Access at all times. Visitors are asked to keep to the seawalls, especially during the breeding season (Apr-Aug).
Facilities: None.
Public transport: None.
Habitat: Saltmarsh, sand-dunes and inter-tidal sandflats.
Key birds: *Summer*: Breeding Little Tern, Redshank, Shelduck. *Winter*: Brent Goose, Common Scoter, Wigeon, Bar-tailed Godwit, Knot, Grey and Golden Plovers. All three harriers recorded on passage. Migrant Whimbrel.
Contact: RSPB, 1 Sirius House, Amethyst Road, Newcastle Business Park NE4 7YL. 0191 256 8200.

8. WHISBY NATURE PARK

Lincolnshire Wildlife Trust.
Location: SK 914 661. W of Lincoln off A46 southern end of Lincoln relief road. Brown tourist signs.
Access: Nature Park open dawn to dusk. Consult notice board at entrance. Car park closed out of hours. Free entry. Natural World Visitor Centre open (10am-5pm), free entry. Some special exhibitions will have a charge. Disabled access. Dogs on leads. Electric buggy for hire by prior arrangement.
Facilities: Coach park, toilets, visitor centre, café, education centre, waymarked routes, interpretation signs, leaflets.
Public transport: No. 65 bus Mon-Sat from Lincoln to Thorpe-on-the-Hill (infrequent), (1/4 mile away).
Habitat: Flooded sand, gravel pits, scrub woodland and grassland.
Key birds: *Summer*: Common Tern on specially-built rafts, Nightingale, Whitethroat, Lesser Whitethroat, Tree Sparrow, Turtle Dove, Sand Martin colony. *Winter*: Wigeon, Teal, Pochard, Tufted Duck, Goldeneye.
Contact: Phil Porter, Whisby Nature Park, Moor Lane, Thorpe-on-the-Hill, Lincoln LN6 9BW. 01522 500676. e-mail: whisby@cix.co.uk
www.lincstrust.co.uk

London, Greater

1. BEDFONT LAKES COUNTRY PARK

Ecology and Countryside Parks Service.
Location: TQ 080 728. OS map sheet 176 (west London). 0.5 miles from Ashford, Middx, 0.5 miles S of A30, Clockhouse Roundabout, on B3003 (Clockhouse Lane).
Access: Open 7.30am-9pm or dusk, whichever is earlier, all days except Christmas Day. Disabled friendly. Dogs on leads. Main nature reserve area only open Sun (2pm-4pm). Keyholder membership available.
Facilities: Toilets, information centre, several hides, nature trail, free parking, up-to-date information.
Public transport: Train to Feltham and Ashford. Bus – H26 and 116 from Hounslow.
Habitat: Lakes, wildflower meadows, woodland, scrub.
Key birds: *Winter*: Water Rail, Bittern, Smew and other wildfowl, Meadow Pipit. *Summer*: Common Tern, Willow, Garden, Reed and Sedge Warblers, Whitethroat, Lesser Whitethroat, hirundines, Hobby, Blackcap, Chiffchaff, Sky Lark. *Passage*: Wheatear, Wood Warbler, Spotted Flycatcher, Ring Ouzel, Redstart, Yellow Wagtail.
Contact: Paul Morgan (Ecology Ranger), BLCP, Clockhouse Lane, Bedfont, Middx TW14 8QA. 01784 423556; Fax: 423451.
e-mail: bedfont_lakes_lnr@hotmail.com

2. BRAMLEY BANK

London Wildlife Trust.
Location: TQ 352 634. Entrance off Riesco Drive, or off Broadcombe, Croydon.
Access: Open all year. No coach access.
Facilities: None.
Public transport: South Croydon British Rail.
Habitat: Oak/sycamore woodland, large pond, acidic grassland and heath.
Key birds: Good range of woodland and parkland birds including woodpeckers and Nuthatch.
Contact: London Wildlife Trust, Harling House, Skyline House, 200 Union Street, London SE1 0LW. 0207 261 0447.
e-mail: enquiries@wildlondon.org.uk
www.wildlondon.org.uk

3. CAMLEY STREET NATURAL PARK

London Wildlife Trust.
Location: From Kings Cross Station drive or walk up Pancras Road between Kings Cross and St Pancras Stations. At the junction under a railway bridge turn R into Goods Way. Turn L into Camley Street and follow the telegraph pole fence to the large wrought-iron gates.
Access: Weekdays (9am-5pm), weekends (11am-5pm). Closed Fri. No coach parking at site. There is wheelchair access.
Facilities: Path.
Public transport: Nearest train/tube: Kings Cross, St Pancras.
Habitat: Woodland scrub, reedbeds, meadow, pond.
Key birds: *Spring/summer*: Warblers. *All year*: Mallard, Tufted Duck, Moorhen, Grey Heron, Sparrowhawk. *Winter*: Siskin, Reed Warbler. Good for dragonflies.
Contact: Trust HQ, 0207 261 0447
e-mail: enquiries@wildlondon.org.uk
www.wildlondon.org.uk

4. CHASE (THE) LNR

London Wildlife Trust.
Location: TQ 515 860. Lies in the Dagenham Corridor, an area of green belt between the London Boroughs of Barking & Dagenham and Havering.
Access: Open throughout the year and at all times. Reserve not suitable for wheelchair access. Eastbrookend Country Park which borders The Chase LNR has surfaced footpaths for wheelchair use.
Facilities: Millennium visitor centre, toilets, ample car parking, Timberland Trail walk.
Public transport: Rail: Dagenham East (District Line) 15 minute walk. Bus: 174 from Romford five minute walk.
Habitat: Shallow wetlands, reedbeds, horse-grazed pasture, scrub and wetland. These harbour an impressive range of animals and plants including the nationally rare black poplar tree. A haven for birds, with approx 190 different species recorded.
Key birds: *Summer*: Breeding Reed Warbler, Lapwing, Water Rail, Lesser Whitethroat and Little

NATURE RESERVES - ENGLAND

Ringed Plover, Kingfisher, Reed Bunting. *Winter*:
Significant numbers of Teal, Shoveler, Redwing,
Fieldfare and Snipe dominate the scene. *Spring/
autumn migration*: Yellow Wagtail, Wheatear, Ruff,
Wood Sandpiper, Sand Martin, Ring Ouzel, Black
Redstart and Hobby regularly seen.
Contact: Gareth Winn/Tom Clarke, Project
Manager/Project Officer, The Millennium Centre,
The Chase, Off Dagenham Road, Rush Green,
Romford, Essex RM7 0SS. 020 8593 8096.
e-mail: lwtchase@cix.co.uk
www.wildlifetrust.org.uk/london/

5. LONDON WETLAND CENTRE

The Wildfowl & Wetlands Trust.
Location: TQ 228 770. Less than 1 mile from South
Circular (A205) at Roehampton. In London, Zone 2,
one mile from Hammersmith.
Access: Winter (9.30am-5pm: last admission 4pm),
summer (9.30am-6pm: last admission 5pm). Charge
for admission.
Facilities: Visitor centre, hides, nature trails, art
gallery. Discovery Restaurant (hot and cold food),
cinema, shop, observatory centre, seven hides (one
with a lift for wheelchair access), three interpretative
buildings.
Public transport: Train: Barnes. Tube:
Hammersmith then Duckbus 283 (comes into centre).
Bus from Hammersmith – 283, 33, 72, 209. Bus
from Richmond 33, 72.
Habitat: Main lake, reedbeds, wader scrape,
mudflats, open water lakes, grazing marsh.
Key birds: Nationally important numbers of

wintering waterfowl, including Gadwall and
Shoveler. Important numbers of wetland breeding
birds, including grebes, swans, a range of duck
species such as Pochard, plus Lapwing, Little
Ringed Plover, Redshank, warblers, Reed Bunting
and Bittern.
Contact: John Arbon (Grounds and Facilities
Manager), Stephanie Fudge (Manager), London
Wetland Centre, Queen Elizabeth Walk, Barnes,
London SW13 9WT. 0208 409 4400.
e-mail: info.london@wwt.org.uk
www.wwt.org.uk

6. SYDENHAM HILL WOOD

London Wildlife Trust.
Location: TQ 335 722. SE London, SE26, between
Forest Hill and Crystal Palace, just off South Circular
(A205).
Access: Open at all times, no permits required. Some
steep slopes make disabled access limited
Facilities: Nature trail, no toilets.
Public transport: Train – Sydenham Hill, Forest
Hill. Bus – 363, 202, 356, 185, 312, 176, P4.
Habitat: Oak and hornbeam woodland, small pond,
meadow and glades.
Key birds: *Resident*: Kestrel, Sparrowhawk, Tawny
Owl, all three woodpeckers, Treecreeper, Nuthatch,
Song Thrush. *Summer*: Chiffchaff, Blackcap.
Winter: Redwing, Fieldfare.
Contact: The Warden, London Wildlife Trust,
Horniman Museum, 100 London Road, London
SE23 3PQ. 020 8699 5698.
e-mail: lwtsydenham@cix.co.uk
www.wildlondon.org.uk

Manchester, Greater

1. AUDENSHAW RESERVOIRS

United Utilities, Bottoms Office.
Location: SJ 915 965. On W side of Manchester.
Access from J24 of M60, then onto Audenshaw
Road (B6390) at N end of site.
Access: Park in Audenshaw Road, no disabled
access.
Facilities: Hide (contact R Travis on 0161 330
2607). Permit (free) from D Tomes, UU Bottoms
Office, Woodhead Road, Tintwistle, Glossop SK13
1HS. **Public transport:** None.

Habitat: A group of three reservoirs.
Key birds: Major migration point; *Winter*: Notable
gull roost inc. regular Mediterranean Gull and
Yellow-legged and white-winged gulls, large
Goosander roost. Many rarities including Red-necked
Grebe, Leach's Petrel, Wryneck.

2. DOVESTONES RESERVOIR

Peak District National Park.
Location: Close to Saddleworth Moor, E of
Oldham. From A635, look for sign on R having

172

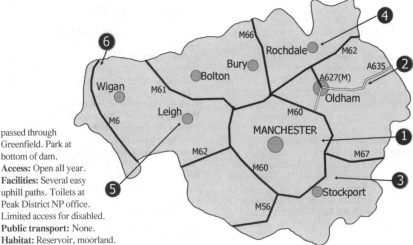

passed through
Greenfield. Park at
bottom of dam.
Access: Open all year.
Facilities: Several easy
uphill paths. Toilets at
Peak District NP office.
Limited access for disabled.
Public transport: None.
Habitat: Reservoir, moorland.
Key birds: *Spring/summer*: Ring Ouzel, Twite,
Peregrine, Raven, Stonechat, Whinchat, Wheatear,
Dipper, Common Sandpiper. Possibility of Redstart
and Crossbill. *All year*: Wildfowl.
Contact: Peak District National Park, 01629
816200.

3. ETHEROW COUNTRY PARK

Stockport Metropolitan Borough Council.
Location: SJ 965 908. From J 27 of M60 head E
on B6104 into Compstall near Romiley, Stockport.
Access: Open at all times; permit required for
conservation area. Keep to paths.
Facilities: Reserve area has SSSI status. Hide,
nature trail, visitor centre, scooters for disabled.
Public transport: None.
Habitat: River Etherow, woodlands, marshy area.
Key birds: Sparrowhawk, Buzzard, Dipper, all
three woodpeckers, Pied Flycatcher, warblers.
Winter: Brambling, Siskin, Water Rail. Frequent
sightings of Merlin and Raven over hills.
Contact: Etherow Country Park, Compstall,
Stockport, Cheshire SK6 5JD. 0161 427 6937; fax
0161 427 3643.

4. HOLLINGWORTH LAKE

Hollingworth Lake – Rochdale MBC.
Location: SD 939 153 (visitor centre). Four miles
NE of Rochdale, signed from A58 Halifax Road
and J21 of M62 – B6225 to Littleborough.
Access: Access open to lake and surroundings at all
times.

Facilities: Cafes, hide, trails and education service,
car parks. Visitor centre open 10.30am-6pm (Mon-
Sun) in summer, 11am-4pm (Mon-Fri), 10.30am-
5pm (Sat & Sun) in winter.
Public transport: Bus Nos 452, 450. Train to
Littleborough or Smithy Bridge.
Habitat: Lake (116 acres 47 ha includes 20 acre
nature reserve), woodland, streams, marsh, willow
scrub.
Key birds: *All year*: Great Crested Grebe,
Kingfisher, Lapwing, Little Owl, Bullfinch,
Cormorant. Occasional Peregrine Falcon, Water Rail,
Snipe. *Spring/autumn*: Passage waders, wildfowl,
Kittiwake. *Summer*: Reed Bunting, Dipper, Common
Sandpiper, Curlew, Oystercatcher, Black Tern,
'Commic' Tern, Grey Partridge, Sedge Warbler,
Blackcap. *Winter*: Goosander, Goldeneye, Siskin,
Redpoll, Golden Plover.
Contact: The Ranger, Hollingworth Lake Visitor
Centre, Rakewood Road, Littleborough, OL15 0AQ.
01706 373421. www.rochdale.gov.uk

5. PENNINGTON FLASH

Wigan Leisure and Culture Trust
Location: SJ 640 990. One mile from Leigh town
centre. Main entrance on A572 (St Helens Road).
Access: Park is signposted from A580 (East Lancs
Road) and is permanently open. Four largest hides,
toilets and information point open 9am-dusk (except
Christmas Day). Main paths flat and suitable for
disabled. Coach parking available if booked in
advance.

Facilities: Toilets including disabled toilet and information point. Total of seven bird hides. Site leaflet available and Rangers based on site. Group visits welcome, guided tours or a site introduction can be arranged subject to staff availability.

Public transport: Only 1 mile from Leigh bus station. Several services stop on St Helens Road near entrance to park. Contact GMPTE 0161 228 7811.

Habitat: Lowland lake, ponds and scrapes, fringed with reeds, rough grassland, scrub and young woodland.

Key birds: Waterfowl all year, waders mainly passage spring and autumn (14-plus species). Breeding Common Tern and Little Ringed Plover. Feeding station attracts Willow Tit and Bullfinch all year.

Contact: Peter Alker, Pennington Flash Country Park, St Helens Road, Leigh WN7 3PA. 01942 605253 (Also fax number). e-mail: pfcp@wlct.org

6. WIGAN FLASHES

Lancashire Wildlife Trust/Wigan Council.
Location: SD 580 035. One mile from J25 of M6.
Access: Free access, open at all times. Areas suitable for disabled but some motor cycle barriers with gates. Paths being upgraded. Access for coaches - contact reserve manager for details.
Facilities: Three hide screens built 2004.
Public transport: 610 bus (Hawkley Hall Circular).
Habitat: Wetland with reedbed.
Key birds: Black Tern on migration. *Summer*: Nationally important for Reed Warbler and breeding Common Tern. Willow Tit, Grasshopper Warbler, Kingfisher. *Winter*: Wildfowl especially diving duck and Gadwall. Bittern (especially winter).
Contact: Mark Champion, Lancashire Wildlife Trust, Clifton Street Community Centre, Clifton Street, Wigan, Lancs WN3 5HN. 01942 233976. e-mail: lwt@wiganflashes.wanadoo.co.uk

Merseyside

1. AINSDALE & BIRKDALE LNR

Sefton Council.
Location: SD 300 115. SD 310 138. Leave the A565 just N of Formby and car parking areas are off the unnumbered coastal road.
Access: Track from Ainsdale or Southport along the shore. Wheelchair access across boardwalks at Ainsdale Sands Lake Nature Trail and the Queen's Jubilee Nature Trail, opposite Weld Road.
Facilities: Ainsdale Visitor Centre open Summer and toilets (Easter-Oct).
Public transport: Ainsdale and Southport stations 20 minute walk. Hillside Station is a 30 minute walk across the Birkdale Sandhills to beach.
Habitat: Foreshore, dune scrub and pine woodland.
Key birds: *Spring/summer*: Grasshopper Warbler, Chiffchaff, waders. *Winter*: Blackcap, Stonechat, Redwing, Fieldfare, waders and wildfowl. *All year*: Sky Lark, Grey Partridge.
Contact: Sefton Council, Southport Town Hall, Lord Street, Southport, PR8 1DA, www.sefton.gov.uk

2. DEE ESTUARY

Metropolitan Borough of Wirral.
Location: SJ 255 815. Leave A540 Chester to Hoylake road at Heswall and head downhill (one mile) to the free car park at the shore end of Banks Road. Heswall is 30 minutes from South Liverpool and Chester by car.
Access: Open at all times. Best viewpoint 600 yards along shore N of Banks Road. No disabled access along shore, but good birdwatching from bottom of Banks Road. Arrive 2.5 hours before high tide. Coach parking available.
Facilities: Toilets in car park and information board. Wirral Country Park Centre three miles N off A540 has toilets, hide, café, kiosk (all accessible to wheelchairs). Birdwatching events programme available from visitor centre.
Public transport: Bus service to Banks Road car park from Heswall bus station. Contact Mersey Travel (tel 0151 236 7676).
Habitat: Saltmarsh and mudflats.
Key birds: *Autumn/winter*: Large passage and winter wader roosts – Redshank, Curlew, Black-tailed

Godwit, Oystercatcher, Golden Plover, Knot, Shelduck, Teal, Red-breasted Merganser, Peregrine, Merlin, Hen Harrier, Short-eared Owl. Smaller numbers of Pintail, Wigeon, Bar-tailed Godwit, Greenshank, Spotted Redshank, Grey and Ringed Plovers, Whimbrel, Curlew Sandpiper, Little Stint, occasional Scaup and Little Egret.
Contact: Martyn Jamieson, Head Ranger, Wirral Country Park Centre, Station Road, Thustaston, Wirral, Merseyside CH61 0HN. 0151 648 4371/ 3884. e-mail: wirralcountrypark@wirral.gov.uk www.wirral.gov.uk/er

3. HILBRE ISLAND LNR

Wirral Country Park Centre (Metropolitan Borough of Wirral).
Location: SJ 184 880. Three tidal islands in mouth of Dee Estuary. Park in West Kirby which is on A540 Chester-to-Hoylake road – 30 minutes from Liverpool, 45 minutes from Chester. Follow brown Marine Lake signs to Dee Lane pay and display car park. Coach parking available at West Kirby but please apply for permit to visit island well in advance as numbers limited.
Access: Two mile walk across the sands from Dee Lane slipway. No disabled access. Do not cross either way within 3.5 hours of high water – tide times and suggested safe route on noticeboard at slipway. Prior booking and permit needed for parties

of six or more – maximum of 50. Book early.
Facilities: Toilets at Marine Lake and Hilbre (primitive!). Permits, leaflets and tide times from Wirral Country Park Centre. Hilbre Bird Observatory.
Public transport: Bus and train station (from Liverpool) within 0.5 mile of Dee Lane slipway. Contact Mersey Travel, tel 0151 236 7676.
Habitat: Sandflats, rocky shore and open sea.
Key birds: *Late summer/autumn*: Seabird passage – Gannets, terns, skuas, shearwaters and after NW gales good numbers of Leach's Petrel. *Winter*: Wader roosts at high tide, Purple Sandpiper, Turnstone, sea ducks, divers, grebes. Passage migrants.
Contact: Martyn Jamieson, Head Ranger, Wirral Country Park Centre, Station Road, Thustaston, Wirral, Merseyside CH61 0HN. 0151 648 4371/ 3884. e-mail: wirralcountrypark@wirral.gov.uk www.wirral.gov.uk/er

4. MARSHSIDE

RSPB (North West England Office).
Location: SD 355 202. On south shore of Ribble Estuary, one mile north of Southport centre on Marine Drive.
Access: Open 8.30am-5pm all year. No toilets. No dogs please. Coach parties please book in advance. No charges but donations welcomed.
Facilities: Two hides and trails accessible to wheelchairs.
Public transport: Bus service to Elswick Road/ Marshside Road half-hourly, bus No 44, from Lord Street. Contact Southport Buses (01704 536137).
Habitat: Coastal grazing marsh and lagoons.
Key birds: *Winter*: Pink-footed Goose, wildfowl, waders, raptors. *Spring*: Breeding waders and wildfowl, Garganey, migrants. *Autumn*: Migrants. *All year*: Black-tailed Godwit.
Contact: Tony Baker, RSPB, 24 Hoghton Street, Southport PR9 0PA. 01704 536378. e-mail: tony.baker@rspb.org.uk

5. SEAFORTH NATURE RESERVE

The Wildlife Trust for Lancashire, Manchester and North Merseyside.
Location: SJ 315 970. Five miles from Liverpool city centre. From M57/M58 take A5036 to docks.
Access: Only organised groups who pre-book are now allowed access. Groups should contact the reserve office

(see below) at least seven days in advance of their planned trip. Coaches welcome.
Facilities: Toilets when visitor centre open, three hides.
Public transport: Train to Waterloo or Seaforth stations from Liverpool. Buses to dock gates from Liverpool.
Habitat: Saltwater and freshwater lagoons, scrub grassland.

Key birds: Little Gull on passage (Apr) plus Roseate, Little and Black Terns. Breeding and passage Common Tern (Apr-Sept) plus Roseate, Little and Black Terns on passage. Passage and winter waders and gulls. Passage passerines, especially White Wagtail, pipits and Wheatear.
Contact: Steve White, Seaforth Nature Reserve, Port of Liverpool, L21 1JD. 0151 9203769.
e-mail: lwildlife@cix.co.uk

Norfolk

1. NWT CLEY MARSHES

Norfolk Wildlife Trust.
Location: TG 054 441. NWT Cley Marshes is situated three miles N of Holt on A149 coast road, half a mile E of Cley-next-the-Sea. Visitor centre and car park on inland side of the road.
Access: Open all year round. Visitor centre open Apr-Oct (10am-5pm daily), Nov-mid Dec (10am-4pm Wed-Sun). Cost: adults £3.75, children under 16 free. NWT members free.
Facilities: Visitor centre, birdwatching hides, wildlife gift shop, refreshments, toilets, coach parking, car park, disabled access to centre, boardwalk and hides and toilets, groups welcome.
Public transport: Bus service from Norwich, Fakenham and Holt Mon-Sat. The Coasthopper service stops outside daily. Connections for train and bus services at Sheringham. Special discounts to visitors arriving by bus.
Habitat: Reedbeds, salt and freshwater marshes, scrapes and shingle ridge with international reputation as one of the finest birdwatching sites in Britain.
Key birds: *Feb:* Brent Goose, Wigeon, Teal, Shoveler, Pintail. *Spring:* Chiffchaff, Wheatear, Sandwich Tern, Reed and Sedge Warblers, Ruff, Black-tailed Godwit. *Jun:* Spoonbill, Avocet, Bittern, Bearded Tit. *Autumn:* Green and Wood Sandpiper, Greenshank, Whimbrel, Little Ringed Plover. Many rarities.
Contact: Dick Bagnall, Oakeley Centre, NWT Cley Marshes, Cley, Holt, Norfolk NR25 7RZ. 01263 740008. e-mail BernardB@nwt.cix.co.uk
www.wildlifetrust.org.uk.Norfolk

2. NWT EAST WRETHAM HEATH

Norfolk Wildlife Trust.
Location: TL 913 887. Site lies in the centre of Breckland, N of Thetford. From A11 take turning for A1075 to Watton and travel for about two miles over the level crossing and pass the lay-by to the left. Car park and entrance to reserve are by the first house on the left.
Access: Open all year dawn to dusk. Limited coach parking in lay by 400m south of reserve entrance.
Facilities: Car park. Trail, hide.
Public transport: None.
Habitat: Breckland grass heath and meres, scrub and woodland.
Key birds: Wildfowl, wading birds, Redstart, Wood Lark.
Contact: Bev Nichols, Norfolk Wildlife Trust, Bewick House, 22 Thorpe Road, Norwich NR1 1RY. 01603 625540.
e-mail BevN@norfolkwildlifetrust.co.uk
www.wildlifetrust.org.uk/Norfolk

3. NWT HICKLING BROAD

Norfolk Wildlife Trust.
Location: TG 428 222. Approx four miles SE of Stalham, just off A149 Yarmouth Road. From Hickling village, follow the brown 'duck' tourist signs into Stubb Road at the Greyhound Inn. Take first turning left to follow Stubb Road for another mile. Turn right at the for the nature reserve. The car park is ahead of you.
Access: Open all year. Visitor centre open Apr-Sept (10am-5pm daily). Cost: adults £3, children under 16 free. NWT members free.

NATURE RESERVES - ENGLAND

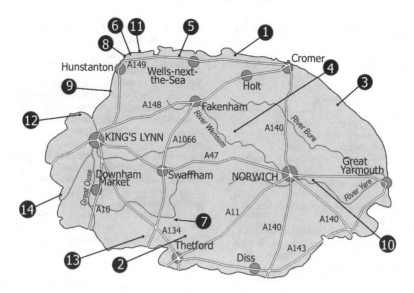

Facilities: Visitor centre, boardwalk trail through reedbeds to open water, birdwatching hides, wildlife gift shop, refreshments, picnic site, toilets, coach parking, car parking, disabled access to broad, boardwalk and toilets. Groups welcome. Water trail (additional charge – booking essential).
Public transport: Morning bus service only Mon-Fri from Norwich (Neaves Coaches) Cromer to North Walsham (Sanders). Buses stop in Hickling village, a 20 minute walk away.
Habitat: Hickling is the largest and wildest of the Norfolk Broads with reedbeds, grazing marshes and wide open skies.
Key birds: Marsh Harriers, Bittern. Swallowtail butterfly, Norfolk hawker (rare dragonfly).
Contact: John Blackburn, Hickling Broad Visitor Centre, Stubb Road, Hickling NR12 0BN.
e-mail johnb@norfolkwildlifetrust.co.uk
www.wildlifetrust.org.uk/norfolk

4. NWT FOXLEY WOOD

Norfolk Wildlife Trust.
Location: TG 049 229. Foxley Wood is situated 12 miles NW of Norwich on A1067, signposted at Foxley village from main road.
Access: Open all year, (10am-5pm). Closed Thursdays. Groups welcome
Facilities: Trails, car park, no toilets.

Public transport: Bus – Norwich to Fakenham services stop at Foxley War Memorial approximately 15 minutes' walk. Sanders Coaches Sun service.
Habitat: Norfolk's largest ancient woodland. Rich woodland flora including early purple orchids, wood anemone, wood sorrel and bluebells.
Key birds: Woodland birds including Great Spotted and Green Woodpeckers, Woodcock, Blackcap, Garden Warbler. White admiral butterflies.
Contact: Trust HQ, 01603 625540.
e-mail admin@norfolkwildlifetrust.org.uk

5. HOLKHAM

English Nature (Norfolk Team).
Location: TF 890 450. From Holkham village turn N off A149 down Lady Ann's Drive; parking.
Access: Access unrestricted, but keep to paths and off grazing marshes and farmland.
Facilities: Two hides. Disabled access.
Public transport: Bus Norbic Norfolk bus information line 0845 3006116 all free of charge.
Habitat: Sandflats, dunes, marshes, pinewoods.
Key birds: *Passage*: Migrants. *Winter*: Wildfowl, inc. Brent, Pink-footed and White-fronted Geese. *Summer*: Breeding Little Tern.
Contact: R Harold, Hill Farm Offices, Main Road, Holkham, Wells-next-the-Sea NR23 1AB.
01328 711183; fax 01328 711893.

NATURE RESERVES - ENGLAND

6. HOLME BIRD OBSERVATORY

Norfolk Ornithologists' Association (NOA).
Location: TF 717 450. E of Hunstanton, signposted from A149. Access from Broadwater Road. The reserve and visitors centre are beyond the White House at the end of the track.
Access: Reserve open daily to members dawn to dusk; non-members (9am-5pm) by permit from the Observatory. Please keep dogs on leads in the reserve. Parties by prior arrangement.
Facilities: Accredited Bird Observatory operating 12 months of the year, bird ringing, MV moth trapping and other scientific monitoring. Visitor centre, car park and several hides (seawatch hide reserved for NOA members) together with access to beach and coastal path.
Public transport: Coastal bus service runs from Hunstanton to Sheringham roughly every 30 mins but is seasonal and times may vary. Phone Norfolk Green Bus, 01553 776 980.
Habitat: In ten acres of diverse habitat: sand dunes, Corsican pines, scrub and reed-fringed lagoon making this a migration hotspot.
Key birds: Species list over 320. Ringed species over 150. Recent rarities have included Red Kite, Common Crane, Red-backed Shrike, Osprey, Pallas', Yellow-browed, Greenish and Barred Warblers.
Contact: Jed Andrews, Holme Bird Observatory, Broadwater Road, Holme, Hunstanton, Norfolk PE36 6LQ. 01485 525406. e-mail: info@noa.org.uk www.noa.org.uk

7. LYNFORD ARBORETUM

Forest Enterprise.
Location: TL 821 943, TL 818 935. NW of Thetford. At Mundford roundabout, follow signs to Swaffham (A1065). Take first R to Lynford Hall. Follow road past hall to car park on the L. Disabled drivers can turn R. Alternatively, from roundabout head SE towards Thetford. Take minor road signed L almost immediately to Lynford Lakes. Follow road to L turn, signed to lakes. There is a car park at bottom of the track.
Access: Open all year.
Facilities: Two car parks, tracks. Suitable for wheelchairs. No dogs in arboretum.
Public transport: None.
Habitat: Plantations, arboretum, lake.
Key birds: *Spring/summer*: Possible Wood Lark, Tree Pipit, possible Nightjar, Kingfisher, waterfowl. *Winter*: Crossbill, possible Hawfinch. *All year*: Usual woodland species, woodpeckers.

Contact: Forest Enterprise, Santon Downham. Brandon, Suffolk IP27 0TJ. 01832 810 271.

8. REDWELL MARSH

Norfolk Ornithologists' Association (NOA).
Location: TF 702 436. In Holme, off A149, E of Hunstanton. Access from Broadwater Road.
Access: View from public footpath from centre of Holme village to Broadwater Road. Open at all times.
Facilities: Member's hide, offering free wheelchair access, (access from Broadwater Road).
Public transport: As for Holme Bird Observatory.
Habitat: Wet grazing marsh with ditches, pond and two large wader scrapes.
Key birds: Wildfowl and waders inc. Curlew/Green/Wood and Pectoral Sandpipers, Greenshank, Spotted Redshank, Avocet and Black-tailed Godwit. Recent rarities have included American Wigeon, Ring Ouzel, Black-winged Stilt, Mediterranean Gull and Grasshopper Warbler. Also a raptor flight path.
Contact: Jed Andrews, Holme Bird Observatory, Broadwater Road, Holme, Hunstanton, Norfolk PE36 6LQ. 01485 525406.
e-mail: info@noa.org.uk
www.noa.org.uk

9. SNETTISHAM

RSPB (East Anglia Office).
Location: TF 630 310. Car park two miles along Beach Road, signposted off A149 King's Lynn to Hunstanton, opposite Snettisham village.
Access: Open at all times. Dogs to be kept on leads. Two hides are suitable for wheelchairs. Disabled access is across a private road. Please phone office number for permission and directions. Coaches welcome, but please book in advance as a height barrier needs to be removed.
Facilities: Four birdwatching hides, connected by reserve footpath. No toilets on site.
Public transport: Nearest over two miles away.
Habitat: Intertidal mudflats, saltmarsh, shingle beach, brackish lagoons, and unimproved grassland/scrub. Best visited on a high tide.
Key birds: *Autumn/winter/spring*: Waders (particularly Knot, Bar and Black-tailed Godwits, Dunlin, Grey Plover), wildfowl (particularly Pink-footed and Brent Geese, Wigeon, Gadwall, Goldeneye), Peregrine, Hen Harrier, Merlin, owls. Migrants in season. *Summer*: Breeding Ringed Plover, Redshank, Avocet, Common Tern. Marsh Harrier regular.

Contact: Jim Scott, RSPB, 43 Lynn Road, Snettisham, King's Lynn, Norfolk PE31 7LR. 01485 542689.

10. STRUMPSHAW FEN

RSPB (East Anglia Office).
Location: TG 33 06. Seven miles ESE of Norwich. Follow signposts. Entrance across level-crossing from car park, reached by turning sharp right and right again into Low Road from Brundall, off A47 to Great Yarmouth.
Access: Open dawn-dusk. RSPB members free, adults £2.50, children 50p, family £5. Guide dogs only. (Wheelchair access not currently promoted due to safety issues with railway crossing).
Facilities: Toilets, reception hide and two other hides, two walks, five miles of trails.
Public transport: Brundall train station about one mile. Bus stop half a mile – contact Norbic(0845 300 6116).
Habitat: Reedbed and reedfen, wet grassland and woodland.
Key birds: *Summer*: Bittern, Bearded Tit, Marsh Harrier, Cetti's Warbler and other reedbed birds. *Winter*: Bittern, wildfowl, Marsh and Hen Harriers. Swallowtail butterflies in Jun.
Contact: Tim Strudwick, Staithe Cottage, Low Road, Strumpshaw, Norwich, Norfolk NR13 4HS. 01603 715191.
e-mail: strumpshaw@rspb.org.uk
www.rspb.org.uk

11. TITCHWELL MARSH

RSPB (East Anglia Office).
Location: TF 749 436. Near Hunstanton. Footpath along sea wall from A149 between Thornham and Titchwell.
Access: All paths and trails suitable for wheelchairs. Reserve and hides open at all times. Coach parking available but pre-booking essential.
Facilities: Visitor centre, shop with large selection of binoculars, telescopes and books, open every day 9.30am to 5pm (Nov 13 - Feb 16, 2005, 9.30 to 4pm). Our new tearoom is open from 9.30am to 4.30pm every date (Nov 13 - Feb 16, 2005, 9.30am to 4pm). Visitor centre and tearoom closed on Christmas day and Boxing day.
Public transport: Phone Norfolk Green Bus 01553 776980.
Habitat: Reedbed, brackish & freshwater pools, saltmarsh, dunes, shingle.
Key birds: *Spring/summer*: Nesting Avocet, Bittern, Bearded Tit, Water Rail, Marsh Harrier, Reed, and Sedge Warblers. *Autumn*: Knot. *Winter*: Brent Geese, Goldeneye, Scoter, Eider, Hen Harrier roost, Snow Bunting and Shore Lark on beach.
Contact: Centre Manager, Titchwell Marsh Reserve, King's Lynn, Norfolk PE31 8BB. Tel/fax 01485 210779.

12. THE WASH NNR

English Nature East Midlands Team.
Location: Between TF 484 280 and TF 598 237. W

of King's Lynn. For main entrance off A17, follow road along East bank of River Nene at Sutton Bridge.
Access: Open access W of River Ouse, though remain on public footpaths along seabank and keep dogs under control.
Facilities: Car parks at Gedney Drove End (TF 481 284), Guy's Head (TF 491 256), Peter Scott's Lighthouse (TF 493 255), Ongar Hill (TF 582 247).
Public transport: None.
Habitat: Saltmarsh and mudflats.
Key birds: *Summer*: Redshank, Oystercatcher. Marsh Harrier, Shelduck, Reed Bunting. Massive numbers of passage and wintering waterfowl including Brent Goose, Knot, Oystercatcher, Lapwing, Redshank, Dunlin, godwits, Shelduck and raptors.
Contact: Simon Cooter, English Nature, 78 High Street, Boston, Lincs PE21 8SX. 01205 311674. e-mail: simon.cooter@english-nature.org.uk www.english-nature.org.uk

13. NWT WEETING HEATH

Norfolk Wildlife Trust.
Location: TL 756 881. Weeting Heath is signposted from the Weeting-Hockwold road, two miles W of Weeting near to Brandon in Suffolk. Nature reserve can be reached via B1112 at Hockwold or B1106 at Weeting.
Access: Open daily from Apr-Sep. Cost: adults £2.50, children free. NWT members free. Disabled access to visitor centre and hides.
Facilities: Visitor centre open daily Apr-Aug, birdwatching hides, wildlife gift shop, refreshments,

toilets, coach parking, car park, groups welcome (book first).
Public transport: Train services to Brandon and bus connections (limited) from Brandon High Street.
Habitat: Breckland, grass heath.
Key birds: Stone Curlew, migrant passerines, Wood Lark.
Contact: Bev Nichols, Trust HQ, 01603 625540. e-mail: BevN@norfolkwildlifetrust.org.uk www.wildlifetrust.org.uk/Norfolk

14. WELNEY

The Wildfowl & Wetlands Trust.
Location: TL 546 944. Ten miles N of Ely, signposted from A10 and A1101.
Access: Open daily (10am-5pm) except Christmas Day. Free admission to WWT members, otherwise £3.50 (adult), £2.75 (senior), £2 (child).
Facilities: Visitor centre, cafe open 10am-4pm daily. Large, heated observatory, plus six other birdwatching hides. Provision for disabled visitors.
Public transport: Poor. Train to Littleport (6 miles away), but from there, taxi is only option. Excellent cycling country!
Habitat: 1,000 acres of washland reserve, spring damp meadows, winter wildfowl marsh.
Key birds: *Winter*: Bewick's Swan, Whooper Swan, winter wildfowl. *Spring*: Breeding/migrant waders, warblers.
Contact: The Warden, WWT, Hundred Foot Bank, Welney, Nr Wisbech PE14 9TN. 01353 860711. e-mail: welney@wwt.org.uk www.wwt.org.uk

Northants

1. CLIFFORD HILL GRAVEL PITS

Location: SP 781 595. Take the A428 Bedford Road from Northampton town centre E to A45 roundabout. Go straight over and turn L to park by Courtyard Hotel.
Access: Open all year. Shooting may take place Tues.
Facilities: Car park, toilets. **Public transport:** None.
Habitat: Reservoir, river, grassland.
Key birds: *Spring*: Wheatear, Sand Martin, Yellow Wagtail. *Winter*: Excellent for wintering wildfowl inc Goosander, Pintail, Red-crested Pochard and Smew.

Geese, Golden Plover, Green Sandpiper, Redshank, thrushes. *All year*: Meadow Pipit, Reed Bunting, Grey Wagtail, Redpoll, Linnet, Green Woodpecker.

2. HOLLOWELL RESERVOIR

Anglian Water.
Location: SP 683 738. From Northampton, take the A5199 NW. After eight miles, turn L. The car park is on the L.
Access: Open all year. Permit required. Keep dogs on lead.

Facilities: None.
Public transport: Bus: No 60 from Northampton, first one arriving at 8.58am. Tel: 01604 676 060.
Habitat: Grass, mature, mixed and conifer plantations.
Key birds: *Autumn/winter*: Dunlin, Greenshank, Redshank, Green Sandpiper, Mediterranean Gull, ducks. Crossbill occurs in invasion years. Bearded Tit and Dartford Warbler occasional.
Contact: Anglian Water, Anglian House, Ambury Road, Huntingdon, Cambs, PE29 3NZ, 01480 323000.

3. PITSFORD RESERVOIR

Beds, Cambs, Northants and Peterborough Wildlife Trust.
Location: SP 787 702. Five miles N of Northampton. On A43 take turn to Holcot and Brixworth. On A508 take turn to Brixworth and Holcot.
Access: Lodge open mid-Mar to mid-Nov from 8am-dusk. Winter opening times variable, check in advance. Permits for reserve available from Lodge on daily or annual basis. Reserve open to permit holders 365 days a year. No dogs. Disabled access from Lodge to first hide.
Facilities: Toilets available in Lodge, 15 miles of paths, eight bird hides, car parking.
Public transport: None.
Habitat: Open water (up to 120 ha), marginal vegetation and reed grasses, wet woodland, grassland and mixed woodland (40 ha).
Key birds: 150 species in 2004. *Summer*: Breeding warblers, terns, Hobby, Tree Sparrow. *Autumn*: Waders only if water levels suitable. *Winter*: Wildfowl, feeding station with Tree Sparrow and Corn Bunting.
Contact: Dave Francis, Pitsford Water Lodge, Brixworth Road, Holcot, Northampton NN6 9SJ. 01604 780148. e-mail: pitsford@cix.compulink.co.uk

4. STANFORD RESERVOIR

Severn Trent Water/Beds, Cambs, Northants and Peterborough Wildilfe Trust.
Location: SP 600 805. One mile SW of South Kilworth off Kilworth/Stanford-on-Avon road.
Access: Daytime, permits from Northants Wildlife Trust. No dogs. Limited access for disabled.
Facilities: Toilets (inc disabled). Two hides, perimeter track. Disabled parking.
Public transport: None.
Habitat: Reservoir with willow, reed and hedgerow edges.
Key birds: *Winter*: Wildfowl, especially Ruddy Duck. *Spring*: Migratory terns and warblers. *Late summer*: Terns, Hobby, waders.
Contact: Trust HQ 01954 712500; fax 01954 710051. e-mail: cambridgeshire@wildlifebcnp.org www.wildlifebcnp.org

5. SUMMER LEYS LNR

Northamptonshire County Council.
Location: SP 886 634. Three miles from Wellingborough, accessible from A45 and A509, situated on Great Doddington to Wollaston

181

40 space car park, small tarmaced circular route suitable for wheelchairs.

Facilities: Three hides, one feeding station. No toilets, nearest are at Irchester Country Park on A509 towards Wellingborough.

Public transport: Nearest main station is Wellingborough. No direct bus service, though buses run regularly to Great Doddington and Wollaston, both about a mile away. Tel: 01604 236712 (24 hrs) for copies of timetables.

Habitat: Scrape, two ponds, lake, scrub, grassland, hedgerow.

Key birds: Hobby, Lapwing, Golden Plover, Ruff, Gadwall, Garganey, Pintail, Shelduck, Shoveler, Little Ringed Plover, Tree Sparrow, Redshank, Green Sandpiper, Oystercatcher, Black-headed Gull colony, terns.

Contact: Chris Haines, Countryside and Tourism, Northamptonshire Council, PO Box 163, County Hall, Northampton NN1 1AX. 01604 237227 – please ring for a leaflet about the reserve. e-mail: countryside@northamptonshire.gov.uk

6. THRAPSTON GRAVEL PITS & TITCHMARSH LNR

Beds, Cambs, Northants and Peterborough Wildlife Trust/English Nature.

Location: TL 008 804.

Access: Public footpath from layby on A605 N of Thrapston. Car park at Aldwincle, W of A605 at Thorpe Waterville.

Facilities: Two hides.

Public transport: Bus service to Thrapston.

Habitat: Alder/birch/willow wood; old duck decoy, series of water-filled gravel pits.

Key birds: *Summer*: Breeding Grey Heron (no access to Heronry), Common Tern, Little Ringed Plover; warblers. Migrants, inc. Red-necked and Slavonian Grebes, Bittern and Marsh Harrier recorded.

Contact: Trust HQ, 01954 713500; fax 01954 710051. e-mail:cambridgeshire@wildlifebcnp.org www.wildlifebcnp.org

Northumberland

1. ARNOLD RESERVE, CRASTER

Northumberland Wildlife Trust.

Location: NU255197. Lies NE of Alnwick and SW of Craster village.

Access: Public footpath from car park in disused quarry.

Facilities: Interpretation boards, information centre open in Summer. Toilets (incl disabled) and picnic site in quarry car park.

Public transport: Arriva Northumberland nos. 501 and 401.

Habitat: Semi-natural woodland and scrub near coast.

Key birds: Good site for migrant passerines to rest and feed. Interesting visitors can inc. Bluethroat, Red-breasted Flycatcher, Barred and Icterine Warblers, Wryneck; moulting site for Lesser Redpoll. *Summer*: Breeding warblers.

Contact: Trust HQ, 0191 284 6884; fax 0191 284 6794; e-mail: northwildlife@cix.co.uk www.wildlifetrust.org.uk/northumberland

2. BOLAM LAKE COUNTRY PARK

Northumberland County Council.

Location: NZ 08 81. 4.8km N of main Jedburgh Road (A696), 27km NW of Newcastle. Signed along a minor road from Belsay.

Access: Open all year, small parking fee.

Facilities: Car parks, leaflets from warden's house in the main car park.

Public transport: None.

Habitat: Parkland, lake, carr, conifers, woodland.

Key birds: *Spring/summer*: Ruddy Duck, Woodcock, Common Sandpiper, Sand Martin, House Martin, Redstart, warblers, inc Grasshopper, Garden, Wood and Willow, possible flycatchers, waders. *Winter*: Whooper Swan, Greylag and Canada Geese, Wigeon, Pintail, Goosander, Water Rail, Woodcock, thrushes, Siskin, Redpoll. *All year*: Possible Grey Partridge, Green and Great Spotted Woodpeckers, Goldcrest, tits, Treecreeper, Nuthatch, Bullfinch.

Contact: Northumberland County Council, County Hall, Morpeth NE61 2EF. 01670 533100.

3. DRURIDGE BAY RESERVES

Northumberland Wildlife Trust.
Location: 1. NU 285 023. S of Amble. Hauxley (67a) approached by track from road midway between High and Low Hauxley. 2. Druridge Pools NZ 272 965. 3. Cresswell Pond NZ 283 945. Half mile N of Cresswell.
Access: Day permits for all three reserves.
Facilities: 1. Visitor centre, five hides (one suitable for disabled). Disabled toilet. Lake with islands behind dunes. 2. Three hides. 3. Hide.
Public transport: None.
Habitat: 2. Deep lake and wet meadows with pools behind dunes. 3. Shallow brackish lagoon behind dunes fringed by saltmarsh and reedbed, some mudflats.
Key birds: 1. *Spring and autumn*: Good for passage birds (inc. divers, skuas). *Summer*: Coastal birds, esp. terns (inc. Roseate). 2. Especially good in spring. Winter and breeding wildfowl; passage and breeding waders. 3. Good for waders, esp. on passage.
Contact: Jim Martin, Hauxley Nature Reserve, Low Hauxley, Amble, Morpeth, Northumberland. 01665 711578.

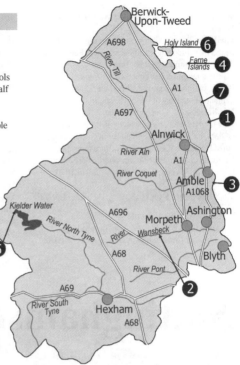

4. FARNE ISLANDS

The National Trust.
Location: NU 230 370. Access by boat from Seahouses Harbour. Access from A1.
Access: Apr, Aug-Sept: Inner Farne and Staple 10.30am-6pm (majority of boats land at Inner Farne). May-Jul: Staple Island 10.30am-1.30pm, Inner Farne: 1.30pm-5pm. Disabled access possible on Inner Farne, telephone Property Manager for details. Dogs allowed on boats – not on islands.
Facilities: Toilets on Inner Farne.
Public transport: Nearest rail stations at Alnmouth and Berwick.
Habitat: Maritime islands between 15-28 depending on tides.
Key birds: 18 species of seabirds/waders, four species of tern (including Roseate), 34,000-plus pairs of Puffin, Rock Pipit, Pied Wagtail, Starling, 1,200 Eider etc.
Contact: John Walton, 8 St Aidans, Seahouses, NorthumberlandNE68 7SR. 01665 720651.

5. KIELDER FOREST

Forest Enterprise.
Location: NY 632 934. Kielder Castle is situated at N end of Kielder Water, NW of Bellingham.

Access: Forest open all year. Toll charge on 12 mile forest drive. Centre opening limited in winter.
Facilities: Visitor centre, exhibition, toilets, shop, access for disabled, licenced café. Local facilities include Youth Hostel, camp site, pub and garage.
Public transport: Bus: 814,815,816 from Hexham and seasonal service 714 from Newcastle.
Habitat: Commercial woodland, mixed species
Key birds: *Spring/summer*: Goshawk, Chiffchaff, Willow Warbler, Redstart, Siskin. *Winter*: Crossbill. *Resident*: Jay, Dipper, Great Spotted Woodpecker, Tawny Owl, Song Thrush, Goldcrest.
Contact: Forest Enterprise, Eals Burn, Bellingham, Hexham, Northumberland, NE48 2HP, 01434 220242. e-mail: richard.gilchrist@forestry.gsi.gov.uk

6. LINDISFARNE NNR

English Nature (Northumbria Team).
Location: NU 090 430. Island access lies two miles E of A1 at Beal, eight miles S of Berwick-on-Tweed.
Access: Via causeway, note high tide times to avoid being stranded. Open all hours. Some restricted

access (refuges). Coach parking available on Holy Island.
Facilities: Toilets, visitor centre in village. Hide on island (new hide with disabled access at Fenham-le-Moor). Self-guided trail on island.
Public transport: Irregular bus service to Holy Island, mainly in summer. Bus route follows mainland boundary of site north-south.
Habitat: Dunes, sand and mudflats.
Key birds: *Passage and winter*: Wildfowl and waders, including pale-bellied Brent Goose, Long-tailed Duck and Whooper Swan. Rare migrants.
Contact: Phil Davey, Site Manager, Beal Station, Berwick-on-Tweed, TD15 2PB. 01289 381470.

7. NEWTON POOL NATURE RESERVE

National Trust (North East).
Location: NU 243 240. Follow signs to High Newton N of Embleton at the junction of the B1339

and B1340. In the village, follow signs to Low Newton. Just before the village is a car park which must be used as no public parking is available further on. A National Trust sign in the village square shows the way to the bird hide along the Craster footpath (about a five minute walk).
Access: Open all year. From May to mid-August parts of the beach may be cordoned off to avoid disturbing nesting birds.
Facilities: Car park, hide.
Public transport: None.
Habitat: Dunes, tidal flats, beach, freshwater pool with artificial islands, scrub.
Key birds: *Spring/summer*: Grasshopper Warbler, Water Rail, Little Grebe, Whitethroat, Yellow Wagtail, Whinchat, Stonechat, Corn Bunting, terns, gulls. *Winter*: Waders, gulls.
Contact: National Trust (North East), Scots' Gap, Morpeth, Northumberland, NE61 4EG, 01670 774691.

Nottinghamshire

1. ATTENBOROUGH GRAVEL PITS

Nottinghamshire Wildlife Trust.
Location: SK 523 343. On A6005, seven miles SW of Nottingham alongside River Trent. Signed from main road.
Access: Open at all times. Dogs on leads. Paths suitable for disabled access.
Facilities: Nature trail (leaflet from Notts WT), one hide (key £2.50 from Notts WT). Visitor centre due to open Spring 2005.
Public transport: Railway station at Attenborough, several buses pass close to reserve (Rainbow 525A from Nottingham every ten minutes).
Habitat: Disused, flooded gravel workings with associated marginal and wetland vegetation.
Key birds: *Spring/summer*: Breeding Common Tern (40-plus pairs), Reed Warbler, Black Tern regular (bred once). *Winter*: Wildfowl (Bittern has wintered for last two years), Grey Heron colony, adjacent Cormorant roost.
Contact: Nottinghamshire Wildlife Trust, The Old Ragged School, Brook Street, Nottingham NG1 1EA. 0115 958 8242. e-mail: nottswt@cix.co.uk www.wildlifetrust.org.uk/nottinghamshire

2. BESTHORPE NATURE RESERVE

Nottinghamshire Wildlife Trust.
Location: SK 817 640 and SK813 646 (access points). Take A1133 N of Newark. Turn into Trent Lane S of Besthorpe village, reserve entrances second turn on left and right turn at end of lane (at River Trent).
Access: Open access to two hides (one with disabled access from car park at present). Open access to SSSI meadows. Limited access to areas grazed with sheep. Dogs on leads.
Facilities: No toilets (pubs etc in Besthorpe village), two hides, paths, nature trail (northern part).
Public transport: Buses (numbers 22, 67, 68, 6, S7L) run by Marshalls, Lincs, Road Car and Travel Wright along A1133 to Besthorpe village (0.75 mile away). Tel: 0115 924 0000 or 01777 710550 for information.
Habitat: Gravel pit with islands, SSSI neutral grasslands, hedges, reedbed, etc.
Key birds: *Spring/summer:* Breeding Grey Heron, Cormorant, Little Ringed Plover, Kingfisher, Grasshopper Warbler. *Winter:* Large numbers of ducks (Pochard, Tufted Duck, Pintail, Wigeon) and Peregrine.

NATURE RESERVES - ENGLAND

Contact: Trust HQ, 0115 958 8242.
e-mail: nottswt@cix.co.uk
www.wildlifetrust.org.uk/nottinghamshire

3. COLWICK COUNTRY PARK

Nottingham City Council.
Location: SK 610 395. Off A612, three miles E of
Nottingham city centre.
Access: Open at all times, but no vehicle access after
dusk or before 7am.
Facilities: Nature trails. Sightings log book in
Fishing Lodge.
Public transport: Call park office for advice
Habitat: Lakes, pools, woodlands, grasslands, new
plantations, River Trent.
Key birds: *Summer*: Warblers, Hobby, Common
Tern (15 plus pairs). *Winter*: Wildfowl and gulls.
Passage migrants.
Contact: Head Ranger, The Fishing Lodge, Colwick
Country Park, River Road, Colwick, Nottingham
NG4 2DW. 0115 987 0785.
www.colwick2000.freeserve.co.uk

4. LOUND GRAVEL PITS

Tarmac, ARC, Nottinghamshire Wildlife Trust.
Location: SK 690 856. Two miles N of Retford off
A638 adjacent to Sutton and Lound villages.
Access: Open at all times. Use public rights of way
only (use OS Map Sheet No 120 - 1:50,000
Landranger Series).
Facilities: Public viewing platform/screen off
Chainbridge lane (overlooking Chainbridge NR
Scrape).
Public transport: Buses from Bawtry (Church
Street), Retford bus station and Worksop (Hardy
Street) on services 27/27A/83/83A/84 to Lound
Village crossroads (Chainbridge Lane).
Habitat: Working sand and gravel quarries, restored
gravel workings, large areas of recently designated
SSSI (Apr 2003), woodland, reedbeds, fishing
ponds, river valley, in-filled and disused fly ash
tanks, farmland, scrub, open water.
Key birds: Over 235 species recorded. *Summer*:
Gulls, terns, wildfowl and waders. Passage waders,
terns, passerines and raptors. *Winter*: Wildfowl,
gulls, raptors. Rarities have inc. Ring-billed Gull,
Caspian, White-winged Black and Whiskered Terns,
Manx Shearwater, Lesser Scaup, Green-winged and
Blue-winged Teal, Richard's Pipit, Baird's, Pectoral
and Buff-breasted Sandpiper, Long-billed Dowitcher,
Killdeer, Spoonbill, Bluethroat, Nightingale, Snow
Bunting, Shore Lark, Great Skua.

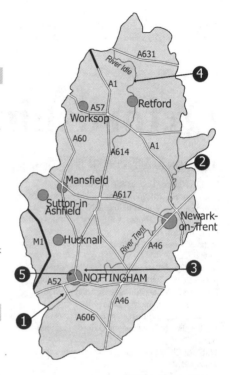

Contact: Lound Bird Club, Gary Hobson
(Secretary), 11 Sherwood Road, Harworth,
Doncaster, South Yorks DN11 8HY. 01302 743654.
e-mail: loundbirdclub@tiscali.co.uk

5. WOLLATON PARK

Wollaton Hall.
Location: Situated approx 5 miles W of Nottingham
City Centre.
Access: Open all year from dawn-dusk.
Facilities: Pay/display car parks. Some restricted
access (deer), leaflets.
Public transport: Trent Buses: no 22, and
Nottingham City Transport: no's 31, and 28 running
at about every 15 mins.
Habitat: Lake, small reedbed, woodland.
Key birds: *All year*: Main woodland species present,
with good numbers of Nuthatch, Treecreeper and all
three woodpeckers. *Summer*: Commoner warblers,
incl Reed Warbler, all four hirundine species, Spotted
Flycatcher. *Winter*: Pochard, Gadwall, Wigeon,
Ruddy Duck, Goosander, occasional Smew and

185

Goldeneye. Flocks of Siskin and Redpoll, often feeding by the lake, Redwing and occasional Fieldfare.

Contact: Wollaton Hall & Park, Wollaton, Nottingham NG8 2AE. 0115 915 3920..
e-mail: wollaton@ncmg.demon.co.uk

Oxfordshire

1. ASTON ROWANT NNR

English Nature (Thames & Chiltern Team).
Location: SU 731 966. From the M40 Lewknor interchange at J6 of M40, travel NE for a short distance and turn R onto A40. After 1.5 miles at the top of the hill, turn R and R again into a narrow, metalled lane. Drive to the end of this road to the car park.
Access: Open all year. Some wheelchair access, please contact site manager for more information.
Facilities: On-site parking, easy access path to viewpoint, seats, interpretation panels.
Public transport: Bus: stops near the reserve (Stokenchurch, Lewknor and Oxford to London bus).
Habitat: Chalk grassland, chalk scrub, beech woodland.
Key birds: *Spring/summer*: Blackcap, warblers, Turtle Dove, Tree Pipit. *Winter*: Possible Short-eared Owl, Brambling, Siskin, winter thrushes. *Passage*: Whinchat, Wheatear, Ring Ouzel. *All year*: Red Kite, Buzzard, Sparrowhawk, Woodcock, Little and Tawny Owls, Green and Great Spotted Woodpeckers, Sky Lark, Meadow Pipit, Marsh Tit.
Contact: English Nature (Thames & Chiltern Team), Aston Rowant Reserve Office, Aston Hill, Lewknor, Watlington OX9 5SG. 01844 351833.
www.english-nature.org.uk

2. BLENHEIM PARK

Blenheim Palace.
Location: SP 440 160. 9.5km from Oxford at Woodstock on A34. Various entrances available for cars and pedestrians.
Access: Open all year. (9am - 4.45pm, last entry). £2 entry fee, £1 children.
Facilities: Car parks, toilets, footpaths. Dogs on leads. Disabled toilet in Pleasure Gardens.
Public transport: None.
Habitat: Lakes, woodland, pastureland.
Key birds: *Spring/summer*: Spotted Flycatcher, Blackcap, Garden Warbler, migrating Garganey.

Winter: Wildfowl, possible Smew, Bittern, Water Rail. *All year*: Gadwall, Barn Owl, Tawny Owl, Little Owl, Kingfisher, Great Spotted and Green Woodpeckers, Jay, other usual woodland species.
Contact: Blenheim Palace, Woodstock, Oxford, 01993 811091

3. FOXHOLES RESERVE

Berks, Bucks & Oxon Wildlife Trust.
Location: SP 254 206. Head N out of Burford on A424 towards Stow-on-theWold. Take third turning on R. Head NE on unclassified road to Bruern for 3.5km. Just before reaching Bruern, turn L along track following Cocksmoor Copse. After 750m, park in car park on R just before farm buildings.
Access: Open all year. Please keep to the paths.
Facilities: Car park, footpaths. Can be very muddy in winter.
Public transport: None.
Habitat: River, woodland, wet meadow.
Key birds: *Spring/summer*: Nightingale, Yellow Wagtail, possible Redstart, Wood Warbler, Spotted Flycatcher. *Winter*: Redwing, Fieldfare, Woodcock. *All year*: Little Owl, all three woodpeckers, possible Hawfinch.
Contact: Trust HQ, 01865 775476.
e-mail: bbowt@cix.co.uk

4. OTMOOR NATURE RESERVE

RSPB (Central England Office).
Location: SP 570 126. Car park seven miles NE of Oxford city centre. From B4027, take turn to Horton-cum-Studley, then first left to Beckley. After 0.67 miles at the bottom of a short hill turn R (before the Abingdon Arms public house). After 200 yards, turn left into Otmoor Lane. Reserve car park is at the end of the lane (approx one mile).
Access: Open dawn-dusk. No permits or fees. No dogs allowed on the reserve visitor trail (except public rights of way). In wet conditions, the visitor route can be muddy and wellingtons are essential.
Facilities: Limited. Small car park with cycle racks,

visitor trail (3 mile round trip) and two screened viewpoints. The reserve is not accessible by coach and is unsuitable for large groups.
Public transport: None.
Habitat: Wet grassland, reedbed and open water.
Key birds: *Summer*: Breeding birds include Cetti's Warbler, Grasshopper Warbler, Lapwing, Redshank, Curlew, Snipe, Yellow Wagtail, Shoveler, Gadwall, Pochard, Tufted Duck, Little Grebe, Great Crested Grebe. Hobby breeds locally. *Winter*: Wigeon, Teal, Shoveler, Pintail, Gadwall, Pochard, Tufted Duck, Lapwing, Golden Plover, Hen Harrier, Peregrine, Merlin. *Autumn and spring passage*: Marsh Harrier, Short-eared Owl, Greenshank, Green Sandpiper, Common Sandpiper, Spotted Redshank and occasional Black Tern.
Contact: Neil Lambert, Site Manager, RSPB, c/o Folly Farm, Common Road, Bexley OX3 9YR. 01865 351163. www.rspb.org.uk

5. SHOTOVER COUNTRY PARK

Oxford City Council.
Location: SP 565 055. Approach from Wheatley or Old Road, Headington.
Access: Open all year, best early morning or late in the evening.
Facilities: Car park, toilets, nature trails, booklets.
Public transport: None.
Habitat: Woodland, farmland, heathland, grassland, scrub.
Key birds: *Spring/summer*: Willow Warbler, Blackcap, Garden Warbler, Spotted Flycatcher, Whitethroat, Lesser Whitethroat, Pied Flycatcher, Redstart, Tree Pipit. *Autumn*: Crossbill, Redpoll, Siskin, thrushes. *All year*: Sparrowhawk, Jay, tits, finches, woodpeckers, Corn Bunting.
Contact: Oxford City Council, PO Box 10, Oxford OX1 1EN. 01865 249811.

Shropshire

1. CHELMARSH RESERVOIR

South Staffordshire Water/Shropshire Wildlife Trust.
Location: SO 726 881.6km south of Bridgnorth, off the B4555. From Chelmarsh village head S towards Highley. Turn L at Sutton and L at the T-junction. From the car park, walk to the other end of the reservoir to the hides.
Access: Open all year.
Facilities: Car park.
Public transport: None.
Habitat: Reservoir, reedbed.
Key birds: *Winter*: Wildfowl, inc Pintail, Smew, geese, swans, Water Rail. *Spring/summer*: Reed and Sedge Warblers, Reed Bunting. *Passage*: Possible Osprey.
Contact: Shropshire Wildlife Trust, 193 Abbey Foregate, Shrewsbury SY2 6AH. 01743 284 280 Fax 01743 284 281.

2. CLUNTON COPPICE

Shropshire Wildlife Trust.
Location: SO 343 806. Take B4385 S from Bishop's Castle. After two miles take road to Brockton, Lower Down and Clunton. Park in Clunton village and walk S into the woodland.
Access: Open at all times. Access along road and public rights of way only.
Facilities: Limited parking in small quarry entrance on R, or opposite The Crown pub.
Public transport: Not known.
Habitat: Oak coppice. Good for ferns, mosses and fungi.
Key birds: Buzzard and Raven regular. *Spring/summer*: Wide range of woodland birds, inc. Redstart, Wood Warbler and Pied Flycatcher, Woodcock.
Contact: Trust HQ, 01743 284280.
e-mail: shropshirewt@cix.co.uk
www.shropshirewildlifetrust.org.uk

3. FENN'S WHIXALL AND BETTISFIELD MOSSES NNR

English Nature (North Mercia Team).
Location: The reserve is located four miles SW of Whitchurch, ten miles SW of Wrexham. It is to the S of the A495 between Fenn's bank, Whixall and Bettisfield. There is roadside parking at entrances, car parks at Morris's Bridge, Roundthorn Bridge, World's End and a large car park at Manor House. Disabled access by prior arrangement along the railway line.
Access: Permit required except on Mosses trail routes.

187

Shropshire

Facilities: There are panels at all of the main entrances to the site, and leaflets are available when permits are applied for. Three interlinking Mosses trails explore the NNR and canal from Morris's and Roundthorn bridges.
Public transport: Bus passes nearby. Railway two miles away.
Habitat: Peatland meres and mosses.
Key birds: *Spring/summer:* Nightjar, Hobby, Curlew, Tree Sparrow. *All year:* Sky Lark, Linnet. Water vole, brown hare. *Winter:* Short-eared Owl.
Contact: English Nature, Manor House, Mosshore, Wixhall, Shropshire SY13 2PD. 01948 880362.
e-mail: jean.daniels@english-nature.org.uk

4. WOOD LANE

Shropshire Wildlife Trust.
Location: SJ 421 331. Turn off A528 at Spurnhill, 1 mile SE of Ellesmere.
Access: Open at all times.
Facilities: Car parks clearly signed. Hides (access by permit).
Public transport: None.
Habitat: Gravel pit.
Key birds: *Summer:* Breeding Sand Martins. Popular staging post for waders (inc. Redshank, Greenshank, Ruff, Dunlin, Little Stint, Green and Wood Sandpiper). *Winter:* Lapwing and Curlew.
Contact: Trust HQ, 01743 284280.
e-mail: shropshirewt@cix.co.uk

Somerset

1. BRANDON HILL NATURE PARK

Bristol City Council/Avon Wildlife Trust.
Location: 578 728. The reserve is in the centre of Bristol, in the SW corner of Brandon Hill Park which overlooks Jacobs Wells Road. Metered parking available in the nearby roads - Great George Street, Berkeley Square, NCP car park on Jacobs Wells Road.
Access: Open all year. Wheelchair access from Great George Street and Berkeley Square only.
Facilities: Woodland walk, butterfly garden, picnic area.
Public transport: Travel line, 0870 6082608.

Habitat: Wildflower meadow, woodland.
Key birds: *Spring/summer:* Blackcap, other warblers. *All year:* Jay, Bullfinch, usual woodland species.
Contact: Trust HQ, 0117 917 7270; fax 0117 929 7273. e-mail: mail@avonwildlifetrust.org.uk www.avonwildlifetrust.org.uk

2. BREAN DOWN

National Trust (North Somerset).
Location: ST 290 590. 182 map. Five miles N of Burnham-on-Sea. J22 of M5, head for Weston-super-Mare on A370 and then head for Brean at Lympsham.

Access: Open all year. Dogs on lead. Steep slope – not suitable for wheelchair-users.
Facilities: Toilets one mile before property, not NT.
Public transport: Call Tourist Information Centre for details 01934 888800 (different in winter/summer).
Habitat: Limestone and neutral grassland, scrub and steep cliffs.
Key birds: *Summer*: Blackcap, Garden Warbler, Whitethroat, Stonechat. *Winter*: Curlew, Shelduck, Dunlin on mudflat. Migrants.
Contact: The National Trust, Barton Rocks, Barton, Winscombe, North Somerset BS25 1DU. 01934 844518.
e-mail: northsomersetstaff@nationaltrust.org.uk

3. BRIDGWATER BAY NNR

English Nature (Somerset & Gloucestershire Team).
Location: ST 270 470. Nine miles N of Bridgwater. Take J23 or 24 off M5. Turn N off A39 at Cannington.
Access: Hides open every day except Christmas Day. Permits needed for Steart Island (by boat only). Dogs on leads – grazing animals/nesting birds. Disabled access to hides by arrangement, other areas accessible.
Facilities: Car park at Steart. Footpath approx 0.5 miles to tower and hides.
Public transport: None.
Habitat: Estuary, intertidal mudflats, saltmarsh.
Key birds: *Winter:* Wildfowl and waders, birds of prey. *Spring/autumn:* Passage migrants.
Contact: Robin Prowse, Dowells Farm, Steart, Bridgwater, Somerset TA5 2PX. 01278 652426. www.english-nature.org.uk

4. CHEW VALLEY LAKE

Avon Wildlife Trust, Bristol Water Plc.
Location: ST 570 600. Reservoir (partly a Trust reserve) between Chew Stoke and West Harptree, crossed by A368 and B3114, nine miles S of Bristol.
Access: Permit for access to hides (five at Chew, two at Blagdon). Best roadside viewing from causeways at Herriott's Bridge (nature reserve) and Herons Green Bay. Day, half-year and year permits from Bristol Water, Recreation Department, Woodford Lodge, Chew Stoke, Bristol BS18 8SH. Tel/fax 01275 332339.
Facilities: Hides.
Public transport: Travel line, 0870 6082608.
Habitat: Reservoir.
Key birds: *Autumn/winter:* Concentrations of wildfowl (inc. Bewick's Swan, Goldeneye, Smew,

Ruddy Duck), gull roost (inc. regular Mediterranean, occasional Ring-billed). Migrant waders and terns (inc. Black). Recent rarities inc. Blue-winged Teal, Spoonbill, Alpine Swift, Citrine Wagtail, Little Bunting, Ring-necked Duck, Kumlien's Gull.
Contact: Trust HQ. 0117 917 7270; fax 0117 929 7273. e-mail: mail@avonwildlifetrust.org.uk www.avonwildlifetrust.org.uk

5. HORNER WOOD NATURE RESERVE

Somerset Wildlife Trust.
Location: From Minehead, take A39 W to a minor road 0.8km E of Porlock signed to Horner. Park in village car park.
Access: Open all year.
Facilities: None.
Public transport: Bus: Porlock.
Habitat: Oak woodland, moorland.
Key birds: *Spring/summer*: Wood Warbler, Pied Flycatcher, Redstart, Stonechat, Whinchat, Tree Pipit, Dartford Warbler possible. *All year*: Dipper, Grey Wagtail, woodpeckers, Buzzard, Sparrowhawk.
Contact: Somerset Wildlife Trust, Fyne Court, Broomfield, Bridgewater, Somerset TA5 2EQ. 01823 451587. e-mail: somwt@cix.co.uk www.wildlifetrust.org.uk/somerset

6. LEIGH WOODS

National Trust (Bristol).
Location: ST 560 736. Two miles W from centre of Bristol. Pedestrian access from North Road, Leigh Woods or via Forestry Commission car park at end of Coronation Avenue, Abbots Leigh. Access to both roads is from A369 which goes from Bristol to J19 of M5.
Access: Open all year. A good network of paths around the plateau. The paths down to the towpath are steep and uneven.
Facilities: Two waymarked trails from Forestry Commission car park: purple trail (1.75 miles) on level ground, red trail (2.5 miles) more undulating.
Public transport: Bristol-Portishead. Bus service (358/658 and 359/659) goes along A369. Leaves Bristol generally at 20 and 50 minutes past the hour. More details from First Badgerline. Tel: 0117 955 3231.
Habitat: Ancient woodland, former wood pasture, two grassland areas, calcareous grassland and scree by towpath.
Key birds: *Summer:* Peregrine, Blackcap, Chiffchaff, Spotted Flycatcher. *Winter:* Great Spotted and Green Woodpeckers, Song Thrush, Long-tailed, Marsh and common tits.

Contact: Bill Morris, Reserve Office, Valley Road, Leigh Woods, Bristol, 01936 429336.
e-mail: wlwbgm@smtp.ntrust.org.uk

7. SHAPWICK HEATH NNR

English Nature (Somerset & Gloucester Team).
Location: Situated between Shapwick and Westhay, near Glastonbury. The nearest car park to the site is at the Willows garden centre, in Westhay.
Access: Open all year. Disabled access to displays, hides.
Facilities: Network of paths, hides. Toilets, leaflets and refreshments available at the garden centre.
Public transport: None.
Habitat: Traditionally managed herb-rich grassland, ferny wet woodland, fen, scrub, ditches, open water, reedswamp and reedbed.
Key birds: *All year*: Ducks, waders.
Contact: English Nature (Somerset and Gloucestershire Team), Roughmoor, Bishop's Hull, Taunton, Somerset, TA1 5AA.01823 283211.
e-mail: somerset@english-nature.org.uk

8. WALBOROUGH

Avon Wildlife Trust.
Location: ST 315 579. On S edge of Weston-super-Mare at mouth of River Axe.
Access: Access from Uphill boatyard. Special access trail suitable for less able visitors.
Facilities: None.
Public transport: Travel line, 0870 6082608.
Habitat: Limestone grassland, scrub, saltmarsh, estuary.
Key birds: The Axe Estuary holds good numbers of migrant and wintering wildfowl (inc. Teal, Shelduck) and waders (inc. Black-tailed Godwit, Lapwing, Golden Plover, Dunlin, Redshank). Other migrants inc. Little Stint, Curlew Sandpiper, Ruff. Little Egret occurs each year, mostly in late summer.
Contact: Trust HQ, 0117 917 7270.
e-mail: mail@avonwildlifetrust.org.uk

9. WEST SEDGEMOOR

RSPB (South West England Office).
Location: ST 361 238. Entrance down by-road off A378 Taunton-Langport road, one mile E of

Fivehead.
Access: Access at all times to woodland car park and heronry hide. Heronry hide and part of the woodland trail are wheelchair accessible.
Facilities: Heronry hide, two nature trails: Woodland Trail and Scarp trail, disabled parking area.
Public transport: Bus from Taunton to Fivehead. Bus company telephone no: First Southern National Ltd 01823 272033.
Habitat: Semi-natural ancient oak woodland and wet grassland. Part of the Somerset Levels and Moors.
Key birds: *Spring/summer*: Breeding Grey Heron, Curlew, Lapwing, Redshank, Snipe, Buzzard, Sedge Warbler, Yellow Wagtail, Sky Lark, Nightingale. Passage Whimbrel and Hobby. *Winter*: Large flocks of waders and wildfowl (including Lapwing, Golden Plover, Shoveler, Pintail, Teal and Wigeon).
Contact: Site Manager, Dewlands Farm, Redhill, Curry Rivel, Langport, Somerset TA10 0PH. 01458 252805, fax 01458 252184.
e-mail: sally.brown@rspb.org.uk
www.rspb.org.uk

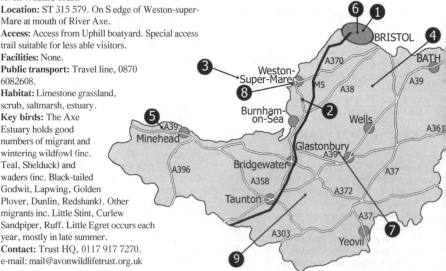

Somerset

Staffordshire

1. BELVIDE RESERVOIR

British Waterways Board and West Midland Bird Club.
Location: SJ865102. Near Brewood, 7 miles NW of Wolverhampton.
Access: Access only by permit from the West Midland Bird Club.
Facilities: Hides.
Public transport: Not known.
Habitat: Canal feeder reservoir with marshy margins and gravel islands.
Key birds: Important breeding, moulting and wintering ground for wildfowl, including Ruddy Duck, Goldeneye and Goosander), passage terns and waders. Night roost for gulls.
Contact: Miss M Surman, 6 Lloyd Square, 12 Nial Close, Edgbaston, Birmingham B15 3LX.
www.westmidlandbirdclub.com/belvide

2. BLACK BROOK

Staffordshire Wildlife Trust.
Location: SK 020 645. N of Leek, W of A53. West of road from Royal Cottage to Gib Tor.
Access: Access is via public footpath from Gib Tor to Newstone Farm; this first passes through a conifer plantation which is outside the reserve.
Facilities: None.
Public transport: None.
Habitat: Heather and bilberry moorland, and upland acidic grassland.
Key birds: Merlin, Kestrel, Red Grouse, Golden Plover, Snipe, Curlew, Dipper, Wheatear, Whinchat, Twite.
Contact: Trust HQ, 01889 880100. e-mail: staffswt@cix.co.uk www.staffs-wildlife.org.uk

3. BLITHFIELD RESERVOIR

South Staffs Waterworks Co.
Location: SK 058 237. View from causeway on B5013 (Rugeley/Uttoxeter).
Access: Access to reservoir and hides by permit from West Midland Bird Club.
Facilities: None.
Public transport: None.
Habitat: Large reservoir.
Key birds: *Winter*: Good populations of wildfowl (inc. Bewick's Swan, Goosander, Goldeneye, Ruddy Duck), large gull roost (can inc. Glaucous, Iceland). Passage terns (Common, Arctic, Black) and waders, esp. in autumn (Little Stint, Curlew Sandpiper, Spotted Redshank regular).
Contact: Miss M Surman, 6 Lloyd Square, 12 Niall Close, Edgbaston, Birmingham B15 3LX.

4. BRANSTON WATER PARK

East Staffordshire Borough Council.
Location: SK 217 207. Follow the brown tourist sign from the A38 N. No access from the A38 S - head to the Barton-under-Needwood exit and return N. The park is 0.5 miles S of the A5121 Burton-upon-Trent exit.
Access: Open all year, flat wheelchair-accessible stone path all round the lake.
Facilities: Disabled toilets, picnic area (some wheelchair accessible tables), modern children's play area.
Public transport: Contact ESBC Tourist information 01283 508111.
Habitat: Reedbed, willow carr woodland, scrub, meadow area.
Key birds: *Spring/summer*: Reed Warbler, Cuckoo, Reed Bunting. Important roost for Swallow and Sand Martin. *Winter*: Waders, Little Ringed Plover occasionally, Pied Wagtail roost.
Contact: East Staffordshire Borough Council, Midland Grain Warehouse, Derby Street, Burton-on-Trent, Staffordshire DE14 2JJ. 01283 508657; (Fax) 01283 508571

5. BROWN END QUARRY

Staffordshire Wildlife Trust/North Staffordshire Group of the Geologists' Association.
Location: SK 090 502. Reserve is at E end of Waterhouses on the A523 Leek-Ashbourne road. The Quarry is on the N side of the road just W of the Manifold Cycle Track.
Access: Open all year. Access to the parking area is over a bridge shared with a cycle hire company.
Facilities: Interpretative trail.
Public transport: None.
Habitat: Scrub, limestone.
Key birds: *Spring/summer*: Warblers.
Contact: Trust HQ, 01889 880100.
e-mail: staffswt@cix.co.uk
www.staffs-wildlife.org.uk

6. BURNT WOOD

Staffordshire Wildlife Trust.
Location: SJ 736 355. From Newcastle on A53 to
Market Drayton. When the main road crosses B5026
turn into Kestrel Drive and then Pheasant Drive.
Access also possible from B5026 Eccleshall Road
0.3 miles from A53. Parking is difficult for this
reserve.
Access: Open all year.
Facilities: None.
Public transport: None.
Habitat: Ancient oak woodland, pond.
Key birds: *All year*: Goshawk, Raven, Woodcock,
all three woodpeckers. Good for butterflies and
moths, adder, grass snake, slow worm and common
lizard.
Contact: Trust HQ, 01889 880100.
e-mail: staffswt@cix.co.uk
www.staffs-wildlife.org.uk

7. COOMBES VALLEY

RSPB (North West England Office).
Location: SK 005 530. Four miles from Leek along
A523 between Leek and Ashbourne and 0.5 miles
down unclassified road – signposted.
Access: No dogs allowed. Most of the trails are
unsuitable for disabled. Open daily – no charge.
Coach groups welcome by prior arrangement.
Facilities: Visitor centre, toilets, two miles of nature
trail, one hide.
Public transport: None.
Habitat: Sessile oak woodland, unimproved pasture
and meadow.
Key birds: *Spring:* Pied Flycatcher, Redstart, Wood
Warbler. *Jan-Mar*: Displaying birds of prey.
Contact: Nick Chambers, Six Oaks Farm, Bradnop,
Leek, Staffs ST13 7EU. 01538 384017.
www.rspb.org.uk

8. CROXALL LAKES

Staffordshire Wildlife Trust.
Location: SK 188 139. From Rugeley follow the
A513 passing through Kings Bromley and Alrewas.
Continue along this road passing over the A38. After
approx one mile, you will pass over the River Tame,
which forms the W boundary of the reserve. The
entrance to the reserve is the second track on the L.
Access: Open all year. Gravel track to the N of the
reserve.
Facilities: Interpretation boards, two bird hides,
parking and access to be improved.

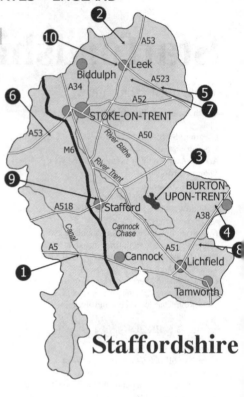

Public transport: None.
Habitat: Two lakes, wader scrapes, pools,
floodplain grassland.
Key birds: *Winter*: Teal, Wigeon, Smew. *Spring/
summer*: Breeding Lapwing, Redshank, Little
Ringed Plover, Oystercatcher.
Contact: Trust HQ, 01889 880100.
e-mail: staffswt@cix.co.uk

9. DOXEY MARSHES

Staffordshire Wildlife Trust.
Location: SJ 903 250. In Stafford. Parking 0.25
miles off M6 J14/A5013 or walk from town centre.
Access: Open at all times. Dogs on leads. Disabled
access being improved. Coach parking off Wooton
Drive.
Facilities: One hide, three viewing platforms, two are
accessible to wheelchairs.
Public transport: Walk from town centre via
Sainsbury's.
Habitat: Marsh, pools, reedbeds, hedgerows, reed
sweet-grass swamp.

Key birds: *Spring/summer*: Breeding Snipe,
Lapwing, Redshank, warblers, buntings, Sky Lark,
Water Rail. *Winter*: Snipe, wildfowl, thrushes,
Short-eared Owl. Passage waders, vagrants.
Contact: Trust HQ, 01889 880100.
e-mail: staffswt@cix.co.uk

10. LONGSDON WOODS

Staffordshire Moorlands District Council
Location: SK 965 555. Part of Ladderedge Country
Park. Reserve lies off A53 Leek road. Can be
approached from Ladderedge near Leek; City Lane,
Longsdon; or Rudyard Station.

Access: Public rights of way only.
Facilities: None.
Public transport: None.
Habitat: Woodland and wet grassland.
Key birds: Heronry, Sparrowhawk, Curlew, Snipe,
Jack Snipe, Woodcock, Little and Tawny Owls, all
three woodpeckers, Redstart, Blackcap, Garden
Warbler.
Contact: Countryside Service, Staffordshire
Moorlands District Council, Moorlands House,
Stockwell Street, Leek, Staffordshire ST13
6HQ.01538 483577.
e-mail: countryside@staffsmoorlands.gov.uk

Suffolk

1. BONNY WOOD

Suffolk Wildlife Trust.
Location: TM 076 520. The reserve is about 0.5
miles from Barking Tye and is an SSSI. From
Needham Market take the B1078 to Barking. Park at
the village hall.
Access: Open all year. Dogs on leads at all times.
Facilities: Nature trail.
Public transport: None.
Habitat: Ancient, semi-natural woodland.
Key birds: *Spring/summer*: Nightingale, Blackcap,
Willow Warbler, Woodcock. *All year*: Treecreeper,
Tawny Owl, usual woodland birds.
Contact: Suffolk Wildlife Trust, Brooke House,
The Green, Ashbocking, Ipswich IP6 9JY. 01473
890089. e-mail: info@suffolkwildlife.cix.co.uk
www.wildlifetrust.org.uk/suffolk

2. CARLTON MARSHES

Suffolk Wildlife Trust.
Location: TM 508 920. SW of Lowestoft, at W
end of Oulton Broad. Take A146 towards Beccles
and turn R after Tesco garage.
Access: Open during daylight hours. Keep to
marked paths. Dogs only allowed in some areas, on
leads at all times. Car park suitable for coaches.
Facilities: Information centre and shop. Snacks
available.
Public transport: Bus and train in walking distance.

Habitat: 100 acres of grazing marsh, peat pools and
fen.
Key birds: Wide range of wetland and Broadland
birds, including Marsh Harrier.
Contact: Catriona Finlayson, Suffolk Broads
Wildlife Centre, Carlton Colville, Lowestoft, Suffolk
NR33 8HU. 01502 564250.
www.wildlifetrust.org.uk/suffolk

3. CASTLE MARSHES

Suffolk Wildlife Trust.
Location: TM 471 904. Head E on the A146 from
Beccles to Lowestoft. Take the first L turn after
Three Horseshoes pub. Continue on the minor road
which bends round to the R. Carry straight on – the
road bends to the R again. The car park is on the L
just after White Gables house.
Access: Public right of way. Unsuitable for
wheelchairs. Stiles where path leaves the reserve.
Unmanned level crossing is gated.
Facilities: None.
Public transport: Bus: nearest bus route is on the
A146 Lowestoft to Beccles road. Tel: 0845 958
3358. Train: Beccles and Oulton Broad South on the
Ipswich to Lowestoft line.
Habitat: Grazing marshes, riverbank.
Key birds: *Spring/summer*: Marsh Harrier, Cetti's
Warbler, occasional Grasshopper Warbler. *Winter*:
Hen Harrier, wildfowl, Snipe.
Contact: Trust HQ, 01473 890089.
e-mail: info@suffolkwildlife.cix.co.uk

4. DINGLE MARSHES, DUNWICH

Suffolk Wildlife Trust/RSPB.
Location: TM 48 07 20. Eight miles from
Saxmundham. Follow brown signs from A12 to
Minsmere and continue to Dunwich. Forest carpark
(hide) – TM 46 77 10. Beach carpark – TM 479
707. The reserve forms part of the Suffolk Coast
NNR.
Access: Open at all times. Access via public rights of
way and permissive path along beach. Dogs on lead
please. Coaches can park on beach car park.
Facilities: Toilets at beach car park, Dunwich. Hide
in Dunwich Forest overlooking reedbed, accessed via
Forest car park. Circular trail waymarked from
carpark.
Public transport: Via Coastlink, Dial a ride service
to Dingle 01728 833546 links to buses and trains.
Habitat: Grazing marsh, reedbed, shingle beach and
saline lagoons
Key birds: *All year:* In reedbed, Bittern, Marsh
Harrier, Bearded Tit. *Winter:* Hen Harrier, White-
fronted Goose, Wigeon, Snipe, Teal on grazing
marsh. *Summer:* Lapwing, Avocet, Snipe, Black-
tailed Godwit, Hobby. Good for passage waders.
Contact: Alan Miller, Suffolk Wildlife Trust, 9
Valley Terrace, Valley Road, Leiston, Suffolk IP16
4AP. 01728 833405.
e-mail: alanm@suffolkwildlife.cix.co.uk

5. HAVERGATE ISLAND

RSPB (East Anglia Office).
Location: TM 425 496. Part of the Orfordness
NNR at the mouth of the River Alde.
Orford is 17km NE
Woodbridge, signposted
off the A12.
Access: Open Apr-Aug
(1st & 3rd weekends and
every Thu), Sep-Mar
(1st Sat every month).
Book in advance
through Minsmere
RSPB visitor centre,
tel 01728 648281.
Park in Orford at the
large pay and display
car park next to the
quay.

Facilities: Toilets, picnic area, five birdwatching
hides, viewing platform, visitor trail (approx 2km).
Public transport: Boat trips from Orford (one mile)
Habitat: Shallow brackish water, lagoons with
islands, saltmarsh, shingle beaches.
Key birds: *Summer:* Breeding Arctic, Common and
Sandwich Terns, migrants. Leading site for Avocet.
Winter: Wildfowl and waders.
Contact: Ian Paradine, Manager, RSPB, c/o Friends
Garage, Front Street, Orford, Suffolk IP12 2LD.
01394 450732.

6. HAZELWOOD MARSHES

Suffolk Wildlife Trust.
Location: TM 435 575. Four miles W of
Aldeburgh. Small car park on A1094. Mile walk
down sandy track.
Access: Open dawn to dusk.
Facilities: Hide.
Public transport: Bus, nearest route on A1094.
Habitat: Estuary, marshes.
Key birds: Marshland and estuary birds; spring and
autumn migrants.
Contact: Trust HQ, 01473 890089.
e-mail: info@suffolkwildilfe.cix.co.uk

7. LACKFORD LAKES

Suffolk Wildlife Trust.
Location: TL 803 708. Via track off N side of
A1101 (Bury St Edmunds to Mildenhall

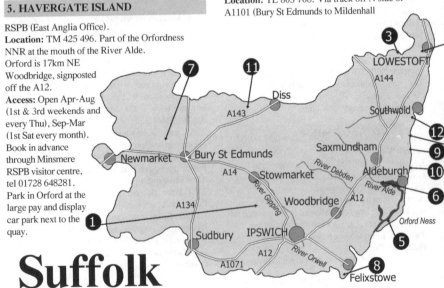

Suffolk

NATURE RESERVES - ENGLAND

road), between Lackford and Flempton. Five miles from Bury.

Access: New reserve centre open winter (10am-4pm), summer (10am-5pm) Wed to Sun, (closed Mon and Tues). Tea and coffee facilities, toilets.

Facilities: New visitor centre with viewing area upstairs. Tea and coffee facilities, toilets. Eight hides. Coaches should pre-book.

Public transport: None.

Habitat: Restored gravel pit with open water, lagoons, islands, willow scrub.

Key birds: *Winter*: Bittern, Water Rail, Bearded Tit. Large gull roost. Wide range of waders and wildfowl (inc. Goosander, Pochard, Tufted Duck, Shoveler. *Spring/autumn*: Migrants, inc. raptors. Breeding Shelduck, Little Ringed Plover and reed warblers.

Contact: Joe Davis, Lackford Lakes Visitor Centre, Lackford Lakes, Lackford, Bury St Edmunds, Suffolk IP28 6HX. 01284 728706.
e-mail: suffolkwildlife@cix.co.uk

8. LANDGUARD BIRD OBSERVATORY

Location: TM 283 317. Road S of Felixstowe to Landguard Nature Reserve and Fort.

Access: Visiting by appointment.

Facilities: Migration watch point and ringing station.

Public transport: Call for advice.

Habitat: Close grazed turf, raised banks with holm oak, tamarisk, etc.

Key birds: Unusual species and common migrants. Seabirds.

Contact: Paul Holmes, Landguard Bird Observatory, View Point Road, Felixstowe, Suffolk IP11 8TW. Ms J Cawston 01473 748463.

9. MINSMERE

RSPB (Eastern England Regional Office)

Location: TM 452 680. Six miles NE of Saxmundham. From A12 head for Westleton, N of Yoxford. Access from Westleton (follow the brown tourist signs).

Access: Open every day, except Tues, Christmas Day and Boxing Day (9am-9pm or dusk if earlier). Visitor centre open 9am-5pm (9am-4pm Nov-Jan). Tea-room 10.30am-4.30pm (10am-4pm Nov-Jan). Free to RSPB members, otherwise £5 adults, £1.50 children, £3 concession. Max two coaches per day (not Bank Holiday weekends).

Facilities: Toilets, visitor centre, hides, nature trails, family activity packs.

Public transport: Train to Saxmundham then taxi.

Habitat: Woodland, wetland – reedbed and grazing marsh, heathland, dunes and beach, farmland – arable

conversion to heath, coastal lagoons, 'the scrape'.

Key birds: *Summer*: Avocet, Bittern, Marsh Harrier, Bearded Tit, Redstart, Nightingale, Nightjar. *Winter*: Wigeon, White-fronted Goose, Bewick's Swan. *Autumn/spring*: Passage migrants, waders etc.

Contact: RSPB Minsmere Nature Reserve, Saxmundham, Suffolk IP17 3BY. 01728 648281.
e-mail: minsmere@rspb.org.uk
www.rspb.org.uk

10. NORTH WARREN & ALDRINGHAM WALKS

RSPB (East Anglia Office).

Location: TM 468 575. Directly N of Aldeburgh on Suffolk coast. Use signposted main car park on beach.

Access: Open at all times. Please keep dogs under close control. Beach area suitable for disabled. Three spaces at Thorpeness Beach Pay&Display Car Park. On a first come first serves basis only.

Facilities: Three nature trails, leaflet available from TIC Aldeburgh or Minsmere RSPB. Toilets in Aldeburgh and Thorpeness.

Public transport: Bus service to Aldeburgh. First Eastern Counties (08456 020121).

Habitat: Grazing marsh, lowland heath, reedbed, woodland.

Key birds: *Winter*: White-fronted Goose, Tundra Bean Goose, Wigeon, Shoveler, Teal, Gadwall, Pintail, Snow Bunting. *Spring/summer*: Breeding Bittern, Marsh Harrier, Hobby, Nightjar, Wood Lark, Nightingale, Dartford Warbler.

Contact: Dave Thurlow, 1 Ness House Cottages, Sizewell, Leiston, Suffolk IP16 4UB. 01728 832719.
e-mail: dave.thurlow@rspb.org.uk

11. REDGRAVE AND LOPHAM FENS

Suffolk Wildlife Trust.

Location: TM 05 07 97. Five miles from Diss, signposted and easily accessed from A1066 and A143 roads.

Access: Open all year, dogs strictly on leads only. Visitor centre open all year at weekends & sometimes during the week in school holidays, call for details on 01379 688333

Facilities: Visitor centre with coffee shop, toilets, including disabled toilets, car park with coach space, wheelchair access to Visitor Centre and viewing platform/short boardwalk. Other general circular trails (no wheelchair access).

Public transport: Buses and trains to Diss town – Simonds coaches to local villages of Redgrave and South Lopham from Diss.

195

Habitat: Calcareous fen, wet acid heath, scrub and woodland
Key birds: *All year*: Water Rail, Snipe, Teal, Woodcock, Sparrowhawk, Kestrel, Great Spotted, Lesser Spotted and Green Woodpeckers, Tawny and Little Owls, Shelduck. *Summer:* Reed, Sedge and Grasshopper Warblers, other leaf and *Sylvia* warblers, Hobby plus large Swallow and Starling roosts. *Winter/ occasionals on passage*: Bearded Tit, Marsh Harrier, Greenshank, Green Sandpiper, Shoveler, Gadwall, Pintail, Garganey, Jack Snipe, Bittern.
Contact: Andrew Excell, Redgrave and Lopham Fens, Low Common Road, South Lopham, Diss, Norfolk IP22 2HX. 01379 687618.
e-mail: redgrave@suffolkwildlife.cix.co.uk
www.wildlifetrust.org.uk/suffolk

12. WALBERSWICK

English Nature (Suffolk Team).
Location: TM 475 733. Good views from B1387 and from lane running W from Walberswick towards Westwood Lodge; elsewhere keep to public footpaths or shingle beach.
Access: Parties and coach parking by prior arrangement.
Facilities: Hide on S side of Blyth estuary, E of A12.
Public transport: Call for advice.
Habitat: Tidal estuary, fen, freshwater marsh and reedbeds, heath, mixed woodland, carr.
Key birds: *Spring/summer*: Marsh Harrier, Bearded Tit, Water Rail, Bittern, Nightjar. *Passage/winter*: Wildfowl, waders and raptors.
Contact: Adam Burrows, English Nature, Regent House, 110 Northgate Street, Bury St Edmunds IP33 1HP. 01502 676171.

Surrey

1. FRENSHAM COMMON

Waverley BC and National Trust .
Location: SU 855 405. Common lies on either side of A287 between Farnham and Hindhead.
Access: Open at all times. Car park (locked 9pm-9am). Keep to paths.
Facilities: Information rooms, toilets and refreshment kiosk at Great Pond.
Public transport: Call Trust for advice.
Habitat: Dry and humid heath, woodland, two large ponds, reedbeds.
Key birds: *Summer*: Dartford Warbler, Wood Lark, Hobby, Nightjar, Stonechat. *Winter*: Wildfowl (inc. occasional Smew), Bittern, Great Grey Shrike.
Contact: Mike Coates, Rangers Office, Bacon Lane, Churt, Surrey GU10 2QB. 01252 792416.

2. LIGHTWATER COUNTRY PARK

Surrey County Council.
Location: SU 921 622. From J3 of M3, take the A322 and follow brown Country Park signs. From the Guildford Road in Lightwater, turn into The Avenue. Entrance to the park is at the bottom of the road.
Access: Open all year dawn-dusk.
Facilities: Car park, visitor centre open most days during summer, toilets, leaflets.
Public transport: Train: Bagshot two miles. Tel 08457 484950. Bus: No 34 from Woking, Guildford and Camberley. Tel: 08706 082608.
Habitat: Reclaimed gravel quarries. Heath, woodland, bog.
Key birds: *Summer*: Dartford Warbler, Stonechat, Wood Lark, Tree Pipit, Hobby, Nightjar, all three woodpeckers. *Autumn*: Ring Ouzel, Crossbill, Siskin, Fieldfare, Redwing, possible Woodcock.
Contact: Surrey County Council, County Hall, Penrhyn Road, Kingston-upon-Thames, Surrey KT1 2DN. 08456 009 009. www.surreycc.gov.uk

3. RIVERSIDE PARK, GUILDFORD

Guildford Borough Council.
Location: TQ 005515 (Guildford BC). From car park at Bowers Lane, Burpham (TQ 011 527). Three miles from town centre.
Access: Open at all times. Follow marked paths. Access to far side of lake and marshland area via boardwalk.
Facilities: Boardwalk.
Public transport: Guildford town centre to Burpham (Sainsburys) No 36 Bus (Arriva timetable information. Tel 0870 608 2608).
Habitat: Wetland, lake, meadow, woodland.
Key birds: *Summer*: Sedge, Reed and Garden Warblers, Common Tern, Lesser Whitethroat, Hobby. *Winter*: Jack Snipe, Chiffchaff, Water Rail. *Passage*: Common Sandpiper, Whinchat, Water Pipit (up to 12 most years).
Contact: Parks Helpdesk, Guildford Borough Council, Millmead House, Millmead, Guildford, Surrey GU2 4BB. 01483 444715.
e-mail: parks@guildford.gov.uk
www.guildford.gov.uk

4. THURSLEY COMMON

English Nature (Sussex & Surrey Team).
Location: SU 900 417. From Guildford, take A3 SW to B3001 (Elstead/Churt road). Use the Moat car park.
Access: Open access. Parties must obtain prior permission.
Facilities: Boardwalk in wetter areas.
Public transport: None.
Habitat: Wet and dry heathland, woodland, bog.
Key birds: *Winter*: Hen Harrier and Great Grey Shrike. *Summer*: Hobby, Wood Lark, Dartford Warbler, Stonechat, Curlew, Snipe, Nightjar.
Contact: Simon Nobes, English Nature, Uplands Stud, Brook, Godalming, Surrey GU8 5LA. 01428 685878.

Sussex, East

1. BEWL WATER

Sussex Wildlife Trust/Southern Water.
Location: TQ 674 320. On B2099 coming from Tilehurst to Wadhurst.
Access: Open all year.
Facilities: Car park, hide. Footpath round reservoir. Southern Water have a Visitor Centre on the N side but the reserve cannot be viewed from there. Call 01892 890661.
Public transport: None.
Habitat: Reservoir, woodland, plantation.
Key birds: *Spring/summer*: Chiffchaff, other warblers, terns. *Passage*: Osprey, Common Sandpiper, Green Sandpiper, Greenshank, stints. *All year*: Pochard, Wigeon, Teal, Gadwall.
Contact: Sussex Wildlife Trust, Woods Mill, Shoreham Road, Henfield, West Sussex, BN5 9SD, 01273 492630. e-mail: enquiries@sussexwt.co.uk www.sussexwt.co.uk

2. FORE WOOD

RSPB (South East England Office).
Location: TQ 758 123. From the A2100 (Battle to Hastings) take lane to Crowhurst at Crowhurst Park

Caravan Park. Park at Crowhurst village hall and walk back up Forwood Lane for 0.5 miles. Entrance to reserve on left at top of hill.
Access: Open all year. Closed Christmas Day. No disabled facilities. No dogs. No coaches.
Facilities: Two nature trails.
Public transport: Station at Crowhurst, about 0.5 mile walk. Charing Cross/Hastings line. No buses within one mile.
Habitat: Semi-natural ancient woodland.
Key birds: Three woodpecker species, Nuthatch, Treecreeper, Sparrowhawk, Marsh Tit. *Spring/Summer:* Blackcap, Nightingale, Spotted Flycatcher.
Contact: Martin Allison, 12 The Grove, Crowborough, East Sussex TN6 1NY. 01273 775333 (South East Regional Office).
e-mail: martin.allison@rspb.org.uk

3. PEVENSEY LEVELS

English Nature (Sussex & Surrey Team).
Location: TQ 665 054. NE of Eastbourne. S of A259, one mile along minor road from Pevensey E to Norman's Bay.
Access: Good views from road.
Facilities: None.
Public transport: Call for advice.
Habitat: Freshwater grazing marsh, subject to light flooding after rains.
Key birds: *Summer:* Breeding Reed and Sedge Warblers, Yellow Wagtail, Snipe, Redshank, Lapwing. *Winter:* Large numbers of wildfowl (inc. some Bewick's and Whooper Swans) and waders (inc. Golden Plover). Birds of prey (inc. Merlin, Peregrine, Hobby, Short-eared Owl).

Contact: Malcolm Emery, English Nature, Phoenix House, 33 North Street, Lewes, E Sussex BN7 2PH. 01273 476595; fax 01273 483063; e-mail sussex.surrey@english-nature.org.uk www.english-nature.org.uk.

4. RYE HARBOUR

Rye Harbour Local Nature Reserve Management Committee.
Location: TQ 941 188. One mile from Rye off A259 signed Rye Harbour. From J10 of M20 take A2070 until it joins A259.
Access: Open at all times by footpaths. Organised groups please book.
Facilities: Car park in Rye Harbour village. Information kiosk in car park. Shop, 2 pubs, Toilets and disabled facilities near car park, four hides (wheelchair access, two with induction sound loop fitted), information centre open most days (10am-4pm) by volunteers.
Public transport: Train (tel: 08457 484950), bus (tel: 0870 608 2608), Rye tourist information (tel: 01797 226696).
Habitat: Sea, sand, shingle, pits and grassland.
Key birds: *Spring:* Passage waders, especially roosting Whimbrel. *Summer:* Breeding terns, waders, Wheatear, Yellow Wagtail. *Winter:* Wildfowl, Water Rail, Bittern.
Contact: Barry Yates, (Manager), 2 Watch Cottages, Winchelsea, East Sussex TN36 4LU. 01797 223862. e-mail: yates@clara.net
www.naturereserve.ryeharbour.org
see also www.rxwildlife.org.uk for latest sightings in area.

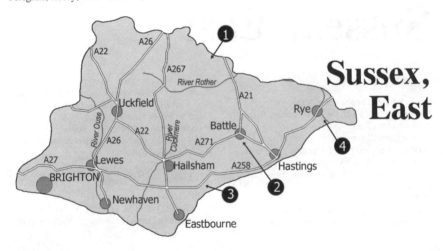

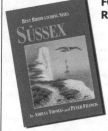
Sussex, West

1. ARUNDEL

The Wildfowl & Wetlands Trust.
Location: TQ 020 081. Clearly signposted from Arundel, just N of A27.
Access: Summer (9.30am-5.30pm) winter (9.30am-4.30pm). Closed Christmas Day. Approx 1.5 miles of level footpaths, suitable for wheelchairs. No dogs except guide dogs.
Facilities: Visitor centre, restaurant, shop, hides, picnic area, seasonal nature trails. Eye of The Wind Wildlife Gallery. Corporate hire facilities.
Public transport: Arundel station, 15-20 minute walk. Tel: 01903 882131.
Habitat: Lakes, wader scrapes, reedbed.
Key birds: *Summer*: Nesting Redshank, Lapwing, Oystercatcher, Common Tern, Sedge, Reed and Cetti's Warblers, Peregrine, Hobby. *Winter*: Teal, Wigeon, Reed Bunting, Water Rail, Cetti's Warbler and occasionally roosting Bewick's Swan.
Contact: James Sharpe, Mill Road, Arundel, West Sussex BN18 9PB. 01903 883355.
e-mail: info.arundel@wwt.org.uk
www.wwt.org.uk

2. KINGLEY VALE

English Nature (Sussex & Surrey Team).
Location: SU 825 088. West Stoke car park. Approx three miles NW of Chichester town centre (as the crow flies). Travel W along B2178 from Chichester, approx three miles, to East Ashling. Immediately after village turn right (on sharp left hand bend). After approx 0.5 miles turn left (off sharp right hand bend) and car park is on the right.
Access: Always open, no permits, no disabled access, no toilets. Dogs on a lead please.
Facilities: There is a nature trail (posts 1-24) and an unmanned information centre – no toilets, plenty of trees and bushes!
Public transport: Nearest railway station approx four miles walking distance. Nearest main bus route just over one mile on A286 at Mid Lavant.
Habitat: Greatest yew forest in Europe (more than 30,000 trees). Chalk grassland, mixed oak/ash woodland and scrub. Chalk heath.
Key birds: *Spring/summer*: Nightingale, Golden Pheasant, Whitethroat, Blackcap, Lesser Whitethroat. *Autumn/winter*: Hen Harrier, Buzzard, Hobby on migration, Red Kite.
Contact: English Nature, Game Keeper's Lodge, West Stoke House Farm, Downs Road, West Stoke, Chichester, West Sussex PO18 9BN. 01243 575353.

3. PAGHAM HARBOUR

West Sussex County Council.
Location: SZ 857 966. Five miles S of Chichester on B2145 towards Selsey.

Access: Open at all times, dogs must be on leads, disabled trail with accessible hide. All groups and coach parties must book in advance.

Facilities: Visitor centre open at weekends (10am-4pm), toilets (including disabled), three hides, one nature trail.

Public transport: Bus stop by visitor centre.

Habitat: Intertidal saltmarsh, shingle beaches, lagoons and farmland.

Key birds: *Spring*: Passage migrants. *Autumn*: Passage waders, other migrants. *Winter*: Brent Goose, Slavonian Grebe, wildfowl. *All year*: Little Egret.

Contact: Sarah Patton, Pagham Harbour LNR, Selsey Road, Sidlesham, Chichester, West Sussex PO20 7NE. 01243 641508.

e-mail: pagham.nr@westsussex.gov.uk

4. PULBOROUGH BROOKS

RSPB (South East England Office).

Location: TQ 054 170. Signposted on A283 between Pulborough (via A29) and Storrington (via A24). Two miles SE of Pulborough.

Access: Open daily. Visitor centre 10am-5pm (Tea-room 4.45pm, 4pm Mon-Fri in winter), closed Christmas Day and Boxing Day. Nature trail and hides (9am-9pm or sunset), closed Christmas Day. Admission fee for nature trail (free to RSPB members). No dogs. All four hides accessible to wheelchair users, although a strong helper is needed.

Facilities: Visitor centre (incl RSPB shop, tea room with terrace, displays, toilets). Nature trail and four hides and two viewpoints. Large car park. Play and picnic areas. An electric buggy is available for free hire, for use on trail.

Public transport: Two miles from Pulborough train station. Connecting bus service regularly passes reserve entrance (not Suns). Compass Travel (01903 233767). Cycle stands.

Habitat: Lowland wet grassland (wet meadows and ditches). Hedgerows and woodland.

Key birds: *Winter*: Wintering waterbirds, Bewick's Swan. *Spring*: Breeding wading birds and songbirds (incl Lapwing and Nightingale). *Summer*: Butterflies and dragonflies, warblers. *Autumn*: Passage wading birds, Redstart, Whinchat.

Contact: Tim Callaway, Site Manager, Upperton's Barn, Wiggonholt, Pulborough, West Sussex RH20 2EL. 01798 875851.

e-mail: pulborough.brooks@rspb.org.uk

5. WARNHAM NATURE RESERVE

Horsham District Council.

Location: TQ 167 324. One mile from Horsham Town Centre. Reserve is located just off the A24 'Robin Hood' roundabout, on the B2237.

Access: Open every day throughout the year and Bank Hols (10am-6pm or dusk). Free access over part of the Reserve, small charge (day/annual permits) for some areas. No dogs allowed. Good disabled access over most of the Reserve.

Facilities: Visitor centre and café open Sat and Sun in summer, Sun only in winter. Public car park with access for coaches at request. Toilets (including disabled), two hides, reserve leaflets, a new nature trail, new bird feeding station, boardwalks, benches and hardstanding paths.

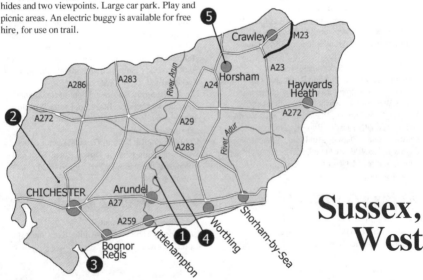

Sussex, West

Public transport: From Horsham Railway Station it is a mile walk along Hurst Road, with a R turn onto Warnham Road. A bus from the 'Carfax' in Horsham Centre can take you to within 150 yards of the reserve. Travel line, 0870 608 2608.
Habitat: 17 acre millpond, reedbeds, marsh, meadow and woodland (deciduous and coniferous).
Key birds: Heronry, Kingfisher, Cetti's Warbler, three woodpeckers, Willow Tit, Goldcrest. *Summer:*

Hirundines, Hobby, Cuckoo, Spotted Flycatcher, warblers. *Winter:* Cormorant, gulls, Little Grebe, Water Rail, Siskin, Lesser Redpoll, thrushes, waders. *Passage:* Waders, Wheatear, Whinchat, Meadow Pipit, terns.
Contact: Countryside Warden, Leisure Services, Park House Lodge, North Street , Horsham, W Sussex RH12 1RL. 01403 256890.
e-mail: sam.bayley@horsham.gov.uk

Tyne & Wear

1. BOLDON FLATS

South Tyneside Metropolitan Council.
Location: NZ 377 614. Take A184 N from Sunderland to Boldon.
Access: View from Moor Lane on minor road NE of East Boldon station towards Whitburn.
Facilities: None.
Public transport: East Boldon Metro Station approx. 800 metres from site.
Habitat: Meadows, part SSSI, managed flood in winter, pond, ditches.
Key birds: *Passage/winter:* Wildfowl and waders. Gull roost may inc. Mediterranean, Glaucous, Iceland; Merlin fairly regular.
Contact: Countryside Officer, South Tyneside Metropolitan Council, Town Hall, Westoe Road, South Shields, Tyne & Wear NE33 2RL. 0191 427 1717; (Fax) 0191 455 0208.
www.southtyneside.info

2. DERWENT WALK COUNTRY PARK

Gateshead Council.
Location: NZ 178 604. Along River Derwent, four miles SW of Newcastle and Gateshead. Several car parks along A694.
Access: Site open all times. Thornley visitor centre open weekends and Bank Holidays (12-5pm). Keys for hides available from Thornley Woodlands Centre (£2).
Facilities: Toilets at Thornley visitor centre. Toilets at Swalwell visitor centre. Hides at Far Pasture Ponds and Thornley feeding station.
Public transport: 45, 46, 46A, M20 and 611 buses from Newcastle/Gateshead to Swalwell/ Rowlands Gill. Bus stop Thornley Woodlands

Centre. (Regular bus service from Newcastle). Information from News Travel Line. Tel: 0191 2325325.
Habitat: Mixed woodland, river, ponds, meadows.
Key birds: *Summer:* Wood Warbler, Pied Flycatcher, Green Sandpiper, Kingfisher, Dipper, Great Spotted and Green Woodpeckers, Blackcap, Garden Warbler, Nuthatch. *Winter:* Brambling, Marsh Tit, Bullfinch, Great Spotted Woodpecker, Nuthatch, Goosander, Kingfisher.
Contact: Stephen Westerberg, Thornley Woodlands Centre, Rowlands Gill, Tyne & Wear NE39 1AU. 01207 545212.
e-mail: countryside@gateshead.gov.uk
www.gatesheadbirders.co.uk

3. RYTON WILLOWS

Gateshead Council.
Location: NZ 155 650. Five miles W of Newcastle. Access along several tracks running N from Ryton.
Access: Open at all times.
Facilities: Nature trail and free leaflet.
Public transport: Regular service to Ryton from Newcastle/Gateshead. Information from Nexus Travelline on 0191 232 5325.
Habitat: Deciduous woodland, scrub, riverside, tidal river.
Key birds: *Winter:* Goldeneye, Goosander, Green Woodpecker, Nuthatch, Treecreeper. *Autumn:* Greenshank. *Summer:* Lesser Whitethroat, Sedge Warbler, Yellowhammer, Linnet, Reed Bunting, Common Sandpiper.
Contact: Andrew McLay, Thornley Woodlands Centre, Rowlands Gill, Tyne & Wear NE39 1AU. 1208 545212. e-mail: countryside@gateshead.gov.uk
www.gatesheadbirders.co.uk

Tyne & Wear

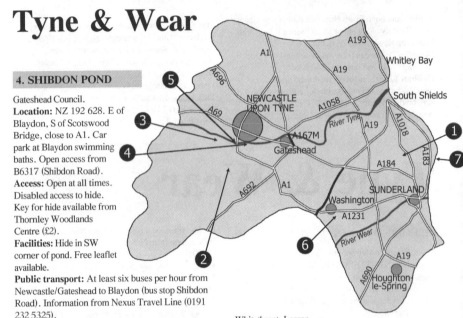

4. SHIBDON POND

Gateshead Council.
Location: NZ 192 628. E of
Blaydon, S of Scotswood
Bridge, close to A1. Car
park at Blaydon swimming
baths. Open access from
B6317 (Shibdon Road).
Access: Open at all times.
Disabled access to hide.
Key for hide available from
Thornley Woodlands
Centre (£2).
Facilities: Hide in SW
corner of pond. Free leaflet
available.
Public transport: At least six buses per hour from
Newcastle/Gateshead to Blaydon (bus stop Shibdon
Road). Information from Nexus Travel Line (0191
232 5325).
Habitat: Pond, marsh, scrub and damp grassland.
Key birds: *Winter*: Wildfowl, Water Rail, white-
winged gulls. *Summer*: Reed Warbler, Sedge
Warbler, Lesser Whitethroat, Grasshopper Warbler,
Water Rail. *Autumn*: Passage waders and wildfowl,
Kingfisher.
Contact: Brian Pollinger, Thornley Woodlands
Centre, Rowlands Gill, Tyne & Wear NE39 1AU.
1209 545212. e-mail: countryside@gateshead.gov.uk
www.gatesheadbirders.co.uk

5. TYNE RIVERSIDE COUNTRY PARK AND THE REIGH

Newcastle City Council
Location: NZ 158 658. From the Newcastle to
Carlisle by-ass on the A69(T) take the A6085 into
Newburn. The park is signposted along the road to
Blaydon. 0.25 miles after this junction, turn due W
(the Newburn Hotel is on the corner) and after 0.5
miles the parking and information area is signed just
beyond the Newburn Leisure Centre.
Access: Open all year.
Facilities: Car park. Leaflets and walk details
available.
Public transport: None.
Habitat: River, pond with reed and willow stands,
mixed woodland, open grassland.
Key birds: *Spring/summer*: Swift, Swallow,
Whitethroat, Lesser
Whitethroat. *Winter*:
Sparrowhawk, Kingfisher, Little Grebe, finches,
Siskin, Fieldfare, Redwing, duck, Goosander and
other wildfowl. *All year*: Grey Partridge, Green and
Great Spotted Woodpecker, Bullfinch,
Yellowhammer.
Contact: Newcastle City Council, The Riverside
Country Park, Newburn, Newcastle upon Tyne
NE15 8BW,

6. WASHINGTON

The Wildfowl & Wetlands Trust.
Location: NZ 331 566. In Washington. On N bank
of River Wear, W of Sunderland. Signposted from
A195, A19, A1231 and A182.
Access: Open 9.30am-5pm (summer), 9.30am-4pm
(winter). Free to WWT members. Admission charge
for non-members. No dogs except guide dogs.
Good access for people with disabilities.
Facilities: Visitor centre, toilets, parent and baby
room, range of hides. Shop and café.
Public transport: Buses to Waterview Park (250
yards walk) from Washington, from Sunderland,
Newcastle-upon-Tyne, Durham and South Shields.
Tel: 0845 6060260 for details.
Habitat: Wetlands, woodland and meadows.
Key birds: *Spring/summer*: Nesting colony of
Grey Heron, other breeders include Common Tern,

202

Oystercatcher, Lapwing. *Winter*: Bird-feeding station visited by Great Spotted Woodpecker, Bullfinch, Jay and Sparrowhawk. Goldeneye and other ducks.
Contact: Dean Heward, (Conservation Manager), Wildfowl & Wetlands Trust, Pottinson, Washington NE38 8LE. 0191 416 5454 ext 231.
e-mail: dean.heward@wwt.org.uk www.wwt.org.uk

7. WHITBURN BIRD OBSERVATORY

National Trust/Durham Bird Club.

Location: NZ 414 633.
Access: Access details from Recorder.
Facilities: None.
Public transport: None.
Habitat: Cliff top location.
Key birds: Esp. seawatching but also passerine migrants inc. rarities.
Contact: Tony Armstrong,39 Western Hill, Durham City, DH1 4RJ. 0191 386 1519;
e-mail: ope@globalnet.co.uk.

Warwickshire

1. ALVECOTE POOLS

Warwickshire Wildlife Trust.
Location: SK 253 034. Located alongside River Anker E of Tamworth. Access via Robey's Lane (off B5000) just past Alvecote Priory car park. Also along towpath via Pooley Hall visitor centre, also number of points along towpath.
Access: Some parts of extensive path system is accessible to disabled. Parking Alvecote Priory car park. Nature trail.
Facilities: Nature trail.
Public transport: Within walking distance of the Avecote village bus stop.
Habitat: Marsh, pools (open and reedbeds) and woodland.
Key birds: *Spring/summer*: Breeding Oystercatcher, Common Tern and Little Ringed Plover. Common species include Great Crested Grebe, Tufted Duck and Snipe. Important for wintering, passage and breeding wetland birds.
Contact: Reserves Team, Brandon Marsh Nature Centre, Brandon Lane, Brandon, Coventry CV3 3GW. 024 7630 2912.
e-mail: reserves@warkswt.cix.co.uk
www.warwickshire-wildlife-trust.org.uk

2. BRANDON MARSH

Warwickshire Wildlife Trust.
Location: SP 386 762. Three miles SE of Coventry, 200 yards SE of A45/A46 junction (Tollbar End). Turn E off A45 into Brandon Lane. Reserve entrance 1.25 miles on right.
Access: Open weekdays (9am-5pm), weekends (10am-4.30pm). Entrance charge currently £2.50 (free to Wildlife Trust members). Wheelchair access to nature trail and Wright hide. No dogs. Parking for 2 coaches.
Facilities: Visitor centre, toilets, tea-room (open at above times), nature trail, seven hides.
Public transport: Bus service from Coventry to Tollbar End then 1.25 mile walk. Tel Travel West Midlands 02476 817032 for bus times.
Habitat: Ten pools, together with marsh, reedbeds, willow carr, scrub and small mixed woodland in 260 acres, designated SSSI in 1972.
Key birds: *Spring/summer*: Garden Warbler, Grasshopper Warbler, Whitethroat, Lesser Whitethroat, Hobby, Little Ringed Plover, Whinchat, Wheatear. *Autumn/winter*: Dunlin, Ruff, Snipe, Greenshank, Green and Common Sandpipers, Wigeon, Shoveler, Pochard, Goldeneye, Siskin, Redpoll. *All year*: Cetti's Warbler, Kingfisher, Water Rail, Gadwall, Little Grebe.
Contact: As Alvecote Pools above.

3. HARTSHILL HAYES

Warwickshire County Council.
Location: SP 317 943. Signposted as 'Country Park' from B4114 W of Nuneaton.
Access: Open all year. Closed Christmas Day.
Facilities: Three waymarked walks (one easy-going). Visitor centre, play area, toilets with facilities for disabled.
Public transport: None.
Habitat: Mixed woodland, grassland hillside.
Key birds: Warblers, woodpeckers, tits, Goldcrest, Sparrowhawk, Tawny Owl.
Contact: Country Park Manager's Office, Kingsbury Water Park, Bodymoor Heath Lane, Sutton Coldfield, West Midlands B76 0DY. 01827 872660;
e-mail parks@warwickshire.gov.uk
www.warwickshire.gov.uk/countryside.

Warwickshire

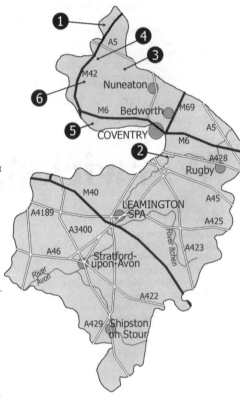

4. KINGSBURY WATER PARK

Warwickshire County Council.
Location: SP 203 960. Signposted 'Water Park'
from J9 M42, A4097 NE of Birmingham.
Access: Open all year except Christmas Day.
Facilities: Four hides, two with wheelchair access.
Miles of flat surfaced footpaths, free loan scheme for
mobility scooters. Cafes, Information Centre with gift
shop.
Public transport: Call for advice.
Habitat: Open water; numerous small pools, some
with gravel islands; gravel pits; silt beds with
reedmace, reed, willow and alder; rough areas and
grassland.
Key birds: *Summer*: Breeding warblers (nine
species), Little Ringed Plover, Great Crested and
Little Grebes. Shoveler, Shelduck and a thriving
Common Tern colony. Passage waders (esp. spring).
Winter: Wildfowl, Short-eared Owl.
Contact: See Hartshill Hayes above.

5. MARSH LANE

Packington Estate Enterprises Limited.
Location: SP 217 804. Equidistant between
Birmingham and Coventry, both approx 7-8 miles
away. Off A452 between A45 and Balsall Common,
S of B4102/A452 junction. Turn right into Marsh
Lane and immediately right onto Old Kenilworth
Road (now a public footpath), to locked gate. Key
required for access.
Access: Only guide dogs allowed. Site suitable for
disabled. Access is by day or year permit only.
Membership rates: annual – adult £22, OAP £16.50,
children (under 16) £11.50 husband/wife £38, OAP
husband/wife £29.50. Contact address below (9am-
5.15pm). Regular newsletter provided to annual
permit holders; day adult £3, OAP £2.75, children
(under 16) £2 obtained from Golf Professional Shop,
Stonebridge Golf Centre, Somers Road, off Hampton
Lane, Meriden, nr Coventry CV7 7PL (tel 01676
522442) only three to four minutes car journey from
site. Open Mon-Sun (7am-7pm). Stonebridge Golf
Centre open to non-members. Visitors can obtain
drinks and meals at Stonebridge Golf Centre. £26.75
deposit required for key. Readily accessible for
coaches.

Facilities: No toilets or visitor centre. Four hides and
hard tracks between hides. Car park behind locked
gates.
Public transport: Hampton-in-Arden railway station
within walking distance on footpath loop. Bus no
194 stops at N end of Old Kenilworth Road one mile
from reserve gate.
Habitat: Two large pools with islands, three small
areas of woodland, five acre field set aside for arable
growth for finches and buntings as winter feed.
Key birds: 177 species. *Summer*: Breeding birds
include Little Ringed Plover, Common Tern, most
species of warbler including Grasshopper. Good
passage of waders in Apr, May, Aug and Sept.
Hobby and Buzzard breed locally.
Contact: Nicholas P Barlow, Packington Hall,
Packington Park, Meriden, Nr Coventry CV7 7HF.
01676 522020. www.packingtonestate.net

NATURE RESERVES - ENGLAND

6. WHITACRE HEATH

Warwickshire Wildlife Trust.
Location: SP 209 931. Three miles N of Coleshill, just W of Whitacre Heath village.
Access: Trust members only.
Facilities: Five hides.
Public transport: Within walking distance of Lea Marston village bus stop.
Habitat: Pools, wet woodland and grassland.

Key birds: *Summer*: Sedge Warbler, Reed Warbler, Lesser Whitethroat, Whitethroat, Garden Warbler, Willow Warbler and Blackcap. Migrant Curlew, Whinchat. Also Snipe, Water Rail and Kingfisher.
Contact: Reserves Team, Brandon Marsh Nature Centre, Brandon Lane, Brandon, Coventry CV3 3GW. 024 7630 2912.
e-mail: reserves@warkswt.cix.co.uk
www.warwickshire-wildlife-trust.org.uk

West Midlands

1. LICKEY HILLS COUNTRY PARK

Birmingham County Council.
Location: Eleven miles SW of Birmingham City Centre.
Access: Open all year.
Facilities: Car park, visitor centre with wheelchair pathway with viewing gallery, picnic site, toilets, café, shop.
Public transport: Bus: West Midlands 62 Rednal (20 mins walk to visitor centre. Rail: Barnt Green (25 mins walk through woods to the centre).
Habitat: Hills covered with mixed deciduous woodland, conifer plantations and heathland.
Key birds: *Spring/summer*: Warblers, Tree Pipit, Redstart. *Winter*: Redwing, Fieldfare. *All year*: Common woodland species.
Contact: The Visitor Centre, Lickey Hills Country Park, Warren Lane, Rednal, Birmingham, B45 8ER, 0121 4477106.
e-mail: visitorcentre@lickeyhills.fsnet.co.uk

2. ROUGH WOOD CHASE

Walsall Metropolitan Borough Council
Location: SJ 987 012. NW of Walsall town centre. From J10 of M6 travelling N, turn L on A454 then R on A462. Turn first R onto Bloxwich Road North then R into Hunts Lane. Car park is on R.
Access: Open at all times.
Facilities: Car park, footpaths, nature trail.
Public transport: WMT bus no 341 from Walsall, Park street and Willenhall.
Habitat: Mature oak and mixed woodland, grassland, open wate.
Key birds: Common woodland and water species.

Contact: Countryside Services, Walsall Metropolitan Borough Council, Dept of Leisure & Community Services, PO box 42, The Civic Centre, Darwall Street, WalsallWS1 1TZ. 01922 650000;(Fax)01922 721862. www.walsall.gov.uk

3. SANDWELL VALLEY 1

Metropolitan Borough Council.
Location: SP 012 918 & SP 028 992.
Access: Access and car park from Dagger Lane or Forge Lane, West Bromwich.
Facilities: Mainly public open space.
Public transport: Call for advice.
Habitat: Nature reserve, lakes, woods and farmland.
Key birds: *Summer*: Breeding Lapwing, Little Ringed Plover, Sparrowhawk. Great Spotted and Green Woodpeckers, Tawny Owl, Reed Warbler. Passage waders.
Contact: Senior Ranger, Sandwell Valley Country Park, Salters Lane, West Bromwich, W Midlands B71 4BG. 0121 553 0220 or 2147.

4. SANDWELL VALLEY 2

RSPB
Location: SP 035 928. Great Barr, Birmingham. Follow signs S from M6 J7 via A34. Take right at 1st junction onto A4041. Take 4th left onto Hamstead Road (B4167), then right at 1st mini roundabout onto Tanhouse Avenue.
Access: 800 metres of paths accessible to assisted and powered wheelchairs with some gradients (please ring centre for further information), centre fully accessible.
Facilities: Visitor centre and car park (open Tue-Sun 9.30am-5pm) with viewing area, small shop and hot

drinks, four viewing screens, one hide. Please ring the centre for details on coach parking.
Public transport: Bus: 16 from Corporation Street (Stand CJ), Birmingham City Centre (ask for Tanhouse Avenue). Train: Hamstead Station, then 16 bus for one mile towards West Bromwich from Hamstead (ask for Tanhouse Avenue).
Habitat: Open water, wet grassland, reedbed, dry grassland and scrub.
Key birds: *Summer*: Lapwing, Reed Warbler, Willow Tit. *Passage*: Sandpipers, Yellow Wagtail, chats, Common Tern. *Winter*: Water Rail, Snipe, Jack Snipe, Goosander, Bullfinch, woodpeckers and wildfowl.
Contact: Lee Copplestone, 20 Tanhouse Avenue, Great Barr, Birmingham B43 5AG. 0121 3577395.

Wiltshire

1. FYFIELD DOWNS NNR

Location: On the Marlborough Downs. From the A345 at the N end of Marlborough a minor road signed Broad Hinton, bisects the downs, dipping steeply at Hackpen Hill to the A361 just before Broad Hinton. From Hackpen Hill walk S to Fyfield Down.
Access: Open all year but avoid the racing gallops. Keep dogs on leads.
Facilities: Car park.
Public transport: None.
Habitat: Downs.
Key birds: *Spring*: Ring Ouzel possible on passage, Wheatear, Cuckoo, Redstart, common warblers. *Summer*: Possible Quail. *Winter*: Occasional Hen Harrier, possible Merlin, Golden Plover, Short-eared Owl, thrushes. *All year*: Sparrowhawk, Buzzard, Kestrel, partridges, Green and Great Spotted Woodpeckers, Goldfinch, Corn Bunting.
Contact: English Nature (Wiltshire team), Prince Maurice Court, Hambleton, Devizes, Wiltshire SN10 2RT. 01380 726344. e-mail: wiltshire@english-nature.org.uk

2. LANGFORD LAKE

Wiltshire Wildlife Trust.
Location: SU 037 370. Nr Steeple Langford, S of A36, approx eight miles W of Salisbury. In the centre of the village, turn S into Duck Street, signposted Hanging Langford. Langford Lakes is the first turning on the L just after a small bridge across the River Wylye.
Access: Opened to the public in Sept 2002. Main gates opening the during the day - ample parking. Advance notice required for coaches. No dogs allowed on this reserve.
Facilities: Four hides, all accessible to wheelchairs. Cycle stands provided (250m from Wiltshire Cycleway between Great Wishford and Hanging Langford).
Public transport: Nearest bus stop 500m - X4 Service between Salisbury and Warminster.
Habitat: Three former gravel pits, with newly created islands and developing reed fringes. 12 ha (29 acres) of open water; also wet woodland, scrub, chalk river.
Key birds: *Summer*: Breeding Coot, Moorhen, Mallard, Tufted Duck, Pochard, Gadwall, Little Grebe, Great Crested Grebe. Also Kingfisher, Common Sandpiper, Grey Wagtail, warblers (8 species). *Winter*: Wildfowl, sometimes also Wigeon, Shoveler, Teal, Water Rail, Little Egret, Bittern. *Passage*: Sand Martin, Green Sandpiper, waders, Black Tern.
Contact: Wiltshire Wildlife Trust, Langford Lakes, Duck Street, Steeple Langford, Salisbury, Wiltshire SP3 4NH. 01722 790770.
e-mail: admin@wiltshirewildlife.org
www.wiltshirewildlife.org

3. SAVERNAKE FOREST

Forest Enterprise.
Location: From Marlborough, A4 Hungerford road runs along N side of forest. Two pillars mark Forest Hill entrance 1.5 miles E of A346/A4 junction. The Grand Avenue leads straight through middle of woodland to join a minor road from Stibb Green on A346 N of Burbage to A4 W of Froxfield.
Access: Open all year.
Facilities: Car park, picnic site at NW end by A346. Fenced-off areas should not be entered unless there is a footpath.

Public transport: None.
Habitat: Ancient woodland, with one of the largest collections of veteran trees in Britain.
Key birds: *Spring/summer*: Garden Warbler, Blackcap, Willow Warbler, Chiffchaff, Wood Warbler, Redstart, occasional Nightingale, Tree Pipit, Spotted Flycatcher. *Winter*: Finch flocks possibly inc

Siskin, Redpoll, Brambling. *All year*: Sparrowhawk, Buzzard, Woodcock, owls, all three woodpeckers, Marsh and Willow Tit, Jay and other woodland birds.
Contact: Forest Enterprise, Postern Hill, Marlborough, 01672 512520.
e-mail: admin@wiltshirewildlife.org
www.wiltshirewildlife.org

Worcestershire

1. MONKWOOD NATURE RESERVE

Worcestershire Wildlife Trust.
Location: SO 804 607. The reserve is about five miles NW of Worcester. On the A443 Worcester to Holt Heath road, take any of the minor roads N of Hallow to Sinton Green. At the village green, take the road to Monkwood by the side of the New Inn pub. About a mile down the road there is a car park on the R.
Access: Open daily excluding Christmas Day. Stoned bridleway. Other paths grass, difficult to negotiate when wet.
Facilities: Two nature trails.
Public transport: None.
Habitat: Ancient woodland.
Key birds: *Spring/summer*: Garden Warbler, Lesser Whitethroat, Cuckoo, Blackcap, Woodcock. *All year*: Tawny Owl, Sparrowhawk, Jay. Good for butterflies, especially Wood White.
Contact: Worcestershire Wildlife Trust, Lower Smite Farm, Smite Hill, Hindlip, Worcester, WR3 8SZ, 01905 754919. e-mail: worcswt@cix.co.uk www.worcswildlifetrust.co.uk

2. TIDDESLEY WOOD

Worcestershire Wildlife Trust.
Location: SO 929 462. Take the A44 from Pershore to Worcester. Turn L near town boundary just before the summit of the hill towards Besford and Croome. The entrance to the reserve is on the L after about 0.75 miles.
Access: Open all year except Christmas Day. Cycles and horses only allowed on the bridleway. Please

keep dogs fully under control. As there is a military firing range at the SW corner of the wood, do not enter the area marked by red flags. The NE plot is private property and visitors should not enter the area. Main ride stoned, with some potholes. Small pathways difficult if wet.
Facilities: Information board. May find numbered posts around the reserve which were described in an old leaflet. Circular trail around small pathways.
Public transport: None.
Habitat: Ancient woodland, conifers.
Key birds: *Spring*: Chiffchaff, Blackcap, Cuckoo, occasional Nightingale. *All year*: Crossbill, Coal Tit, Goldcrest, Sparrowhawk, Willow Tit, Marsh Tit.

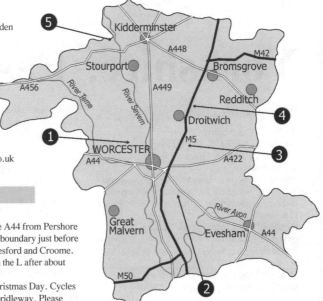

Winter: Redwing, Fieldfare.
Contact: Trust HQ, 01905 754919.
e-mail: worcswt@cix.co.uk

3. TRENCH WOOD

Worcestershire Wildlife Trust.
Location: SO 931 585. NE of Worcester.
Access: Open daily exc Christmas Day. Pathways difficult if wet, all grass or exposed mud.
Facilities: Car park.
Public transport: None.
Habitat: (NB mature trees to SW and SE of wood are not part of reserve). Young broadleaved woodland, mixed scrub.
Key birds: Very good for warblers; Woodcock.
Contact: Trust HQ, 01905 754919.
e-mail: worcswt@cix.co.uk
www.worcswildlifetrust.co.uk

4. UPTON WARREN

Worcestershire Wildlife Trust.
Location: SO 936 675. Two miles S of Bromsgrove on A38.
Access: Always open except Christmas Day. Trust membership gives access, or day permit from sailing centre. Disabled access to west hide at moors only. Dogs on leads.
Facilities: Seven hides, maps at entrances, can be very muddy.

Public transport: Birmingham/Worcester bus passes reserve entrance.
Habitat: Fresh and saline pools with muddy islands, some woodland and scrub.
Key birds: *Winter*: Wildfowl. *Spring/autumn*: Passage waders, Common Tern, Cetti's Warbler, Oystercatcher and Little Ringed Plover, many breeding warblers.
Contact: A F Jacobs, 3 The Beeches, Upton Warren, Bromsgrove, Worcs B61 7EL. 01527 861370.

5. WYRE FOREST

English Nature/Worcs Wildlife Trust.
Location: SO 750 760. A456 out of Bewdley.
Access: Observe reserve signs and keep to paths. Forestry Commission visitor centre at Callow Hill. Fred Dale Reserve is reached by footpath W of B4194 (parking at SO776763).
Facilities: Facilities for disabled (entry by car) if Warden telephoned in advance.
Public transport: None.
Habitat: Oak forest, conifer areas, birch heath, stream.
Key birds: Buzzard, Pied Flycatcher, Wood Warbler, Redstart, all three woodpeckers, Woodcock, Crossbill, Siskin, Hawfinch, Kingfisher, Dipper, Grey Wagtail, Tree Pipit.
Contact: Tim Dixon, 01531 638500.

Yorkshire, East

1. BEMPTON CLIFFS

RSPB (North of England Office).
Location: TA 197 738. Near Bridlington. Take cliff road N from Bempton Village off B1229 to car park and visitor centre
Access: Visitor centre open Mar-Nov and weekends in Dec and Feb. Public footpath along cliff top with observation points. Four miles of chalk cliffs, highest in the county.
Facilities: Visitor centre, toilets. Viewing platforms.Picnic area.
Public transport: Railway 1.5 miles - irregular bus service to village 1.25 miles.
Habitat: Seabird nesting cliffs, farmland, scrub.
Key birds: Best to visit May to mid-July for eg

Puffin, Gannet (only colony on English mainland), Fulmar, Kittiwake; also nesting Tree Sparrow, Corn Bunting; good migration watchpoint for skuas, shearwaters and terns.
Contact: Site Manager, RSPB Visitor Centre, Cliff Lane, Bempton, Bridlington, E Yorks YO15 1JF. 01262 851179.

2. BLACKTOFT SANDS

RSPB (North of England Office).
Location: SE 843 232. Eight miles E of Goole on minor road between Ousefleet and Adlingfleet.
Access: Open 9am-9pm or dusk if earlier. RSPB members free, £3 permit for non-members, £2 concessionary, £1 children, £6 family.
Facilities: Car park, toilets, visitor centre, six hides,

footpaths suitable for wheelchairs.
Public transport: Goole/Scunthorpe bus (Sweynes' Coaches stops outside reserve entrance).
Habitat: Reedbed, saline lagoons, lowland wet grassland, willow scrub.
Key birds: *Summer*: Breeding Avocet, Marsh Harrier, Bittern, Bearded Tit, passage waders (exceptional list inc many rarities). *Winter*: Hen Harrier, Merlin, Peregrine, wildfowl.
Contact: Pete Short (Warden) & Mike Pilsworth (Asst Warden), Hillcrest, Whitgift, Nr Goole, E Yorks DN14 8HL. 01405 704665.
e-mail: pete.short@rspb.org.uk
mike.pilsowrth@rspb.org.uk
www.rspb.org

3. FLAMBOROUGH CLIFFS

Yorkshire Wildlife Trust
Location: TA 240 722. The reserve is part of Flamborough headland approx 4 miles NE of Bridlington. From Bridlington take B1255 to Flamborough and follow signs for North Landing.
Access: Open all year. There is a public pay and display car park at North Landing which gives access to both parts of the reserve. Paths not suitable for wheelchairs.
Facilities: Car park (pay and display), trails, refreshments available at café at North Landing (open Apr-Oct 10am-5pm), toilets.
Public transport: Flamborough is served by buses from Bridlington and Bempton. Phone 01482 222 222 for details.
Habitat: Coastal cliffs, rough grassland and scrub, farmland.
Key birds: *Summer*: Puffin, Guillemot, Razorbill, Kittiwake, Shag, Fulmar, Sky Lark, Meadow Pipit, Linnet, Whitethroat, Yellowhammer, Tree Sparrow, occasional Corn Bunting. *Passage migrants*: Fieldfare, Redwing and occasional rarities such as Wryneck and Red-backed Shrike.
Contact: Trust HQ, 01904 659570
e-mail: info@yorkshirewt.cix.co.uk
www.yorkshire-wildlife-trust.org.uk

4. NORTH CAVE WETLANDS

Yorkshire Wildlife Trust
Location: SE 887 328. North Cave Wetlands is located to NW of North Cave village, approx 10 miles W of Hull. From junction 28 of M62, follow signs to North Cave on B1230. Turn L in village and follow road to next crossroads. Follow road round to L, take next L onto Dryham Lane. If coming from N, follow minor road direct from Market Weighton. One mile after the turning for Hotham, take next R (Dryham Lane).
Access: Open all year with car parking on Dryham Lane. Some of the footpaths are suitable for all abilities.
Facilities: Three bird-viewing hides, two of which are accessible to wheelchair users. The nearest toilet and refreshment facilities are in the village of North Cave, one mile away,
Public transport: Buses serve North Cave from Hull and Goole, telephone 01482 222 222 for details.
Habitat: Six former gravel pits have been converted into various lagoons for wetland birds, including one reedbed. There are also grasslands, scrub and hedgerows.
Key birds: Over 150 different species have been recorded including Great Creasted Grebe, Gadwall, Pochard, Sparrowhawk, Avocet, Ringed Plover, Golden Plover, Dunlin, Ruff, Redshank, Green and Common Sandpipers, Tree Sparrow.
Contact: Trust HQ, 01904 659570
e-mail: info@yorkshirewt.cix.co.uk
www.yorkshire-wildlife-trust.org.uk

5. SPURN NNR

Yorkshire Wildlife Trust.
Location: Entrance Gate TA 417 151. 26 miles from Hull. Take A1033 from Hull to Patrington then B1445 to Easington and unclassed roads on to Kilnsea and Spurn Head.
Access: Normally open at all times. Vehicle admission fee (at present £3). No charge for pedestrians. No dogs allowed under any circumstances, not even in cars. Coaches by permit only (must be in advance).
Facilities: Chemical portaloos next to information centre. Information centre open weekends, Bank Holidays, school holidays. Three hides. Cafe at point open weekends Apr to Oct 10am to 5pm.
Public transport: Nearest bus service is at Easington (3.5 miles away).
Habitat: Sand dunes with marram and sea buckthorn scrub. Mudflats around Humber Estuary.
Key birds: *Spring*: Many migrants on passage and often rare birds such as Red-backed Shrike, Bluethroat etc. *Autumn*: Passage migrants and rarities like Wryneck, Pallas's Warbler. *Winter*: Waders and Brent Goose.
Contact: Spurn Reserves Officer, Spurn NNR, Blue Bell, Kilnsea, Hull HU12 0UB.
e-mail: spurnywt@ukonline.co.uk

Yorkshire, North

1. COATHAM MARSH

Tees Valley Wildlife Trust.
Location: NZ 585 250. Located on W edge of
Redcar. Access from minor road to Warrenby from
A1085/A1042.
Access: Reserve is open throughout daylight hours.
Please keep to permissive footpaths only.
Facilities: Two hides. Key required for one of these
– available to Tees Valley Wildlife Trust members for
£10 deposit. No toilets or visitor centre.
Public transport: Frequent bus service between
Middlesbrough and Redcar. Nearest stops are in
Coatham 0.25 mile from reserve (Arriva tel 0870
6082608). Redcar Central Station one mile from site.
Frequent trains from Middlesbrough and Darlington.
Habitat: Freshwater wetlands, lakes, reedbeds.
Key birds: *Spring/autumn*: Wader passage (including
Wood Sandpiper and Greenshank). *Summer:*
Passerines (including Sedge Warbler, Yellow
Wagtail). *Winter:* Ducks (including Smew).
Occasional rarities, Water Rail, Great White Egret,
Avocet, Bearded Tit and Bittern.
Contact: Bill Ashton-Wickett, Trust HQ. 01642
759900. e-mail: info@teesvalleywt.cix.co.uk
www.wildlifetrust.org.uk/teesvalley

2. FILEY BRIGG ORNITHOLOGICAL GROUP BIRD OBSERVATORY

FBOG and Yorkshire Wildlife Trust (The Dams).
Location: TA 10 68 07. Two access roads into Filey
from A165 (Scarborough to Bridlington road). Filey
is ten miles N of Bridlington and eight miles S of
Scarborough.
Access: Opening times – no restrictions. Dogs only
in Parish Wood and The Old Tip (on lead). Coaches
welcome. Park in the North Cliff Country Park.
Facilities: No provisions for disabled at present. Two
hides at The Dams, one on The Brigg (open most
weekends from late Jul-Oct, key can be hired from
Country Park café). Toilets in Country Park (Apr-
Nov 1) and town centre. Nature trails at The Dams,
Parish Wood/Old Tip. Cliff-top walk for seabirds
along Cleveland Way.
Public transport: All areas within a mile of Filey
railway station. Trains into Filey tel. 08457 484950;
buses into Filey tel. 01723 503020.
Habitat: The Dams – two freshwater lakes, fringed
with some tree cover and small reedbeds. Parish
Wood – a newly created wood which leads to the Old

Tip. The latter has been fenced (for stock and crop
strips) though there is a public trail. Carr Naze has a
pond and can produce newly arrived migrants.
Key birds: *The Dams*: Breeding and wintering water
birds, breeding Sedge Warbler, Reed Warbler and
Tree Sparrow. *The Tip*: Important for breeding Sky
Lark, Meadow Pipit, common warblers and Grey
Partridge. *Winter:* Area for buntings including
Lapland. *Seawatch Hide*: Jul-Oct. All four skuas,
shearwaters, terns. *Winter*: Divers and grebes. *Totem
Pole Field:* A new project should encourage breeding
species and wintering larks, buntings etc. Many sub-
rare/rare migrants possible at all sites.
Contact: Lez Gillard, Recorder, 12 Sycamore
Avenue, Filey, N Yorks YO14 9NU. 01723
516383. e-mail: lezgillard@tiscali.co.uk
www.fbog.co.uk

3. HUNTCLIFF

Tees Valley Wildlife Trust.
Location: NZ 674 215. Along Cleveland Way
footpath from Saltburn or Skinningrove.
Access: Open all year.
Facilities: None.
Public transport: None.
Habitat: Strip of coastal grassland and cliff face.
Key birds: Large colony of Kittiwakes and Fulmars
with some Cormorants.
Contact: Trust HQ, 01642 759900.

4. LOWER DERWENT VALLEY

English Nature (North & East Yorks).
Location: SE 691 447. Six miles SE of York,
stretching 12 miles S along the River Derwent from
Newton-on-Derwent to Wressle and along the
Pocklington Canal. Visitor facilities at Bank Island,
Wheldrake Ings YWT (SE 691 444), Thorganby (SE
693 422) and North Duffield Carrs (SE 698 366).
Access: Open all year. No dogs. Disabled access at
North Duffield Carrs (two hides and car park). 500m
path.
Facilities: North Duffield Carrs – two hides,
wheelchair access. Wheldrake Ings (YWT) – five
hides. Bank Island – two hides, viewing tower.
Thorganby – viewing platform.
Public transport: Bus from York/Selby – contact
First (01904 622992). Bicycle stands provided in car
parks at Bank Island and North Duffield Carrs.
Habitat: Hay meadow and pasture, swamp, open

NATURE RESERVES - ENGLAND

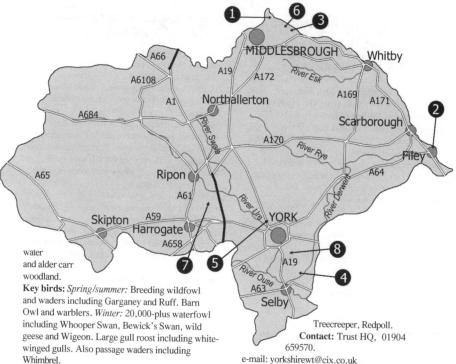

water
and alder carr
woodland.
Key birds: *Spring/summer:* Breeding wildfowl
and waders including Garganey and Ruff. Barn
Owl and warblers. *Winter:* 20,000-plus waterfowl
including Whooper Swan, Bewick's Swan, wild
geese and Wigeon. Large gull roost including white-
winged gulls. Also passage waders including
Whimbrel.
Contact: Peter Roworth, Site Manager English
Nature, Genesis 1, Heslington Road, York YO10
5ZQ. 01904 435500.
e-mail: york@english-nature.org.uk
Leaflet available SAE please or check website
www.english-nature.org.uk

5. MOORLANDS WOOD RESERVE

Yorkshire Wildlife Trust.
Location: From the York ringroad, take the A19
Thirsk/Northallerton road. After 1.6km turn R at The
Blacksmith's Arms pub for Skelton. Go through the
village. After 3.2km, look for an open parking area
at a wide verge on the L alongside the wood and just
before the entrance gate.
Access: Open all year. Entry free.
Facilities: Car park, woodland paths. Suitable for
wheelchairs.
Public transport: None.
Habitat: Mature mixed woodland, ponds.
Key birds: *Spring/summer:* Cuckoo, Curlew,
Garden Warbler, Blackcap, Spotted Flycatcher.
Winter: Thrushes, Siskin. *All year:* Tawny Owl,
Great Spotted Woodpecker, tits inc Marsh Tit, Jay,

Treecreeper, Redpoll.
Contact: Trust HQ, 01904
659570.
e-mail: yorkshirewt@cix.co.uk

6. SALTBURN GILL

Tees Valley Wildlife Trust.
Location: NZ 674 205. From Middlesbrough take
A174 to Saltburn-by-the-Sea. Access from Cat Nab
car park on seafront.
Access: Public footpaths through the site, therefore
open at all times. Paths can be muddy after rain.
Some steep gradients.
Facilities: No visitor centre or hides. Parking at Cat
Nab car park, Saltburn (charges). Toilets in car
park.
Public transport: Bus – frequent to Saltburn town
centre from Middlesbrough. 0.75 mile walk to site.
Hourly 62 bus from Middlesbrough to Loftus stops
at seafront (Arriva tel 0870 6082608). Train –
frequent services to Saltburn from Middlesbrough
and Darlington. Reserve is 0.75 mile from station.
Habitat: Semi-ancient natural deciduous woodland.
Some grassland and scrub, stream.
Key birds: Typical woodland species, Marsh Tit,
Woodcock, Grey Wagtail.
Contact: Trust HQ, 01642 759900.
e-mail: info@teesvalleywt.cix.co.uk

7. STAVELEY NATURE RESERVE

Yorkshire Wildlife Trust
Location: SE 365 634. The reserve is just N of
Staveley village, miles SW of Boroughbridge.
Access: Open all year. Public footpath runs through
the reserve. Parking limited. Park near village green
opposite phone box. Walk along Minskip Road for
150 yeards past Spellow Crescent on R. Turn L
between bungalow and Ceres House into an
unmarked lane. At the end of lane bear R through a
kissing gate onto reserve.
Facilities: An information panel overlooks the main
lagoon. There is a locked bird hide. Keys available
from the reserve chairman for a small fee.
Public transport: Staveley is served by bus.
Telephone: 01423 537300.
Habitat: Shallow lagoon edged with reedswamp, fen
and flower-rich grassland with scrub and pasture.
Ponds and small area of woodland.
Key birds: *Summer*: Reed, Sedge, Garden and
Grasshopper Warblers and Reed Bunting all breed on
site. *Winter*: Wildfowl are attracted to main lagoon.
Jack Snipe and Short-eared Owl frequently seen.
Contact: Bob Evison (reserve chairman), Trust HQ,
01904 659570 or 01423 865342.
e-mail: info@yorkshirewt.cix.co.uk

8. WHELDRAKE INGS/LOWER DERWENT VALLEY NNR

Yorkshire Wildlife Trust.
Location: Leave York ring road to S onto A19 Selby
road. After one mile turn L to Wheldrake signed
Wheldrake 4 and Thorganby 6.5. After 3.5 miles
pass through Wheldrake and continue towards
Thorganby. After a sharp R bend turn L after 0.5
miles onto an unsigned tarmac track. Look for two
old stone gateposts with pointed tops. Car park is
about 0.25 miles down track. To reach reserve, cross
the bridge over the river and turn R over a stile.
Access: Open all year. Please keep to the riverside
path. From Apr-Sep.
Facilities: Car park, four hides.
Habitat: Water meadows, river, scrub, open water.
Key birds: *Spring/summer*: Wildfowl species, Grey
Partridge, Turtle Dove, some waders, Spotted
Flycatcher, warblers. *Winter*: Occasional divers and
scarce grebes. wildfowl inc. Pintail, Pochard,
Goshawk, Hen Harrier, Water Rail, Short-eared Owl,
thrushes, good mix of other birds.
Contact: Trust HQ, 01904 659570.

Yorkshire, South & West

1. BRETTON COUNTRY PARK AND OXLEY BANK WOOD

Yorkshire Wildlife Trust/Wakefield MDC.
Location: SE 295 125. 15 miles S of Leeds and N
of Sheffield. Leave motorway at J38. Take A637
Huddersfield road to N. After 0.5 miles, entrance to
Park is on L. Can also park in Sculpture Park's car
park, which is first L off A637 in West Bretton.
Access: Open all year. Permit required for the
Yorkshire Wildlife Trust area.
Facilities: Car park, visitor centre, information
leaflets.
Public transport: None.
Habitat: Landscaped park, mature woodland, two
lakes.
Key birds: *Spring/summer*: Cuckoo, warblers,
Spotted Flycatcher, Sand Martin, Swallow. *Winter*:
Fieldfare, Redwing, Brambling, Redpoll, Siskin,
Hawfinch. *All year*: Kingfisher, all three
woodpeckers, Little and Tawny Owls, Linnet,
Bullfinch, Yellowhammer, usual woodland birds.
Contact: Trust HQ, 01904 659570.
e-mail: info@yorkshirewt.cix.co.uk

2. DENABY INGS NATURE RESERVE

Yorkshire Wildlife Trust.
Location: Reserve is on A6023 from Mexborough.
Look for a L fork signed Denaby Ings Nature
Reserve. Proceed along Pastures Road for 0.5 miles
and watch for 2nd sign on R marking entrance to car
park. From car park, walk back to road to set of
concrete steps on R which lead up to a small visitor
centre and a hide.
Access: Open all year.
Facilities: Car park, visitor centre, hide, nature trail.
Public transport: None.
Habitat: Water, deciduous woodland, marsh,
willows.
Key birds: *Spring/summer*: Waterfowl, Little
Ringed Plover, Turtle Dove, Cuckoo, Little and
Tawny Owl, Sand Martin, Swallow, Whinchat,

possible Grasshopper Warbler, Lesser Whitethroat, Whitethroat, other warblers, Spotted Flycatcher, Red-legged and Grey Partridges, Kingfisher. *Passage*: Waders, Common, Arctic and Black Terns, Redstart, Wheatear. *Winter*: Whooper Swan, wildfowl, Jack Snipe, waders, Grey Wagtail, Short-eared Owl, Stonchat, Fieldfare, Redwing, Brambling, Siskin. *All year*: Corn Bunting, Yellowhammer, all three woodpeckers possible, Willow Tit.
Contact: Trust HQ, 01904 659570.
e-mail: info@yorkshirewt.cix.co.uk
www.yorkshire-wildlife-trust.org.uk

3. FAIRBURN INGS

RSPB (North West England Office).
Location: SE 452 277. 12.5 miles from Leeds, six miles from Pontefract, 3.5 miles from Castleford situated next to A1 at Fairburn turnoff.
Access: Reserve and hides open every day (9am-dusk). Centre with shop open weekdays (11am-4pm) and weekends (10am-5pm) and Bank Holidays. Hot and cold drinks available. Dogs on leads at all times. Boardwalk leading to Pickup Pool and feeding station and paths to centre wheelchair-friendly.
Facilities: Reserve hides include three open at all times with one locked at dusk. Toilets open when centre open or 9am-5pm. Disabled access to toilets. All nature trails follow public paths and are open at all times.

Public transport: Nearest train stations are Castleford or Pontefract. Buses approx every hour from Pontefract and Tadcaster. Infrequent from Castleford and Selby.
Habitat: Open water due to mining subsidence, wet grassland, marsh and willow scrub, reclaimed colliery spoil heaps.
Key birds: *Winter*: A herd of Whooper Swan usually roost. Normally up to five Smew including male, Wigeon, Gadwall, Goosander, Goldeney. Spring: Osprey, Wheatear, Little Gull and five species of tern pass through. *Summer*: Breeding birds include Reed and Sedge Warblers, Shoveler, Gadwall, Cormorant.
Contact: Chris Drake, Information Warden, Fairburn Ings Visitor Centre, Newton Lane, Fairburn, Castleford WF10 2BH. 01977 603796.

4. HARDCASTLE CRAGS

National Trust.
Location: From Halifax, follow A646 W for five miles to Hebden Bridge and pick up National Trust signs in town centre. Continue to A6033 Keighley Road. Follow this for 0.75 miles. Turn L at National Trust sign to car parks.
Access: Open all year.
Facilities: 2 small pay and display car parks, cycle racks and several way marked trails. Gibson mill has toilets, café, exhibitions. The Mill is not connected to any mains services, in extreme conditions may be closed for health and safety reasons.
Public transport: Good public transport links. Trains to Hebden Bridge every half hour. Regular buses from major towns in the area. Catch bus C from station, becomes bus H from town centre, for hourly service to the Crags. Pleasant 1.5 mile riverside walk from town to Hardcastle Crags, well signed.
Habitat: Wooded valleys, ravines, streams, hay meadows and moorland edge.
Key birds: *Spring/summer*: Cuckoo, Redstart, Lesser Whitethroat, Garden Warbler, Blackcap, Wood Warbler, Chiffchaff, Spotted Flycatcher, Pied Flycatcher, Curlew, Lapwing, Meadow Pipit. *All year*: Sparrowhawk, Kestrel, Green and Greater Spotted Woodpeckers, Tawny Owl, Barn Owl, Little Owl, Jay, Coal Tit and other woodland species.
Contact: National Trust, Hardcastle Crags, Hollin Hall Office, Hebden Bridge, West Yorks HX7 7AP. 01422 844518

5. INGBIRCHWORTH RESERVOIR

Yorkshire Water.
Location: Leave M1 at J37, take A628 to Manchester and Penistone. After five miles turn R at roundabout onto A629 Huddersfield road. After 2.5 miles you reach Ingbirchworth. At a sign for The Fountain Inn, turn L. Pass a pub. Road bears L to cross dam, go straight forward onto track leading to car park. From car park, follow footpath to reservoir.
Access: Open all year. One of the few reservoirs in the area with access.
Facilities: Car park, picnic tables.
Habitat: Reservoir, small strip of deciduous woodland.
Key birds: *Spring/summer*: Whinchat, warblers, woodland birds, House Martin. *Spring/autumn passage*: Little Ringed Plover, Ringed Plover, Dotterel, waders, Common Tern, Arctic Tern, Black Tern, Yellow Wagtail, Wheatear. *Winter*: Wildfowl, Golden Plover, waders, occasional rare gull such as Iceland or Glaucous, Grey Wagtail, Fieldfare, Redwing, Brambling, Redpoll.
Contact: Yorkshire Water, PO Box 52, Bradford BD3 7YD.

6. RSPB OLD MOOR

RSPB North West Office.
Location: SE 422 011. From M1 J36, then A6195. From A1 J37, then A635 and A6195 – follow brown signs.
Access: Open Apr 1-Oct 31 (Wed-Sun 9am-5pm), Nov 1-Mar 31 (Wed/Thu-Sat/Sun 10am-4pm). Entry fee with Annual Membership available.
Facilities: Toilets (including disabled), large visitor centre and shop, five superb hides. All sites including hides fully accessible for disabled.
Public transport: Buses – information from South Yorkshire Passenger Transport 01709 589200.
Habitat: Lakes and flood meadows, wader scrape and reedbeds.
Key birds: *Winter*: Large numbers of wildfowl. *Summer*: Breeding waders and wildfowl. Rare vagrants recorded annually.
Contact: The Warden, RSPB Old Moor, Old Moor Lane, Wombwell, Barnsley, South Yorkshire S73 0YF. 01226 751593 Fax: 01226 341078. www.rspb.org.uk

7. POTTERIC CARR

Yorkshire Wildlife Trust.
Location: SE 589 007. From M18 junction 3 take A6182 (Doncaster) and at first roundabout take third exit; entrance and car park are on R after 50m.
Access: Access by permit only. Parties must obtain prior permission.
Facilities: Field Centre (refreshments, toilet) open (9.30 a.m.-4 p.m.) Thursdays to Sundays all year, Bank Holiday Mondays open, also Tues 9.30-1.30. Eight hides (three for disabled). Ticketing 7 days/week. Approx 12 km of footpaths including 8km suitable for disabled unaided. 10 new/refurbished hides, 6 suitable for wheelchairs. See website for more details and events programme.
Public transport: Buses from Doncaster to new B&Q store travel within easy reach of entrance.
Habitat: Reed fen, subsidence ponds, artificial pools, grassland, woodland.
Key birds: 96 species have bred. Nesting waterfowl (inc. Shoveler, Gadwall, Pochard), Water Rail, Kingfisher, all three woodpeckers, Lesser Whitethroat, Reed and Sedge Warblers, Willow Tit. *Passage/winter*: Bittern, Marsh Harrier, Black Tern, waders, wildfowl.
Contact: For further contact information, www.potteric-carr.org.uk (01302 364152).

8. SPROTBOROUGH FLASH AND THE DON GORGE

Yorkshire Wildlife Trust.
Location: From A1, take A630 to Rotherham, 3 miles W of Doncaster. After 1/4 mile, turn R at lights to Sprotborough. After road drops down, cross bridge over river then another over canal. Turn immediately L. Park in small roadsidearea on L beside canal. Walk along canal bank, past The Boat Inn. Reserve entrance is approx 1/2 mile further.
Access: Open all year.
Facilities: Three hides, footpaths.
Public transport: River bus from Doncaster in summer months.
Habitat: River, reed, gorge, woodland.
Key birds: *Summer*: Turtle Dove, Cuckoo, hirundines, Lesser Whitethroat, Whitethroat, Garden Warbler, Blackcap, Chiffchaff, Willow Warbler, Spotted Flycatcher. *Spring/autumn passage*: Little Ringed Plover, Dunlin, Greenshank, Green Sandpiper, waders, Yellow Wagtail. *Winter/all year*: wildfowl, Water Rail, Snipe, Little Owl, Tawny Owl, all three woodpeckers, thrushes, Siskin, possible Corn Bunting.
Contact: Trust HQ, 01904 659570.
e-mail: info@yorkshirewt.cix.co.uk www.yorkshire-wildlife-trust.org.uk

SCOTLAND

Border counties

Borders

1. BEMERSYDE MOSS

Scottish Wildlife Trust.
Location: NT 614 340. 7.9 miles E of Melrose on minor road, between Melrose and Smailholm.
Access: Permit required.
Facilities: Hide with parking nearby.
Public transport: None.
Habitat: Shallow loch and marsh with large Black-headed Gull colony and wintering wildfowl.
Key birds: *Summer*: Black-necked Grebe, Grasshopper Warbler. *Winter:* Wildfowl, waders on migration, raptors.
Contact: Trust HQ, 0131 312 7765.

2. DUNS CASTLE

Scottish Wildlife Trust.
Location: NT 778 550. Located N of the centre of Duns (W of Berwick-upon-Tweed).
Access: Access from Castle Street or at N end of reserve from B6365. Alternatively take A6112 North.
Facilities: Deposit £2. Leaflet available from Trust.
Public transport: None.
Habitat: Loch and woodland.
Key birds: Woodland birds, including Marsh Tit, Chiffchaff and Pied Flycatcher, waterfowl. Also occasional otter and red squirrel.
Contact: Trust HQ, 0131 312 7765.

3. GUNKNOWE LOCH AND PARK

Borders Council.
Location: NT 523 51. two miles from Galashiels on A6091. Park at Gunknowe Loch.
Access: Open all year.
Facilities: Car park, paths.
Public transport: None.
Habitat: River, parkland, scrub, woodland.
Key birds: *Spring/summer*: Grey Wagtail, Kingfisher, Sand Martin, Blackcap, Sedge and Grasshopper Warblers. *Passage*: Yellow Wagtail, Whinchat, Wheatear. *Winter*: Thrushes, Brambling, Wigeon, Tufted Duck, Pochard, Goldeneye.

All year: Great Spotted and Green Woodpeckers, Redpoll, Goosander, possible Marsh Tit.
Contact: Borders Council, Newton Street, Boswells, Melrose TD6 0SA. 01835 824000.
e-mail: enquiries@scotborders.gov.uk

4. ST ABB'S HEAD

National Trust for Scotland.
Location: NT 914 693. Lies five miles N of Eyemouth. Follow A1107 from A1.
Access: Reserve open all year. Keep dogs on lead. Cliff path is not suitable for disabled visitors.
Facilities: Visitor centre and toilets open daily Apr-Oct.
Public transport: Nearest rail station is Berwick-upon-Tweed. Bus service from Berwick, tel 018907 81533.
Habitat: Cliffs, coastal grasslands and freshwater loch.
Key birds: *Apr-Aug*: Seabird colonies with large numbers of Kittiwake, auks, Shag, Fulmar, migrants. *Apr-May and Sept-Oct*: Good autumn seawatching.
Contact: Kevin Rideout, Rangers Cottage, Northfield, St Abbs, Borders TD14 5QF. 018907 71443. e-mail: krideout@nts.org.uk
www.nts.org.uk

5. THE HIRSEL

The Estate Office, The Hirsel.
Location: NT 827 403. Signed off A698 Kelso-Coldstream road on outskirts of Coldstream.
Access: Open all year. Private estate so please keep to the public paths.
Facilities: Car parks, visitor centre, leaflets, trails.
Public transport: Bus: Coldstream, Kelso, Berwick-upon-Tweed, Edinburgh.
Habitat: Freshwater loch, reeds, woods.
Key birds: *Spring/summer*: Redstart, Garden Warbler, Blackcap, flycatchers, possible Water Rail, wildfowl. *Autumn*: Wildfowl, Goosander, possible Green Sandpiper. *Winter*: Whooper Swan, Pink-footed Goose, Wigeon, Goldeneye, Pochard, occasional Smew, Scaup, Slavonian Grebe.
Contact: The Estate Office, The Hirsel, Coldstream TD12 4LF. 01890 882834.

Dumfries and Galloway

6. CAERLAVEROCK

The Wildfowl & Wetlands Trust.
Location: NY 051 656. From Dumfries take B725 towards Bankend.
Access: Open daily except Christmas Day.
Facilities: 20 hides, heated observatory, four towers, Salcot Merse Observatory, sheltered picnic area. Self-catering accommodation and camping facilities. Nature trails in summer. Old Granary visitor building with fair-trade coffee shop serving light meals and snacks; natural history bookshop; binoculars & telescopes for sale. Theatre/conference room. Binoculars for hire. Parking for coaches.
Public transport: Bus 371 from Dumfries stops 30 mins walk from reserve. Western Buses 01837 253496.
Habitat: Saltmarsh, grassland.
Key birds: *Winter*: Wildfowl esp. Barnacle Goose (max 25,000) and Whooper Swan.
Contact: The Wildfowl & Wetlands Trust, Eastpark Farm, Caerlaverock, Dumfries DG1 4RS. 01387 770200.

7. KEN/DEE MARSHES

RSPB (South & West Scotland Office).
Location: NX 699 684. Six miles from Castle Douglas – good views from A762 and A713 roads to New Galloway.
Access: From car park at entrance to farm Mains of Duchrae. Open during daylight hours. No dogs.
Facilities: Hides, nature trails. Three miles of trails available, nearer parking for elderly and disabled, but phone warden first. Part of Red Kite trail.
Public transport: None.
Habitat: Marshes, woodlands, open water.
Key birds: *All year*: Mallard, Grey Heron, Buzzard. *Spring/summer*: Pied Flycatcher, Redstart, Tree Pipit, Sedge Warbler. *Winter*: Greenland White-fronted and Greylag Geese, birds of prey (Hen Harrier, Peregrine, Merlin, Red Kite).

Contact: Paul Collin, Gairland, Old Edinburgh Road, Minnigaff, Newton Stewart DG8 6PL. 01671 402861.

8. MERSEHEAD

RSPB (South & West Scotland Office).
Location: NX 925 560. From Dalbeattie, take B793 or A710 SE to Caulkerbush.
Access: Open at all times.
Facilities: Hide, nature trails, information centre and toilets.
Public transport: None.
Habitat: Wet grassland, arable farmland, saltmarsh, inter-tidal mudflats.
Key birds: *Winter*: Up to 9,500 Barnacle Geese, 4,000 Teal, 2,000 Wigeon, 1,000 Pintail, waders (inc. Dunlin, Knot, Oystercatcher). *Summer*: Breeding birds include Lapwing, Redshank, Sky Lark.
Contact: Eric Nielson, Mersehead, Southwick, Mersehead, Dumfries DG2 8AH. 01387 780298.

9. MULL OF GALLOWAY

RSPB (South & West Scotland Office).
Location: NX 156 304. Most southerly tip of Scotland – five miles from village of Drummore, S of Stranraer.
Access: Open at all times. Access suitable for disabled. Disabled parking by centre. Centre open summer only (Apr-Oct).
Facilities: Visitor centre, toilets, nature trails, CCTV on cliffs.
Public transport: None.

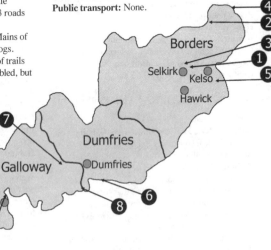

Habitat: Sea cliffs, coastal heath.
Key birds: *Spring/summer*: Guillemot, Razorbill, Kittiwake, Black Guillemot, Puffin, Fulmar, Raven, Wheatear, Rock Pipit, Twite. Migrating Manx Shearwater. *All year*: Peregrine.
Contact: Paul Collin, Gairland, Old Edinburgh Road, Minnigaff, Newton Stewart DG8 6PL. 01671 402851.

10. WIGTOWN BAY

Dumfries & Galloway Council.
Location: NX 465 545. Between Wigtown and Creetown. It is the largest LNR in Britain at 2,845 ha. The A75 runs along E side with A714 S to Wigtown and B7004 providing superb views of the LNR.
Access: Open at all times. The hide is disabled friendly. Main accesses: Roadside lay-bys on A75 near Creetown and parking at Martyr's Stake and Wigtown Harbour. All suitable for coaches. The visitor facility in Wigtown County Building has full disabled access including lift and toilets.
Facilities: A hide at Wigtown Harbour with views over the River Bladnoch, saltmarsh and fresh water wetland has disabled access from harbour car park. Walks and interpretation in this area. Visitor room in Wigtown County Buildings has interpretation facilities and a commanding view of the bay. CCTV of Ospreys breeding in Galloway during summer and wetland birds in winter. Open Mon-Sat (10am-5pm, later some days). Sun (2pm-5pm).
Public transport: Travel Information Line 08457 090510 (local rate, 9am-5pm Mon-Fri). Bus No 415 for Wigtown and west side. Bus No 431 or 500 X75 for Creetown and E side.
Habitat: Estuary with extensive saltmarsh/merse and mudflats with newly developed fresh water wetland at Wigtown Harbour.
Key birds: *Winter*: Internationally important for Pink-footed Goose, nationally important for Curlew, Whooper Swan and Pintail, with major gull roost and other migratory coastal birds. *Summer*: Breeding waders and duck.
Contact: Elizabeth Tindal, County Buildings, Wigtown, DG8 9JH. 01988 402 401, (M)07702 212 728. e-mail: Elizabeth.Tindal@dumgal.gov.uk www.dgcommunity.net

Central Scotland

Argyll

1. COLL RSPB RESERVE

RSPB (South & West Scotland Office).
Location: NM 168 561. By ferry from Oban. Take B8070 W from Arinagour for five miles. Turn R at Arileod. Continue for about one mile. Park at end of road. Reception point at Totronald.
Access: Open all year. Please avoid walking through fields and crops.
Facilities: Car park, information bothy at Totronald, guided walks in summer. Corn Crake viewing bench.
Public transport: None.
Habitat: Sand dunes, beaches, machair grassland, moorland, farmland.
Key birds: *Spring/summer*: Corn Crake, Redshank, Lapwing, Snipe. *Winter*: Barnacle and Greenland White-fronted Geese.

Contact: RSPB Coll Nature Reserve, Totronald, Isle of Coll, Argyll, PA78 6TB, 01879 230301.

2. LOCH GRUINART, ISLAY

RSPB Scotland (South & West Scotland Office).
Location: Sea loch on N coast, seven miles NW from Bridgend.
Access: Hide open all hours, no dogs, visitor centre open (10am-5pm), disabled access to hide, toilets.
Facilities: Toilets, visitor centre, hide, trail.
Public transport: None.
Habitat: Low wet grasslands, moorland.
Key birds: *Oct-Apr*: Barnacle and Greenland White-fronted Goose. *May-Aug*: Corn Crake. *Sept-Nov*: Migrating wading birds.
Contact: Liz Hathaway, RSPB Scotland, Bushmills Cottage, Gruinart, Isle of Islay PA44 7PP. 01496 850505. e-mail: liz.hathaway@rspb.org.uk www.rspb.org.uk

3. MACHRIHANISH SEABIRD OBSERVATORY

Eddie Maguire and John McGlynn.
Location: NR 628 209. Southwest Kintyre, Argyll. Six miles W of Campbeltown on A83, then B843.
Access: Daily May-Oct. Wheelchair access. Dogs welcome. Parking for three cars.
Facilities: Seawatching hide, toilets in nearby village.
Public transport: Regular buses from Campbeltown (West Coast Motors, tel. 01586 552319).
Habitat: Marine, rocky shore and upland habitats.
Key birds: *Summer:* Golden Eagle, Peregrine and Twite. *Autumn:* Passage seabirds and waders. Gales often produce inshore movements of Leach's Petrel and other scarce seabirds including Balearic Shearwater and Grey Phalarope. *Winter:* Great Northern Diver.
Contact: Eddie Maguire, 25B Albyn Avenue, Campbeltown, Argyll PA28 6LX. 07919 660292. www.mso.1c24.net

Ayrshire

4. AYR GORGE WOODLANDS

Scottish Wildlife Trust.
Location: NS 457 249. From Ayr take A719 NE for about three miles to A768. Go straight over the roundabout onto B473 and continue to Failford. Park in the lay-by in village.
Access: Open all year. Access by well-maintained path along west bank of River Ayr.
Facilities: Footpaths, interpretation boards, leaflets and information board.
Public transport: None.
Habitat: Woodland.
Key birds: Woodland and riverside birds.
Contact: Trust HQ. 0131 312 7765.
e-mail: scottishwt@cix.co.uk
www.swt.org.uk

5. CULZEAN CASTLE & COUNTRY PARK

National Trust for Scotland.
Location: NS 234 103.
Access: Open all year.
Facilities: Car park, visitor centre.
Public transport: None.
Habitat: Shoreline, parkland, woodland, streams, ponds.
Key birds: *Spring/summer:* Blackcap and other warblers. *Winter:* Waders, wildfowl. *All year:* Tits, finches.
Contact: National Trust for Scotland, Wemyss House, 28 Charlotte Square, Edinburgh EH2 4ET. 0131 243 9300.

6. GARNOCK FLOODS

Scottish Wildlife Trust.
Location: NS 306 417. N of Irvine. From A779 N, take the one-way road to Bogside, just beyond A779 interchange. Park on roadside. Best viewed from cycle track along E boundary.
Access: Open all year.
Facilities: None.
Public transport: Bus service on A737, 1 mile distant.
Habitat: River, ponds, fields, wood.
Key birds: *Spring/summer:* Sedge and Willow Warblers, Lesser Whitethroat, Sand Martin. *Winter:* wildfowl inc Goldeneye, Mute Swan, occasional Garganey, waders inc Ruff and Snipe, number of ducks. Winter thrushes. *All year:* Kestrel, Sparrowhawk, Buzzard.
Contact: Trust HQ. 0131 312 7765.
e-mail: scottishwt@cix.co.uk www.swt.org.uk

Clyde

7. FALLS OF CLYDE

Scottish Wildlife Trust.
Location: NS 88 34 14. Approx one mile S of Lanark. Directions from Glasgow – travel S on M74 until J7 then along A72, following signs for Lanark and New Lanark.
Access: Open daylight hours all year. Partial disabled access.
Facilities: Visitor centre open 11am-5pm (Mar-Dec), 12-4pm (Jan-Feb). Toilets and cafeteria on site. Seasonal viewing facility for Peregrines. Numerous walkways and ranger service offers comprehensive guided walks programme.
Public transport: Scotrail trains run to Lanark (0845 7484950). Local bus service from Lanark to New Lanark.
Habitat: River Clyde gorge, waterfalls, mixed riparian/conifer woodlands, meadow, pond.
Key birds: More than 100 species of bird recorded on the reserve, including unrivalled views of breeding Peregrine. Others include Goshawk, Barn Owl, Kingfisher, Dipper, Lapwing, Pied Flycatcher and Sky Lark.

NATURE RESERVES - SCOTLAND

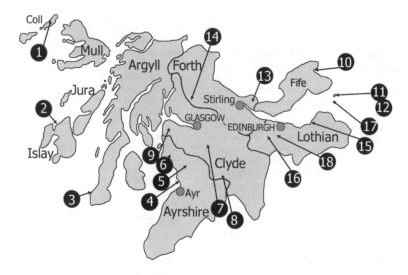

Contact: Miss Lindsay Cook, The Scottish Wildlife Trust Visitor Centre, The Falls of Clyde Reserve & Visitor Centre, New Lanark, South Lanark ML11 9DB. 01555 665262. e-mail: fallsofclyde@swt.co.uk www.swt.org.uk

8. KNOCKSHINNOCK LAGOONS

Scottish Wildlife Trust.
Location: NS 608 137. Car park off B741 (New Cumnock to Dalmellington road), or from Kirkbrae in New Cumnock.
Access: Open all year. Access from Church Lane, New Cumnock.
Facilities: None.
Public transport: None.
Habitat: Mining spoil, lagoons, meadows, woodland.
Key birds: *Summer*: Breeding Redshank, Lapwing, Snipe, Curlew, Shoveler, Teal, Pochard and Garganey. *Winter*: Whooper Swans and large number of ducks. *Spring/autumn*: Good for migrating waders.
Contact: Trust HQ, 0131 312 7765

9. LOCHWINNOCH

RSPB (South & West Scotland Office).
Location: NS 358 582. 18 miles SW of Glasgow, adjacent to A760.
Access: Open every day except Christmas and

Boxing Day, Jan 1 and Jan 2. (10am-5pm).
Facilities: Special facilities for schools and disabled. Refreshments available. Visitor centre, hides.
Public transport: Rail station adjacent, bus services nearby.
Habitat: Shallow lochs, marsh, mixed woodland.
Key birds: *Winter*: Wildfowl (esp. Whooper Swan, Greylag, Goosander, Goldeneye). Occasional passage migrants inc. Whimbrel, Greenshank. *Summer*: Breeding Great Crested Grebe, Water Rail, Sedge and Grasshopper Warblers.
Contact: RSPB Nature Centre, Largs Road, Lochwinnoch, Renfrewshire PA12 4JF. 01505 842663; (fax) 01505 843026.
e-mail lochwinnoch@rspb.org.uk.

Fife

10. EDEN ESTUARY

Fife Council.
Location: NO 470 195. The reserve can be accessed from Guardbridge, St Andrews (one mile) on A91, and from Leuchars via Tentsmuir Forest off A919 (four miles).
Access: The Eden Estuary Centre is open (9am-5pm) every day except Christmas Day, New Year's Day and the day of the Leuchars airshow. Reserve is open all year, but a permit (from Ranger service) is

required to access the N shore. Limited coach access and coach charge if using Kinshaldy car park.
Facilities: Visitor centre at Guardbridge. Information panels at Outhead. Hide at Balgove Bay (key from Ranger Service).
Public transport: Leuchars train station. Regular buses Cupar-Dundee-St Andrews. Tel: 01334 474238.
Habitat: Saltmarsh, river, tidal flats, sand dunes.
Key birds: *Winter*: Main interest is wildfowl and waders, best place in Scotland to see Black-tailed Godwit. Other species include Grey Plover, Shelduck, Bar-tailed Godwit. Offshore Common and Velvet Scoter occur and Surf Scoter is regularly seen. Peregrine, Merlin and Short-eared Owl occur in winter.
Contact: Les Hatton, Fife Ranger Service, Craigtown Country Park, St Andrews, Fife KY16 8NX. 01334 473047. (M)/07985 707593.
e-mail: refrs@craigtoun.freserve.co.uk

11. ISLE OF MAY BIRD OBSERVATORY

Location: Firth of Forth
Access: Public boat from Anstruther, observatory has a contract with local firm Smilven Divers.
Facilities: Hostel accommodation in ex-lighthouse (the Low Light) for up to six, (Apr-Oct); usual stay is one week. No supplies on island; visitors must take own food and sleeping bag. Five Heligoland traps used for ringing migrants when qualified personnel present. SNH Warden usually resident Apr-Sep.
Key birds: Breeding seabirds and migrants including a number of 'firsts' for Britain. Bookings: Mike Martin, 2 Manse Park, Uphall, W Lothian EH52 6NX. 01506 855285;
e-mail: mwa.martin@btinternet.com,

12. ISLE OF MAY NNR

Scottish Natural Heritage.
Location: NT 655 995. Small island lying six miles off Fife Ness in Firth of Forth.
Access: Boats run from Anstruther and North Berwick. Contact SNH for details 01334 654038. Keep to paths. Fishing boat from Anstruther arranged by the Observatory for those using Observatory accommodation. Delays are possible, both arriving and leaving, because of weather.
Facilities: No dogs; no camping; no fires. Prior permission required if scientific work or filming is to be carried out.

Public transport: Regular bus service to Anstruther and North Berwick harbour.
Habitat: Sea cliffs, rocky shoreline.
Key birds: *Early Summer*: Breeding auks and terns, Kittiwake, Shag, Eider, Fulmar. Over 68,000 pairs of Puffins. *Autumn/spring:* Weather-related migrations include rarities each year.
Contact: For Observatory accomodation: David Thorne, Craigurd House, Blyth Bridge, West Linton, Peeblesshire EH46 7AH. For all other enquiries: SNH, 46 Crossgate, Cupar, Fife KY15 5HS.

Forth

13. GARTMORN DAM

Clackmannanshire Council.
Location: NS 912 940. Approx one mile NE of Alloa, signposted from A908 in Sauchie.
Access: Open at all times. No charge. Access road not suitable for large coaches.
Facilities: Visitor centre with toilets. Open 8.30am-8.00pm daily (Apr-Sept inclusive) and 1pm-4pm (weekends only Oct-Mar). One hide, key obtainable from visitor centre. Provision for disabled.
Public transport: Bus service to Sauchie. First Bus, 01324 613777. Stirling Bus Station, 01786 446 474.
Habitat: Open water (with island), woodland – deciduous and coniferous, farmland.
Key birds: *Summer*: Great Crested Grebe, breeding Sedge and Reed Warblers, occasional Osprey. *Autumn*: Migrant waders. *Winter*: Wildfowl (regionally important site), Kingfisher, Water Rail.
Contact: Clackmannanshire Ranger Service, Lime Tree House, Alloa, Clackmannanshire FK10 1EX. 01259 450000. e-mail: rangers@clacks.gov.uk www.clacksweb.org.uk

14. INVERSNAID

RSPB (South & West Scotland Office).
Location: NN 337 088. On E side of Loch Lomond. Via B829 W from Aberfoyle, then along minor road to car park by Inversnaid Hotel.
Access: Open all year.
Facilities: None.
Public transport: None.
Habitat: Deciduous woodland rises to craggy ridge and moorland.
Key birds: *Summer*: Breeding Buzzard, Blackcock, Grey Wagtail, Dipper, Wood Warbler, Redstart, Pied Flycatcher, Tree Pipit. The loch is on a migration route, especially for wildfowl and waders.

Contact: RSPB South & West Scotland Office, 10 Park Quadrant, Glasgow, G3 6BS. 0141 331 0993.

Lothian

15. ABERLADY BAY

East Lothian Council (LNR).
Location: NT 472 806. From Edinburgh take A198 E to Aberlady. Reserve is 1.5 miles E of Aberlady village.
Access: Open at all times. Please stay on footpaths to avoid disturbance. Disabled access from reserve car park. No dogs.
Facilities: Small car park and toilets. Notice board with recent sightings at end of footbridge.
Public transport: Edinburgh to N Berwick bus service stops at reserve (request), service no 124. Railway 4 miles away at Longniddry.
Habitat: Tidal mudflats, saltmarsh, freshwater marsh, dune grassland, scrub, open sea.
Key birds: *Summer*: Breeding birds include Shelduck, Eider, Reed Bunting and up to eight species of warbler. Passage waders inc. Green, Wood and Curlew Sandpipers, Little Stint, Greenshank, Whimbrel, Black-tailed Godwit. *Winter*: Divers (esp. Red-throated), Red-necked and Slavonian Grebes and geese (large numbers of Pink-footed roost); sea-ducks, waders.
Contact: Ian Thomson, 4 Craigielaw, Longniddry, East Lothian EH32 0PY. 01875 870588.

16. ALMONDELL AND CALDERWOOD

West Lothian Council.
Location: NT 077 670. Several entrances but this is closest to visitor centre – signposted off A89, two miles S of Broxburn.
Access: Open all year. Parking available off Bank Street in Mid Calder. Walk down the footpath beside the Masonic Hall or park in the Oakbank car park on A71 between the two shale bings, then along the roadside to the crash barrier and in.
Facilities: Car park, café, picnic area, toilets,
pushchair access, partial access for wheelchairs, visitor centre (open Sat-Thu), shop, countryside ranger service.
Public transport: None.
Habitat: Woodland, marshland.
Key birds: *Spring/summer*: Woodcock, Tawny Owl, Grasshopper Warbler, Yellowhammer, Blackcap, Garden Warbler. *Winter*: Goldcrest, Redpoll, Willow Tit. *All year*: Dipper, Grey Wagtail, Sparrowhawk.
Contact: Almondell and Calderwood Country Park, Visitor Centre, Broxburn, West Lothian, EH52 5PE, 01506 882254.

17. BASS ROCK

Location: NT602873. Island in Firth of Forth, lying E of North Berwick.
Access: Private property. Regular daily sailings from N Berwick around Rock; local boatman has owner's permission to land individuals or parties by prior arrangement. For details contact Fred Marr, N Berwick on 01620 892838.
Facilities: None.
Public transport: None.
Habitat: Sea cliffs.
Key birds: The spectacular cliffs hold a large Gannet colony, (up to 9,000 pairs), plus auks, Kittiwake, Shag and Fulmer.

18. GLADHOUSE RESERVOIR LNR

Scottish Water.
Location: NT 295 535. S of Edinburgh off A703.
Access: Open all year although there is no access to the reservoir itself. Most viewing can be done from the road (telescope required).
Facilities: Small car park on north side. Not suitable for coaches.
Public transport: None.
Habitat: Reservoir, grassland, farmland.
Key birds: *Spring/summer*: Oystercatcher, Lapwing, Curlew. Possible Black Grouse. *Winter*: Geese, including Pinkfeet, Twite, Brambling, Hen Harrier.
Contact: Scottish Water, PO Box 8855, Edinburgh, EH10 6YQ, 0845 601 8855.
e-mail: customer.service@scottishwater.co.uk
www.scottishwater.co.uk

Eastern Scotland

Angus & Dundee

1. LOCH OF KINNORDY

RSPB (East Scotland).
Location: NO 351 539. Car park on B951 one mile
W of Kirriemuir. Perth 45 minutes drive, Dundee 30
minutes drive, Aberdeen one hour drive.
Access: Open dawn-dusk. Disabled access to two
hides via short trails.
Facilities: Three birdwatching hides.
Public transport: Nearest centre is Kirriemuir.
Habitat: Freshwater loch, fen, carr, marsh.
Key birds: *Spring/summer*: Osprey, Black-necked
Grebe, Blacked-headed Gull. *Winter*: Wildfowl
including Goosander, Goldeneye and Whooper Swan.
Contact: Alan Leitch, RSPB, 1 Atholl Crescent,
Perth PH1 5NG. 01738 639783.
e-mail: alan.leitch@rspb.org
www.rspb.org

2. LOCH OF LINTRATHEN

Scottish Wildlife Trust.
Location: NO 27 54. Seven miles W of Kirriemuir.
Take B951 and choose circular route on unclassified
roads round loch.
Access: Public hide planned but to date, Scottish
Wildlife Trust hide (members' permit system only).
Good viewing points from several places along
unclassified roads.
Facilities: None.
Public transport: None.
Habitat: Oligatrophic/mesotrophic loch. Surrounded
by mainly coniferous woodland.
Key birds: *Summer*: Osprey. *Winter*: Greylag
Goose, Goosander, Whooper Swan, Wigeon, Teal.
Contact: Rick Goater, SWT, Annat House, South
Anag, Ferryden, Montrose, Angus DD10 9UT.
01674 676555. e-mail: swtnero@cix.co.uk

3. MONTROSE BASIN

Scottish Wildlife Trust on behalf of Angus Council.
Location: NO 690 580 Centre of Basin. NO 702
565 Wildlife SWT Centre on A92. 1.5 miles from
centre of Montrose.
Access: Apr 1-Oct 31 (10.30am-5pm). Nov 1-Mar
31, (10.30am-4pm).

Facilities: Visitor centre, shop, vending machine,
toilets, disabled access to centre, two hides on
western half of reserve.
Public transport: Train 1.5 miles in Montrose.
Buses same as above.
Habitat: Estuary, saltmarsh, reedbeds, farmland.
Key birds: Pink-footed Goose – up to 35,000 arrive
Oct. Wintering wildfowl and waders. Breeding Eider
Ducks.
Contact: Karen van Eeden, Scottish Wildlife Trust,
Montrose Basin Wildlife Centre, Rossie Braes,
Montrose DD10 9TJ. 01674 676336.
e-mail: montrosebasin@swt.org.uk
www.swt.org.uk

4. SEATON CLIFFS

Scottish Wildlife Trust.
Location: NO 667 416. 30 acre cliff reserve, nearest
town Arbroath 2.2 miles. Car parking at N end of
promenade at Arbroath.
Access: Open all year.
Facilities: Nature trail with interpretation boards.
Public transport: None.
Habitat: Red sandstone cliffs with nature trail.
Key birds: Seabirds inc. Eider, auks, terns; Rock
Dove and House Martin breed.
Contact: Trust HQ, 0131 312 7765.

Moray & Nairn

5. CULBIN SANDS

RSPB (North Scotland Office).
Location: NH 900 580. Approx ½ mile from Nairn.
Access to parking at East Beach car park, signed off
A96.
Access: Open at all times. Path to reserve suitable for
all abilities.
Facilities: Toilets at car park. Track along dunes and
saltmarsh.
Public transport: Buses stop in Nairn, half mile W
of site. Train station in Nairn three-quarters mile W
of reserve.
Habitat: Saltmarsh, sandflats, dunes.
Key birds: *Winter*: Flocks of Common Scoter,
Long-tailed Duck, Knot, Bar-tailed Godwit, Red-
breasted Merganser. Raptors like Peregrine, Merlin

and Hen Harrier attracted by wader flocks. Roosting geese. *Summer*: Breeding Ringed Plover, Oystercatcher and Common Tern.
Contact: RSPB North Scotland Office., Etive House, Beechwood Park, Inverness IV2 3BW. 01463 715000. e-mail: nsro@rspb.org.uk www.rspb.org.uk

6. SPEY BAY LEIN (THE)

Scottish Wildlife Trust.
Location: NJ 325 657. Eight miles NE of Elgin. From Elgin take A96 and B9015 to Kingston. Reserve is immediately E of village. Car parks at Kingston and Tugnet.
Access: Open all year.
Facilities: Wildlife centre, car park, information board.
Public transport: None.
Habitat: Shingle, rivermouth and coastal habitats.
Key birds: *Summer*: Osprey, waders, wildfowl. *Winter*: Seaduck and divers offshore (esp. Long-tailed Duck, Common and Velvet Scoters, Red-throated Diver).
Contact: Trust HQ, 0131 312 7765.

NE Scotland

7. FORVIE NNR

Scottish Natural Heritage.
Location: NK 034 289.
Access: Dogs on leads only. Reserve open at all times but ternery closed Apr 1-end of Aug annually. Stevenson Forvie Centre open every day (Apr-Sept) and, when staff are available, outside those months. Wheelchair access to the centre.
Facilities: Interpretive display and toilets in Stevenson Forvie Centre. Bird hide, waymarked trail. Space is available for coach parking at the Stevenson Forvie Centre or at Waterside Car Park.
Public transport: Bluebird No 263 to Cruden Bay. Ask for the Newburgh or Collieston Crossroads stop. Tel: 01224 591381.
Habitat: Estuary, dunes, coastal heath.
Key birds: *Spring/summer*: Eider and

terns nesting. *Winter*: Waders and wildfowl on estuary.
Contact: Alison Matheson (Area Officer), Scottish Natural Heritage, Stevenson Forvie Centre, Little Collieston Croft, Collieston, Aberdeenshire AB41 8RU. 01358 751330.
www.snh.org.uk

8. FOWLSHEUGH

RSPB (East Scotland).
Location: NO 879 80. Cliff top path N from Crawton, signposted from A92, three miles S of Stonehaven.
Access: Unrestricted. Boat trips (May-Jul) from Stonehaven Harbour. Booking essential. Contact East Scotland regional office. Tel: 01224 624824.
Facilities: Car park with limited number of spaces 200 yards from reserve.
Public transport: None.
Habitat: Sea cliffs.
Key birds: Spectacular seabird colony, mainly Kittiwake and auks. Bottle-nosed dolphins seen regularly.
Contact: The Warden, Starnafin, Crimond, Fraserburgh AB43 8QN. 01346 532017. e-mail: strathbeg@rspb.org.uk

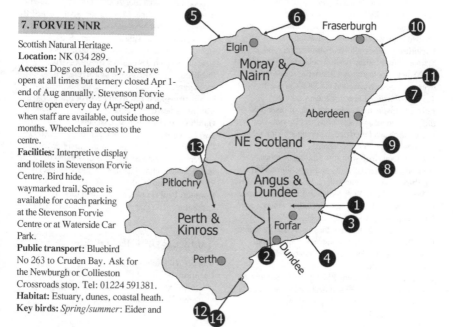

9. GLEN TANAR NNR

Glen Tanar Estate.
Location: 47 96. On the A93 Banchory-Braemar road. From Aboyne turn S across the river, then W on the B976. At Bridge O'Ess, turn L to Braeloine to the car park.
Access: Open all year.
Facilities: Car parks, visitor centre (open Apr-Sep), trails. Do not attempt the summits unless you are properly prepared.
Public transport: None.
Habitat: Remnant of old Caledonian forest, heather moorland, mountains.
Key birds: *All year*: Black Grouse, Woodcock, Siskin, Grey Wagtail, Dipper. *Winter*: Possible Golden Eagle.
Contact: Ranger, Glen Tanar Estate, Brooks House, Glen Tanar, Aboyne, Aberdeenshire AB34 5EU. 01339 880047. e-mail: office@glentanar.co.uk

10. LOCH OF STRATHBEG

RSPB (East Scotland).
Location: NK 057 581. Near Crimond on the A90, nine miles S of Fraserburgh.
Access: Access to Fen, Bay and Bank hides is limited to between 8am and 4pm. Starnafin visitor centre and Tower Pool hide open at all times dawn-dusk. (Loch hides, access restricted to between 8.00am and 4.00pm daily.) Visitor centre now fully accessible to wheelchairs and disabled visitors.
Facilities: Visitor centre and observation room at Starnafin, four hides, Tower Pool hide, accessible via 1,000 metre footpath from Starnafin, (three hides overlooking Loch accessed via MOD airfield). Toilets (with disabled access), car parking.
Public transport: (Access to whole of reserve is difficult without a vehicle.) Bus service runs between Fraserburgh and Peterhead, stopping at Crimond just over one mile from visitor centre.
Habitat: Dune loch with surrounding marshes, reedbeds, grasslands, dunes and agricultural land.
Key birds: *Winter*: Internationally important numbers of Whooper Swan, Pink-footed and Barnacle Geese, large numbers of winter duck including Smew. *Spring/summer*: Waders, Black-headed Gull, Common Tern, Water Rail, farmland birds including Corn Bunting. *Spring/autumn*: Spoonbill, Little Egret, Marsh Harrier, passage waders including Black-tailed Godwit.
Contact: RSPB Warden, RSPB Loch of Strathbeg, Starnafin, Crimond, Fraserburgh, AB43 8QN. 01346 532017. e-mail: strathbeg@rspb.org.uk
www.rspb.org.uk

11. LONGHAVEN CLIFFS

Scottish Wildlife Trust.
Location: NK 116 394. 6.2 miles S of Peterhead. Take A952 S from Peterhead and then A975 to Bullers of Buchan (gorge).
Access: Access from car park at Blackhills quarry or Bullers of Buchan.
Facilities: Leaflet available from Mark Young, Mechlepark, Oldmeldrum, Inverurie, Aberdeenshire, AB5 0DC.
Public transport: None.
Habitat: Rugged red granite cliffs and cliff-top vegetation.
Key birds: *May-July:* Nine species of breeding seabird inc. Kittiwake, Shag, Guillemot, Razorbill, Puffin.
Contact: Trust HQ, 0131 312 7765.

Perth & Kinross

12. LOCH LEVEN NNR

SNH, Loch Leven Laboratory.
Location: NO 150 010. Head S from Perth and leave M90 at exit 6, S of Kinross.
Access: Traditional shoreline access areas at three stretches of shoreline. Most birders go to the RSPB-owned part of the NNR at Vane Farm. New local access guidance is in place at the site. See www.snh.org.uk for details, pick up leaflet locally. New paths and hides planned fpr the reserve.
Facilities: Extensive ornithological research programme.
Public transport: Bus from Perth or Edinburgh to Kinross section of shoreline.
Habitat: Lowland loch with islands.
Key birds: *Winter*: Flocks of geese (over 20,000 Pinkfeet), ducks, Whooper Swan. *Summer*: Greatest concentration of breeding ducks in Britain (10 species) and grebes. *Passage*: Waders (Golden Plover flocks up to 500).
Contact: Paul Brooks, SNH, Loch Leven Laboratory, The Pier, Kinross KY13 8UF. 01577 864439.

13. LOCH OF THE LOWES

Scottish Wildlife Trust.
Location: NO 042 435. Sixteen miles N of Perth, two miles NE of Dunkeld – just off A923 (signposted).

Access: Visitor centre open Apr-Sept inclusive (10am-5pm), mid-Jul to mid-Aug (10am-6pm). Observation hide open all year – daylight hours. No dogs allowed. Partial access for wheelchairs.
Facilities: Visitor centre with toilets, observation hide.
Public transport: Railway station – Birnam/Dunkeld – three miles from reserve. Bus from Dunkeld – two miles from reserve.
Habitat: Freshwater loch with fringing woodland.
Key birds: Breeding Ospreys(Apr-end Aug). Nest in view, 200 metres from hide. Wildfowl and woodland birds. Greylag roost (Oct-Mar).
Contact: Uwe Stoneman, (Manager), Scottish Wildlife Trust, Loch of the Lowes, Visitor Centre, Dunkeld, Perthshire PH8 0HH. 01350 727337.

14. VANE FARM NNR

RSPB (East Scotland).
Location: NT 160 993. By Loch Leven. Take exit 5 from M90 onto B9097.
Access: Open daily (10am-5pm) except Christmas Day, Boxing Day, Jan 1and Jan 2. Cost £3 adults,

£2 concessions, 50p children, £6 family. Free to members. No dogs except guide dogs. Disabled access to shop, coffee shop, observation room and toilets. Coach parking available for up to 2 coaches. Free car parking.
Facilities: Shop, coffee shop and observation room overlooking Loch Leven and the reserve. There is a 1.25 mile hill trail through woodland and moorland. Wetland trail with three observation hides. Toilets, including disabled.
Public transport: Nearest train station Cowdenbeath (nine miles away). Nearest bus station Kinross at Green Hotel (five miles away).
Habitat: Wet grassland and flooded areas by Loch Leven. Arable farmland. Native woodland and heath moorland.
Key birds: *Spring/summer*: Breeding and passage waders (including Lapwing, Redshank, Snipe, Curlew). Farmland birds (including Sky Lark and Yellowhammer). *Winter*: Whooper Swan, Bewick's Swan, Pink-footed Goose.
Contact: Ken Shaw, Senior Site Manager, Vane Farm Nature Centre, Kinross, Tayside KY13 9LX. 01577 862355. e-mail: vanefarm@rspb.co.uk

Highlands & Islands

Highland & Caithness

1. ABERNETHY FOREST RESERVE – LOCH GARTEN

RSPB (North Scotland Office).
Location: NH 981 184. 2.5 miles from Boat of Garten, eight miles from Aviemore. Off B970, follow 'RSPB Ospreys' road signs (between Apr - Aug only).
Access: Osprey Centre open daily 10am-6pm (Apr to end Aug). Disabled access. No dogs (guide dogs only). No charge to RSPB members. Non-members: adults £3, senior citizens £2, children 50p.
Facilities: Osprey Centre overlooking nesting Ospreys, toilets, optics and CCTV live pictures.

Public transport: Bus service to Boat of Garten from Aviemore, 2.5 mile footpath to Osprey Centre. Steam railway to Boat of Garten from Aviemore.
Habitat: Caledonian pine wood.
Key birds: Ospreys nesting from Apr to Aug, Crested Tit, Crossbill, red squirrel. In 2004 hide provided views of lekking Capercaillies Apr to mid-May.
Contact: R W Thaxton, RSPB, Forest Lodge, Nethybridge, Inverness-shire PH25 3EF. 01479 821894.

2. BEN MORE COIGACH

Scottish Wildlife Trust.
Location: NC 075 065. 10 miles N of Ullapool, W of A835.
Access: Access at several points from minor road off A835 to Achiltibuie. Open all year, do not enter croftland without permission.
Facilities: None.

Public transport: None.
Habitat: Loch, bog, mountain and moorland.
Key birds: Upland birds (inc. Ptarmigan, Raven, Ring Ouzel, Golden Plover, Twite). *Winter:* Grazing Barnacle Goose.
Contact: John Smith, North Keanchullish, Ullapool, Wester Ross IV26 2TW. 01854 612531.

3. FORSINARD

RSPB (North Scotland).
Location: NC 89 04 25. 30 miles SW of Thurso on A897. Turn off A9 at Helmsdale from S (24 miles) or A836 at Melvich from N coast road (14 miles).
Access: Open at all times, but few birds Sept-Feb. Contact visitor centre during breeding season (mid-Apr to end Jun) and during deerstalking season (Jul 1-Feb 15) for advice. Self-guided trail open all year, no dogs, not suitable for wheelchairs.
Facilities: Visitor centre open Apr 1-Oct (9am-6pm), seven days per week. Static and AV displays, live CCTV and webcam link to Hen Harrier nest in breeding season. Wheelchair access to centre and toilet. Guided walks Tue and Thu, May-Aug. Hotel nearby.
Public transport: Train from Inverness and Thurso (0845 484950) visitor centre in Forsinard Station building.
Habitat: Blanket bog, upland hill farm.
Key birds: Golden Plover, Greenshank, Dunlin, Hen Harrier, Merlin, Short-eared Owl.
Contact: RSPB, Forsinard Peatlands Reserve, Reserve Office, Forsinard, Sutherland KW13 6YT. 01641 571225. e-mail: forsinard@rspb.org.uk www.rspb.org.uk

4. INSH MARSHES

RSPB (North Scotland Office).
Location: NN 775 999. In Spey Valley, two miles NE of Kingussie on B970 minor road.
Access: Open at all times. No disabled access. Coach parking available along access road to car park.
Facilities: Information viewpoint, two hides, three nature trails. Not suitable for disabled. No toilets.
Public transport: Nearest rail station Kingussie (two miles).
Habitat: Marshes, woodland, river, open water.
Key birds: *Spring/summer:* Waders (Lapwing, Curlew, Redshank, Snipe), wildfowl (including Goldeneye and Wigeon), Wood Warbler, Redstart, Tree Pipit. *Winter:* Hen Harrier, Whooper Swan, other wildfowl.
Contact: Pete Moore, Ivy Cottage, Insh, Kingussie,

Inverness-shire PH21 1NT. 01540 661518.
e-mail: pete.moore@rspb.org.uk
www.kincraig.com/rspb.htn

5. ISLE OF EIGG

Scottish Wildlife Trust.
Location: NM 38 48. Small island S of Skye, reached by ferry from Maillaig or Arisaig (approx 12 miles).
Access: Ferries seven days per week (weather permitting) during summer. Four days per week (weather permitting) Sept-Apr. Coach parties would need to transfer to ferries for visit to Eigg. Please contact ferry companies prior to trip.
Facilities: Pier centre – shops/Post Office, tea-room, craftshop, toilets.
Public transport: Caledonian MacBrayne Ferries NE from Mallaig (tel: 01687 462403), *MV Sheerwater* from Arisaig (tel: 01678 450 224).
Habitat: Moorland (leading to sgurr pitchstone ridge), wood and scrub, hay fields, shoreline. Marsh and bog.
Key birds: Red-throated Diver, Golden Eagle, Buzzard, Raven. *Summer:* Manx Shearwater, Arctic Tern, various warblers, Twite, etc.
Contact: John Chester, Millers Cottage, Isle of Eigg, Small Isles PH42 4RL. 01687 482477.
www.isleofeigg.org

6. ISLE OF RUM

SNH (North West Region).
Location: NM 370 970. Island lying S of Skye. Passenger ferry from Mallaig, take A830 from Fort William.
Access: Contact Reserve Office for details of special access arrangements relating to breeding birds, deer stalking and deer research.
Facilities: Kinloch Castle Hostel, 01687 462037, Bayview Guest House, 01687 462023. General store and post office, guided walks in summer.
Public transport: Caledonian MacBrayne ferry from Mallaig, 01687 450224, www.arisaig.co.uk
Habitat: Coast, moorland, woodland restoration, montane.
Key birds: *Summer:* Large Manx Shearwater colonies on hill tops; breeding auks (inc. Black Guillemot), Kittiwake, Fulmar, Eider, Golden Plover, Merlin, Red-throated Diver, Golden Eagle.
Contact: SNH Reserve Office, Isle of Rum PH43 4RR, 01687 462026; (fax) 01687 462805.

7. NIGG BAY RSPB NATURE RESERVE

RSPB (North Scotland Office).
Location: NH 805 731. One mile (1.6km) N of Nigg village on the B9175.
Access: Open at all times.
Facilities: Hide, car park.
Public transport: Bus: for info call Stagecoach on 01463 239292. Train station at Fearn, four miles.
Habitat: Extensive area of mudflat, saltmarsh and wet grassland.
Key birds: *Winter*: Large flocks of wintering waders, waterfowl and Pink-footed Geese. *Summer*: Breeding Lapwing, Oystercatcher, Redshank. Osprey feeding in the bay.
Contact: RSPB North Scotland Office, 01463 715000. e-mail: nsro@rspb.org.uk

8. UDALE BAY RSPB RESERVE

RSPB North Scotland Office.
Location: One mile W of Jemimaville on the B9163.
Access: Open all year.
Facilities: Hide, large lay-by. No coach parking.
Public transport: No 26 bus stops in Jemimaville six times a day (approx 5 min walk to reserve). Bus Info contact: Rapsons, 01463 710555.
Habitat: Mudflat, saltmarsh and wet grassland.
Key birds: *Spring/summer*: wildfowl, Oystercatcher, Redshank, waders. Possible Osprey fishing. *Autumn/winter*: Large flocks of wildfowl, geese, waders.
Contact: RSPB North Scotland Office, 01463 715000. e-mail: nsro@rspb.org.uk

Orkney

9. BIRSAY MOORS

RSPB (East Scotland).
Location: Access to hide at Burgar Hill (HY 344 257), signposted from A966 at Evie. Birsay Moors viewed from layby on B9057 NW of Dounby (HY 347 245).
Access: Open access all year round.
Facilities: One hide at Burgar Hill very good for watching breeding Red Throated Divers Apr to Aug. Wheelchair access.
Public transport: Orkney Coaches. Service within 0.5 mile of reserve. Tel: 01856 877500.
Habitat: Diverse example of Orkney moorland - wet and dry heath, bog, mire, scrub and some farmland.
Key birds: *Spring/summer*: Nesting Hen Harrier, Merlin, Great and Arctic Skuas, Short-eared Owl,

Golden Plover, Curlew, Red-throated Diver. *Winter*: Hen Harrier roost.
Contact: The Warden, 12/14 North End Road, Stromness, Orkney KW16 3AG. 01856 850176. e-mail: orkney@rspb.org.uk www.rspb.co.uk

10. BRODGAR

RSPB (East Scotland).
Location: HY 296 134. Reserve surrounds the Ring of Brodgar, part of the Heart of Neolithic Orkney World Heritage Site on the B9055 off the Stromness-Finstown Road.
Access: Open all year.
Facilities: Footpath, circular route approx one mile.
Public transport: Orkney Coaches. Service within 0.5 mile of reserve. Tel: 01856 877500. Occasional service past reserve.
Habitat: Wetland and farmland including species-rich grassland, loch shores.
Key birds: *Spring/summer*: Breeding waterfowl on farmland and nine species of waders breed here. The farmed grassland is suitable for Corn Crake and provides water, food and shelter for finches, larks and buntings. *Winter*: Large numbers of Golden Plover, Curlew and Lapwing.
Contact: (See Birsay Moors above), 01856 850 176.

11. COPINSAY

RSPB (East Scotland).
Location: HY 610 010. Access by private boat or hire boat from mainland Orkney.
Access: Open all year round.
Facilities: House on island open to visitors, but no facilities. No toilets nor hides.
Public transport: None.
Habitat: Sea cliffs, farmland.
Key birds: *Summer*: Stunning seabird-cliffs with breeding Kittiwake, Guillemot, Black Guillemot, Puffin, Razorbill, Shag, Fulmar, Rock Dove, Eider, Twite, Raven and Greater Black-backed Gull. Passage migrants esp. during periods of E winds.
Contact: (See Birsay Moors above), 01856 850 176. S Foubisher (boatman) 01856 741252 - cannot sail if wind is in the east

12. HOBBISTER

RSPB (East Scotland).
Location: HY 396 070 or HY 381 068. Near Kirkwall.
Access: Open access between A964 and the sea. Dogs on leads please.
Facilities: A council maintained footpath to

Waulkmill Bay, two car parks, walks along peat-cutters' tracks.
Public transport: Orkney Coaches. Tel: 01856 877500.
Habitat: Orkney moorland, bog, fen, saltmarsh, coastal cliffs, scrub.
Key birds: *Summer*: Breeding Hen Harrier, Merlin, Short-eared Owl, Red Grouse, Red-throated Diver, Eider, Merganser, Black Guillemot. Wildfowl and waders at Waulkmill Bay. *Autumn/winter*: Waulkmill for sea ducks, divers, auks and grebes (Long-tailed Duck, Red, Black and Great Northern Divers, Slavonian Grebe).
Contact: The Warden, 12/14 North End Road, Stromness, Orkney KW16 3AG. 01856 850176.
e-mail: orkney@rspb.org.uk

13. HOY

RSPB (East Scotland).
Location: HY 210 025. Located in NW of Hoy, a large island S of mainland Orkney. Car ferry from Houten to Lyness.
Access: Open all year round. Keep dogs on lead. Not suitable for wheelchairs – rough terrain.
Facilities: Toilet facilities at Moaness Pier and at Rackwick. Nature trail – circular route from Moaness Pier to Old Man of Hoy via Old Rackwick Post Road. Leaflets available from 2003.
Public transport: Foot passenger ferry service from Stromness to Moaness Pier. Minibus taxis
Habitat: Coastal heath, moorland, fellfield, woodland and cliffs.
Key birds: *Spring/summer:* Red-throated Diver, Merlin, Peregrine, Golden Plover, Dunlin, Great Skua, Arctic Skua, Short-eared Owl, Guillemot, Razorbill, Puffin, Fulmar, Kittiwake, Stonechat, Wheatear. *Autumn/winter:* Redwing, Fieldfare, Snow Bunting. *Migration species:* Whimbrel, Brambling plus almost anything is possible.
Contact: The Warden, Ley House, Hoy, Orkney KW16 3NJ. 01856 791298.
e-mail: orkney@rspb.org.uk

14. MARWICK HEAD

RSPB (East Scotland).
Location: HY 229 242. On W coast of mainland Orkney, near Dounby. Path N from Marwick Bay, or from council car park at Cumlaquoy at HY 232 252.
Access: Open all year. Rough terrain not suitable for wheelchairs.
Facilities: Cliff top path.
Public transport: Orkney Coaches (01856 877500).

Habitat: Rocky bay, sandstone cliffs. Beach path good place for Great Yellow Bumble Bee in Aug.
Key birds: May-Jul best. Huge numbers of Kittiwakes and auks, inc. Puffins, also nesting Fulmar, Rock Dove, Raven, Rock Pipit.
Contact: See Hobbister above, 01856 850176.
e-mail: orkney@rspb.org.uk www.rspb.co.uk

15. NORTH HILL, PAPA WESTRAY

RSPB (East Scotland).
Location: HY 496 538. Small island lying NE of Westray, reserve at N end of island's main road.
Access: Access at all times. During breeding season report to summer warden at Rose Cottage, 650 yards S of reserve entrance (Tel 01857 644240.) or use trail guide.
Facilities: Nature trails, hide/info hut.
Public transport: Orkney Ferries (01856 872044), Loganair (01856 872494).
Habitat: Sea cliffs, maritime heath.
Key birds: *Summer*: Close views of colony of Puffin, Guillemot, Razorbill and Kittiwake. Black Guillemot nest under flagstones around reserve's coastline. One of UK's largest colonies of Arctic Tern, also Arctic Skua.
Contact: Apr-Aug, The Warden at Rose Cottage, Papa Westray DW17 2BU. 01857 644240.,
2. RSPB Orkney Office 12/14 North End Road, Stromness, Orkney KW16 3AG. 01856 850176.
e-mail: orkney@rspb.org.uk www.rspb.co.uk

16. NORTH RONALDSAY BIRD OBSERVATORY

Location: HY 64 52. 35 miles from Kirkwall, Orkney mainland
Access: Open all year except Christmas.
Facilities: Accommodation, display room, meals, snacks etc for non-residents, fully licenced, toilets, croft walk.
Public transport: Twice daily (Mon-Sat) subsidised flights from Kirkwall (Loganair 01856 872494). Sunday flights in Summer. Once weekly ferry from Kirkwall (Fri or Sat), some Sun sailings in summer (Orkney Ferries Ltd 01856 872044).
Habitat: Crofting island with a number of eutrophic and oligotrophic wetlands. Coastline has both sandy bays and rocky shore. Walled gardens concentrate passerines.
Key birds: *Spring/Autumn*: Prime migration site in including regular BBRC species. Wide variety of breeding seabirds, wildfowl and waders. *Winter:* Waders and wildfowl include Whooper Swan and hard weather movements occur.

NATURE RESERVES - SCOTLAND

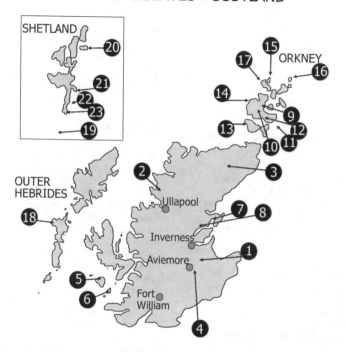

Contact: Alison Duncan, North Ronaldsay Bird Observatory, Twingness, North Ronaldsay, Orkney KW17 2BE. 01857 633200.
e-mail: alison@nrbo.prestel.co.uk
www.nrbo.f2s.com

17. NOUP CLIFFS, WESTRAY

RSPB (East Scotland).
Location: HY 392 499. Westray lies NE of Mainland and Rousay. From Pierowall take minor road to Noup Farm (HY 492 488) then track NW to lighthouse. Track not suitable for coaches. Rough terrain.
Access: No dogs, even on a lead.
Facilities: Trail guide leaflet and information board.
Public transport: Flights from Kirkwall daily (Loganair 01856 872494). Daily ferry (Orkney Ferries 01856 872044).
Habitat: 2.5km of old red sandstone cliff.
Key birds: *Summer*: May-Jul best. Huge seabird colony with breeding Kittiwake, Guillemot, Razorbill, Fulmar, Puffin and Gannet. Also breeding Shag, Raven, Rock Dove, Rock Pipit. Adjacent to reserve breeding Arctic Skua and Arctic Tern.
Contact: In Summer - RSPB Warden, Rose Cottage, Papa Westray, Orkney, KW17 2BU. 01857 644 240.

Other times -RSPB Orkney Office, 12/14 North End Road, Stromness, Orkney KW16 3AG. 01856 850176. e-mail: orkney@rspb.org.uk
www.rspb.co.uk

Outer Hebrides

18. BALRANALD

RSPB (North Scotland Office).
Location: NF 705 707. From Skye take ferry to Lochmaddy, North Uist. Drive W on A867 for 20 miles to reserve. Turn off main road three miles NW of Bayhead at signpost to Houghharry.
Access: Open at all times, no charge. Dogs on leads. Disabled access.
Facilities: Visitor centre and toilets – disabled access. Marked nature trail.
Public transport: Bus service (tel 01876 560244).
Habitat: Freshwater loch, machair, coast and croft lands.
Key birds: *Summer*: Corn Crake, Corn Bunting, Lapwing, Oystercatcher, Dunlin, Ringed Plover, Redshank, Snipe. *Winter*: Twite, Greylag Goose, Wigeon, Teal, Shoveler. *Passage*: Barnacle Goose,

Pomarine Skua, Long-tailed Skua.
Contact: Jamie Boyle, 9 Grenitote, Isle of North
Uist, H56 5BP01876 560287.
e-mail: james.boyle3@btinternet.com

Shetland

19. FAIR ISLE BIRD OBSERVATORY

Fair Isle Bird Observatory.
Location: HZ 2172.
Access: Open from end Apr-end Oct. Free to roam
everywhere except one croft (Lower Leogh).
Facilities: Public toilets at Airstrip and Stackhoull
Stores (shop). Accommodation at Fair Isle Bird
Observatory (phone/e-mail: for brochure/details).
Guests can be involved in observatory work and get
to see birds in the hand. Slide shows, guided walks
through Ranger Service.
Public transport: Tue, Thurs, Sat – ferry (12
passengers) from Grutness, Shetland. Tel: Jimmy or
Florrie Stout 01595 760222. Mon, Wed, Fri, Sat –
air (7 seater) from Tingwall, Shetland. Tel: Loganair
01595 840246.
Habitat: Heather moor and lowland pasture/crofting
land. Cliffs.
Key birds: Large breeding seabird colonies (auks,
Gannet, Arctic Tern, Kittiwake, Shag, Arctic Skua
and Great Skua). Many common and rare migrants
Apr/May/early Jun, late Aug-Nov.
Contact: Deryk Shaw (Warden), Hollie Shaw
(Administrator), Fair Isle Bird Observatory, Fair Isle,
Shetland ZE2 9JU. 01595 760258.
e-mail: fairisle.birdobs@zetnet.co.uk
www.fairislebirdobs.co.uk

20. FETLAR

RSPB (East Scotland).
Location: HU 603 917. Lies E of Yell. Take car
ferry from Gutcher, N Yell. Booking advised. Tel:
01957 722259.
Access: Part of RSPB reserve (Vord Hill) closed
mid-May-end Jul. Entry during this period is only by
arrangement with warden.
Facilities: Hide at Mires of Funzie. Displays etc at
interpretive centre, Houbie. Toilets at ferry terminal,
shop and interpretive centre.
Public transport: None.
Habitat: Serpentine heath, rough hill lane, upland
mire.
Key birds: *Summer*: Breeding Red-throated Diver,
Eider, Shag, Whimbrel, Golden Plover, Dunlin,

skuas, Manx Shearwater, Storm Petrel. Red-necked
Phalarope on Loch of Funzie (HU 655 899) viewed
from road or RSPB hide overlooking Mires of
Funzie.
Contact: RSPB North Isles Officer, Bealance,
Fetlar, Shetland ZE2 9DJ. Tel/Fax: 01957 733246.
e-mail: malcolm.smith@rspb.org.uk

21. ISLE OF NOSS NNR

Scottish Natural Heritage (Shetland Office).
Location: HU 531 410. Four miles by car ferry and
road to the E of Bressay. Take ferry to Bressay and
follow signs for Noss. Park at end of road and walk
to shore (600 yards) where ferry to island will collect
you (if red flag is flying, island is closed due to sea
conditions). Freephone 0800 107 7818 for daily ferry
information.
Access: Access (Tue, Wed, Fri, Sat, Sun) 10am-
5pm, May-late Aug. Access by zodiac inflatable. No
dogs allowed on ferry. Steep rough track down to
ferry.
Facilities: Visitor centre, toilets.
Public transport: None. Post car available, phone
Royal Mail on 01595 820200. Cycle hire in Lerwick.
Habitat: Dune grassland, moorland, blanket bog, sea
cliffs.
Key birds: *Spring/summer:* Fulmar, Shag, Gannet,
Arctic Tern, Kittiwake, Great Black-backed Gull,
Great Skua, Arctic Skua, Guillemot, Razorbill,
Puffin, Black Guillemot, Eider.
Contact: Simon Smith, Scottish Natural Heritage,
Stewart Building, Alexandra Wharf, Lerwick,
Shetland ZE1 0LL. 01595 693345.
e-mail: northern_isles@snh.gov.uk

22. MOUSA

RSPB (Shetland Office).
Location: HU 460240. Small uninhabited island east
of Sandwick in South Mainland of Shetland.
Access: By ferry from Leebitton Pier, Sandwick,
Shetland – mid-Apr-mid-Sept.
Facilities: The Mousa Broch is the best preserved
Iron Age tower in the world (World Heritage Site).
Public transport: Buses run to Sandwick from
Lerwick. Details of ferry available from Tom
Jamieson (01950 431367) or his web site
(www.mousaboattrips.co.uk.).
Habitat: A small uninhabited island with maritime
grassland and a small area of shell sand.
Key birds: *Summer*: Storm Petrels can be seen on
the special night trips run by Tom Jamieson. Arctic
Tern, Arctic and Great Skuas, Black Guillemot and
Puffin.

Contact: Tom Jamieson, RSPB Shetland Office, East House, Sumburgh Head Lighthouse, Virkie, Shetland ZE3 9JN. 01950 460800.

23. SUMBURGH HEAD

RSPB (Shetland Office).
Location: HU 407 079. S tip of mainland Shetland.
Access: Open all year, but seabirds best May-mid Aug.

Facilities: View points.
Public transport: None.
Habitat: Sea cliffs.
Key birds: Breeding Puffin, Guillemot, Razorbill, Kittiwake, Shag, also minke and killer whales. Humpback whale seen occasionally.
Contact: RSPB Shetland Office, East House, Sumburgh Head Lighthouse, Virkie, Shetland ZE3 9JN. 01950 460 800.

East Wales

1. BRECHFA POOL

Brecknock Wildlife Trust.
Location: SO 118 377. Travelling NE from Brecon
look for lane off A470, 1.5 miles SW of Llyswen;
on Brechfa Common, pool is on R after cattle grid.
Access: Open dawn to dusk.
Facilities: None.
Public transport: None.
Habitat: Marshy grassland, large shallow pool.
Key birds: Teal, Wigeon, Bewick's Swan,
Redshank, Lapwing, Dunlin, Lesser Black-backed
Gull.
Contact: Trust HQ, 01874 625708.

2. BWLCHCOEDIOG

W K and Mrs J Evans.
Location: SH 878 149. From Machynlleth take
A489 NE for 14 miles. Half mile east of Mallwyd,
turn left into Cwm Cewydd, then 1.25 miles up the
valley. Park at Bwlchcoediog House.
Access: Open all year round. No dogs in fenced
areas. Please keep to paths.
Facilities: None.
Public transport: None.
Habitat: Farmland, woodland, streams, two ponds,
lake.
Key birds: *Summer:* Breeding Tree Pipit, Redstart,
Garden and Wood Warblers, Pied and Spotted
Flycatchers. *Winter:* Woodcock, Brambling, Raven,
Buzzard, Sparrowhawk. *All Year:* Siskin, occasional
Peregrine, Red Kite, Dipper.
Contact: W K and Mrs J Evans, Bwlchcoediog Isaf,
Cwm Cewydd, Mallwyd, Machynlleth, Powys SY20
9EE. 01650 531243.

3. CORS DYFI

Montgomeryshire Wildlife Trust.
Location: SN701985. 3.5 miles SW of Machynlleth
on the A487 S of Morben Isaf Caravan Park.
Access: Open all year. Coach parking by the
entrance. Wheelchair access by prior arrangement.
Facilities: Footpaths.
Public transport: None.
Habitat: Open water, swamp, wet woodland, bog.
Key birds: Nightjar, Snipe, Grasshopper Warbler,
Reed Bunting, Stonechat.

Contact: Trust HQ, 01983 555654
e-mail: info@montwt.co.uk
www.wildlifetrust.org.uk/montgomeryshire

4. DOLYDD HAFREN

Montgomeryshire Wildlife Trust.
Location: SJ 208 005. W of B4388. Go through
Forden village and on about 1.5 miles. Turn R at
sharp L bend at Gaer Farm and down farm track to
car park at other end.
Access: Open at all times – dogs to be kept on lead
at all times.
Facilities: Two bird hides.
Public transport: None.
Habitat: Riverside flood meadow – bare shingle,
permanent grassland, ox-bow lakes and new pools.
Key birds: Goosander, Redshank, Lapwing, Snipe,
Oystercatcher, Little Ringed Plover, Osprey. *Winter:*
Curlew.

Contact: Trust HQ, 01983 555654
e-mail: info@montwt.co.uk
www.wildlifetrust.org.uk/montgomeryshire

5. ELAN VALLEY

Dwr Cymru Welsh Water & Elan Valley Trust.
Location: SN 928 646 (visitor centre). Three miles
SW of Rhayader, off B4518.
Access: Mostly open access.
Facilities: Visitor centre and toilets (open mid Mar-
end Oct), nature trails all year and hide at SN 905
617.
Public transport: Post bus from Rhayader and
Llandrindod Wells.
Habitat: 45,000 acres of moorland, woodland, river
and reservoir.
Key birds: *Spring/summer*: Birds of prey, upland
birds including Golden Plover and Dunlin. Woodland
birds include Redstart and Pied Flycatcher.
Contact: Pete Jennings, Rangers Office, Elan Valley
Visitor Centre, Rhayader, Powys LD6 5HP. 01597
810880. e-mail: pete@elanvalley.org.uk
www.elanvalley.org.uk

6. GILFACH

Radnorshire Wildlife Trust.
Location: SN 952 714. Two miles NW from
Rhayader/Rhaeadr-Gwy. Take minor road to St
Harmon from A470 at Marteg Bridge.
Access: Centre open Easter-Sept 31 (10am-5pm).
Apr (every day). May/Jun (Fri-Mon). Jul/Aug
(every day). Sept (Fri-Mon). Reserve open every day
all year. Dogs on leads only. Disabled access and
trail.
Facilities: Visitor centre – open as above. Way-
marked trails.
Public transport: None.
Habitat: Upland hill farm, river, oak woods,
meadows, hill-land.
Key birds: *Spring/summer*: Pied Flycatcher,
Redstart. *All year*: Dipper, Red Kite.
Contact: Tim Thompson, Gilfach, St Harmon,
Rhaeadr-Gwy, Powys LD6 5LF. 01597 870 301.
e-mail: tim@ratgilfoelfisnet.co.uk
http//westwales.co.uk/gilfach.htm

7. GLASLYN, PLYNLIMON

Montgomeryshire Wildlife Trust.
Location: SN 826 941. Nine miles SE of
Machynlleth. Off minor road between the B4518 near
Staylittle and A489 at Machynlleth. Go down track
for about a mile to car park.

Access: Open at all times – dogs on a lead at all
times.
Facilities: Footpath.
Public transport: None.
Habitat: Heather moorland and upland lake.
Key birds: Red Grouse, Short-eared Owl, Meadow
Pipit, Sky Lark, Wheatear and Ring Ouzel, Red Kite,
Merlin, Peregrine. Goldeneye – occasional.
Contact: Trust HQ, 01983 555654
e-mail: info@montwt.co.uk
www.wildlifetrust.org.uk/montgomeryshire

8. LLANDINAM GRAVELS

Montgomeryshire Wildlife Trust.
Location: SO022876. From Newton, upon entering
Llandinam, turn R by statue, over the bridge, turn
first L along track to car park.
Access: Open all year. Coach parking in village car
park.
Facilities: Footpaths.
Public transport: From Newton.
Habitat: Hay meadow, wet woodland, river, shingle
beds.
Key birds: Redstart, Pied Flycatcher, Common
Sandpiper.
Contact: Trust HQ, 01983 555654
e-mail: info@montwt.co.uk
www.wildlifetrust.org.uk/montgomeryshire

9. LLYN MAWR

Montgomeryshire Wildlife Trust.
Location: SO 009 971. From Newtown, head NW
on A470 and then take minor 'no through' road N of
Clatter. After Pontdolgoch take road towards Bwlch
y Gasreg, park in small car park in front of white
cottage.
Access: Footpath, dogs on lead at all times.
Facilities: None.
Public transport: None.
Habitat: Upland lake, wetland, scrub.
Key birds: *Summer*: Breeding Great Crested Grebe,
Black-headed Gull, Snipe, Curlew, Whinchat,
Redkite. *Winter*: Occasional Goldeneye, Goosander,
Whooper Swan.
Contact: Trust HQ, 01938 555654.
e-mail: info@montwt.cix.co.uk
www.wildlifetrust.org.uk/montgomeryshire

10. PWLL PENARTH

Montgomeryshire Wildlife Trust.
Location: SO 137 926. Take B4568 from Newtown
to Llanllwchaiarn, turn down by church and follow

lane for a mile to sewage works gates. Park in small car park down track to right of gates.
Access: Open at all times. Wheelchair access by prior arrangement. Dogs to be kept on lead at all times.
Facilities: Two hides.
Public transport: None.
Habitat: Lake, Sand Martin bank, arable crops.
Key birds: *Late spring/summer*: Sand Martin, Lapwing, Sky Lark, Grey Wagtail. *Winter*: Buntings, finches. *All year*: Kingfisher, Mallard, Coot, Canada Goose, Ruddy Duck.
Contact: Trust HQ, 01938 555654.
e-mail: info@montwt.cix.co.uk
www.wildlifetrust.org.uk/montgomeryshire

11. PWLL-Y-WRACH

Brecknock Wildlife Trust.
Location: SO 165 327. Between Hay-on-Wye and Brecon at foot of Black Mountains. Half mile SE of Talgarth. At T-junction in centre of Talgarth (Tourist Information Centre on R) turn R and take an almost immediate L round a very sharp ninety degree bend. After 20m turn L opposite Bell Hotel and follow minor road for 1.5 miles past The Prya Centre (formerly Mid Wales Hospital) on L. Road narrows and nature reserve car park is a few hundred yards on R.
Access: A small car park which can take about six cars. A coach could be accommodated provided the car park is not already full.
Facilities: A network of footpaths, including a disabled access path.
Public transport: Bus 39 to Talgarth (Brecon to Hay-on-Wye).
Habitat: Steep valley woodland, stream and waterfall.
Key birds: Dipper, Grey Wagtail, woodland species (inc. Pied Flycatcher, Wood Warbler). Dormouse colony.
Contact: Trust HQ, 01938 555654.
e-mail: info@montwt.cix.co.uk
www.wildlifetrust.org.uk/montgomeryshire

12. ROUNDTON HILL

Montgomeryshire Wildlife Trust.
Location: SO 293 947. SE of Montgomery. From Churchstoke on A48, take minor road to Old Churchstoke, R at phone box, then first R.

Access: Open access. Tracks rough in places. Dogs on lead at all times.
Facilities: Car park. Waymarked trails.
Public transport: None.
Habitat: Ancient hill grassland, woodland, streamside wet flushes, scree, rock outcrops.
Key birds: Buzzard, Raven, Wheatear, all three woodpeckers, Tawny Owl, Redstart, Linnet, Goldfinch.
Contact: Trust HQ, 01938 555654.
e-mail: info@montwt.cix.co.uk
www.wildlifetrust.org.uk/montgomeryshire

13. TALYBONT RESERVOIR

(Dwr Cymru) Welsh Water.
Location: SO 100 190. Take minor road off B4558 S of Talybont, SE of Brecon.
Access: No access to reservoir area, view from road.
Facilities: Displays at the Glyn Collwm information centre at Aber, between the reservoir and Talybont. Bird hide.
Public transport: None.
Habitat: Reservoir, woodland.
Key birds: *Winter*: Wildfowl (inc. Goldeneye, Goosander, Whooper Swan), Redpoll, Siskin. Migrant waders.

14. VYRNWY (LAKE)

RSPB (North Wales Office).
Location: SJ 020 193. Located WSW of Oswestry. Nearest village is Llanfyllin on A490. Take B4393 to lake.
Access: Reserve open all year. Visitor centre open Apr-Dec (10.30am-4.30pm), Dec-Apr weekends only (10.30am-4.30pm).
Facilities: Toilets, visitor centre, hides, nature trails, coffee shop, RSPB shop, craft workshops.
Public transport: Train and bus Welshpool (25 miles away).
Habitat: Heather moorland, woodland, meadows, rocky streams and large reservoir.
Key birds: Dipper, Kingfisher, Pied Flycatcher, Wood Warbler, Redstart, Peregrine and Buzzard.
Contact: Jo Morris, Centre Manager, RSPB Lake Vyrnwy Reserve, Bryn Awel, Llanwddyn, Oswestry, Salop SY10 0LZ. 01691 870278.
e-mail: lake.vyrnwy@rspb.org.uk

North Wales

1. BARDSEY BIRD OBSERVATORY

Bardsey Bird Observatory.
Location: SH 11 21. Private 444 acre island. One hour boat journey from Pwllheli (18 miles SW of Bangor).
Access: Mar-Nov. No dogs. Visitor accommodation in 150-year-old farmhouse (two single, two double, two x four dorms). To stay at the Observatory contact Alicia Normand (tel 01758 760667, e-mail bob&lis@solfach.freeserve.co.uk). Day visitors by Bardsey Ferries (01758 730326).
Facilities: Public toilets available for day visitors. Three hides, one on small bay, two seawatching.
Public transport: Trains from Birmingham to Pwllheli. Tel: 0345 484950. Arriva bus from Bangor to Pwllheli. Tel: 0870 6082608.
Habitat: Sea-birds cliffs viewable from boat only. Farm and scrubland, Spruce plantation, willow copses and gorse-covered hillside.
Key birds: *All Year*: Chough, Peregrine. *Spring/ summer*: Manx Shearwaters (16,000 pairs), other seabirds. Migrant warblers, chats, Redstart, thrushes. *Autumn*: Many rarities including Eye-browed Thrush, Lanceolated Warbler, American Robin, Yellow throat, Scarlet Tanager.
Contact: Steven Stansfield, Cristin, Ynys Enlli (Bardsey), off Aberaron, via Pwllheil, Gwynedd LL53 8DE. 07855 264151.
e-mail: warden@bbfo.org.uk
www.bbfo.org.uk

2. CEIRIOG FOREST

Forest Enterprise.
Location: SJ 166 384. From Glyn Ceiriog take road to Nantyr. Turn R at a white cottage called Bryn Awel, through a gate marked Glyndyfrdwy into forest. Park at Forest Enterprise picnic site.
Access: Open all year. Long walk on metalled track.
Facilities: None.
Public transport: None.
Habitat: Heather moor, woodland.
Key birds: *Spring/summer*: Redstart, Pied Flycatcher, Wood Warbler, Whinchat, Tree Pipit, Ring Ouzel, Wheatear. *All year*: Black Grouse, Dipper, Grey wagtail, Red Grouse, Raven, Chaffinch, Redpoll, Siskin, Crossbill, all three woodpeckers, Sparrowhawk, Buzzard.

Contact: Forest Enterprise Wales, Victoria Terrace, Aberystwyth, Ceredigion, SY23 2DQ, 01970 612367.

3. CEMLYN

North Wales Wildlife Trust.
Location: SH 337 932. Ten miles from Holyhead, Anglesey, minor roads from A5025 at Tregele.
Access: Open all the time. Dogs on leads. No wheelchair access. During summer months walk on seaward side of ridge and follow signs. Car park suitable for mini buses. Coaches to park on roadside at grid ref. SH 336 929.
Facilities: None.
Public transport: None.
Habitat: Brackish lagoon, shingle, ridge.
Key birds: *Summer*: Breeding terns. *Winter*: Waders/ducks (including Little Grebe, Goldeneye and Shoveler).
Contact: Chris Wynne, North Wales Wildlife Trust, 376 High Street, Bangor, Gwynedd LL57 1YE. 01248 351541. e-mail: info@nwwt.cix.co.uk www.wildlifetrust.org.uk/northwales

4. CONNAHS QUAY

Deeside Naturalists' Society .
Location: SJ 275 715. NW of Chester. Take B5129 from Queensferry towards Flint, two miles.
Access: Advance permit required.
Facilities: Field studies centre, four hides.
Public transport: None.
Habitat: Saltmarsh, mudflats, grassland scrub, open water, wetland meadow.
Key birds: High water roosts of waders, inc. Black-tailed Godwit, Oystercatcher, Redshank, Spotted Redshank. Passage waders. *Winter*: Wildfowl (inc. Teal, Pintail, Goldeneye), Merlin, Peregrine.
Contact: Secretary, Deeside Naturalist's Society, 38 Kelsterton Road, Connahs Quay, Flints CH5 4BJ.

5. CONWY

RSPB (North Wales Office).
Location: SH 799 771. On E bank of Conwy Estuary. Access from A55 at exit signed to Conwy and Deganwy.
Access: Open daily (10am-5pm) or dusk if earlier. Closed for Christmas Day.
Facilities: Visitor centre, toilets including disabled.

235

One hide, accessible to wheelchairs. Trails firm and level, though a little rough in places. Two further hides accessible to pedestrians.
Public transport: Train service to Llandudno Junction. Bus service to Tesco supermarket, Llandudno Junction. Tel: 08706 082 608.
Habitat: Open water, islands, reedbeds, grassland, estuary.
Key birds: *Spring/summer:* Breeding Reed and Sedge Warblers, Lapwing, Redshank, Little Ringed Plover, Sky Lark, Reed Bunting and rarities. *Autumn:* Passage waders, spectacular Starling roost and rarities. *Winter:* Kingfisher, Goldeneye, Water Rail, Red-breasted Merganser, wildfowl.
Contact: Alan Davies, Conwy RSPB Nature Reserve, Llandudno Junction, Conwy, North Wales LL31 9XZ. 01492 584091.

6. FORYD

Gwynedd Council.
Location: SW of Caernarfon, at W end of Menai Straits. Either take minor road along shore from A487 S of Caernarfon (SH 483 617), and also minor road S to Llanfaglan, llandwrog and Dinas Dinlle from above, or W from A499 (SH 452 555)
Access: Several minor roads lead to the bay. The picnic site and hide are wheelchair accessible. Public transport available.
Facilities: Car park and picnic site on E shore (SH 453 604), bird hide on E shore (SH 452 586) phone Countryside Warden to arrange access to hide.
Public transport: Buses available to neighbouring villages, with access on foot to bay within 1 mile, contact 0870 60 82 608 or www.gwynedd.gov.uk
Habitat: Partially enclosed intertidal bay, sand and mud flats, salt marsh. SSSI, SAC, and LNR.
Key birds: Large wintering populations of Wigeon, Shelduck, Oystercatcher, Curlew, Lapwing, Dunlin, and Redshank, with smaller numbers of Golden and Grey Plover, Knot, Greenshank, Black and Bar-tailed Godwits, along with other ducks and divers.
Contact: Gwynedd Council, Countryside Wardens. 01286 672255.
e-mail: CefnGwlad@gwynedd.gov.uk
www.gwynedd.gov.uk

7. GORS MAEN LLWYD

North Wales Wildlife Trust.
Location: SH 975 580. Follow A5 to Cerrigydrudion (seven miles S of site), then take B4501 and go past Llyn Brennig Visitor Centre. Approx two miles beyond centre, turn right (still on

B4501). First car park on right approx 300 yards after cattle grid.
Access: Open all the time. Dogs on leads. Keep to the paths. Rare breeding birds on the heather so keep to paths.
Facilities: In second car park by lake shore there are toilets and short walk to bird hide. Paths are waymarked, but can be very wet and muddy in poor weather.
Public transport: None.
Habitat: Heathland. Heather and grass overlooking large lake.
Key birds: *Summer:* Red and Black Grouse, Hen Harrier, Merlin, Sky Lark, Curlew. *Winter:* Wildfowl on lake.
Contact: Neil Griffiths, Reserves Officer, NWWT, 376 High Street, Bangor, Gwynedd LL57 1YE. 01248 351541. e-mail: info@nwwt.cix.co.uk www.wildlifetrust.org.uk/northwales

8. LLYN ALAW

Welsh Water/United Utilities.
Location: SH 390 865. Large lake five miles from Amlwch in northern part of Anglesey. Signposted from A55/A5/B5112/B5111/B5109.
Access: Open all year. No dogs to hides or sanctuary area but dogs allowed (maximum two per adult) other areas. Limited wheelchair access.
Facilities: Toilets (including disabled), two hides, two nature trails, information centre, car parks, network of mapped walks, picnic sites, information boards.
Public transport: Not to within a mile.
Habitat: Large area of standing water, shallow reedy bays, hedges, scrub, woodland, marsh, grassland.
Key birds: *Summer:* Lesser Whitethroat, Sedge and Grasshopper Warblers, Little and Great Crested Grebes, Tawny Owl, Barn Owl, Buzzard. *Winter:* Whooper Swan, Goldeneye, Hen Harrier, Short-eared Owl, Redwing, Fieldfare, Peregrine, Raven. *All year:* Bullfinch, Siskin, Redpoll, Goldfinch, Stonechat. *Passage waders:* Ruff, Spotted Redshank, Curlew Sandpiper, Green Sandpiper.
Contact: Jim Clark, Llyn Alaw, Llantrisant, Holyhead LL65 4TW. 01407 730762.
e-mail: llynalaw@amserve.net

9. LLYN CEFNI

Welsh Water/United Utilities.
Location: SH 440 775. A reservoir located two miles NW of Llangefni, in central Anglesey. Follow B5111 or B5109 from the village.

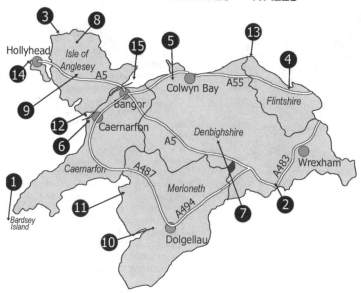

Access: Open at all times. Dogs allowed except in sanctuary area. Good wheelchair access most of Southern (Langefni) side.
Facilities: Toilets (near waterworks), picnic site, hide, information boards.
Public transport: Bus 32, 4 (45 Sat only, 52 Tue and Thu only). Tel 0870 6082608 for information.
Habitat: Large area of open water, reedy bays, coniferous woodland, carr, scrub.
Key birds: *Summer*: Sedge and Grasshopper Warblers, Whitethroat, Buzzard, Tawny Owl, Little Grebe, Gadwall, Shoveler. *Winter*: Waterfowl (Whooper Swan, Goldeneye), Crossbill, Redpoll, Siskin, Redwing. *All year*: Stonechat, Treecreeper, Song Thrush.
Contact: Jim Clark, Llyn Alaw, Llantrisant, Holyhead LL65 4TW. 01407 730762.
e-mail: llynalaw@amserve.net

10. MAWDDACH VALLEY

RSPB (North Wales Office).
Location: SH 696 185 (information centre). Two miles W of Dolgellau on A493. Next to toll bridge at Penmaenpool.
Access: Reserve open at all times. Information centre open daily during Easter week and from Whitsun to first weekend of Sept (11am-5pm). Between Easter week and Whitsun, weekends only (noon-4pm).
Facilities: Toilets and car park at information centre.
Public transport: Buses run along A493. Morfa

Mawddach railway halt four miles from information centre.
Habitat: Oak woodlands of Coed Garth Gell and willow/alder scrub at Arthog Bog SSSI.
Key birds: *Spring/summer*: Pied Flycatcher, Redstart and Tree Pipit. *Winter*: Raven, roving flocks of Siskin, Redpoll. Goosander and Goldeneye on the estuary.
Contact: The Warden, Mawddach Valley Nature Reserves, Ynys-Hir Reserve, Eglwys-Fach, Machynlleth, Powys SY20 8TA. 01654 700222.
e-mail: mawddach@rspb.org.uk
www.rspb.org.uk

11. MORFA HARLECH NNR

Countryside Council for Wales.
Location: SH 574 317. On A496 Harlech road.
Access: Open all year.
Facilities: Car park.
Public transport: The site is served by both bus and train. Train stations are at Harlech and Ty Gwyn (Ty Gwyn is near the saltmarsh wintering birds.)
Habitat: Shingle (no shingle at Harlech), coast, marsh, dunes. Also forestry plantation, grassland, swamp.
Key birds: *Spring/summer*: Whitethroat, Spotted Flycatcher, Grasshopper Warbler, migrants. *Passage*: Waders, Manx Shearwater, ducks. *Winter*: Divers, Whooper Swan, Wigeon, Teal, Pintail, Scaup, Common Scoter, Hen Harrier, Merlin, Peregrine,

Short-eared Owl, Little Egret, Water Pipit, Snow Bunting, Twite. *All year/breeding*: Redshank, Lapwing, Ringed Plover, Snipe, Curlew, Shelduck, Oystercatcher, Stonechat, Whinchat, Wheatear, Linnet, Reed Bunting, Sedge Warbler. Also Red-breasted Merganser, Kestrel, gulls.
Contact: Countryside Council for Wales, Maes y Ffynnon, Ffordd, Bangor, Gwynedd, LL57 2DN, 0845 1306229. e-mail: enquiries@ccw.gov.uk

12. NEWBOROUGH WARREN

CCW (North West Area).
Location: SH 406 670/430 630. In SE corner of Anglesey. From Menai Bridge head SW on A4080 to Niwbwrch or Malltraeth.
Access: Permit required for places away from designated routes.
Facilities: None.
Public transport: None.
Habitat: Sandhills, estuaries, saltmarshes, dune grasslands, rocky headlands.
Key birds: Wildfowl and waders at Malltraeth Pool (visible from road), Braint and Cefni estuaries (licensed winter shoot on marked areas of Cefni estuary administered by CCW); waterfowl at Llyn Rhosddu (public hide).
Contact: W Sandison, CCW North West Area, Tel/fax 01248 716422; (M)0468 918572.

13. POINT OF AYR

RSPB (Dee Estuary Office).
Location: SJ 140 840. At mouth of the Dee Estuary. Three miles E of Prestatyn. Access from A548 coast road to Talacre village. Park at end of Station Road.
Access: Open at all times.
Facilities: Car park, public hide overlooks saltmarsh and mudflats. No visitor centre. Toilets in Talacre village. Group bookings, guided walks and events. Track is 10min walk (0.5 mile).
Public transport: Bus – Prestatyn (Arriva 11, 11A/Crossville). Rail – Prestatyn.
Habitat: Intertidal mud/sand, saltmarsh, shingle.
Key birds: *Spring/summer*: Breeding Sky Lark, Meadow Pipit, Reed Bunting. *Late summer*: Pre-migratory roost of Sandwich and Common Terns. *Autumn*: Passage waders. *Winter*: Roosting waterfowl (eg Shelduck, Pintail), Oystercatcher, Curlew, Redshank, Merlin, Peregrine, Short-eared Owl. Rarities have occurred.
Contact: Colin E Wells, Burton Point Farm, Station Road, Burton, Nr Neston, Cheshire CH64 5SB. 0151 3367681. e-mail: colin.wells@rspb.org.uk

14. SOUTH STACK CLIFFS

RSPB (North Wales Office).
Location: RSPB Car Park SH 211 818, Ellins Towerr information centre SH 206 820. Follow A55 to W end in Holyhead, proceed straight on at roundabout, continue straight on through traffic lights. After another 1/2 mile turn L and follow Brown Tourist signs for RSPB Ynys Lawd South Stack.
Access: RSPB car park with disabled parking, 'Access for all' track leading to a viewing area overlooking the lighthouse adjacent to Ellins Tower Visitor centre. Access to Ellins Tower gained via staircase. Access to other areas of the reserve via an extensive network of paths some of which are steep and uneven. Coach parking by prior arrangement at The South Stack Kitchen, Tel:01407 762181 (privately owned). Access to Ellins Tower from here either by steep path with steps or by alighting at the RSPB car park and walking down 'Access for all' track.
Facilities: Free access to Ellin's Tower which has windows overlooking main auk colony open daily (10am-5.30pm Easter-Sep).
Public transport: Mainline station Holyhead. Infrequent local bus service, Holyhead-South Stack. Tel. 0870 608 2608.
Habitat: Sea cliffs, maritime grassland, maritime heath, lowland heath.
Key birds: Peregrine, Chough, Fulmar, Puffin, Guillemot, Razorbill, Rock Pipit, Sky Lark, Stonechat, Linnet, Shag, migrant warblers and passage seabirds.
Contact: Dave Bateson, Plas Nico, South Stack, Holyhead, Anglesey LL65 1YH. 01407 764973.

15. TRAETH LAFAN

Gwynedd Council.
Location: NE of Bangor, stretching to Llanfairfechan. 1) Minor road from old A55 near Tal-y-Bont (SH 610 710) to Aber Ogwen car park by coast (SH 614 723). 2) Also access from minor road from Abergwyngregyn village to Morfa Aber Reserve (SH 646 731) 3) Access on foot to Morfa Madryn Reserve 1 mile W from Llanfairfechan promenade (SH 679 754).
Access: Open access from 1, 2, and 3. All sites are wheelchair accessible. Public transport available.
Facilities: Public paths. 1) Car park, hides 200m away at Spinnies Reserve. 2) Car park and hide. 3) Car and coach park with toilets and café, hides at reserve.
Public transport: For local bus and train timetables

call 0870 60 82 608 or log on to
www.gwynedd.gov.uk or www.conwy.gov.uk
Habitat: Intertidal sands and mudflats, wetlands,
streams. SPA SAC SSSI and LNR.
Key birds: Third most important area in Wales for
wintering waders; of national importance for moulting
Great Crested Grebe and Red-breasted Merganser;
internationally important for Oystercatcher and
Curlew; passage waders; winter concentrations of
Goldeneye and Greenshank, and of regional
significance for wintering populations of Black-
throated, Red-throated & Great Northern Divers and

Black-necked and Slavonian Grebes and breeding
Lapwings at Morfa Madryn.
Contact: Countryside Wardens, Countryside and
Access Unit, Gwynedd Council, Council Offices,
Caernarfon LL55 1SH. Gwynedd council
Countryside Wardens on 01286 672255;
e-mail: CefnGwlad@gwynedd.gov.uk
www.gwynedd.gov.uk Conwy County Borough
Council Countryside Service on 01492 575200.
e-mail: cg.cs@conwy.gov.uk
www.conwy.gov.uk

South Wales

1. CWM CLYDACH

RSPB (South Wales Office).
Location: SN 584 026. Three miles N of J45 on
M4, through village of Clydach on B4291.
Access: Open at all times along public footpaths and
waymarked trails. Coach parking not available.
Facilities: Nature trails, car park, information
boards.
Public transport: Buses from Swansea stop at
reserve entrance. Nearest railway station is eight
miles away in Swansea.
Habitat: Oak woodland on steep slopes lining the
banks of the fast-flowing Lower Clydach River.
Key birds: *Spring/summer*: Nesting Buzzard,
Sparrowhawk and Raven. Nestboxes are used by
Pied Flycatcher, Redstart and tits while Wood
Warbler, all three species of woodpecker, Nuthatch,
Treecreeper and Tawny Owl also nest. Dipper and
Grey Wagtail frequent the river.
Contact: Martin Humphreys, 2 Tyn y
Berllan, Craig Cefn
Parc, Clydach,
Swansea SA6
5TL. 01792
842927.

2. KENFIG NNR

Bridgend County Borough Council.
Location: SS 802 811. Seven miles W of Bridgend.
From J37 on M4, drive towards Porthcawl, then
North Cornelly, then follow signs.
Access: Open at all times.
Facilities: Toilets, hides, free car parking and nature
trail for visually impaired. Visitor centre open
weekends and holidays (10am-4.30pm), weekdays
(2pm-4.30pm).
Public transport: None.
Habitat: Sand dunes, dune slacks, Kenfig Pool, Sker
Beach.

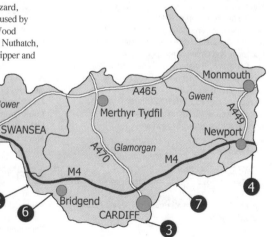

Key birds: *Summer*: Warblers including Cetti's, Grasshopper, Sedge, Reed, Willow and Whitethroat. One of the UK's best sites for orchids. *Winter*: Wildfowl, Water Rail, Bittern, grebes.
Contact: David Carrington, Ton Kenfig, Bridgend, CF33 4PT. 01656 743386.
e-mail: carridg@bridgend.gov.uk

3. LAVERNOCK POINT

The Wildlife Trust of South and West Wales.
Location: ST 182 680. Public footpaths S of B4267 between Barry & Penarth.
Access: No restrictions.
Facilities: None.
Public transport: Call Trust for advice.
Habitat: Cliff top, unimproved grassland, dense scrub.
Key birds: Seawatching in late summer; Glamorgan's best migration hotspot in autumn.
Contact: Trust HQ, 01656 724100.
e-mail: information@wtsww.cix.co.uk

4. MAGOR MARSH

Gwent Wildlife Trust.
Location: ST 427 867. S of Magor. Leave M4 at exit 23, turning R onto B4245. Follow signs for Redwick in Magor village. S of Magor village, look for gate on Whitewall Common on E side of reserve.
Access: Open all year. Keep to path.
Facilities: Hide. Car park, footpaths and boardwalks.
Public transport: Bus service to Magor village. Reserve is approx 10 mins walk along Redwick road.
Habitat: Sedge fen, reedswamp, willow carr, damp hay meadows and open water.
Key birds: Important for wetland birds. *Spring*: Reed, Sedge and Grasshopper Warbler, occasional Garganey and Green Sandpiper on passage. Hobby and Peregrine. *Winter*: Teal, Bittern records in two recent years. *All year*: Snipe, Reed Bunting, Cetti's Warbler and Water Rail.
Contact: Trust HQ, 01600 715501.
e-mail: gwentwildlife@cix.co.uk
www.wildlifetrust.org.uk/gwent

5. OXWICH

CCW (Swansea Office).
Location: SS 872 773. 12 miles from Swansea, off A4118.
Access: NNR open at all times. No permit required for access to foreshore. Dunes, woodlands and facilities.
Facilities: Private car park, summer only. Toilets

summer only. Marsh boardwalk and marsh lookout. No visitor centre, no facilities for disabled visitors.
Public transport: Bus service Swansea/Oxwich. First Cymru, tel. 01792 580580.
Habitat: Freshwater marsh, saltmarsh, foreshore, dunes, woodlands.
Key birds: *Summer*: Breeding Reed, Sedge and Cetti's Warblers, Treecreeper, Nuthatch, woodpeckers. *Winter*: Wildfowl.
Contact: Countryside Council for Wales, RVB House, Llys Felin Newydd, Phoenix Way, Swansea SA7 9FG. 01792 763500.

6. PARC SLIP NATURE PARK

The Wildlife Trust of South and West Wales.
Location: SS 880 840. Tondu, half mile W of Aberkenfig. From Bridgend take A4063 N, turning L onto B4281 after passing M4. Reserve is signposted from this road.
Access: Open dawn to dusk. Space for coach parking.
Facilities: Three hides, nature trail, interpretation centre.
Public transport: None.
Habitat: Restored opencast mining site, wader scrape, lagoons.
Key birds: *Summer*: Breeding Tufted Duck, Lapwing, Sky Lark. Migrant waders (inc. Little Ringed Plover, Green Sandpiper), Little Gull. Kingfisher, Green Woodpecker.
Contact: Trust HQ, 01656 724100.
e-mail: information@wtsww.cix.co.uk

7. PETERSTONE WENTLOOGE

Gwent Wildlife Trust.
Location: ST 269 800. Reserve overlooks Severn Estuary, between Newport and Cardiff. Take B4293 to Peterstone Wentlooge village.
Access: Park in large lay-by opposite the church and take path to the sea wall.
Facilities: None.
Public transport: None.
Habitat: Foreshore, inter-tidal mudflats, grazing.
Key birds: Passage waders and winter wildfowl.
Contact: Trust HQ, 01600 715501.

8. WHITEFORD NNR

Location: SS 450 960. Pass through Llanmadoc village, downhill, turn R at church to Cwm Ivy. Lane leads from here downhill to Whiteford Plantation. Follow footpath through Plantation, across Burrows to hide on Berges Island.

Access: Free access. Best to get to hide before a.m. high water for waders and wildfowl.
Facilities: None. Area not recommended for those with restricted mobility but excellent view of Whiteford Marsh from hillside road above Britannia Inn at Cheriton.
Public transport: None.

Habitat: Conifer plantation, marsh, mudflats.
Key birds: *Autumn/winter*: Divers, Red-necked, Slovinian and Black-necked Grebes, Brent Goose, Wigeon, Teal, Pintail and Eider. Common and Jack Snipe occur, with Whimbrel and Spotted Redshank on passage. Turnstone and Purple Sandpiper at Whiteford Point

West Wales

1. CASTLE WOODS

The Wildlife Trust of South and West Wales.
Location: SN 615 217. About 60 acres of woodland overlooking River Tywi, W of Llandeilo town centre.
Access: Open all year by footpath from Tywi Bridge, Llandeilo (SN 627 221).
Facilities: Call for advice.
Public transport: None.
Habitat: Old mixed deciduous woodlands.
Key birds: All three woodpeckers, Buzzard, Raven, Sparrowhawk. *Summer:* Pied and Spotted Flycatchers, Redstart, Wood Warbler. *Winter:* On water meadows below, look for Teal, Wigeon, Goosander, Shoveler, Tufted Duck and Pochard.
Contact: Steve Lucas, Area Officer, 35 Maesquarre Road, Betws, Ammanford, Carmarthenshire SA18 2LF. 01269 594293.
e-mail: information@wtsww.cix.co.uk
www.wildlifetrust.org.uk/wtsww

2. CORS CARON

CCW (West Wales Area).
Location: SN 697 632 (car park). Reached from B4343 N of Tregaron.
Access: Open access to S of car park along railway to boardwalk, out to SE bog. Access to rest of reserve by permit. Dogs on lead. Access for coaches.
Facilities: None at present.
Public transport: None.
Habitat: Raised bog, river, fen, wet grassland, willow woodland, reedbed.
Key birds: *Summer*: Lapwing, Redshank, Curlew, Red Kite, Grasshopper Warbler, Whinchat. *Winter:* Teal, Wigeon, Whooper Swan, Hen Harrier, Red Kite.

Contact: Paul Culyer, CCW, Neuaddlas, Tregaron, Ceredigion. 01974 298480. e-mail: p.culyer@ccw.gov.uk www.ccw.gov.uk

3. DINAS & GWENFFRWD

RSPB (South Wales Office).
Location: SN 788 472. Dinas car park off B road to Llyn Brianne Reservoir.
Access: Public nature trail at Dinas open at all times.
Facilities: None
Public transport: Nearest station at Llandovery.
Habitat: Hillside oakwoods, streams, bracken slopes and moorland.
Key birds: Buzzard, Pied Flycatcher, Redstart, Wood Warbler, Tree Pipit, Red Kite and Peregrine in area. Dipper, Goosander, Raven.
Contact: Mr M Humphreys, 2 Tyn y Berllan, Craig Cefn Par, Craig Cefn Parc, Clydach, Swansea SA6 5TL. 01792 842927. www.rspb.org.uk

4. DYFI

CCW (West Wales Area).
Location: SN 610 942. Large estuary area W of Machynlleth. Public footpaths off A493 E of Aberdyfi, and off B4353 (S of river); minor road from B4353 at Ynyslas to dunes and parking area.
Access: Ynyslas dunes and the estuary have unrestricted access. No access to Cors Fochno (raised bog) for casual birdwatching; permit required for study and research purposes. Good views over the bog and Aberleri marshes from W bank of Afon Leri.
Facilities: Public hide overlooking marshes beside footpath at SN 611 911.
Public transport: None.
Habitat: Sandflats, mudflats, saltmarsh, creeks, dunes, raised bog, grazing marsh.

Key birds: *Winter:* Greenland White-fronted Goose, wildfowl, waders and raptors. *Summer:* Breeding wildfowl and waders (inc. Teal, Shoveler, Merganser, Lapwing, Curlew, Redshank).
Contact: Mike Bailey, CCW Warden, Plas Gogerddan, Aberystwyth, Ceredigion SY23 3EE. 01970 821100.

5. THE NATIONAL WETLANDS CENTRE, WALES

The Wildfowl & Wetlands Trust.
Location: SS 533 984. Leave M4 at junction 48. Signposted from A484, E of Llanelli.
Access: Open daily (9.30am-6.00pm summer, earlier in winter) except Christmas Eve and Christmas Day.
Facilities: Visitor centre, restaurant, hides, education facilities, disabled access. Overlooks Burry Inlet.
Public transport: None.
Habitat: Inter-tidal mudflats, reedbeds, pools, marsh, waterfowl collection.
Key birds: Large flocks of Curlew, Oystercatcher, Redshank on saltmarsh. *Winter:* Pintail, Wigeon, Teal. Also Little Egret, Short-eared Owl, Peregrine.
Contact: Mr Nigel Williams, Centre Manager, The National Wetlands Centre, Llwynhendy, Llanelli SA14 9SH. 01554 741087; (Fax)01554 744101.
e-mail: info.llanelli@wwt.org.uk
www.wwt.org.uk

6. RAMSEY ISLAND

RSPB (South Wales Office).
Location: SM 706237. One mile offshore St Justinians, slipway, two miles W of St Davids.
Access: Open every day, Easter-Oct 31. For boat bookings contact: Thousand Island expeditions: 01437 721686 or sales@thousandislands.co.uk
Facilities: Toilets, small RSPB shop, tuck

shop, hot drinks and snacks, self-guiding trail.
Public transport: Trains to Haverfordwest Station. Hourly buses to St Davids, bus to St Justinians.
Habitat: Acid grassland, maritime heath, seacliffs.
Key birds: *Spring/summer:* Cliff-nesting auks (Guillemot, Razorbill). Kittiwake, Lesser, Great Black-backed and Herring Gulls, Shag, Peregrine, Raven, Chough, Lapwing, Wheatear, Stonechat.
Contact: RSPB Wales Headquarters, Sutherland House, Castle Bridge, Cowbridge Rd East, Cardiff CF11 9AB. 02920 353000. www.rspb.org.uk

7. SKOKHOLM ISLAND

The Wildlife Trust of South and West Wales.
Location: SM 738 037. Island lying S of Skomer.
Access: Day visits, Mon only Jun-Aug from Martinshaven. Weekly accomm. Apr-Sep, tel 01239 621212 for details and booking.
Facilities: Call for details.
Public transport: None.
Habitat: Cliffs, bays and inlets.
Key birds: *Summer:* Large colonies of Razorbill, Puffin, Guillemot, Manx Shearwater, Storm Petrel,

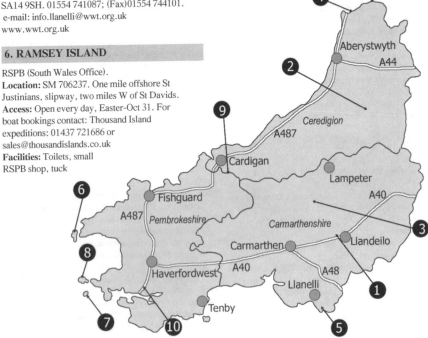

NATURE RESERVES - WALES

Lesser Black-backed Gull. Migrants inc. rare species.
Contact: Trust HQ, 01239 621212.

8. SKOMER ISLAND

The Wildlife Trust of South and West Wales.
Location: SM 725 095. Fifteen miles from
Haverfordwest. Take B4327 turn-off for Marloes,
embarkation point at Martin's Haven, two miles past
village.
Access: Apr 1-Oct 31. Boats sail at 10am, 11am and
noon every day except Mon (Bank Holidays
excluded). Closed four days beginning of Jun for
seabird counts. Not suitable for infirm (steep landing
steps and rough ground).
Facilities: Information centre, toilets, two hides,
wardens, booklets, guides, nature trails.
Public transport: None.
Habitat: Maritime cliff, bracken, bluebells and red
campion, heathland, freshwater ponds.
Key birds: Largest colony of Manx Shearwater in
the world (overnight). Puffin, Guillemot, Razorbill
(Apr-end Jul). Kittiwake (until end Aug), Fulmar
(absent Oct), Short-eared Owl (during day Jun and
Jul), Chough, Peregrine, Buzzard (all year),
migrants.
Contact: Juan Brown, Skomer Island, Marloes,
Pembs SA62 2BJ. 07971 114302.
e-mail: skomer@wtww.co.uk

9. WELSH WILDLIFE CENTRE

The Wildlife Trust of South and West Wales.
Location: SN 188 451. Two miles SE of Cardigan.
River Teifi is N boundary. Sign posted from
Cardigan to Fishguard Road.
Access: Open 10am-5pm all year. Free parking for
WTWW members, £5 non-members. Dogs welcome
– on a lead. Disabled access to visitor centre, paths,
four hides.
Facilities: Visitor centre, restaurant, network of paths
and seven hides.
Public transport: Train station, Haverfordwest (23
miles). Bus station in Cardigan. Access on foot from
Cardigan centre, ten mins.
Habitat: Wetlands, marsh, swamp, reedbed, open
water, creek (tidal), river, saltmarsh, woodland.
Key birds: Cetti's Warbler, Kingfisher, Water Rail,
Greater Spotted Woodpecker, Dipper, gulls, Marsh

Harrier, Sand Martin, Hobby, Redstart, occasional
Bittern and Red Kite.
Contact: The Welsh Wildlife Centre, Cilgerran,
Cardigan SA43 2TB. 01239 621212.
e-mail: information@wtsww.cix.co.uk

10. WESTFIELD PILL

The Wildlife Trust of South and West Wales.
Location: SM 958 073.
Access: Open all year.
Facilities: Car park, cycle track.
Public transport: None.
Habitat: Freshwater lagoons, disused railway
embankment, scrub, woodland margins.
Key birds: *Spring/summer*: Hirundines, Whitethroat,
Blackcap, Spotted Flycatcher. *Passage*: Waders.
Winter: Little Grebe, Peregrine, Water Rail,
Woodcock, Fieldfare, Redwing, Siskin, Redpoll. *All
year*: Sparrowhawk, Kingfisher, Tawny Owl, Grey
Wagtail, Dunnock, Raven, Bullfinch.
Contact: Trust HQ, 01239 621212.

11. YNYS-HIR

RSPB (CYMRU).
Location: SN 68 29 63. Off A487 Aberystwyth -
Machynlleth road in Eglwys-fach village. Six miles
SW of Machynlleth.
Access: Open every day (9am-9pm or dusk if
earlier). Visitor centre open daily Apr-Oct (10am-
5pm), Wed-Sun Nov-Mar (10am-4pm). Coaches
welcome but please call for parking information.
Facilities: Visitor centre and toilets, both with
disabled access. Numerous trails, seven hides, drinks
machine.
Public transport: Bus service to Eglwys-fach from
either Machynlleth or Aberystwyth, tel. 01970
617951. Rail service to Machynlleth.
Habitat: Estuary, freshwater pools, woodland and
wet grassland.
Key birds: *Winter*: Greenland White-fronted Goose,
Wigeon, Hen Harrier, Barnacle Goose. *Spring/
summer*: Wood Warbler, Redstart, Pied Flycatcher.
All year: Peregrine, Red Kite, Buzzard, Goshawk.
Contact: Frances Hazell, Ynys-Hir RSPB Nature
Reserve, Eglwys-fach, Machynlleth, Powys SY20
8TA. 01654 700222.
e-mail: ynyshir@rspb.org.uk

Northern Ireland

Co Antrim

BOG MEADOWS

Ulster Wildlife Trust.
Location: J 315 726. Two miles SW of Belfast city centre. Signposted from the Falls Road. (OS map 15).
Access: Open at all times. Coach parties welcome.
Facilities: Car park, bird hide, disabled access, high quality paths.
Public transport: City bus from city centre or taxi. For more details contact Translink on 028 9066 6630.
Habitat: Wet grassland, scrub, ponds, reedbed.
Key birds: *Summer:* Breeding Sedge Warbler, Grasshopper Warbler, Stonechat, Grey Wagtail, Blackcap, Reed Bunting. *Winter:* Water Rail, Snipe, Teal, occasional Long-eared Owl.
Contact: Ross Towers, Ulster Wildlife Trust, c/o Colin Glen Forest Park Centre, 163 Stewartstown Road, Dunmurry, Belfast BT17 0HW. 028 9062 8647. e-mail: ross.towers@ulsterwildlifetrust.org

ECOS NATURE RESERVE

Ulster Wildlife Trust.
Location: D 118 036. 0. 5 miles NE of Ballymena town centre. OS map 9.
Access: Open at all times. Coaches welcome. Disabled access.
Facilities: Environmental centre, car park, toilets, bird hide, high-quality paths.
Pulbic transport: Bus from Ballymena town centre or within easy walking distance. For more details contact Translink on 028 9066 6630.
Habitat: Lake, wet meadows, willow coppice, scrub.
Key birds: *Summer:* Breeding Snipe, Sedge and Grasshopper Warblers, Reed Bunting. *Winter:* Teal, Goldeneye, Lapwing, Curlew. Rarities have included, White-winged Black Tern and Hoopoe in recent years.
Contact: Trust HQ, 028 4483 0282.
e-mail: andrew.upton@ulsterwildifetrust.org

GLENDUN FARM

Ulster Wildlife Trust.
Location: D201317. Situated just off A2, 4 miles NW of Cushendun in the Glens of Antrim (OS Map 5).
Access: Open at all times. Not suitable for coach parties. Unsuitable for wheelchair users.
Facilities: Car park, waymarked trail
Public transport: Bus from Cushendall to Ballycastle. Contact Translink on 028 9066 6630 for more details
Habitat: Working hill farm with traditional stock, unimproved grassland, upland moorland
Key birds: *Summer:* Red Grouse, Hen Harrier, Whinchat, Spotted Flycatcher, Dipper, Grey Wagtail, Buzzard, Raven. *Winter:* Snow Bunting.
Contact: Trust HQ, 028 4483 0282.
e-mail: andrew.upton@ulsterwildifetrust.org
www.ulsterwildlifetrust. org

ISLE OF MUCK

Ulster Wildlife Trust.
Location: D 464 024. Situated off NR tip of Island Magee, Co Antrim (OS Map 9).
Access: Wildlife trust members only. Permit required to land on island from UWT. Island only accessible by boat. Not suitable for coaches or disabled.
Facilities: None.
Public transport: None.
Habitat: Offshore island with cliffs and stack
Key birds: *Summer:* Fulmar, Manx Shearwater, Peregrine, Kittiwake, terns, Guillemot, Razorbill, Black Guillemot, Puffin. *Winter:* Red-throated Diver.
Contact: Trust HQ, 028 4483 0282.
e-mail: andrew.upton@ulsterwildifetrust.org
www.ulsterwildlifetrust. org

KEBBLE

Department of the Environment NI.
Location: D 095 515. W end of Rathlin Island, off coast from Ballycastle.
Access: Scheduled ferry service from Ballycastle.
Facilities: None.
Public transport: None.
Habitat: Sea cliffs, grass, heath, lake, marsh.
Key birds: Major cliff nesting colonies of auks (inc. Puffin), Fulmar and Kittiwake; also Buzzard and Peregrine. Manx Shearwater and other seabirds on passage.
Contact: Dept of the Environment NI, Portrush Countryside Centre, 8 Bath Road, Portrush, Co Antrim BT56 8AP. 028 7082 3600.

PORTMORE LOUGH

RSPB (Northern Ireland Office).
Location: J 107 685. Eight miles from Lurgan. Signposted from Aghalee village.
Access: Open every day, unmanned. No access to meadows during winter. Limited disabled facilities.
Facilities: Car park, toilets, information shelter and one hide.
Public transport: None.
Habitat: Lowland wet grassland, scrub and reedbed.
Key birds: *Spring/summer*: Breeding Curlew, Snipe, Lapwing. *Winter*: Greylag Goose, Whooper Swan and a variety of wildfowl.
Contact: John Scovell, 02897 510097; (M)07736 792516. e-mail: john.scovell@rspb.org.uk

RATHLIN ISLAND

RSPB (Northern Ireland Office).
Location: Of NE coast of County Antrim. Five mile ferry journey from Ballycastle.
Access: Apr-Aug by appointment with warden only. Four miles from harbour, approx 100 steps. No toilets. Small shelter.
Facilities: RSPB viewpoint at the West Lighthouse.
Public transport: Caledonian MacBrayne ferry service from Ballycastle to Rathlin, tel 028 207 69299. Minibus tel 028 207 63451.
Habitat: Sea cliffs and offshore stacks.
Key birds: *Spring/summer*: Puffin, Guillemot, Razorbill, Fulmar, Kittiwake.
Contact: Liam McFaul/Alison McFaul, RSPB Northern Ireland Office, South Cleggan, Rathlin Island, Ballycastle, Co Antrim, 028 207 63948.

SLIEVENACLOY

Ulster Wildlife Trust.
Location: J 255 712. Situated in the Belfast Hills, take Ballycolin Road off A01 (OS Map 15).
Access: Permit currently required from project officer. Not suitable for coaches.
Facilities: None at present, currently being developed.
Public transport: Bus service from Belfast to Glenavy.
Habitat: Unimproved grassland, scrub.
Key birds: *Summer*: Snipe, Curlew, Sky Lark, Grey Wagtail, Wheatear, Grasshopper Warbler, Reed Bunting. *Winter*: Hen Harrier, Merlin, Fieldfare, Snow Bunting.
Contact: Mark Edgar, Slievenacloy Project Officer, c/o Colin Glen Forest Park Centre, 163 Stewartstown Road, Dunmurry, Belfast BT17 0HW. 028 9062 8647. e-mail: slievenacloy@btopenworld.com

STRAIDKILLY

Ulster Wildlife Trust.
Location: D302165. Situated midway between Glenarm and Carnlough in the Glens of Antrim (OS Map 9)
Access: Open at all times. Not suitable for coach parties. Unsuitable for wheelchair users.
Facilities: Waymarked trail, picnic areas and interpretation viewpoint
Public transport: 128 Ballymena-Carnlugh or 162 Larne-Cushendun. For more details contact Translink on 028 9066 6630.
Habitat: Semi-natural woodland
Key birds: *Summer*: Buzzard, Raven, Blackcap, Bullfinch. *Winter*: Woodcock.
Contact: Trust HQ, 028 4483 0282.
e-mail: andrew.upton@ulsterwildifetrust.org
www.ulsterwildlifetrust. org

Co Armagh

OXFORD ISLAND

Craigavon Borough Council.
Location: J 061 608. On shores of Lough Neagh, 2. 5 miles from Lurgan, Co Armagh. Signposted from J10 of M1.
Access: Site open at all times. Car parks are locked at varying times depending on seasons (see signs). Coach parking available. Lough Neagh Discovery Centre open every day Apr-Sept (10am-6pm Mon-Sat, 10am-7pm Sun), Oct-Mar (10am-5pm every day) Closed Christmas Day. Dogs on leads please. Most of site and all of Centre accessible for wheelchairs. Wheelchairs and mobility scooters available for visitors.
Facilities: Public toilets, Lough Neagh Discovery Centre with exhibitions, loop system for hard-of-hearing, shop and café. Four miles of footpaths, five birdwatching hides, children's play area, picnic tables, public jetties. Guided walks available (pre-booking essential). Varied programme of events.
Public transport: Ulsterbus Park'n Ride at Lough Road, Lurgan is 0. 5 miles from reserve entrance. Tel 028 9033 3000. Lurgan Railway Station, 3 miles from reserve entrance.
Habitat: Freshwater lake, ponds, wet grassland, reedbed, woodland.

Key birds: *Winter*: Large flocks of wildfowl, especially Pochard, Tufted Duck, Goldeneye and Scaup (mainly Dec/Jan). Whooper and Bewick's Swans (Oct-Apr). *Summer*: Sedge and Grasshopper Warblers and Great Crested Grebe.
Contact: Rosemary Mulholland, Conservation Officer, Lough Neagh Discovery Centre, Oxford Island NNR, Lurgan, Co Armagh, N Ireland BT66 6NJ. 028 383 322205.
e-mail: oxford.island@craigavon.gov.uk
www.oxfordisland.com

Co Down

BELFAST LOUGH RESERVE

RSPB (Northern Ireland Office).
Location: Take A2 N from Belfast and follow signs to Belfast Harbour Estate. Both entrances to reserve have checkpoints. From Dee Street 2 miles to reserve, from Tillysburn entrance 1 mile.
Access: Dawn to dusk.
Facilities: Lagoon overlooked by observation room (check for opening hours), two view points.
Public transport: None.
Habitat: Mudflats, wet grassland, freshwater lagoon.
Key birds: Noted for Black-tailed Godwit numbers and excellent variety of waterfowl in spring, autumn and winter, with close views. Rarities have included Buff-breasted, Pectoral, White-rumped and Semi-palmated Sandpipers, Spotted Crake, American Wigeon, Laughing Gull.
Contact: Anthony McGeehan, 028 9147 9009.

CASTLE ESPIE

The Wildfowl & Wetlands Trust.
Location: J 474 672. On Strangford Lough 10 miles E of Belfast, signposted from A22 in Comber area.
Access: Open daily except Christmas Day (10. 30am Mon-Sat, 11. 30am Sun).
Facilities: Visitor centre, educational facilities, views over lough, three hides, woodland walk.
Public transport: Call for advice.
Habitat: Reedbed filtration system with viewing facilities.
Key birds: *Winter*: Wildfowl esp. pale-bellied Brent Goose, Scaup. *Summer*: Warblers. Wader scrape has attracted Little Egret, Ruff, Long-billed Dowitcher, Killdeer.
Contact: James Orr, Centre Manager, The Wildfowl & Wetlands Trust, Castle Espie, Ballydrain Road, Comber, Co Down BT23 6EA. 028 9187 4146.

COPELAND BIRD OBSERVATORY

Location: Situated on a 40-acre island on outer edge of Belfast Lough, four miles N of Donaghadee.
Access: Access is by chartered boat from Donaghadee.
Facilities: Observatory open Apr-Oct most weekends and some whole weeks. Hostel-type accommodation for up to 20. Daily ringing, bird census, sea passage recording. General bookings: Neville McKee, 67 Temple Rise, Templepatrick, Co. Antrim BT39 0AG (tel 028 9443 3068).
Public transport: None.
Habitat: Grassy areas, rock foreshore.
Key birds: Large colony of Manx Shearwaters; Black Guillemot, Eider, Water Rail also nest. *Summer*: Visiting Storm Petrels. Moderate passage of passerine migrants.
Contact: Dr Peter Munro, Talisker Lodge, 54B Templepatrick Road, Ballyclare, Co Antrim BT39 9TX. 028 9332 3421.

CRAWFORDSBURN COUNTRY PARK

Department of the Environment NI.
Location: J 467 826. Signposted off A2 Belfast-Bangor road.
Access: Open at all times.
Facilities: Car parks.
Public transport: None.
Habitat: Sea, shore (rocky and sandy), woodland, glen, open fields.
Key birds: Woodland and grassland species; Dipper; Eider, gulls, divers, terns, shearwaters can all be seen offshore, esp. in autumn.
Contact: Ciaran McLarnon, Crawfordsburn Country Park, Bridge Road South, Helen's Bay, Co Down BT19 1LD. 028 9185 3621.

MURLOUGH

National Trust.
Location: J 394 338. Ireland's first nature reserve, between Dundrum and Newcastle, close to Mourne Mountains.
Access: Permit needed except on marked paths.
Facilities: Visitor centre.
Public transport: Local bus service from Belfast-Newcastle passes reserve entrances.
Habitat: Sand dunes, heathland.
Key birds: Waders and wildfowl occur in Inner Dundrum Bay adjacent to the reserve; divers and large numbers of Scoter (inc. regular Surf Scoter) and Merganser in Dundrum Bay.

Contact: Head Warden, Murlough NNR, The Stable Yard, Keel Point, Dundrum, Newcastle, Co Down BT33 0NQ. Tel/fax 028 437 51467; e-mail umnnrw@smtp.ntrust.org.uk.

NORTH STRANGFORD LOUGH

National Trust.
Location: J 510 700. View from adjacent roads and car parks; also from hide at Castle Espie (J 492 675).
Access: Call for advice.
Facilities: Hide.
Public transport: None.
Habitat: Extensive tidal mudflats, limited saltmarsh.
Key birds: Major feeding area for pale-bellied Brent Goose, also Pintail, Wigeon, Whooper Swan. Waders (inc. Dunlin, Knot, Oystercatcher, Bar-tailed Godwit).
Contact: Head Warden, National Trust, Strangford Lough Wildlife Scheme, Strangford Lough Wildlife Centre, Castle Ward, Strangford, Co Down BT30 7LS. Tel/fax 028 4488 1411; e-mail uslwcw@smtp. ntrust.org.uk.
e-mail: strangford@nationaltrust.org.uk

QUOILE PONDAGE

c/o Department of the Environment NI.
Location: J 500 478. One mile N of Downpatrick on road to Strangford, at S end of Strangord Lough.
Access: Open all year. No permit, but keep to trails.
Facilities: Large modern hide, visitor centre. Nature trail.
Public transport: None.
Habitat: Freshwater pondage to control flooding, with many vegetation types on shores.
Key birds: Many wildfowl species, woodland birds; migrant and wintering waders including Spotted Redshank, Ruff and Black-tailed Godwit.
Contact: Warden, Quoile Countryside Centre, 5 Quay Road, Downpatrick, Co Down BT30 7JB. 028 4461 5520.

Co Londonderry

LOUGH FOYLE

RSPB (Northern Ireland Office).
Location: C 545 237. Large sea lough NE of Londonderry. Take minor roads off Limavady-Londonderry road to view-points (choose high tide) at Longfield Point, Ballykelly, Faughanvale.

Access: Open all year. No permit, but keep to trails.
Facilities: None.
Public transport: None.
Habitat: Beds of eel-grass, mudflats, surrounding agricultural land.
Key birds: Staging-post for migrating wildfowl (eg. 15, 000 Wigeon, 4, 000 Pale-bellied Brent Geese in Oct/Nov). *Winter*: Slavonian Grebe, divers, Bewick's and Whooper Swans, Bar-tailed Godwit, Golden Plover, Snow Bunting. *Autumn*: Waders (inc. Ruff, Little Stint, Curlew Sandpiper, Spotted Redshank).
Contact: RSPB N Ireland HQ (01232 491547),

UMBRA

Ulster Wildlife Trust.
Location: C 725 355. Ten miles W of Coleraine on A2 – entrance beside automatic railway crossing about 1. 5 miles W of Downhill. OS 1:50 000 sheet 4.
Access: Wildlife Trust members. Not suitable for coaches. Unsuitable for wheelchair users.
Facilities: Informal paths.
Public transport: Ulsterbus service to Downhill from Coleraine. For more details contact Translink on 028 9066 6630.
Habitat: Sand dunes.
Key birds: *Summer*: Breeding Sky Lark. *Winter*: Woodcock, Peregrine, plus Great Northern Diver offshore.
Contact: Trust HQ, 028 4483 0282.
e-mail: andrew.upton@ulsterwildifetrust.org
www.ulsterwildlifetrust.org

Co. Tyrone

BLESSINGBOURNE

Ulster Wildlife Trust.
Location: H448487. Situated immediately NE of Fivemiletown (OS Map 18)
Access: Open at all times. Not suitable for coach parties. Unsuitable for wheelchair users.
Facilities: Paths
Public transport: For more details contact Translink on 028 9066 6630.
Habitat: Lake, reedbeds, mixed woodland
Key birds: *Summer*: Water Rail, Kingfisher, Sedge and Grasshopper Warblers, Blackcap.
Contact: Trust HQ, 028 4483 0282.
e-mail: andrew.upton@ulsterwildifetrust.org

Isle of Man

BREAGLE GLEN

Manx Wildlife Trust/Castletown Town
Commissioners Habitats.
Location: SC 196 688. In Port Erin from St Georges
Crescent, which forms the whole N and W
boundary.
Access: Open all year.
Facilities: None.
Public transport: Regular bus service from Douglas
to Port Erin and then short walk.
Habitat: Small woodland area, shrubs.
Key birds: *Passage:* Yellow-browed Warbler,
Barred Warbler, Firecrest, Red-breasted Flycatcher
have been recorded.
Contact: Tricia Sayle, Reserves Officer, Manx
Wildlife Trust, Tynwald Mills, St John's, Isle of
Man IM4 3AE. 01624 801985.
e-mail: manxwt@cix.co.uk
www.wildlifetrust.org.uk/manxwt

CALF OF MAN BIRD OBSERVATORY

Administration Department, Manx National Heritage.
Location: SC 15 65. Small island off SW of Isle of
Man.
Access: Landings by private craft all year. No dogs,
fires, camping or climbing.
Facilities: Accommodation for 8 people plus 2
volunteers sharing from Apr-Sept. Bookings: at
contact address. Bird ringers welcome to join in
ringing activities with prior notice.
Public transport: Local boat from Port Erin (Apr-
Sept) or Port St Mary all year.
Habitat: Heather/bracken moor and seabird cliffs.
Key birds: *All year:* Hen Harrier, Peregrine and
Chough. Breeding seabirds including Shag, Razorbill,
Manx Shearwater etc. Excellent spring and autumn
migration, seabird passage best in autumn.
Contact: Ben Jones, (Warden), Manx National
Heritage, Douglas, Isle of Man IM1 3LY.

CLOSE SARTFIELD

Manx Wildlife Trust.
Location: SC 361 956. From Ramsey drive W on
A3. Turn on to B9, take third R and follow this road
for nearly a mile. Reserve entrance is on R.
Access: Open all year round. No dogs. Path and
boardwalk suitable for wheelchairs from car park
through wildflower meadow and willow scrub to
hide.
Facilities: Car park, hide, reserve leaflet (50p,
available from office) outlines circular walk.
Public transport: None.
Habitat: Wildflower-rich hay meadow, marshy
grassland, willow scrub/developing birch woodland,
bog.
Key birds: *Winter:* Large roost of Hen Harrier.
Summer: Corn Crake (breeding 1999 and 2000 after
11 years' absence), Curlew, warblers.
Contact: Tricia Sayle, Reserves Officer, Manx
Wildlife Trust, Tynwald Mills, St John's, Isle of
Man IM4 3AE. 01624 801985.
e-mail: tricia@manxwt.cix.co.uk

CRONK Y BING

Manx Wildlife Trust.
Location: NX 381 017. Take A10 coast road N
from Jurby. Approx two miles along there is a sharp
right hand turn over a bridge. Before bridge there is a
track to L. Parking area is available at end of track.
Access: Open all year round. Dogs to be kept on a
lead. Not suitable for the disabled.
Facilities: None.
Public transport: None.
Habitat: Open dune and dune grassland.
Key birds: *Summer:* Terns. *Winter:* Divers, grebes,
skuas, gulls.
Contact: Tricia Sayle, Reserves Officer, Manx
Wildlife Trust, Tynwald Mills, St John's, Isle of
Man IM4 3AE. 01624 801985.
e-mail: tricia@manxwt.cix.co.uk

COUNTY DIRECTORY

Great Crested Grebe by Nick Williams

ENGLAND

THE INFORMATION in the directory has been obtained either from the persons listed or from the appropriate national or other bodies. In some cases, where it has not proved possible to verify the details directly, alternative responsible sources have been relied upon. When no satisfactory record was available, previously included entries have sometimes had to be deleted. Readers are requested to advise the editor of any errors or omissions.

AVON

See Somerset.

BEDFORDSHIRE

Bird Atlas/Avifauna
An Atlas of the Breeding Birds of Bedfordshire 1988-92 by R A Dazley and P Trodd (Bedfordshire Natural History Society, 1994).

Bird Recorder
Dave Odell, 43 Tyne Crescent, Brickhill, Bedford, MK41 7UN. e-mail: daveodell@tiscali.co.uk

Bird Report
BEDFORDSHIRE BIRD REPORT (1946-), from Mary Sheridan, 28 Chestnut Hill, Linslade, Leighton Buzzard, Beds LU7 2TR. 01525 378245.

BTO Regional Representative & Regional Development Officer
RR. Phil Cannings, 30 Graham Gardens, Luton, Beds, LU3 1NQ. H:01582 400394; W:01234 842203; e-mail: philcannings@btopenworld.com

RDO. Judith Knight, 381 Bideford Green, Linslade, Leighton Buzzard, Beds, LU7 2TY. Home 01525 378161; e-mail: judy.knight@tinyonline.co.uk

Club
BEDFORDSHIRE BIRD CLUB. (1992; 252). Miss Sheila Alliez, Flat 61 Adamson Court, Adamson Walk, Kempston, Bedford, MK42 8QZ. e-mail: alliezsec@peewit.freeserve.co.uk www.bedsbirdcub.org.uk
Meetings: 8.00pm, last Tuesday of the month (Sep-Mar), Maulden Village Hall, Maulden, Beds.

Ringing Groups
IVEL RG. Errol Newman, 29 Norse Road, Goldington, Bedford, MK41 0NR. 01234 312787; e-mail: lew.n1@ntlworld.com

RSPB Local Groups
BEDFORD. (1970; 80). Barrie Mason, 6 Landseer Walk, Bedford, MK41 7LZ. 01234 262280.
Meetings: 7.30pm, 3rd Thursday of the month, A.R.A. Manton Lane, Bedford.

LUTON AND SOUTH BEDFORDSHIRE. (1973; 120). Mick Price, 120 Common Road, Kensworth, Beds LU6 3RG. 01582 873268.
Meetings: 7.45pm, 2nd Wednesday of the month, Houghton Regis Social Cnetre, Parkside Drive, Houghton Regis, LU5 5QN.

Wildlife Trust
See Cambridgeshire.

BERKSHIRE

BirdAtlas/Avifauna
The Birds of Berkshire by P E Standley et al (Berkshire Atlas Group/Reading Ornithological Club, 1996).

Bird Recorder
RECORDER (Records Committee and rarity records).Chris DR Heard, 3 Waterside Lodge, Ray Mead Road, Maidenhead, Berkshire SL6 8NP. 01628 633828; e-mail: chris.heard@virgin.net

ASSISTANT RECORDER (Rare breeding records, bird survey data). Derek J Barker, 40 Heywood Gardens, Woodlands Park, Maidenhead, Berkshire SL6 3LZ. 01628 820125.

Bird Reports
BERKSHIRE BIRD BULLETIN (Monthly, 1986-), from Brian Clews, 118 Broomhill, Cookham, Berks SL6 9LQ. 01628 525314; e-mail: brian.clews@btconnect.com

BIRDS OF BERKSHIRE (1974-), from Secretary of the Reading Ornithological Club,

BIRDS OF THE THEALE AREA (1988-), from Secretary, Theale Area Bird Conservation Group.

NEWBURY DISTRICT BIRD REPORT (1959-), from Secretary, Newbury District Ornithological Club.

BTO Regional Representative & Regional Development Officer
RR. Chris Robinson, 2 Beckfords, Upper Basildon, Reading, RG8 8PB. 01491 671420; e-mail: berks_bto_rep@btinternet.com

Clubs
BERKSHIRE BIRD BULLETIN GROUP. (1986; 100). Berkshire Bird Bulletin Group, PO Box 680, Maidenhead, Berks SL6 9ST. 01628 525314; e-mail: brian.clews@btconnect.com

ENGLAND

NEWBURY DISTRICT ORNITHOLOGICAL CLUB. (1959; 110). Trevor Maynard, 15 Kempton Close, Newbury, Berks RG14 7RS. 01635 36752; info@ndoc.org.uk www.ndoc.org.uk.

READING ORNITHOLOGICAL CLUB. (1945; 200). Renton Righelato, 63 Hamilton Road, Reading RG1 5RA. 0787 981 2564; e-mail: renton@righelato.net www.theroc.org.uk
Meetings: 8pm, alternate Wednesdays (Oct-Mar). University of Reading.

THEALE AREA BIRD CONSERVATION GROUP. (1988; 75). Brian Uttley, 65 Omers Rise, Burghfield Common, Reading RG7 3HH. 0118 983 2894.
Meetings: 8pm, 1st Tuesday of the month, Englefield Social Club.

Ringing Groups
NEWBURY RG. J Legg, 1 Malvern Court, Old Newtown Road, Newbury, Berks, RG14 7DR.
e-mail: janlegg@btinternet.com

RUNNYMEDE RG. D G Harris, 22 Blossom Waye, Hounslow, TW5 9HD.
e-mail: daveharris@tinyonline.co.uk

RSPB Local Groups
EAST BERKSHIRE. (1974; 200). Ken Panchen, 7 Knottocks End, Beaconsfield, Bucks, HP9 2AN. 01494 675779; e-mail: ken.panchen@care4free.net www.eastberksrspb.org.uk

READING. (1986; 80). Carl Feltham, 39 Moriston Close, Reading, RG30 2PW. 0118 941 1713. www.reading-rspb.org.uk
Meetings: 2nd Tuesday of the month (Sep-Jun), Pangbourne Village Hall, Pangbourne, READING.

WOKINGHAM & BRACKNELL. (1979; 200). Les Blundell, Folly Cottage, Buckle Lane, Warfield RG42 5SB. 01344 861964; e-mail: les@folly-cottage.fsnet.co.uk www.wbrspb.btinternet.co.uk
Meetings: 8.00pm, 2nd Tuesday of the month (Sep-Jun), Finchampstead Memorial Hall, Wokingham.

Wildlife Hospitals
KESTREL LODGE. D J Chandler, 101 Sheridan Avenue, Caversham, Reading, RG4 7QB. 01189 477107. Birds of prey, ground feeding birds, waterbirds, seabirds. Temporary homes for all except large birds of prey. Veterinary support. Small charge.

LIFELINE. Wendy Hermon, Treatment Centre Co-ordinator, Swan Treatment Centre, Cuckoo Weir Island, South Meadow Lane, Eton, Windsor, Berks, SL4 6SS. 01753 859397; fax 01753 622709; www.swanlifeline.org.uk

Registered charity. Thames Valley 24-hour swan rescue and treatment service. Veterinary support and hospital unit. Operates membership scheme.

Wildlife Trust
Director, See Oxfordshire,

BUCKINGHAMSHIRE

BirdAtlas/Avifauna
The Birds of Buckinghamshire ed by P Lack and D Ferguson (Buckinghamshire Bird Club, 1993).

Bird Recorder
Andy Harding, 15 Jubilee Terrace, Stony Stratford, Milton Keynes, MK11 1DU. H:01908 565896; W:01908 653328; e-mail: a.v.harding@open.ac.uk

Bird Reports
AMERSHAM BIRDWATCHING CLUB ANNUAL REPORT (1975-), from Secretary,

BUCKINGHAMSHIRE BIRD REPORT (1980-), from John Gearing, Valentines, Dinton, Aylesbury, Bucks, HP17 8UW. e-mail: john_gearing@hotmail.com

NORTH BUCKS BIRD REPORT (10 pa), from Recorder,

BTO Regional Representative & Regional Development Officer
RR. Mick A'Court, 6 Chalkshire Cottages, Chalkshire road, Butlers Cross, Bucks ,HP17 0TW. H:01296 623610; W:01494 462246; e-mail: mick@focusrite.com or a.arundinaceous@virgin.net

RDO. Peter Hearn, 160 High Street, Aylesbury, Bucks HP20 1RE. Home & fax 01296 581520; Work 01296 424145.

Clubs
BUCKINGHAMSHIRE BIRD CLUB. (1981; 340). Roger S Warren, 59 Glade Road, Marlow, Bucks SL7 1DQ. 01628 484807. www.hawfinches.freeserve.co.uk

NORTH BUCKS BIRDERS. (1977; 50). Andy Harding, 15 Jubilee Terrace, Stony Stratford, Milton Keynes, MK11 1DU. H:01908 565896; W:01908 653328.
Meetings: Last Tuesday of the month (Nov, Jan, Feb, Mar), The Cock, High Street, Stony Stratford.

Ringing Groups
HUGHENDEN RG. Peter Edwards, 8 The Brackens, Warren Wood, High Wycombe, Bucks, HP11 1EB. 01494 535125.

RSPB Local Groups
See also Herts: Chorleywood,

AYLESBURY. (1981; 220). Barry Oxley, 3 Swan Close, Station Road, Blackthorn, Bicester, Oxon, OX25 1TU. 01869 247780.

NORTH BUCKINGHAMSHIRE. (1976; 430). Jim Parsons, 8 The Mount, Aspley Guise, Milton Keynes, MK17 8EA. 01908 582450.
Meetings: 8.00pm, 2nd Tuesday of the month, Jennie Lee Theatre, Bletchley Leisure Centre.

Wildlife Hospitals

MILTON KEYNES WILDLIFE HOSPITAL. Mr & Mrs V Seaton, 150 Bradwell Common Boulevard, Milton Keynes, MK13 8BE. 01908 604198; www-tec.open.ac.uk/staff/robert/mkwh.htm Registered charity. All species of British birds and mammals. Veterinary support.

WILDLIFE HOSPITAL TRUST. St Tiggywinkles, Aston Road, Haddenham, Aylesbury, Bucks HP17 8AF. 01844 292292; fax 01844 292640; e-mail: mail@sttiggywinkles.org.uk www.sttiggywinkles.org.uk Registered charity. All species. Veterinary referrals and helpline for vets and others on wild bird treatments. Full veterinary unit and staff. Pub: *Bright Eyes* (free to members - sae).

Wildlife Trust

Director, See Oxfordshire,

CAMBRIDGESHIRE

BirdAtlas/Avifauna

An Atlas of the Breeding Birds of Cambridgeshire (VC 29) P M M Bircham et al (Cambridge Bird Club, 1994).

The Birds of Cambridgeshire: checklist 2000 (Cambridge Bird Club)

Bird Recorders

CAMBRIDGESHIRE. John Oates, 7 Fassage Close, Lode, Cambridge CB5 9EH. 01223 812546, (M)07860 132708. e-mail: joates9151@aol.com

HUNTINGDON & PETERBOROUGH. John Clark, 7 West Brook, Hilton, Huntingdon, Cambs, PE28 9NW. 01480 830472.

Bird Reports

CAMBRIDGESHIRE BIRD REPORT (1925-), from Secretary, Cambridge Bird Club,

PAXTON PITS BIRD REPORT (1994-) £3 inc postage, from Trevor Gunton, 15 St James Road, Little Paxton, Cambs PE19 6QW. (Tel/fax)01480 473 562.

PETERBOROUGH BIRD CLUB REPORT (1999-), from Secretary, Peterborough Bird Club,

BTO Regional Representatives

CAMBRIDGESHIRE RR. John Le Gassick, 17 Acacia Avenue, St Ives, Cambs PE27 6TN. 01480 391991; e-mail: jclegassick@ntlworld.com

HUNTINGDON & PETERBOROUGH. Phillip Todd, 01733 810832; e-mail: huntspbororr@yahoo.co.uk

Clubs

CAMBRIDGESHIRE BIRD CLUB. (1925; 310). Bruce Martin, 178 Nuns Way, Cambridge, CB4 2NS. 01223 700656; e-mail: bruce.s.martin@ntlworld.com www.cambridgeshirebirdclub.org.uk
Meetings: 2nd Friday of the month, St John's Church Hall, Hills Road, Cambridge/Milton CP Visitors Centre, Milton, Cambridge.

PETERBOROUGH BIRD CLUB. (1999; 210)Janet Darke (membership secretary), 34 High Street, Stilton, Peterborough PE7 3RA. 01733 243556; e-mail: janet@jdarke.freeserve.co.uk www.pbc.codehog.co.uk
Meetings: Indoor last Tuesday of each month from Sep-Apr inclusive at 7.30pm at Burghley Club, Burghley Square, Peterborough. Outdoor meeting monthly throughout most of year.

ST NEOTS BIRD & WILDLIFE CLUB. (1993; 150). Stuart Elsom, 117 Andrew Road, Eynesbury, St Neots, Cambridgeshire PE19 2PP.
e-mail: stuart.elsom@tringa.co.uk www.paxton-pits.org.uk
Meetings: 7.30pm, various Tuesdays, St Neots Bowling Club, St Anselm Place, check website for details.

Ringing Group

WICKEN FEN RG. Dr C J R Thorne, 17 The Footpath, Coton, Cambs, CB3 7PX. 01954 210566; e-mail: cjrt@cam.ac.uk

RSPB Local Groups

CAMBRIDGE. (1977; 150). Colin Kirtland, 22 Montgomery Road, Cambridge, CB4 2EQ. 01223 363092.
Meetings: 3rd Wednesday of every month Jan-May and Sept-Dec 8pm. Chemistry Labs, Lensfield Road, Cambridge.

HUNTINGDONSHIRE. (1982; 200). Pam Peacock, Old Post Office, Warboys Road, Pidley, Huntingdon, Cambs, PE28 3DA. 01487 840615; e-mail: pamp@oldpostoffice.fsnet.co.uk www.huntsrspb.co.uk
Meetings: 7.30pm, last Wednesday of the month (Sep-Apr), Free Church, St Ives.

Wetland Bird Survey Organisers

CAMBRIDGESHIRE OLD COUNTY. Bruce Martin, 178 Nuns Way, Cambridge, CB4 2NS(H)01223 363656; (W)01223 246644; e-mail: bruce.s.martin@ntlworld.com

NENE WASHES. Charlie Kitchin, RSPB Nene Washes, 21a East Delph, Whittlesey, Cambs PE7 1RH. 01733 205140.

Wildlife Trust
BEDS, CAMBS, NORTHANTS & PETERBOROUGH WILDLIFE TRUST. (1990; 12,000). The Manor House, Broad Street, Great Cambourne, Cambridgeshire CB3 6DH. 01954 713500; fax 01954 710051; e-mail: cambridgeshire@wildlifebcnp.org www.wildlifebcnp.org

CHESHIRE

BirdAtlas/Avifauna
The Birds of Sandbach Flashes 1935-1999 by Andrew Goodwin and Colin Lythgoe (The Printing House, Crewe, 2000).

Bird Recorder (inc Wirral)
Tony Broome, 4 Larchwood Drive, Wilmslow, Cheshire, SK9 2NU. 01625 540434; e-mail: tonybroome@cawos.org

Bird Report
CHESHIRE & WIRRAL BIRD REPORT (1969-), from David Cogger, 113 Nantwich Road, Middlewich, Cheshire, CW10 9HD. 01606 832517; e-mail: memsec@cawos.org www.cawos.org

SOUTH EAST CHESHIRE ORNITHOLOGICAL SOCIETY BIRD REPORT (1985-), from Secretary, South East Cheshire Ornithol Soc. 01270 582642.

BTO Regional Representatives & Regional Development Officer
MID RR. Paul Miller, 01928 787535; e-mail: huntershill@worldline.co.uk

NORTH & EAST RR. Charles Hull, Edleston Cottage, Edleston Hall Lane, Nantwich, Cheshire, CW5 8PL. 01270 628194; e-mail: edleston@yahoo.co.uk

SOUTH RR & RDO. Charles Hull, Edleston Cottage, Edleston Hall Lane, Nantwich, Cheshire, CW5 8PL. 01270 628194; e-mail: edleston@yahoo.co.uk

Clubs
CHESHIRE & WIRRAL ORNITHOLOGICAL SOCIETY. (1988; 375). David Cogger, 113 Nantwich Road, Middlewich, Cheshire, CW10 9HD. 01606 832517; e-mail: memsec@cawos.org www.cawos.org
Meetings: 7.45pm, 1st Friday of the month, Knutsford Civic Centre.

CHESTER & DISTRICT ORNITHOLOGICAL SOCIETY. (1967; 50). David King, 13 Bennett Close, Willaston,

South Wirral, CH64 2XF. 0151 327 7212.
Meetings: 7.30pm, 1st Thursday of the month (Oct-Mar), Caldy Valley Community Centre.

KNUTSFORD ORNITHOLOGICAL SOCIETY. (1974; 45). Roy Bircumshaw, 267 Longridge, Knutsford, Cheshire, WA16 8PH. 01565 634193. www.10x50.com
Meetings: 7.30pm, 4th Friday of the month (not Dec), Jubilee Hall.

LANCASHIRE & CHESHIRE FAUNA SOCIETY. (1914; 140). Dave Bickerton, 64 Petre Crescent, Rishton, Lancs, BB1 4RB. 01254 886257; e-mail: bickertond@aol.com www.lacfs.org.uk

LYMM ORNITHOLOGY GROUP. (1975; 60). Mrs Ann Ledden, 4 Hill View, Widnes, WA8 9AL. 0151 424 0441; e-mail: secretary-log@tiscali.co.uk
Meetings: 8.00pm, last Friday of the month (Aug-May), Lymm Village Hall.

MID-CHESHIRE ORNITHOLOGICAL SOCIETY. (1963; 80). Les Goulding, 7 Summerville Gardens, Stockton Heath, Warrington WA4 2EG. 01925 265578; e-mail: les@goulding7.fsnet.co.uk www.midcheshireos.co.uk
Meetings: 7.30pm, 2nd Friday of the month (Oct-Mar), Hartford Village Hall.

NANTWICH NATURAL HISTORY SOCIETY. (1979; 40). Mike Holmes, 4 Tenchers Field, Stapeley, Nantwich, Cheshire CW5 7GR.01270 611577; e-mail: mike@mimprove.com www.nantnats.fsnet.co.uk

SOUTH EAST CHESHIRE ORNITHOLOGICAL SOCIETY. (1964; 140). Colin Lythgoe, 11 Waterloo Road, Haslington, Crewe, CW1 5TF. 01270 582642. www.secos.freeuk.com
Meetings: 2nd Friday (Sept-Apr), 7.30pm, St Mathews Church Hall, Elworth.

WILMSLOW GUILD BIRDWATCHING GROUP. (1965; 67). Tom Gibbons, Chestnut Cottage, 37 Strawberry Lane, Wilmslow, Cheshire, SK9 6AQ. 01625 520317.
Meetings: 7.30pm last Friday of the month, Wilmslow Guild, Bourne St, Wilmslow.

Ringing Groups
MERSEYSIDE RG. P Slater, 45 Greenway Road, Speke, Liverpool, L24 7RY.

SOUTH MANCHESTER RG. C M Richards, Fairhaven, 13 The Green, Handforth, Wilmslow, Cheshire, SK9 3AG. 01625 524527; e-mail: cliveandkay.richards@care4free.net

RSPB Local Groups
CHESTER. (1987; 250). Roger Nutter, Group Leader, 2

Lower Farm Court, Duckington, Malpas, Cheshire SY14 8LQ. 01829 782237;
e-mail: lexeme@onetel.com
Meetings: 7.30pm, 3rd Wednesday of the month (Sep-Apr), St Mary's Centre, Chester.

MACCLESFIELD. (1979; 394). Ray Evans, 01625 432635; e-mail: chair@macclesfieldrspb.org.uk
www.macclesfieldrspb.org.uk
Meetings: 7.45pm, Tuesdays, Senior Citizens Hall, Duke Street Car Park, Macclesfield. Doors open from 7.00 pm.

NORTH CHESHIRE. (1976; 100) Paul Grimmet. 01925 268770; e-mail: paulw@ecosse.net
Meetings: 7.30pm for 7.45pm start, St. Matthews C of E Primary School, Stretton Road, Stretton.

Wildlife Hospitals
RSPCA STAPELEY GRANGE WILDLIFE HOSPITAL. London Road, Stapeley, Nantwich, Cheshire, CW5 7JW. 0870 442 7102. All wild birds. Oiled bird wash facilities and pools. Veterinary support.

Wildlife Trust
CHESHIRE WILDLIFE TRUST. (1962; 3600). Grebe House, Reaseheath, Nantwich, Cheshire CW5 6DG. 01270 610180; fax 01270 610430;
e-mail: info@cheshirewt.cix.co.uk
www.wildlifetrust.org.uk/cheshire

CORNWALL

Bird Recorders
CORNWALL. K Wilson, No.1 Tol-pedn House, School Hill Road, St Levan, Penzance, Cornwall, TR19 6LP. 01736 871800; e-mail: kesteraw@yahoo.co.uk

ISLES OF SCILLY. Nigel Hudson, Post Office Flat, Hugh Street, St Mary's, Isles of Scilly TR21 0JE. 01720 422267; e-mail: nigel-hudson@tiscali.co.uk

Bird Reports
BIRDS IN CORNWALL (1931-), from Colin Boyd, 4 Henliston Drive, Helston, Cornwall, TR13 8BW.

ISLES OF SCILLY BIRD REPORT and NATURAL HISTORY REVIEW 2000 (1969-), from Recorder, Isles of Scilly Bird Group.

BTO Regional Representatives & Regional Development Officers
CORNWALL RR & RDO.

ISLES OF SCILLY RR & RDO. Will Wagstaff, 42 Sally Port, St Mary's, Isles of Scilly, TR21 0JE. 01720 422212; e-mail: william.wagstaff@virgin.net

Clubs
CORNWALL BIRDWATCHING & PRESERVATION

SOCIETY. (1931; 990). Darrell Clegg, 55 Lower Fore Street, Saltash, Cornwall PL12 6JQ.
www.cbwps.org.uk

CORNWALL WILDLIFE TRUST PHOTOGRAPHIC GROUP. (40). David Chapman, 41 Bosence Road, Townshend, Nr Hayle, Cornwall TR27 6AL. 01736 850287; e-mail: david@ruralimages.freeserve.co.uk
www.ruralimages.freeserve.co.uk
Meetings: Mixture of indoor and outdoor meetings, please phone for details.

ISLES OF SCILLY BIRD GROUP (2000; 510). Katharine Sawyer, Membership Secretary, Alegria, High Lanes, St Mary's, Isles of Scilly TR21 0NW. e-mail: katharine.sawyer@which.net
www.scillybirding.co.uk

Ringing Group
SCILLONIA SEABIRD GROUP. Peter Robinson, 19 Pine Park Road, Honiton, Devon, EX14 2HR. (Tel/fax) 01404 549873; e-mail: pjrobinson2@compuserve.com

RSPB Local Group
CORNWALL. (1972; 600). Gordon Mills, 11 Commercial Square, Camborne, Cornwall TR14 8JZ. 01209 713144 (eve).
Meetings: Indoor meetings (Oct-Apr), outdoor throughout the year.

Wetland Bird Survey Organisers
CORNWALL (Excl. Graham Hobin, Lower Drift Farmhouse, Drift, Buryas Bridge, Penzance TR19 6AA; e-mail: graham@birdbrain.freeserve.co.uk

TAMAR COMPLEX. Gladys Grant, 18 Orchard Crescent, Oreston, Plymouth, PL9 7NF. 01752 406287

Wildlife Hospital
MOUSEHOLE WILD BIRD HOSPITAL & SANCTUARY ASSOCIATION LTD. Raginnis Hill, Mousehole, Penzance, Cornwall, TR19 6SR. 01736 731386. All species. No ringing.
E-mail: mouseholebirdhospital@hotmail.com
www.chycor.co.uk/mousehole-sanctuary/index.htm

Wildlife Trust
CORNWALL WILDLIFE TRUST. (1962; 6000). Five Acres, Allet, Truro, Cornwall, TR4 9DJ. 01872 273939; fax 01872 225476; e-mail: info@cornwt.demon.co.uk
www.cornwallwildlifetrust.org.uk

THE ISLES OF SCILLY WILDLIFE TRUST. Carn Thomas, Hugh Town, St Marys, Isles of Scilly TR21 0PT. 01720 422153;(fax) 01720 422153;
e-mail: enquiries@ios-wildlifetrust.org.uk
www.ios-wildlifetrust.org.uk

CUMBRIA

BirdAtlas/Avifauna
The Breeding Birds of Cumbria by Stoff, Callion, Kinley, Raven and Roberts (Cumbria Bird Club, 2002).

Bird Recorders
COUNTY. Colin Raven, 18 Seathwaite Road, Barrow-in-Furness, Cumbria, LA14 4LX; e-mail: colin@walneyobs.fsnet.co.uk

NORTH EAST (Carlisle & Eden). Michael F Carrier, Lismore Cottage, 1 Front Street, Armathwaite, Cumbria, CA4 9PB. 01697 472218.

NORTH WEST (Allerdale & Copeland). J K Manson, Fell Beck, East Road, Egremont, Cumbria, CA22 2ED. 01946 822947; e-mail: jake@jakemanson.freeserve.co.uk

SOUTH (South Lakeland & Furness). Ronnie Irving, 24 Birchwood Close, Kendal, Cumbria, LA9 5BJ. 01539 727523; e-mail: ronnie@fenella.fslife.co.uk

Bird Reports
BIRDS AND WILDLIFE IN CUMBRIA (1970-), from D Clarke, Tullie House Museum, Castle Street, Carlisle, Cumbria, CA3 8TP; e-mail: DavidC@carlisle-city.gov.uk

WALNEY BIRD OBSERVATORY REPORT, from Warden, see Reserves.

BTO Regional Representatives
NORTH RR. Clive Hartley, 01539 532856; e-mail: clivehartley@marshcott.freeserve.co.uk

SOUTH RR. Clive Hartley, 01539 532856; e-mail: clivehartley@marshcott.freeserve.co.uk

Clubs
ARNSIDE & DISTRICT NATURAL HISTORY SOCIETY. (1967; 221). Mrs GM Smith, West Wind, Orchard Road, Arnside, via Carnforth, Cumbria, LA5 0DP. 01524 762522.
Meetings: 7.30pm, 2nd Tuesday of the month (Sept-Apr). WI Hall, Arnside. (Also summer walks).

CUMBRIA BIRD CLUB. (1989; 230). Dave Piercy, Derwentwater Youth Hostel, Borrowdale, Keswick CA12 5UR. 017687 77246; e-mail: daveandkathypiercy@tiscali.co.uk www.cumbriabirdclub.freeserve.co.uk
Meetings: Various evenings and venues (Oct-Mar) check on website for further details. £2 for non-members.

CUMBRIA RAPTOR STUDY GROUP. (1992). P N Davies, Snowhill Cottage, Caldbeck, Wigton, Cumbria, CA7 8HL. 016973 71249; e-mail: pete.caldbeck@virgin.net

Ringing Groups
EDEN RG. G Longrigg, Mere Bank, Bleatarn, Warcop, Appleby, Cumbria, CA16 6PX.

MORECAMBE BAY WADER RG. J Sheldon, 415 West Shore Park, Barrow-in-Furness, Cumbria, LA14 3XZ. 01229 473102.

WALNEY BIRD OBSERVATORY. K Parkes, 176 Harrogate Street, Barrow-in-Furness, Cumbria, LA14 5NA. 01229 824219.

RSPB Local Groups
CARLISLE. (1974; 400). Bob Jones, 130 Greenacres, Wetheral, Carlisle.
Meetings: 7.30pm, Wednesday monthly, Tythe Barn, Carlisle.

SOUTH LAKELAND. (1973; 340). Ms Kathleen Atkinson, 2 Langdale Crescent, Windermere, Cumbria, LA23 2HE. 01539 444254.

WEST CUMBRIA. (1986; 230). Neil Hutchin, 3 Camerton Road, Gt Broughton, Cockermouth, Cumbria, CA13 0YR. 01900 825231; e-mail: neil@hutchin50.fsnet.co.uk
Meetings: 7.30pm, Tuesdays (Sept-Apr), United Reformed Church, Main St, Cockermouth

Wetland Bird Survey Organiser
DUDDON ESTUARY. Bob Treen, 5 Rydal Close, Dalton-in-Furness, Cumbria LA15 8QU. 01229 464789.

Wildlife Trust
CUMBRIA WILDLIFE TRUST. (1962; 5000). Plumgarths, Crook Road, Kendal, Cumbria LA8 8LX. 01539 816300; fax 01539 816301; e-mail: mail@cumbriawildlifetrust.org.uk www.wildlifetrust.org.uk/cumbria

DERBYSHIRE

Bird Recorders
1. Rare breeding records. Roy A Frost, 66 St Lawrence Road, North Wingfield, Chesterfield, Derbyshire, S42 5LL. 01246 850037; e-mail: frostra66@lepidoptera.force9.co.uk.

2. Records Committee & rarity records. Rodney W Key, 3 Farningham Close, Spondon, Derby, DE21 7DZ. 01332 678571; e-mail: r.key3@ntlworld.com

3. JOINT RECORDER and Annual Report editor. Richard M R James, 10 Eastbrae Road, Littleover, Derby, DE23 1WA. 01332 771787; e-mail: rmrjames@yahoo.co.uk

Bird Reports
BENNERLEY MARSH WILDLIFE GROUP ANNUAL REPORT, from Secretary.

CARSINGTON BIRD CLUB ANNUAL REPORT, from Secretary.

DERBYSHIRE BIRD REPORT (1954-), from Bryan Barnacle, Mays, Malthouse Lane, Froggatt, Hope Valley, Derbyshire S32 3ZA. 01433 630726; e-mail: barney@mays1.demon.co.uk

OGSTON BIRD CLUB REPORT (1970-), from Secretary.

BTO Regional Representatives
NORTH RR. Dave Budworth, 121 Wood Lane, Newhall, Swadlincote, Derbys, DE11 0LX. 01283 215188; e-mail: dbud01@aol.com

SOUTH RR. Dave Budworth, 121 Wood Lane, Newhall, Swadlincote, Derbys, DE11 0LX. 01283 215188; e-mail: dbud01@aol.com

Clubs
BENNERLEY MARSH WILDLIFE GROUP. (1995; 135). Richard Rogers, 19 Arundel Drive, Beeston, Nottingham, NG9 3FX. email: rtnr@breathe.com

BAKEWELL & DISTRICT BIRD STUDY GROUP. (1987; 70).Bill Millward, Dale House, The Dale, Hope Valley, Derbys S32 1AQ.
Meetings: 7.30pm, 2nd Monday of the month, Friends Meeting Room, Bakewell.

BUXTON FIELD CLUB. (1946; 68). B Aries, 1 Horsefair Avenue, Chapel-en-le-Frith, High Peak, Derbys, SK23 9SQ. 01298 815291;
e-mail: brian.aries@horsefair.ndo.co.uk
Meetings: 7.30pm, Saturdays fortnightly (Oct-Mar), Methodist Church Hall, Buxton.

CARSINGTON BIRD CLUB. (1992; 257). Mrs Dorothy Evans, 0775 992 4259;
e-mail: Secretary@carsingtonbirdclub.co.uk
www.carsingtonbirdclub.co.uk
Meetings: 7.30pm, 3rd Tuesday of the month, Hognaston Village Hall, Hognaston, 2 miles from Carsington.

 **Derbyshire Ornithological Society** (1954; 550). Steve Shaw, 84 Moorland View Road, Walton, Chesterfield, Derbys, S40 3DF. 01246 236090;
e-mail: steveshaw@ornsoc.freeserve.co.uk
www.derbyshireOS.org.uk
Meetings: 7.30pm, last Friday of the winter months, various venues.

OGSTON BIRD CLUB. (1969; 1126). Mrs Ann Hill, 2 Sycamore Avenue, Glapwell, Chesterfield, S44 5LH. 01623 812159. www.ogstonbirdclub.co.uk

SOUTH PEAK RAPTOR STUDY GROUP. (1998; 12). M E Taylor, 76 Hawksley Avenue, Newbold, Chesterfield, Derbys, S40 4TL. 01246 277749.

Ringing Groups
DARK PEAK RG. W M Underwood, Ivy Cottage, 15 Broadbottom Road, Mottram-in-Longdendale, Hyde, Cheshire SK14 6JB.
e-mail: w.m.underwood@talk21.com

SORBY-BRECK RG. Dr Geoff Mawson, Moonpenny Farm, Farwater Lane, Dronfield, Sheffield, S18 1RA. 01246 415097; e-mail: gpmawson@hotmail.com

SOUDER RG. Dave Budworth, 121 Wood Lane, Newhall, Swadlincote, Derbys, DE11 0LX. 0121 695 3384.

RSPB Local Groups
CHESTERFIELD. (1987; 274). Sue Ottowell, 01246 569431; e-mail: sue@ottowell.wanadoo.co.uk
Meetings: 7.15pm, Winding Wheel, New Exhibition Centre, 13 Holywell Street, Chesterfield.

DERBY LOCAL GROUP. (1973; 560). Paul Highman, 12 Ford Lane, Allestree, Derby, DE22 2EW. 01332 551636; e-mail: paul.highman@ntlworld.com.
Meetings: 7.30pm, 1st Wednesday of the month (Sep-Apr), Lund Pavilion, Derbyshire County Cricket Ground.

Wildlife Trust
DERBYSHIRE WILDLIFE TRUST. (1962; 5000). East Mill, Bridgefoot, Belper, Derbyshire DE56 1XH. 01773 881188; fax 01773 821826;
e-mail: enquiries@derbyshirewt.co.uk
www.derbyshirewildlifetrust.org.uk

DEVON

BirdAtlas/Avifauna
Tetrad Atlas of Breeding Birds of Devon by H P Sitters (Devon Birdwatching & Preservation Society, 1988).

Bird Recorder
Mike Tyler, The Acorn, Shute Road, Kilmington, Axminster, Devon EX13 7ST. 01297 34958; e-mail: mike@mwtyler.freeserve.co.uk

Bird Reports
DEVON BIRD REPORT (1928-), from H Kendall, 33 Victoria Road, Bude, Cornwall, EX23 8RJ. 01288 353818; e-mail: harvey.kendall@btopenworld.com

LUNDY FIELD SOCIETY ANNUAL REPORT (1946-), from Secretary. Index to Report is on Society's website.

BTO Regional Representative & Regional Development Officer
RR. John Woodland, Glebe Cottage, Dunsford, Exeter, EX6 7AA. Tel/fax 01647 252494;
e-mail: jwoodland@btodv.fsnet.co.uk

Clubs
DEVON BIRDWATCHING & PRESERVATION SOCIETY.

(1928; 1200). Mrs Joy Vaughan, 28 Fern Meadow, Okehampton, Devon, EX20 1PB. 01837 53360; e-mail: joy@vaughan411.freeserve.co.uk

KINGSBRIDGE & DISTRICT NATURAL HISTORY SOCIETY. (1989; 130). Martin Catt, Migrants Rest, East Prawle, Kingsbridge, Devon, TQ7 2DB. 01548 511443; e-mail: martin.catt@btinternet.com
Meeting: 7.30pm, 4th Monday of month (Sept-Apr), phone for venue.

LUNDY FIELD SOCIETY. (1946; 450). Frances Stuart, 3 Lower Linden Road, Clevedon, North Somerset BS21 7SU. 01275 871434; e-mail: fs@ifrc.co.uk www.lundy.org.uk

TOPSHAM BIRDWATCHING & NATURALISTS' SOCIETY. (1969; 140). Mrs Janice Vining, 2 The Maltings, Fore Street, Topsham, Exeter, EX3 0HF. 01392 873514; e-mail: tbnsociety@hotmail.com www.members.tripod.co.uk/tbns
Meetings: 7.30pm, 2nd Friday of the month (Sep-May), Matthews Hall, Topsham.

Ringing Groups
DEVON & CORNWALL WADER RG. R C Swinfen, 72 Dunraven Drive, Derriford, Plymouth, PL6 6AT. 01752 704184.

LUNDY FIELD SOCIETY. A M Taylor, 26 High Street, Spetisbury, Blandford, Dorset, DT11 9DJ. 01258 857336; e-mail: ammataylor@yahoo.co.uk www.lundy.org.uk/lfs/

SLAPTON BIRD OBSERVATORY. Peter Ellicott, 10 Chapel Road, Alphington, Exeter, EX2 8TB. 01392 277387.

RSPB Local Groups
EXETER & DISTRICT. (1974; 466). John Allan, Coxland-by-Sigford, Sigford, Near Newton Abbot, TQ12 6LE. 01626 821344; e-mail: john-allan@coxland.fsnet.co.uk www.exeter-RSPB.org.uk
Meetings: 7.30p, various evenings, Southernhay United Reformed Church Rooms, Dix's Field, EXETER.

NORTH DEVON. (1976; 68). David Gayton, 29 Merrythorne Road, Fremington, Barnstaple, Devon, EX31 3AL. 01271 371092. e-mail: hevdav@aol.com
Meetings: 7 for 7.30pm, last Friday of the month, The Civic Centre, Barnstaple.

PLYMOUTH. (1974; 600). Mrs Eileen Willey, 11 Beverstone Way, Roborough, Plymouth, PL6 7DY. 01752 208996; e-mail: willey@willey.eurobell.co.uk

Wetland Bird Survey Organiser
TAMAR COMPLEX. Gladys Grant, 18 Orchard Crescent, Oreston, Plymouth, PL9 7NF. 01752 406287; e-mail: gladysgrant@onetel.com.

Wildlife Hospitals
BIRD OF PREY CASUALTY CENTRE. Mrs J E L Vinson, Crooked Meadow, Stidston Lane, South Brent, Devon, TQ10 9JS. 01364 72174. Birds of prey, with emergency advice on other species. Aviaries, releasing pen. Veterinary support.

BONDLEIGH BIRD HOSPITAL. Manager, Samantha Hart, The Old Forge, Bondleigh, North Tawton, Devon, EX20 2AJ. 01837 82328. All species. 14 aviaries, 2 aquapens. Veterinary support available, if requested, with payment of full charges.

HURRELL, Dr L H. Dr Hurrell, 201 Outland Road, Peverell, Plymouth, PL2 3PF. 01752 771838. Birds of prey only. Veterinary support.

TORBAY WILDLIFE RESCUE CENTRE. Malcolm Higgs, 6A Gerston Place, Paignton, S Devon, TQ3 3DX. 01803 557624. All wild birds, inc. oiled. Pools, aviaries, intensive care, washing facilities. Open at all times. 24-hr veterinary support. Holding areas off limits to public as all wildlife must be returned to the wild.

Wildlife Trust
DEVON WILDLIFE TRUST. (1962; 10,600). Shirehampton House, 35-37 St David's Hill, Exeter, EX4 4DA. 01392 279244; fax 01392 433221; e-mail: contactus@devonwt.cix.co.uk www.devonwildlifetrust.org

DORSET

BirdAtlas/Avifauna
Dorset Breeding Bird Atlas (working title). In preparation.

Bird Recorder
James Lidster, 35 Napier Road, Poole, Dorset BH15 4LX. 01202 672406; e-mail: dorsetbirds@btopenworld.com

Bird Reports
DORSET BIRDS (1987-), from Miss J W Adams, 16 Sherford Drive, Wareham, Dorset, BH20 4EN. 01929 552299.

THE BIRDS OF CHRISTCHURCH HARBOUR (1956-), from Ian Southworth, 1 Bodowen Road, Burton, Christchurch, Dorset BH23 7JL. e-mail: ianbirder@aol.com

PORTLAND BIRD OBSERVATORY REPORT, from Warden, see Reserves,

BTO Regional Representatives
Mike Pleasants, 10 Green Lane, Bournemouth, BH10 5LB. 01202 593500 or 07762 987115; e-mail: mike-pleasants@hotmail.com

Clubs

CHRISTCHURCH HARBOUR ORNITHOLOGICAL GROUP. (1956; 150). Mr. I.H. Southworth, 1 Bodowen Road, Burton, Christchurch, Dorset BH23 7JL. 01202 478093. www.chog.org.uk

DORSET BIRD CLUB. (1987; 550). Mrs Eileen Bowman, 53 Lonnen Road, Colehill, Wimborne, Dorset, BH21 7AT. 01202 884788. www.dorsetbirdclub.org.uk
Meetings: Usually 7.30pm, no set day or venue.

DORSET NATURAL HISTORY & ARCHAEOLOGICAL SOCIETY. (1845; 2188). Dorset County Museum, High West Street, Dorchester, Dorset, DT1 1XA. 01305 262735; www.dor-mus.demon.co.uk e-mail: dorsetcountymuseum @dor-mus.demon.co.uk

Ringing Groups

CHRISTCHURCH HARBOUR RS. E C Brett, 3 Whitfield Park, St Ives, Ringwood, Hants, BH24 2DX. e-mail: ed_brett@lineone.net

PORTLAND BIRD OBSERVATORY. Martin Cade, Old Lower Light, Portland Bill, Dorset, DT5 2JT. 01305 820553; e-mail: obs@btinternet.com www.portlandbirdobs.btinternet.co.uk

STOUR RG. R Gifford, 62 Beacon Park Road, Upton, Poole, Dorset, BH16 5PE.

RSPB Local Groups

BLACKMOOR VALE. (1981; 106). Mrs Margaret Marris, 15 Burges Close, Marnhull, Sturminster Newton, Dorset, DT10 1QQ. 01258 820091.
Meetings: 7.30pm, 3rd Friday in the month, Gillingham Primary School.

EAST DORSET. (1974; 435). Tony Long (Group Leader: S.Cresswell), 93 Wimborne Road, Corfe Mullen, Wimborne, BH21 3DS. 01202 880508; e-mail: tony@joan1206.fsnet.co.uk
Meetings: 7.30pm, 2nd Wednesday of the month, St Mark's church Hall, Talbot Village, Wallisdown, Eastbourne.

POOLE. (1982; 305). John Derricott, 51 Dacombe Drive, Upton, Poole, Dorset, BH16 5JJ. 01202 776312.John Derricott Tel: 01258 450927; e-mail: johnmal@derricott.fslife.co.uk

SOUTH DORSET. (1976; 400). Marion Perriss, Old Barn Cottage, Affpuddle, Dorchester, Dorset, DT2 7HH. 01305 848268; e-mail: affpuddle@btinternet.com; www.southdorset-rspb.org.uk.
Meetings: 3rd Thursday of each month (Sep-Apr), Dorchester Town Hall.

Wildlife Hospital

SWAN RESCUE SANCTUARY. Ken and Judy Merriman, The Wigeon, Crooked Withies, Holt, Wimborne, Dorset, BH21 7LB. 01202 828166; mobile 0385 917457; e-mail: ken@swan-rescue.fsnet.co.uk www.swan-rescue.co.uk
Swans. Hospital unit with indoor ponds and recovery pens. Outdoors: 35 ponds and lakes, and recovery pens. 24-hr veterinary support. Viewing by appointment only.

Wetland Bird Survey Organisers

THE FLEET & PORTLAND HARBOUR. Steve Groves, Abbotsbury Swannery, New Barn Road, Abbotsbury, Dorset, DT3 4JG. (W)0305 871684; e-mail: abbotsbury.swannery@btinternet.com

RADIPOLE & LODMOOR. Keith Ballard, RSPB Visitor Centre, Swannery Carpark, Weymouth, Dorset, DT4 7TZ. 01305 778313.

Wildlife Trust

DORSET WILDLIFE TRUST. (1961; 8000). Brooklands Farm, Forston, Dorchester, Dorset, DT2 7AA. 01305 264620; fax 01305 251120; www.dorsetwildlife.co.uk e-mail: dorsetwt@cix.co.uk

DURHAM

Bird Atlas/Avifauna

A Summer Atlas of Breeding Birds of County Durham by Stephen Westerberg/Kieth Bowey. (Durham Bird Club, 2000)

Bird Recorders

Tony Armstrong, 39 Western Hill, Durham City, DH1 4RJ. 0191 386 1519; e-mail: ope@globalnet.co.uk

CLEVELAND. Rob Little, 5 Belgrave Court, Seaton Carew, Hartlepool TS25 1BF. 01429 428940; e-mail: rob.little@ntlworld.com

Bird Reports

BIRDS IN DURHAM (1971-), from D Sowerbutts, 9 Prebends Fields, Gilesgate, Durham, DH1 1HH. H:0191 386 7201; e-mail: d16lst@tiscali.co.uk

CLEVELAND BIRD REPORT (1974-), from Mr J Sharp, 10 Glendale, Pinehills, Guisborough, TS14 8JF. 01287 633976.

BTO Regional Representatives

David L Sowerbutts, 9 Prebends Field, Gilesgate Moor, Durham, DH1 1HH. H:0191 386 7201; e-mail: d16lst@tiscali.co.uk

CLEVELAND RR. Position vacant.

Clubs

DURHAM BIRD CLUB. (1975; 263). Steve Evans, 07979 601231; StevieEvans@btinternet.com www.durhambirdclub.org

ENGLAND

TEESMOUTH BIRD CLUB. (1960; 260). Chris Sharp, 20 Auckland Way, Hartlepool, TS26 0AN. 01429 865163. www.teesmouthbc.freeserve.co.uk
Meetings: 7.30pm, 1st Wednesday of the month (Sep-Apr), Billingham Arms Hotel, Billingham.

Ringing Groups
DURHAM RG. S Westerberg, 9 Coal Fell Tce, Hallbank Gate, Brampton, Cumbria, CA8 2PY.

DURHAM DALES RG. J R Hawes, Fairways, 5 Raby Terrace, Willington, Crook, Durham, DL15 0HR.

RSPB Local Group
DURHAM. (1974; 125). Lo Brown, 4 Ann's Place, Langley Moor, Durham, DH7 8JY. 0191 378 2433. www.afbp16.freewire.co.uk
Meetings: 7.30pm, 2nd Tuesday of the month (Oct-Mar), Room CG83, adjacent to Scarborough Lecture Theatre, University Science Site, Stockton Road entrance.

Wetland Bird Survey Organisers
TEES ESTUARY. Mike Leakey, c/o English Nature, British Energy, Tees Road, Hartlepool, TS25 2BZ. 01429 853325; e-mail: mike.leakey@english-nature.org.uk

Wildlife Trust
DURHAM WILDLIFE TRUST. (1971; 3500). Rainton Meadows, Chilton Moor, Houghton-le-Spring, Tyne & Wear, DH4 6PU. 0191 584 3112; (fax)0191 584 3934; e-mail: info@durhamwt.co.uk
www.wildlifetrust.org.uk/durham

ESSEX

Bird Atlas/Avifauna
Birds of Essex (provisional title) by Simon Woods (Essex Birdwatching Society, date to be announced).

The Breeding Birds of Essex by M K Dennis (Essex Birdwatching Society, 1996).

Bird Recorder
JOINT RECORDER. Roy Ledgerton (joint), 25 Bunyan Road, Braintree, Essex CM7 2PL. 01376 326103; e-mail: r.ledgerton@virgin.net

JOINT RECORDER. Paul Levey (joint), 5 Hedingham Place, Rectory Road, Rochford Essex SS4 1UP. 01702 549070; e-mail: essex.birds@btopenworld.com

SENIOR RECORDER. Howard Vaughan, 103 Darnley Road, Strood, Rochester, Kent ME2 2EY. 01634 325864; e-mail: howardebs@blueyonder.co.uk

JOINT RECORDER. Bob Flindall, 60 Lady Lane, Chelmsford, , Essex CM2 0TH. 01245 344206; e-mail: robert.flindall@btinternet.com

Bird Report
ESSEX BIRD REPORT (inc Bradwell Bird Obs records) (1950-), from Peter Dwyer, Sales Officer, 48 Churchill Avenue, Halstead, Essex, CO9 2BE. Tel/fax 01787 476524; e-mail: petedwyer@petedwyer.plus.com or pete@northessex.co.uk

BTO Regional Representatives & Regional Development Officer
NORTH-EAST RR & RDO. Peter Dwyer, 48 Churchill Avenue, Halstead, Essex, CO9 2BE. Tel/fax 01787 476524; e-mail: petedwyer@petedwyer.plus.com or pete@northessex.co.uk

NORTH-WEST RR. Roy Ledgerton, 25 Bunyan Road, Braintree, Essex, CM7 2PL. 01376 326103; e-mail: r.ledgerton@virgin.net

SOUTH RR. Position vacant.

Club

ESSEX BIRDWATCHING SOCIETY. (1949; 700). Roy Ledgerton, 25 Bunyan Road, Braintree, Essex, CM7 2PL. 01376 326103; e-mail: r.ledgerton@virgin.net
www.essexbirdwatchsoc.co.uk
Meetings: 1st Friday of the month (Oct-Mar), Friends' Meeting House, Rainsford Road, Chelmsford.

Ringing Groups
ABBERTON RG. C P Harris, Wylandotte, Seamer Road, Southminster, Essex, CM0 7BX.

BASILDON RG. B J Manton, 72 Leighcliff Road, Leigh-on-Sea, Essex, SS9 1DN. 01702 475183; e-mail: bjmanton@lineone.net

BRADWELL BIRD OBSERVATORY. C P Harris, Wyandotte, Seamer Road, Southminster, Essex, CM0 7BX.

RSPB Local Groups
CHELMSFORD AND CENTRAL ESSEX. (1976; 5500). Mike Logan Wood, Highwood, Ishams Chase, Wickham Bishops, Essex, CM8 3LG. 01621 892045; e-mail: mike.lw@tiscali.co.uk
www.chelmsfordrspb.org.uk
Meetings: 8pm, Thursdays, eight times a year. The Cramphorn Theatre, Chelmsford.

COLCHESTER. (1981; 250). Mrs V Owen, Tawnies, Hall Lane, Langenhoe, Colchester, CO5 7NA.
Meetings: 7.45pm, 2nd Thursday of the month (Sep-Apr), Shrub End Community Hall, Shrub End Road, Colchester.

SOUTHEND. (1983; 200). Graham Mee, 24 Sunbury Court, North Shoebury, Essex SS3 8TB. 01702 297554; e-mail: grahamm@southendrspb.co.uk
www.southendrspb.co.uk

ENGLAND

Meetings: 8pm, usually 1st Monday of the month (Sep-May), Sports Centre, Wellstead Gardens, Westcliff.

Wetland Bird Survey Organisers
STOUR ESTUARY. Rick Vonk, RSPB, Unit 3 Court Farm, 3 Stutton Road, Brantham, Suffolk CO11 1PW. (Day) 01473 328006.

LEE VALLEY. Ian Kendall, 18 North Barn, Broxbourne, Herts EN10 6RR; Tel: 01992 709964; e-mail: ikendall@leevalleypark.org.uk

ESSEX (Other Sites). Howard Vaughan, 103 Darnley Road, Strood, Rochester, Kent ME2 2EY. 01634 325864 (after 7pm); e-mail; howardebs@blueyonder.co.uk

Wildlife Trust
ESSEX WILDLIFE TRUST. (1959; 15500). Abbots Hall Farm, Great Wigborough, Colchester, CO5 7RZ. 01621 862960; fax 01621 862990; www.essexwt.org.uk e-mail: admin@essexwt.org.uk

GLOUCESTERSHIRE

Bird Atlas/Avifauna
Atlas of Breeding Birds of the North Cotswolds. (North Cotswold Ornithological Society, 1990)

Bird Recorder
GLOUCESTERSHIRE EXCLUDING S.GLOS (AVON). Richard Baatsen, 1 Prestwick Terrace, Bristol Road, Whitminster, Glos GL2 7PA. e-mail: baatsen@surfbirder.com

Bird Reports
CHELTENHAM BIRD CLUB BIRD REPORT (1998-), from Secretary.

GLOUCESTERSHIRE BIRD REPORT (1953-), from David Cramp, 2 Ellenor Drive, Alderton, Tewkesbury, GL20 8NZ. e-mail: david.cramp@care4free.net

NORTH COTSWOLD ORNITHOLOGICAL SOCIETY ANNUAL REPORT (1983-), from T Hutton (Chairman), 15 Green Close, Childswickham, Broadway, Worcs, WR12 7JJ. 01386 858511.

BTO Regional Representative
Mike Smart, 143 Cheltenham Road, Gloucester, GL2 0JH. Home/work 01452 421131; e-mail: smartmike@smartmike.fsnet.co.uk

Clubs
CHELTENHAM BIRD CLUB. (1976; 85). Mrs Frances Meredith, 14 Greatfield Drive, Charlton Kings, Cheltenham, GL53 9BU. 01242 516393; e-mail: chelt.birds@virgin.net www.beehive.thisisgloucestershire.co.uk/cheltbirdclub

Meetings: 7.15pm, Mondays (Oct-Mar), Bournside School, Warden Hill Road, Cheltenham.

DURSLEY BIRDWATCHING & PRESERVATION SOCIETY. (1953; 500). Roger Pedley, Brookside Cottage, Waterley Bottom, Dursley, Glos GL11 6EF.01453 543334; e-mail: rogerxpedley@hotmail.com http://beehive.thisisgloucestershire.co.uk/dbwps
Meetings: 7.45pm, 2nd and 4th Monday (Sept-Mar), Dursley Community Centre.

GLOUCESTERSHIRE NATURALISTS' SOCIETY. (1948; 600). Mike Smart, 143 Cheltenham Road, Gloucester, GL2 0JH. 01452 421131; e-mail: smartmike.fsnet.co.uk www.glosnats.org.uk

NORTH COTSWOLD ORNITHOLOGICAL SOCIETY. (1982; 70). T Hutton, 15 Green Close, Childswickham, Broadway, Worcs, WR12 7JJ. 01386 858511.
Meetings: Monthly field meetings, usually Sunday 9.30pm.

Ringing Groups
COTSWOLD WATER PARK RG. R Hearn, Wildfowl & Wetlands Trust, Slimbridge, Glos, GL2 7BT. 01453 891900 ext 185; e-mail: richard.hearn@wwt.org.uk

SEVERN ESTUARY GULL GROUP. M E Durham, 6 Glebe Close, Frampton-on-Severn, Glos, GL2 7EL. 01452 741312.

SEVERN VALE RG. R Hearn, Wildfowl & Wetlands Trust, Slimbridge, Glos, GL2 7BT. 01453 891900 ext 185; e-mail: richard.hearn@wwt.org.uk

WILDFOWL & WETLANDS TRUST. R Hearn, Wildfowl & Wetlands Trust, Slimbridge, Glos, GL2 7BT. 01453 891900 ext 185; e-mail: richard.hearn@wwt.org.uk

RSPB Local Group
GLOUCESTERSHIRE. (1972; 740). David Cramp, 2 Ellenor Drive, Alderton, Tewkesbury, GL20 8NZ. 01242 620281.
Meetings: 7.30pm, 3rd Tuesday of the month, Sir Thomas Rich's School, Gloucester.

Wildlife Hospital
GLOUCESTER WILDLIFE RESCUE CENTRE. Alan and Louise Brockbank, 1 Moorend Lodge, Moorend, Hartpury, Glos GL19 3DG. 01452 700038; e-mail: info@gloswildliferescue.fsnet.co.uk www.gloswildliferescue.fsnet.co.uk Intensive care, treatment and rehabilitation facilities. Vetinary support. No restrictions or conditions.

VALE WILDLIFE RESCUE - WILDLIFE HOSPITAL + REHABILITATION CENTRE. Ms Caroline Gould, Station Road, Beckford, Tewkesbury, Glos, GL20 7AN. 01386 882288; (Fax)01386 882299; e-mail: info@vwr.org.uk www.vwr.org.uk
All wild birds. Intensive care. Registered charity. Veterinary support.

Wetland Bird Survey Organisers
SEVERN ESTUARY – GLOUCESTERSHIRE. Colette Hall, Slimbridge, Gloucester, GL2 7BT. 01453 890333.

GLOUCESTERSHIRE (Inland). Les Jones, Chestnut House, Water Lane, Somerford Keynes, Cirencester, Glos GL7 6DS; 01285 861545
e-mail: leslie@somerfordk.freeserve.co.uk

Wildlife Trust
GLOUCESTERSHIRE WILDLIFE TRUST. (1961; 6200). Dulverton Building, Robinswood Hill Country Park, Reservoir Road, Gloucester, GL4 6SX. 01452 383333; fax 01452 383334;
e-mail: info@gloucestershirewildlifetrust.co.uk
www.gloucestershirewildlifetrust.co.uk

HAMPSHIRE

Bird Atlas/Avifauna
Birds of Hampshire by J M Clark and J A Eyre (Hampshire Ornithological Society, 1993).

Bird Recorder
John Clark, 4 Cygnet Court, Old Cove Road, Fleet, Hants, GU51 2RL. Tel/fax 01252 623397; e-mail: johnclark@cygnetcourt.demon.co.uk

Bird Reports
HAMPSHIRE BIRD REPORT (1955-), from Mrs Margaret Boswell, 5 Clarence Road, Lyndhurst, Hants, SO43 7AL. 023 8028 2105;
e-mail: mag.bos@btinternet.com
2002 edition £9 including p&p.

BTO Regional Representative & Regional Development Officer
RRGlynne C Evans, Waverley, Station Road, Chilbolton, Stockbridge, Hants, SO20 6AL. H:01264 860697; e-mail: hantsbto@hotmail.com

Clubs
HAMPSHIRE ORNITHOLOGICAL SOCIETY. (1979; 955). Barrie Roberts, 149 Rownhams Lane, North Baddesley, Southampton, SO52 9LU. 023 8073 7023;
e-mail: robertsbarrie@hotmail.com

SOUTHAMPTON & DISTRICT BIRD GROUP. (1994; 68). Vic Short, 20 Westbroke Gardens, Romsey, SO51 7RQ. 01794 511843; e-mail: vicshort@btopenworld.com
Meetings: Programme available.

Ringing Groups
FARLINGTON RG. Pete Potts, Solent Court Cottage, Chilling Lane, Warsash, Southampton, Hampshire SO31 9HF;
e-mail: ppotts@compuserve.com

ITCHEN RG. W F Simcox, 10 Holdaway Close, Kingsworthy, Winchester, SO23 7QH.
e-mail: wsimcox@sparsholt.ac.uk

LOWER TEST RG. J Pain, Owlery Holt, Nations Hill, Kingsworthy, Winchester, SO23 7QY. 023 8066 7919;
e-mail: jessp@hwt.org.uk

RSPB Local Groups
BASINGSTOKE. (1979; 90). Peter Hutchins, 35 Woodlands, Overton, Whitchurch, RG25 3HN. 01256 770831; e-mail: fieldfare@jaybry.gotadsl.co.uk
Meetings: 7.30pm 3rd Wednesday of the month (Sept-May), The Barn, Church Cottage, St Michael's Church, Church Square, Basingstoke.

NORTH EAST HAMPSHIRE. (1976; 250). The Group leader, 4 Buttermer Close, Farnham, Surrey, GU10 4PN. 01252 724093;
e-mail: Mailto:northeasthantsRSPB.org.uk
www.northeasthantsrspb.org.uk

PORTSMOUTH. (1974; 201). Gordon Humby, 19 Charlesworth Gardens, Waterlooville, Hants, PO7 6AU. 02392 353949.
Meetings: 4th Saturday of the month, programme for members, St Colmans Church Hall, Cosham.

WINCHESTER & DISTRICT. (1974; 175). Maurice Walker, Jesmond, 1 Compton Way, Olivers Battery, Winchester, SO22 4EY. 01962 854033.
Meetings: 7.30pm, 1st Tuesday of the month (not Jul or Aug), Cromwell Suite, The Stanmore, Winchester.

Wetland Bird Survey Organisers
HAMPSHIRE (Estuaries/Coastal). Dave Unsworth, 5 Nelson Road, Bishopstoke, Eastleigh, Hampshire, SO59 7BR. 01703 329191;
e-mail: David_Unsworth@mcga.gov.uk

HAMPSHIRE (Inland - excluding Avon Valley). Keith Wills, 51 Peabody Road, Farnborough, GU14 6EB(H)01252 548408;
e-mail: keithb.wills@ukgateway.net

AVON VALLEY. John Clark, 4 Cygnet Court, Old Cove Road, Fleet, Hants . 01252 623397;
e-mail johnclark@cygnetcourt.demon.co.uk

Wildlife Hospital
NEW FOREST OWL SANCTUARY. Bruce Berry, New Forest Owl Sanctuary, Crow Lane, Crow, Ringwood, Hants, BH24 1EA. 01425 476487;
e-mail: nfosowls@aol.com www.owlsanctuary.co.uk
A selection of Owls, Hawks and Falcons from around the world with flying demonstrations at set times throughout the day. An opportunity to observe birds of prey at close range, an enjoyable day for the whole family. Open daily from Feb to Nov.

Wildlife Trust
HAMPSHIRE & ISLE OF WIGHT WILDLIFE TRUST. (1960; 11,295). Beechcroft House, Vicarage Lane,

ENGLAND

Curdridge, Hampshire SO32 2DP. 01489 774 400; fax 023 8068 8900; e-mail: feedback@hwt.org.uk www.hwt.org.uk

WIGHT WILDLIFE. 2 High Street, Newport, Isle of Wight PO30 1SS01983 533 180; e-mail: feedback@hwt.org.uk

HEREFORDSHIRE

Bird Recorder
Steve Coney, 5 Springfield Road, Withington, Hereford, HR1 3RU. 01432 850068; e-mail: coney@bluecarrots.com

Bird Report
THE YELLOWHAMMER - Herefordshire Ornithological Club annual report, (1951-), from Mr I Evans, 12 Brockington Drive, Tupsley, Hereford, HR1 1TA. 01432 265509; e-mail: iforelaine@care4free.net

BTO Regional Representative
Steve Coney, 5 Springfield Road, Withington, Hereford, HR1 3RU. 01432 850068; e-mail: coney@bluecarrots.com

Club
HEREFORDSHIRE ORNITHOLOGICAL CLUB. (1950; 385). TM Weale, Foxholes, Bringsty Common, Worcester, WR6 5UN. 01886 821368; e-mail: weale@tinyworld.co.uk www.herefordshirebirding.net
Meetings: 7.30pm, 2nd Thursday of the month (Autumn/winter), Holmer Parish Centre, Holmer, Hereford.

Ringing Group
LLANCILLO RG. Dr G R Geen, 6 The Copse, Bannister Green, Felsted, Dunmow, Essex, CM6 3NP. 01371 820189; e-mail: thegeens@aol.com

Wildlife Hospital
ATHENE BIRD SANCTUARY. B N Bayliss, 61 Chartwell Road, Hereford, HR1 2TU. 01432 273259. Birds of prey, ducks and waders, seabirds, pigeons and doves. Heated cages, small pond. Veterinary support.

Wildlife Trust
HEREFORDSHIRE NATURE TRUST. (1962; 1450). Lower House Farm, Ledbury Road, Tupsley, Hereford, HR1 1UT. 01432 356872; fax 01432 275489; e-mail: herefordwt@cix.co.uk www.wildlifetrust.org.uk/hereford

HERTFORDSHIRE

Bird Atlas/Avifauna
Birds at Tring Reservoirs by R Young et al (Hertfordshire Natural History Society, 1996).

Mammals, Amphibians and Reptiles of Hertfordshire by Hertfordshire NHS in association with Training Publications Ltd, 3 Finway Court, Whippendell Road, Watford WD18 7EN, (2001).

The Breeding Birds of Hertfordshire by K W Smith et al (Herts NHS, 1993).

Bird Recorder
Mike Ilett, 14 Cowper Crescent, Bengeo, Hertford, Herts, SG14 3DY. e-mail: michael.ilett@uk.tesco.com

Bird Report
HERTFORDSHIRE BIRD REPORT (1908-1998), from Hon Secretary, Herts Bird Club, 46 Manor Way, Boreham Wood, Herts, WD6 1QY.

BTO Regional Representative & Regional Development Officer
RR & RDO. Chris Dee, 26 Broadleaf Avenue, Thorley Park, Bishop's Stortford, Herts, CM23 4JY. H:01279 755637; e-mail: chris_w_dee@hotmail.com

Clubs
FRIENDS OF TRING RESERVOIRS. (1993; 350). Rose Barr, (Membership Secretary), P.O. Box 1083, Tring, Herts HP23 5WU. www.fotr.org.uk

HERTFORDSHIRE BIRD CLUB. (1971; 290) Part of Hertfordshire NHS. Jim Terry, 46 Manor Way, Borehamwood, Herts WD6 1QY. 020 8905 1461; e-mail: jim@jayjoy.fsnet.co.uk www.lanius.co.uk

HERTFORDSHIRE NATURAL HISTORY SOCIETY. Christine Shepperson, 63 Station Road, Smallford, HertsAL4 0HB; e-mail: shepperson@waitrose.com www.hnhs.org

Ringing Groups
AYLESBURY VALE RG (main activity at Marsworth). S M Downhill, 12 Millfield, Berkhamsted, Herts, HP4 2PB. 01442 865821; e-mail: smdjbd@waitrose.com

MAPLE CROSS RG. P Delaloye, e-mail: pdelaloye@tiscali.co.uk

RYE MEADS RG. Chris Dee, 26 Broadleaf Avenue, Thorley Park, Bishop's Stortford, Herts, CM23 4JY. H:01279 755637; e-mail: ringingsecretary@rmrg.org.uk

TRING RG. Mick A'Court, 6 Chalkshire Cottages, Chalkshire road, Butlers Cross, Bucks ,HP17 0TW. H:01296 623610; W:01494 462246; e-mail: mick@focusrite.com a.arundinaceous@virgin.net

RSPB Local Groups
CHORLEYWOOD & DISTRICT. (1977; 142). Sam Thomas, 36 Field Way, Rickmansworth, Herts WD3 2EJ. 01923 449917.

262

ENGLAND

Meetings: 8pm, last Thursday of the month (Sept-May).

HARPENDEN. (1974; 1000). Geoff Horn, 01582 765443; e-mail: geoffrhorn@yahoo.co.uk

HEMEL HEMPSTEAD. (1972; 150). Paul Green, 207 Northridge Way, Hemel Hempstead, Herts, HP1 2AU. 01442 266637; paul@310nrwhh.freeserve.co.uk www.hemelrspb.org.uk
Meetings: 8pm, 1st Monday of the month (Sep-Jun),The Cavendish School.

HITCHIN & LETCHWORTH. (1973; 106). Dr Martin Johnson, 1 Cartwright Road, Royston, , Herts, SG8 9ET. 01763 249459;
e-mail: martin.2.johnson@gsk.com
Meetings: 7.30pm, 1st Friday of the month, The Settlement, Nevells Road, Letchworth.

POTTERS BAR & BARNET. (1977; 1800). Stan Bailey, 23 Bowmans Close, Potters Bar, Herts, EN6 5NN. 01707 646073.
Meetings: 2pm, 2nd Tuesday of the month, St Johns URC Hall, Mowbray Road, Barnet, 8.00pm, changeable Mondays, Wyllyotts Centre, Potters Bar.

ST ALBANS. (1979; 1550 in catchment area). John Maxfield, 46 Gladeside, Jersey Farm, St Albans, Herts, AL4 9JA. 01727 832688;
e-mail: peter@antram.demon.co.uk
www.antram.demon.co.uk/
Meetings: 8pm, 2nd Tuesday of the month (Sep-May), St Saviours Church Hall, Sandpit Lane, St Albans.

SOUTH EAST HERTS. (1971; 150). Terry Smith, 31 Marle Gardens, Waltham Abbey, Essex EN9 2DZ. 01992 715634; e-mail: se_herts_rspb@yahoo.co.uk http://uk.geocities.com/seherts_rspb
Meetings: 8.00pm, last Tuesday of the month, URC Church Hall, Mill Lane, Broxbourne.

STEVENAGE. (1982; 1300 in the catchment area). Mrs Ann Collis, 16 Stevenage Road, Walkern, Herts, 01483 861547.
Meetings: 7.30pm, 3rd Tuesday of the month, Friends Meeting House, Cuttys Lane, Stevenage.

WATFORD. (1974; 590). Philip and Marilyn McGovern, 65 Harford Drive, Watford WD17 3DQ. 01923 243761. www.members.lycos.co.uk/watford_RSPB
Meetings: 7.30pm, 2nd Wednesday of the month (Sep-Jun), St Thomas' Church Hall, Langley Road, Watford.

Wildlife Hospital
SWAN CARE. Secretary, Swan Care, 14 Moorland Road, Boxmoor, Hemel Hempstead, Herts, HP1 1NH. 01442 251961. Swans. Sanctuary and treatment centre. Veterinary support.

Wildlife Trust
HERTS & MIDDLESEX WILDLIFE TRUST. (1964; 8500). Grebe House, St Michael's Street, St Albans, Herts, AL3 4SN. 01727 858901; fax 01727 854542; e-mail: info@hmwt.org www.wildlifetrust.org.uk/herts/

ISLE OF WIGHT

Bird Recorder
G Sparshott, Leopards Farm, Main Road, Havenstreet, Isle of Wight, PO33 4DR. 01983 882549;
e-mail: grahamspa@aol.com

Bird Reports
ISLE OF WIGHT BIRD REPORT (1986-) (Pre-1986 not available), from Mr DJ Hunnybun, 40 Churchill Road, Cowes, Isle of Wight, PO31 8HH. 01983 292880; email: davehunnybun@hotmail.com

BTO Regional Representative
James C Gloyn, 3 School Close, Newchurch, Isle of Wight, PO36 0NL. 01983 865567;
e-mail: gloynjc@yahoo.com

Clubs
ISLE OF WIGHT NATURAL HISTORY & ARCHAEOLOGICAL SOCIETY. (1919; 500). The Secretary, Salisbury Gardens, Dudley Road, Ventnor, Isle of Wight, PO38 1EJ. 01983 855385.

ISLE OF WIGHT ORNITHOLOGICAL GROUP. (1986; 155). Mr DJ Hunnybun, 40 Churchill Road, Cowes, Isle of Wight, PO31 8HH. 01983 292880;
email: davehunnybun@hotmail.com

Wildlife Trust
See Hampshire.

KENT

Bird Atlas/Avifauna
Birding in Kent by D W Taylor et al (Kent Ornithological Society, 1981). Pica Press

Kent Ornithological Society Winter Bird Survey by N Tardivel (KOS, 1984).

Bird Recorder
Don Taylor, 1 Rose Cottages, Old Loose Hill, Loose, Maidstone, Kent, ME15 0BN. 01622 745641; e-mail: don@collared.free-online.co.uk

Bird Reports
DUNGENESS BIRD OBSERVATORY REPORT (1989-), from Warden, see Reserves.

KENT BIRD REPORT (1952-), from Dave Sutton, 61 Alpha Road, Birchington, Kent, CT7 9ED. 01843 842541; e-mail: dave@suttond8.freeserve.co.uk

SANDWICH BAY BIRD OBSERVATORY REPORT, from Warden, see Reserves,

BTO Regional Representative & Regional Development Officer
RR. Martin Coath, 77 Oakhill Road, Sevenoaks, Kent, TN13 1NU. 01732 460710;
e-mail: marron2@onetel.com

Club

KENT ORNITHOLOGICAL SOCIETY. (1952; 720). Mrs Ann Abrams, 4 Laxton Way, Faversham, Kent ME13 8LJ. e-mail: marron2@onetel.com www.kentos.org.uk
Meetings: Indoor: October-April at various venues; the AGM in April is at St Paul's Church Hall, Boxley Road, Maidstone. See website for details: www.kentos.org.uk

Ringing Groups
DARTFORD RG. R Taylor, 21 Dallin Road, Plumstead, London SE18 3NY.

DUNGENESS BIRD OBSERVATORY. David Walker, Dungeness Bird Observatory, Dungeness, Romney Marsh, Kent, TN29 9NA. 01797 321309; e-mail: dungeness.obs@tinyonline.co.uk www.dungenessbirdobs.org.uk

RECULVER RG. Chris Hindle, 42 Glenbervie Drive, Herne Bay, Kent, CT6 6QL. 01227 373070; e-mail: christopherhindle@hotmail.com

SANDWICH BAY BIRD OBSERVATORY. K Thornton, Sandwich Bay Bird Observatory, Guilford Road, Sandwich, Kent, CT13 9PF. 01304 617341; e-mail: sbbot@talk21.com

SWALE WADER GROUP. Rod Smith, 67 York Avenue, Chatham, Kent, ME5 9ES. 01634 865863; e-mail: rodandmarg@tiscali.co.uk

RSPB Local Groups
CANTERBURY. (1973; 216). Jean Bomber, St Heliers, 30a Castle Road, Tankerton, Whitstable, Kent, CT5 2DY. 01227 277725.
http://cantrspb.members.easyspace.com/
Meetings: 8.00pm, 2nd Wednesday of the month (Sept-Apr), Chaucer Technology School, Spring Lane, Canterbury.

GRAVESEND & DISTRICT. (1977; 260). Malcolm Jennings, 206 Lower Higham Road, Gravesend, Kent DA12 2NN. 01474 322171.
Meetings: 7.45pm, 2nd Wednesday of the month (Sep-May), St Botolph's Hall, North Fleet, Gravesend.

MAIDSTONE. (1973; 250). Dick Marchese, 11 Bathurst Road, Staplehurst, Tonbridge, Kent, TN12 0LG. 01580 892458

www.vidler23.freeserve.co.uk/
Meetings: 7.30pm, 3rd Thursday of the month, Grove Green Community Hall, Penhurst Close, Grove Green, opposite Tesco's.

MEDWAY. (1974; 230). Sue Carter, 31 Ufton Lane, Sittingbourne, ME10 1JB. 01795 427854
www.medway-rspb.pwp.blueyonder.co.uk
Meetings: 7.45pm 3rd Tuesday of the month (except Aug), Strood Library, Bryant Road, Strood.

SEVENOAKS. (1974; 300). Bernard Morris, New House, Kilkhampton, Bude, Cornwall, EX23 9RZ. 01288 321727; or 07967 564699;(Fax)01288 321838; e-mail: bernard.morris5@btinternet.com
Meetings: 7.45pm 2nd Thursday of the month, Playhouse, Stag Theatre.

SOUTH EAST KENT. (1981; 165). Pauline McKenzie-Lloyd, Hillside, Old Park Avenue, Dover, Kent CT16 2DY. 01304 826529; email: pauline.mcklloyd@tiscali.co.uk
Meetings: 7.30pm, 3rd Wednesday of the month (Sep-May), United Reform Church, Folkestone.

THANET. (1976; 170). Hazel Johnson, Flat 5, Abbeygate, Pegwell Road, Ramsgate CT11 0JR. 01843 596336; e-mail: hdjbird@aol.com
Meetings: 7.30pm last Tuesday of the month (Jan-Nov), Holy Trinity Church Hall.

TONBRIDGE. (1975; 150 reg attendees/1700 in catchment). Ms Gabrielle Sutcliffe, 1 Postern Heath Cottages, Postern Lane, Tonbridge, Kent, TN11 0QU. 01732 365583.
Meetings: 7.45pm 3rd Wednesday of the month (Sept-Apr), St Phillips Church, Salisbury Road.

Wetland Bird Survey Organisers
EAST KENT. Ken Lodge, 14 Gallwey Avenue, Birchington, Kent CT7 9PA. 01843 843105; e-mail: kenlodge@minnisbay15.freeserve.co.uk

MEDWAY ESTUARY & NORTH KENT MARSHES. Sally Jennings, RSPB, Bromhey Farm, Cooling, Rochester, Kent ME3 8DS. 01634 222480 .

SWALE ESTUARY. Sally Jennings, RSPB, Bromhay Farm, Cooling, Rochester, Kent ME3 8DS. 01634 222480.

Wildlife Hospital
RAPTOR CENTRE. Eddie Hare, Ivy Cottage, Groombridge Place, Groombridge, Tunbridge Wells, Kent, TN3 9QG. 01892 861175; fax 01892 863761
www.raptorcentre.co.uk
Birds of prey. Veterinary support.

Wildlife Trust
KENT WILDLIFE TRUST. (1958; 10500). Tyland Barn, Sandling, Maidstone, Kent, ME14 3BD. 01622 662012; fax 01622 671390; e-mail: info@kentwildlife.org.uk
www.kentwildlife.org.uk

ENGLAND

LANCASHIRE

Bird Atlas/Avifauna
An Atlas of Breeding Birds of Lancaster and District by Ken Harrison (Lancaster & District Birdwatching Society, 1995).

Breeding Birds of Lancashire and North Merseyside (2001), sponsored by North West Water. Contact: Bob Pyefinch, 12 Bannistre Court, Tarleton, Preston PR4 6HA.

Bird Recorder
(See also Manchester).

Inc North Merseyside. Steve White, 102 Minster Court, Crown Street, Liverpool, L7 3QD. 0151 707 2744; e-mail: stephen.white2@tesco.net

Bird Reports
BIRDS OF LANCASTER & DISTRICT (1959-), from Secretary, Lancaster & District BWS, 01524 734462.

EAST LANCASHIRE ORNITHOLOGISTS' CLUB BIRD REPORT (1982-), from Secretary, 01282 617401; e-mail: doug.windle@care4free.net

BLACKBURN & DISTRICT BIRD CLUB ANNUAL REPORT (1992-), from Doreen Bonner, 6 Winston Road, Blackburn, BB1 8BJ. Tel/fax;01254 261480; www.blackburnbirds.freeuk.com

FYLDE BIRD REPORT (1983-), from Secretary, Fylde Bird Club,

LANCASHIRE BIRD REPORT (1914-), from Secretary, Lancs & Cheshire Fauna Soc,

ROSSENDALE ORNITHOLOGISTS' CLUB BIRD REPORT (1977-) from Secretary, Rossendale Ornithologists Club.

BTO Regional Representatives & Regional Development Officer
EAST RR. Tony Cooper, 28 Peel Park Avenue, Clitheroe, Lancs, BB7 1ET. 01200 424577; e-mail: tonycooper@beeb.net

NORTH & WEST RR. Keith Woods, 2 Oak Drive, Halton, Lancaster, LA2 6QL. 01524 811478; e-mail: woods.keith@btopenworld.com

SOUTH RR. Philip Shearwood, Netherside, Green Lane, Whitestake, Preston, PR4 4AH. 01772 745488

Clubs
BLACKBURN & DISTRICT BIRD CLUB. (1991; 134). Jim Bonner, 6 Winston Road, Blackburn, BB1 8BJ. Tel/fax;01254 261480. www.blackburnbirds.freeuk.com

Meetings: Normally 7.30pm, 1st Monday of the month, (Sept-Apr), Church Hall, Preston New Road.

CHORLEY & DISTRICT NATURAL HISTORY SOCIETY. (1979; 170). Phil Kirk, Millend, Dawbers Lane, Euxton, Chorley, Lancs, PR7 6EB. 01257 266783; e-mail: philipdkirk@clara.co uk www.philkirk.clara.net/cdnhs/
Meetings: 7.30pm, 3rd Thursday of the month (Sept-Apr), St Mary's Parish Centre, Chorley

EAST LANCASHIRE ORNITHOLOGISTS' CLUB. (1955; 45). Doug Windle, 39 Stone Edge Road, Barrowford, Nelson, Lancs, BB9 6BB. 01282 617401; e-mail: doug.windle@care4free.net www.eastlancashireornithologists.org.uk
Meetings: 7.30pm, 1st Monday of the month, St Anne's Church Hall, Feuce, Nr Burnley.

FYLDE BIRD CLUB. (1982; 60). Paul Ellis, 18 Staining Rise, Blackpool, FY3 0BU. 01253 891281; e-mail: paul.ellis24@btopenworld.com or kinta.beaver@man.ac.uk www.fyldebirdclub.org
Meetings: 7.45pm, 4th Tuesday of the month, River Wyre Hotel, Breck Road, Poulton le Fylde.

FYLDE NATURALISTS' SOCIETY. (1946; 140). Gerry Stephen, 10 Birch Way, Poulton-le-Fylde, Blackpool, FY6 7SF. 01253 895195.
Meetings: 7.30pm, fortnightly Wednesdays (Sep-Mar), Fylde Coast Alive, Church Hall, Raikes Parade, Blackpool unless otherwise stated in the Programme.

LANCASHIRE & CHESHIRE FAUNA SOCIETY. (1914; 140). Dave Bickerton, 64 Petre Crescent, Rishton, Lancs, BB1 4RB. 01254 886257; www.lacfs.org.uk e-mail: bickertond@aol.com

LANCASHIRE BIRD CLUB. (1996). Dave Bickerton, 64 Petre Crescent, Rishton, Lancs, BB1 4RB. 01254 886257; e-mail: bickertond@aol.com www.lacfs.org.uk

LANCASTER & DISTRICT BIRD WATCHING SOCIETY. (1959; 200). Andrew Cadman, 57 Greenways, Over Kellet, Carnforth, Lancs, LA6 1DE. 01524 734462; e-mail: ldbws@yahoo.co.uk http://libweb.lancs.ac.uk/ldbws.htm
Meetings: 7.30pm, last Monday of the month (Sep-Apr, not Dec), Unitarian Church, Scotforth, Lancaster.

ROSSENDALE ORNITHOLOGISTS' CLUB. (1976; 35). Ian Brady, 25 Church St, Newchurch, Rossendale, Lancs, BB4 9EX. 01706 222120.
Meetings: 7.30pm, 3rd Monday of the month, Weavers Cottage, Bacup Road, Rawtenstall.

Ringing Groups
FYLDE RG. G Barnes, 17 Lomond Avenue, Marton, Blackpool, FY3 9QL.

MORECAMBE BAY WADER RG. J Sheldon, 415 West

ENGLAND

Shore Park, Barrow-in-Furness, Cumbria, LA14 3XZ. 01229 473102.

NORTH LANCS RG. John Wilson BEM, 40 Church Hill Avenue, Warton, Carnforth, Lancs, LA5 9NU.

SOUTH WEST LANCASHIRE RG. I H Wolfenden, 35 Hartdale Road, Thornton, Merseyside. 01519 311232.

RSPB Local Groups
BLACKPOOL AND FYLDE. (1983; 170). Alan Stamford, 6 Kensington Road, Cleveleys, FY5 1ER. 01253 859662.
Meetings: 7.30pm, 2nd Friday of the month (Sept-June), Frank Townend Centre, Beach Road, Cleveleys.

LANCASTER. (1972; 176). John Wilson BEM, 40 Church Hill Avenue, Warton, Carnforth, Lancs, LA5 9NU.

Wetland Bird Survey Organisers
RIBBLE ESTUARY. Mr Mike Gee, Ribble Estuary National Nature Reserve, Reserve Office, Old Hollow, Marsh Road, Banks PR9 8DU. 01704 225624; e-mail: english-nature@ribble-nnr.freeserve.co.uk

NORTH LANCASHIRE (Inland). Mr Pete Marsh, Leck View Cottage, Ashle's farm, High Tatham, Lancaster, LA2 8PH. 01524 264944; e-mail: pbmarsh@btopenworld.com

Wildlife Trust
THE WILDLIFE TRUST FOR LANCASHIRE, MANCHESTER AND NORTH MERSEYSIDE. (1962; 3500). The Barn, Berkeley Drive, Bamber Bridge, Preston, PR5 6BY. 01772 324129; fax: 01772 628849; e-mail: lancswt@cix.co.uk
www.wildlifetrust.org.uk/lancashire

LEICESTERSHIRE & RUTLAND

Bird Recorder
Rob Fray, 5 New Park Road, Aylestone, Leicester, LE2 8AW. 0116 223 8491;
e-mail: robfray@fray-r.freeserve.co.uk

Bird Reports
LEICESTERSHIRE & RUTLAND BIRD REPORT (1941-), from Mrs S Graham, 5 Brading Road, Leicester, LE3 9BG. 0116 262 5505; e-mail: jsgraham83@aol.com

RUTLAND NAT HIST SOC ANNUAL REPORT (1965-), from Secretary, 01572 747302.

BTO Regional Representative
LEICESTER & RUTLAND RR. Tim Grove, 35 Clumber Street, Melton Mowbray, Leicestershire LE13 0ND. 01664 850766; e-mail: k.grove1@ntlworld.com

Clubs
BIRSTALL BIRDWATCHING CLUB. (1976; 50). Mr KJ

Goodrich, 6 Riversdale Close, Birstall, Leicester, LE4 4EH. 0116 267 4813.
Meetings: 7.30pm, 2nd Tuesday of the month (Oct-Apr), Longslade Community College, Martin Luther Building.

LEICESTERSHIRE & RUTLAND ORNITHOLOGICAL SOCIETY. (1941; 580). Mrs Marion Vincent, 48 Templar Way, Rothley, Leicester, LE7 7RB. 0116 230 3405
www.lros.org.uk
Meetings: 7.30pm, 1st Friday of the month, Leicester Adult Education College, Wellington St, Leicester. Additional meeting at Rutland Water Birdwatching Cntr.

MARKET HARBOROUGH & DISTRICT NATURAL HISTORY SOCIETY. (1971; 40). Mrs Marion Mills, 36 Nelson Street, Market Harborough, LeicsLE16 9AY. 01858 462346.
Meetings: 7.30pm, 2nd Monday in the month, Welland Park College

RUTLAND NATURAL HISTORY SOCIETY. (1964; 256). Mrs L Worrall, 6 Redland Close, Barrowden, Oakham, Rutland, LE15 8ES. 01572 747302. www.rnhs.org.uk
Meetings: 7.30pm, 1st Tuesday of the month (Oct-Apr), Oakham C of E School, Burley Road, Oakham.

Ringing Groups
STANFORD RG. M J Townsend, 87 Dunton Road, Broughton Astley, Leics, LE9 6NA.

RSPB Local Groups
LEICESTER. (1969; 1600 in catchment area). Chris Woolass, 136 Braunstone Lane, Leicester, LE3 2RW. 0116 299 0078;
e-mail: chris@jclwoolass.freeserve.co.uk

LOUGHBOROUGH. (1970; 300). Robert Orton, 12 Avon Road, Barrow-on-Soar, Leics, LE12 8LE. 15094 13936.
Meetings: Monthly Friday nights, Loughborough University.

Wetland Bird Survey Organisers
LEICESTERSHIRE & RUTLAND (excl Rutland Water). Mr Andrew Harrop, 30 Dean Street, Oakham LE15 6AF. (H)01572 757134; e-mail: andrew.harrop@virgin.net

RUTLAND WATER. Tim Appleton, Fishponds Cottage, Stamford Road, Oakham, LE15 8AB. (Day)01572 770651 e-mail: awbc@rutland water.org.uk

Wildlife Trust
LEICESTERSHIRE & RUTLAND WILDLIFE TRUST. (1956; 3000). Brocks Hill Environment Centre, Washbrook Lane, Oadby, Leicestershire LE2 5JJ. 0116 272 0444; fax 0116 272 0404;
e-mail: info@lrwt.org.uk
www.lrwt.org.uk

LINCOLNSHIRE

Bird Recorders
NORTH. Covered temporarily by South Lincs Recorder.

SOUTH. Steve Keightley, Redclyffe, Swineshead Road, Frampton Fen, Boston PE20 1SG. 01205 290333; e-mail: s.keightley@btinternet.com

Bird Reports
LINCOLNSHIRE BIRD REPORT inc Gibraltar Point Bird Obs (1979-), from R K Watson, 8 High Street, Skegness, Lincs, PE25 3NW. 01754 763481.

SCUNTHORPE & NORTH WEST LINCOLNSHIRE BIRD REPORT (1973-), from Secretary, Scunthorpe Museum Society, Ornithological Section,

BTO Regional Representatives & Regional Development Officer
EAST RR. Position vacant.

NORTH RR. John Turner, 01652 650119; e-mail: johnturner17@onetel.com

SOUTH RR. Richard & Kay Heath, 56 Pennytoft Lane, Pinchbeck, Spalding, Lincs, PE11 3PQ. 01775 767055; e-mail: heathsrk@ukonline.co.uk

WEST RR. Peter Overton, Hilltop Farm, Welbourn, Lincoln, LN5 0QH. Work 01400 273323; e-mail: nyika@biosearch.org.uk

RDO. Nicholas Watts, Vine House Farm, Deeping St Nicholas, Spalding, Lincs, PE11 3DG. 01775 630208.

Club
LINCOLNSHIRE BIRD CLUB. (1979; 220). Janet Eastmead, 3 Oxeney Drive, Langworth, Lincoln LN3 5DD. 01522 754522; e-mail: jee@freeuk.com
Meetings: Local groups hold winter evening meetings (contact Secretary for details).

SCUNTHORPE MUSEUM SOCIETY (Ornithological Section). (1973; 50). Keith Parker, 7 Ryedale Avenue, Winterton, Scunthorpe, Lincs DN15 9BJ.
Meetings: 7.15pm, 3rd Monday of the month (Sep-Apr), Scunthorpe Museum, Oswald Road.

Ringing Groups
GIBRALTAR POINT BIRD OBSERVATORY. Mark Grantham, 12 Sybill Wheeler Close, Thetford, Norfolk, IP24 1TG. 01842 750050; (M)07818 497470.

MID LINCOLNSHIRE RG. J Mawer, 2 The Chestnuts, Cwmby Road, Searby, Lincolnshire DN38 6EH. 01652 628583.

WASH WADER RG. P L Ireland, 27 Hainfield Drive, Solihull, W Midlands, B91 2PL. 0121 704 1168; e-mail: enquiries@wwrg.org.uk

RSPB Local Groups
GRIMSBY AND CLEETHORPES. (1986; 2200). Andy Downes, Meadowcroft, 17 Bulwick Avenue, Scartho, Grimsby, DN33 3BH. 01472 319257; e-mail: andrewandlynndownes@ntlworld.com
Meetings: 7.30pm, 1st Monday of the month (Sept-May), Cromwell Banqueting Suite, Cromwell Road, Cleethorpes.

LINCOLN. (1974; 250). Peter Skelson, 26 Parksgate Avenue, Lincoln, LN6 7HP. 01522 695747; e-mail: peter.skelson@lincolnrspb.org.uk www.lincolnrspb.org.uk
Meetings: 7.30pm, 2nd Thursday of the month (not Jun, Jul, Aug, Dec), The Lawn, Union Road, Lincoln.

SOUTH LINCOLNSHIRE. (1987; 350). Barry Hancock, The Limes, Meer Booth Road, Antons Gowt, Boston, Lincs, PE22 7BG. 01205 280057; e-mail: info@southlincsrspb.org.uk www.southlincsrspb.org.uk

Wetland Bird Survey Organisers
HUMBER ESTUARY - MID SOUTH. Ian Shepherd, 38 Lindsey Road, Cleethorpes, Lincolnshire DN35 8TN. (H)01472 697142.

WASH – LINCOLNSHIRE. Lewis James, RSPB Snettisham, 13 Beach Road, Snettisham, King's Lynn, Norfolk PE31 7RA. 01485 542689.

Wildlife Trust
LINCOLNSHIRE WILDLIFE TRUST. (1948; 10800). Banovallum House, Manor House Street, Horncastle, Lincs, LN9 5HF. 01507 526667; fax 01507 525732; e-mail: info@lincstrust.co.uk   www.lincstrust.org.uk

LONDON, GREATER

Bird Atlas/Avifauna
The Breeding Birds Illustrated magazine of the London Area, 2002. ISBN 0901009 121 ed Jan Hewlett(London Natural History Society).

Bird Recorder see also Surrey
Andrew Self, 16 Harp Island Close, Neasden, London, NW10 0DF. e-mail: andrewself@lineone.net www.lnhs.org.uk

Bird Report
CROYDON BIRD SURVEY (1995), from Secretary, Croydon RSPB Group, 020 8777 9370.

LONDON BIRD REPORT (20-mile radius of St Paul's Cath) (1936-), from Catherine Schmitt, 4 Falkland Avenue, London, N3 1QR.

BTO Regional Representative & Regional Development Officer
LONDON & MIDDLESEX RR. Derek Coleman, 23c Park Hill, Carshalton, Surrey, SM5 3SA. 020 8669 7421.

Clubs
LONDON NATURAL HISTORY SOCIETY (Ornithology Section). (1858; 1000). Ms N Duckworth, 9 Abbey Court, Cerne Abbas, Dorchester, Dorset, DT2 7JH. 01300 341 195. www.lnhs.org.uk

LNHS

MARYLEBONE BIRDWATCHING SOCIETY. (1981; 86). Judy Powell, 7 Rochester Terrace, London, NW1 9JN. 020 7485 0863; e-mail: birdsmbs@yahoo.com www.geocities.com/birdsmbs **Meeting:**2nd Friday of month, 7.15pm Gospel Oak Methodist Chapel, Lisburne Road, London NW3

Ringing Groups
LONDON GULL STUDY GROUP - (SE including Hampshire, Surrey, Susex, Berkshire and Oxfordshire).. (Also includes Hampshire, Surrey, Sussex, Berkshire and Oxfordshire). No longer in operation but able to give information on gulls.. Mark Fletcher, 24 The Gowans, Sutton-on-the-Forest, York, YO61 1DJ. e-mail: m.fletcher@csl.gov.uk

RUNNYMEDE RG. D G Harris, 22 Blossom Waye, Hounslow, TW5 9HD. e-mail: daveharris@tinyonline.co.uk

RSPB Local Groups
BEXLEY. (1979; 180). Tony Banks, 15 Boundary Road, Sidcup, Kent DA15 8SS. 020 8859 3518; e-mail: Tony@banks76.freeserve.co.uk www.bexleyrspb.org.uk **Meetings:** 7.30pm, 3rd Friday of the month, Hurstmere School Hall, Hurst Road, Sidcup.

BROMLEY. (1972; 285). Bob Francis, 2 Perry Rise, Forest Hill, London, SE23 2QL. 020 8669 9325 www.bromleyrspb.org.uk **Meetings:** 2nd Wednesday of the month (Sep-Jun), Large Hall, Bromley Central Library Building, Bromley High Street.

CENTRAL LONDON. (1974; 330). Miss Annette Warrick, 12 Tredegar Sq, London, E3 5AD. 020 8981 9624; e-mail: annette@warricka.freeserve.co.uk www.janja@dircon.co.uk/rspb **Meetings:** 2nd Thursday of the month (Sep-May), St Columba's Church Hall, Pont St, London SW1.

CROYDON. (1973; 4000 in catchment area). Sheila Mason, 5 Freshfields, Shirley, Croydon, CR0 7QS. 020 8777 9370 www.croydon-rspb.org.uk **Meetings:** 2nd Monday of each month at 2pm-4pm and again at 8pm-10pm at St Peter's Hall, Ledbury Road, South Croydon.

ENFIELD. (1971; 2700). Norman G Hudson, 125 Morley Hill, Enfield, Middx, EN2 0BQ. 020 8363 1431.

HAVERING. (1972; 270). David Coe, 8 The Fairway, Upminster, Essex, RM14 1BS. 01708 220710. **Meetings:** 8.00pm, 2nd Friday of the month, Hornchurch Library, North Street, Hornchurch.

NORTH LONDON. (1974; 3000). John Parsons, 65 Rutland Gardens, Harringay, London, N4 1JW. 020 8802 9537.

NORTH WEST LONDON RSPB GROUP. (1983; 2000 in catchment area). Bob Husband, The Firs, 49 Carson Road, Cockfosters, Barnet, Herts, EN4 9EN. 020 8441 8742. **Meetings:** 8.00pm, last Tuesday of the month (Sept-Apr), Union Church Hall, Eversfield Gardens, Mill Hill, NW7 (new for 2005).

PINNER & DISTRICT. (1972; 300). Dennis Bristow, 118 Crofts Road, Harrow, Middx, HA1 2PJ. 020 8863 5026. **Meetings:** 8pm, 2nd Thursday of the month, Church Hall, St john The Baptist parish church, Pinner.

RICHMOND & TWICKENHAM. (1979; 375). Steve Harrington, 93 Shaftesbury Way, Twickenham, TW2 5RW. 020 8898 4539. **Meetings:** 8.00pm, 1st Wednesday of the month, York House, Twickenham.

WEST LONDON. (1973; 400). Alan Bender, 020 8841 1952; e-mail: westlondonRSPB@lineone.net. **Meetings:** Ealing Town Hall, 22 Uxbridge Road, LONDON, W5 2BU

Wildlife Hospitals
WILDLIFE RESCUE & AMBULANCE SERVICE. Barry and June Smitherman, 19 Chesterfield Road, Enfield, Middx, EN3 6BE. 020 8292 5377. All categories of wild birds. Emergency ambulance with full rescue equipment, boats, ladders etc. Own treatment centre and aviaries. Veterinary support. Essential to telephone first.

Wildlife Trust
LONDON WILDLIFE TRUST. (1981; 7500). Skyline House, 200 Union Street, London SE1 0LW. 020 7261 0447; fax: 020 7633 0811. e-mail: enquiries@wildlondon.org.uk www.wildlondon.org.uk

MANCHESTER, GREATER

Bird Atlas/Avifauna
Breeding Birds in Greater Manchester by Philip Holland et al (1984).

Bird Recorder
RECORDER AND REPORT EDITOR. Mrs A Judith Smith,

ENGLAND

12 Edge Green Street, Ashton-in-Makerfield, Wigan, WN4 8SL. 01942 712615; e-mail: judith@gmbirds.freeserve.co.uk www.gmbirds.freeserve.co.uk

ASSISTANT RECORDER. Ian McKerchar, 42 Green Ave, Astley, Manchester, M29 7EH . 01942 701758; e-mail: ian@mckerchar1.freeserve.co.uk

ASSISTANT RECORDER. Antony Wainwright, 17 Hobart St, Halliwell, Bolton BL1 3PY. 01204 456415; e-mail: AtnWain@aol.com

Bird Reports
BIRDS IN GREATER MANCHESTER (1976-), from Mrs M McCormick, 91 Sinderland Road, Altrincham WA14 5JJ. (only editions up to year 2000. Year 2001 onwards from County Recorder).

LEIGH ORNITHOLOGICAL SOCIETY BIRD REPORT (1971-), from J Critchley, 2 Albany Grove, Tyldesley, Manchester, M29 7NE. 01942 884644.

BTO Regional Representative & Regional Development Officer
RR. Steve Sutthill, 01457 836 360; e-mail: steve@marctheprinters.org

RDO. Jim Jeffery, 20 Church Lane, Romiley, Stockport, Cheshire, SK6 4AA. H:0161 494 5367; W:01625 522107 ext 112; e-mail: jim_jeffery1943@yahoo.co.uk

Clubs
MANCHESTER ORNITHOLOGICAL SOCIETY. (1954; 70). Dr R Sandling, Maths Dept, University of Manchester, 8 Sackville Street, PO Box 88Manchester M60 1QD. e-mail: rsandling@manchester.ac.uk
Meetings: 7.30pm, Tuesdays, St James Church Hall, off Church Street, Gatley

GREATER MANCHESTER BIRD RECORDING GROUP. (2002: 40) Restricted to contributors of the county bird report.. Mrs A Judith Smith,01942 712615; e-mail: judith@gmbirds.freeserve.co.uk www.gmbirds.freeserve.co.uk

HALE ORNITHOLOGISTS. (1968; 82). Ms Diana Grellier, 8 Apsley Grove, Bowdon, Altrincham, Cheshire, WA14 3AH. 0161 928 9165.
Meetings: 7.30pm, 2nd Wednesday of the month (Sept-July), St Peters Assembly Rooms, Hale.

LEIGH ORNITHOLOGICAL SOCIETY. (1971; 120). Mr D Shallcross, 10 Holden Brook Close, Leigh, Lancs, WN7 2HL. 01942 607206; e-mail: chairman@leighos.org.uk www.leighos.org.uk
Meetings: 7.15pm, Fridays, Leigh Library (check website for details).

ROCHDALE FIELD NATURALISTS' SOCIETY. (1970; 90).

Mrs J P Wood, 196 Castleton Road, Thornham, Royton, Oldham, OL2 6UP. 0161 345 2012.

STOCKPORT BIRDWATCHING SOCIETY. (1972; 80). Dave Evans, 36 Tatton Road South, Stockport, Cheshire, SK4 4LU. 0161 432 9513; e-mail: dave.36tatton@ntlworld.com
Meetings: 7.30pm, last Wednesday of the month, Tiviot Dale Church.

Ringing Groups
LEIGH RG. A J Gramauskas, 21 Elliot Avenue, Golborne, Warrington, WA3 3DU. 0151 929215.

SOUTH MANCHESTER RG. C M Richards, Fairhaven, 13 The Green, Handforth, Wilmslow, Cheshire, SK9 3AG. 01625 524527; e-mail: cliveandray.richards@care4free.net

RSPB Local Groups
BOLTON. (1978; 320). Mrs Alma Schofield, 29 Redcar Road, Little Lever, Bolton, BL3 1EW. 01204 791745.
Meetings: 7.30pm, Thursdays (dates vary), Main Hall, Smithills School, Smithills Dean Road, Bolton.

MANCHESTER. (1972;3600 in catchment area). Peter Wolstenholme, 31 South Park Road, Gatley, Cheshire, SK8 4AL. 0161 428 2175.

STOCKPORT. (1979; 250). Brian Hallworth, 69 Talbot Street, Hazel Grove, Stockport, SK7 4BJ. 0161 456 5328; e-mail: brian.hallworth@ntlworld.com
Meetings: 7.30pm, 2nd Monday of the month (Sep-Apr), Stockport College of Technology, Lecture Theatre B.

WIGAN. (1973; 80). Graham Tonge, 01942 248238; e-mail: derek.dsc@btinternet.com.
Meetings: 7.45pm. Wigan Council For Voluntary Youth Service, Penson Street, Wigan Lane, near Swinley Labour Club, WIGAN.

Wildlife Hospital
THREE OWLS BIRD SANCTUARY AND RESERVE. Trustee, Nigel Fowler, Wolstenholme Fold, Norden, Rochdale, OL11 5UD. 01706 642162; Emergency helpline 07973 819389; e-mail: info@threeowls.co.uk www.threeowls.co.uk Registered charity. All species of wild bird. Rehabilitation and release on Sanctuary Reserve. Open every Sunday 1200-1700, otherwise visitors welcome by appointment. Bi-monthly newsletter. Veterinary support.

Wildlife Trust
Director, See Lancashire.

MERSEYSIDE & WIRRAL

Bird Atlas see Cheshire

Bird Recorders see Cheshire; Lancashire

ENGLAND

Bird Reports see also Cheshire
HILBRE BIRD OBSERVATORY REPORT, from Warden, see Reserves.

NORTHWESTERN BIRD REPORT (1938- irregular), from Secretaries, Merseyside Naturalists' Assoc.

BTO Regional Representatives
MERSEYSIDE RR and RDO. Bob Harris, 2 Dulas Road, Wavertree Green, Liverpool, L15 6UA. Work 0151 706 4311; e-mail: harris@liv.ac.uk

WIRRAL RR. Paul Miller, 01928 787535; e-mail: huntershill@worldline.co.uk

Clubs
MERSEYSIDE NATURALISTS' ASSOCIATION. (1938; 260). Steven Cross, (W)0151 478 4291, (H)0151 920 5718.
www.geocities.com/mnahome

WIRRAL BIRD CLUB. (1977; 150). Mrs Hilda Truesdale, Cader, 8 Park Road, Meols, Wirral, CH47 7BG. 0151 632 2705; info@wirralbirdclub.com
www.wirralbirdclub.com

Ringing Groups
MERSEYSIDE RG. P Slater, 45 Greenway Road, Speke, Liverpool, L24 7RY.

SOUTH WEST LANCASHIRE RG. J D Fletcher, 4 Hawksworth Drive, Freshfield, Formby, Merseyside, L37 7EZ. 01704 877837.

RSPB Local Groups
LIVERPOOL. (1966; 162). Chris Tynan, 10 Barker Close, Huyton, Liverpool, L36 0XU. 0151 480 7938; e-mail: christtynan@aol.com
www.livbird.pwp.blueyonder.co.uk
Meetings: 7 for 7.30pm, 3rd Monday of the month (Sep-Apr), Mossley Hill Parish Church, Junc. Rose Lane and Elmswood Rd.

SEFTON COAST. (1980; 150). Peter Taylor, 26 Tilston Road, Walton, Liverpool, L9 6AJ. 0151 524 1905; e-mail: ptaylor@liv.ac.uk   www.scmg.freeserve.co.uk
Meetings: 7.30pm, 2nd Tuesday of the month, St Lukes Church Hall, Liverpool Road, Crosby.

SOUTHPORT. (1974; 250). Roy Ekins, 01704 875898; e-mail: royekins@yahoo.co.uk
Meetings: 7.45pm, Lord Street West Church Hall, Duke Street, SOUTHPORT.

WIRRAL. (1982; 120). Jeremy Bradshaw, 0151 632 2364; e-mail: bradshaw@woodbank82.fsnet.co.uk

Wetland Bird Survey Organiser
DEE ESTUARY. Colin Wells, Burton Farm Point, Station Road, Nr Neston, South Wirra CH64 5SB. e-mail: colin.wells@rspb.org.uk

Wildlife Trust
Director, See Lancashire.

NORFOLK

Bird Atlas/Avifauna
The Birds of Norfolk by Moss Taylor, Michael Seago, Peter Allard & Don Dorling (Pica Press, 1999).

Bird Recorder
Giles Dunmore, 49 NelsonRoad, Sheringham, Norfolk, NR26 8DA. 01263 822550; e-mail:jdunmore@ukgateway.net

Bird Reports
CLEY BIRD CLUB 10-KM SQUARE BIRD REPORT (1987-), from Peter Gooden, 45 Charles Road, Holt, Norfolk, NR25 6DA. 01263 712368

NAR VALLEY ORNITHOLOGICAL SOCIETY ANNUAL REPORT (1976-), from Secretary.

NORFOLK BIRD & MAMMAL REPORT (1953-), from Secretary, Norfolk and Norwich Naturalists Society, Castle Museum, Norwich, NR1 3JU.

NORFOLK ORNITHOLOGISTS' ASSOCN ANNUAL REPORT (1961-), from Secretary.

BTO Regional Representatives
NORTH-EAST RR. Chris Hudson, Cornerstones, Ringland Road, Taverham, Norwich, NR8 6TG. 01603 868805; e-mail: chris.hudson@osb.uk.net

NORTH-WEST RR. Position vacant.

SOUTH-EAST RR. Position vacant.

SOUTH-WEST RR. Vince Matthews, Rose's Cottage, The Green, Merton, Thetford, Norfolk, IP25 6QU.

Clubs
CLEY BIRD CLUB. (1986; 400). Peter Gooden, 45 Charles Road, Holt, Norfolk, NR25 6DA. 01263 712368.
Meetings: 8pm, Wednesdays, monthly (Dec-Feb), George Hotel, Cley.

GREAT YARMOUTH BIRD CLUB. (1989; 25). Keith R Dye, 104 Wolseley Road, Great Yarmouth, Norfolk, NR31 0EJ. 01493 600705; e-mail: keith@dye3833.freeserve.co.uk
Meetings: 7.45pm, 4th Monday of the month, Rumbold Arms, Southtown Road.

NAR VALLEY ORNITHOLOGICAL SOCIETY. (1976; 125). Ian Black, Three Chimneys, Tumbler Hill, Swaffham, Norfolk, PE37 7JG. 01760 724092; e-mail: ian_a_black@hotmail.com

ENGLAND

NORFOLK & NORWICH NATURALISTS' SOCIETY.
(1869; 490). The Secretary, c/o The Castle Museum,
Norwich, Norfolk NR1 3JU. 01263 712282;
e-mail: leecha@dialstart.net

 NORFOLK ORNITHOLOGISTS'
ASSOCIATION. (1962; 1350). Jed
Andrews, Broadwater Road, Holme-
next-Sea, Hunstanton, Norfolk, PE36
6LQ. 01485 525406.

WENSUM VALLEY BIRDWATCHING SOCIETY. (2003;
93). Colin Wright, 7 Hinshalwood Way, Old Costessey,
Norwich, Norfolk NR8 5BN. 01603 740548;
e-mail: atomic.colin@virgin.net
www.wvbs.co.uk/
Meetings: 7.30pm, 3rd Thursday of the month,
Lenwade village hall.

Ringing Groups
BTO NUNNERY RG. Dawn Balmer, c/o BTO, The
Nunnery, Thetford, Norfolk IP24 2PU. e-mail:
dawn.balmer@bto.org

HOLME BIRD OBSERVATORY. J M Reed, 143 Daniells,
Panshanger, Welwyn Garden City, Herts, AL7 1QP.
01707 336351.

NORTH WEST NORFOLK RG. J M Reed, 143 Daniells,
Panshanger, Welwyn Garden City, Herts, AL7 1QP.
01707 336351.

SHERINGHAM RG. D Sadler, Denver House, 25 Holt
Road, Sheringham, Norfolk, NR26 8NB. 01263 821904;
e-mail: dhsadler@onetel.net.uk

UEA RG. D Thomas, 15 Grant Street, Norwich, NR2
4HA.

WASH WADER RG. P L Ireland, 27 Hainfield Drive,
Solihull, W Midlands, B91 2PL. 0121 704 1168; e-mail:
enquiries@wwrg.org.uk

WISSEY RG. Dr S J Browne, End Cottage, 24
Westgate Street, Hilborough, Norfolk, IP26 5BN. e-
mail: sjbathome@aol.com

RSPB Local Groups
NORWICH. (1971; 360). Paul Knapp, 31 Jubilee
Terrace, Norwich, NR1 2HT. 01603 631841;
email: paul@deetours.freeserve.co.uk
www.norwichRSPB.org uk
Meetings: 7.30pm, 2nd Monday of the month (except
Aug), Hellesdon Community Centre, Middletons Lane,
Hellesdon, Norwich (entrance of Woodview Road).

WEST NORFOLK. (1977; 247). Mr R Gordon, 3 Rectory
Close, King's Lynn, Norfolk, PE32 1AS. 01485 600937;
e-mail: robanngordon@btopenworld.com
Meetings: 7.30pm, 3rd Wednesday of the month
(Sep-Apr), South Wootton Village Hall, Church Lane,
South Wootton, King's Lynn.

Wetland Bird Survey Organisers
BREYDON WATER. Peter Allard, 39 Mallard Way,
Bradwell, Great Yarmouth, Norfolk NR3 8JY . 01493
657798

NORTH NORFOLK COAST. Michael Rooney, English
Nature, Hill Farm Offices, Main Road, Well-next-the –
Sea, Norfolk NR23 1AB. 01328 711866

INLAND. Tim Strudwick, RSPB Strumpshaw Fen,
Staithe Cottage, Low Road, Strumpshaw, Norfolk
NR13 4HS. 01603 715191

Wildlife Trust
NORFOLK WILDLIFE TRUST. (1926; 17,500). Bewick
House, 22 Thorpe Road, Norwich, Norfolk NR1 1RY.
01603 625540; fax 01603 598300;
e-mail: admin@norfolkwildlifetrust.org.uk
www.wildlifetrust.org.uk/norfolk/

NORTHAMPTONSHIRE

Bird Recorder
Gary Pullan, 21 Banbury Lane, Byfield, Daventry,
Northants NN11 6UX; 01327 262671.

Bird Report
NORTHAMPTONSHIRE BIRD REPORT (1969-), from
Alan Coles, 99 Rickyard Road, The Abours, Northants
NN3 3RR.

**BTO Regional Representative & Regional
Development Officer**
RR. Bill Metcalfe, Blendon, Rockingham Hills, Oundle,
Peterborough, PE8 4QA. 01832 274797.

RDO. Bill Metcalfe, Blendon, Rockingham Hills, Oundle,
Peterborough, PE8 4QA. 01832 274797.

Clubs
DAVENTRY NATURAL HISTORY SOCIETY. (1970; 18).
Leslie G Tooby, The Elms, Leamington Road, Long
Itchington, Southam, Warks, CV47 9PL. 0192 681
2269.
Meetings: 7.30pm, last Wednesday of the month,
United Reformed Church Rooms, Foundry Place,
Daventry.

NORTHAMPTONSHIRE BIRD CLUB. (1973; 100). Mrs
Eleanor McMahon, Oriole House, 5 The Croft, Hanging
Houghton, Northants, NN6 9HW. 01604 880009.
http://homepage.ntlworld.com/northantsbirds
Meetings: 7.30pm, 1st Wednesday of the month.
Village Hall, Pound Lane, Moulton, Northants.

Ringing Group
NORTHANTS RG. D M Francis, 2 Brittons Drive, Billing
Lane, Northampton, NN3 5DP.

RSPB Local Groups
MID NENE. (1975; 350). Michael Ridout, Melrose, 140

271

Northampton Road, Rushden, Northants, NN10 6AN.
01933 355544;
e-mail: mridout@smithchamberlain.co.uk.
Meetings: 7.30pm, 2nd or 3rd Thursday of the month
(Sep-Apr), The Saxon Hall, Thorpe Street/Brook
Street, Raunds.

NORTHAMPTON. (1978; 3000). Liz Wicks, 6 Waypost
Court, Lings, Northampton, NN3 8LN. 01604 513991;
e-mail: lizydrip@ntlworld.com
Meetings: 7.30pm, 2nd Thursday of the month,
Kingsthorpe Methodist Church Hall, Kingsthorpe,
Northampton.

Wetland Bird Survey Organiser
Robert Ratcliffe, 173 Montague Road, Bilton, Rugby,
Warks CV22 6LG. 01788 336983.

Wildlife Trust
Director, See Cambridgeshire.

NORTHUMBERLAND

Bird Atlas/Avifauna
The Atlas of Wintering Birds in Northumbria edited by
J C Day et al (Northumberland and Tyneside Bird Club,
2003). Available from 31 Uplands, Whitley Bay,
Northumberland NE25 9AG. £42 including P&P.

Bird Recorder
Ian Fisher, 74 Benton Park Road, Newcastle upon
Tyne, NE7 7NB. 0191 266 7900;
e-mail: ian@hauxley.freeserve.co.uk
www.ntbc.org.uk

Bird Reports
BIRDS IN NORTHUMBRIA (1970-), from Muriel
Cadwallender, 22 South View, Lesbury, NE66 3PZ.
01665 830884;
e-mail: tomandmurielcadwallender@hotmail.com

BIRDS ON THE FARNE ISLANDS (1971-), from
Secretary, Natural History Society of Northumbria,
0191 2326386; e-mail: NHSN@ncl.ac.uk

**BTO Regional Representative & Regional
Development Officer**
RR. Tom Cadwallender, 22 South View, Lesbury,
Alnwick, Northumberland, NE66 3PZ. H:01665 830884;
W:01670 533039;
e-mail: tomandmurielcadwallender@hotmail.com

RDO. Muriel Cadwallender, 22 South View, Lesbury,
Alnwick, Northumberland, NE66 3PZ. 01665 830884;
e-mail: tomandmurielcadwallender@hotmail.com

Clubs
NATURAL HISTORY SOCIETY OF NORTHUMBRIA.
(1829; 850). David C Noble-Rollin, Hancock Museum,
Barras Bridge, Newcastle upon Tyne, NE2 4PT. 0191
232 6386; e-mail: nhsn@ncl.ac.uk

www.NHSN.ncl.ac.uk
Meetings: 7.00pm, every Friday in the winter, The
Hancock Museum.

NORTH NORTHUMBERLAND BIRD
CLUB. (1984; 210). Mrs RA
Skinner, North Cottage, West
Fleetham, Chathill,
Northumberland NE67 5JU. 01665
589383.
Meetings: 7.30pm, 2nd Friday of the
month (Sep-Jun), Bamburgh Pavilion (below castle).

NORTHUMBERLAND & TYNESIDE BIRD CLUB. (1958;
270). Sarah Barratt, 18 Frances Ville, Scotland Gate,
Northumberland, NE62 5ST. 01670 827465:
e-mail: sarah.barratt@btopenworld.com
www.ntbc.org.uk

Ringing Groups
BAMBURGH RS. Mike S Hodgson, 31 Uplands,
Monkseaton, Whitley Bay, Tyne & Wear, NE25 9AG.
0191 252 0511.

NATURAL HISTORY SOCIETY OF NORTHUMBRIA. Dr C
P F Redfern, Westfield House, Acomb, Hexham,
Northumberland, NE46 4RJ.

NORTHUMBRIA RG. Secretary. B Galloway, 34 West
Meadows, Stamfordham Road, Westerhope,
Newcastle upon Tyne, NE5 1LS. 0191 286 4850.

Wetland Bird Survey Organisers
NORTHUMBERLAND COAST. Miss J Roper, 1 Long Row,
Howick, Alnwick, Northumberland, NE66 3LQ.

NORTHUMBERLAND (Inland). Steve Holliday, 2
Larriston Place, Cramlington, Northumberland NE23
8ER. 01670 731063;
e-mail: steve@sjjholliday.freeserve.co.uk

Wildlife Hospitals
BERWICK SWAN & WILDLIFE TRUST. The Honourable
Secretary, North Road Industrial Estate, Berwick upon
Tweed, TD15 1UN. 01289 302882; e-mail: mail@
swan-trust.org.uk www.swan-trust.org.uk
Registered charity. All categories of birds. Pools for
swans and other waterfowl. Veterinary support.

Wildlife Trust
NORTHUMBERLAND WILDLIFE TRUST. (1962; 5000).
The Garden House, St Nicholas Park, Jubilee Road,
Newcastle upon Tyne, NE3 3XT. 0191 284 6884; fax
0191 284 6794; e-mail: mail@northwt.org.uk
www.wildlifetrust.org.uk/northumberland

NOTTINGHAMSHIRE

Bird Recorders
Andy Hall, 10 Staverton Road, Bilborough,
Nottingham, NG8 4ET. 0115 916 9763;
e-mail: andy.h11@ntlworld.com

ENGLAND

Bird Reports
LOUND BIRD REPORT (1990-), from Mr G Hobson, 11 Sherwood Road, Harworth, Doncaster, DN11 8HY. 01302 743654

BIRDS OF NOTTINGHAMSHIRE (1943-), from Davis, 3 Windrush Close, Bramcote View, Nottingham, NG9 3LN. 0115 922 8547; e-mail: prlg@talk21.com

BTO Regional Representative & Regional Development Officer
RR. Mrs Lynda Milner, 6 Kirton Park, Kirton, Newark, Notts, NG22 9LR. 01623 862025; e-mail: lyndamilner@hotmail.com

Clubs
LOUND BIRD CLUB. (1991; 50). Gary Hobson, 11 Sherwood Road, Harworth, Doncaster, South Yorkshire DN11 8HY. (m)0771 2244469; e-mail: loundbirdclub:tiscali.co.uk

NETHERFIELD WILDLIFE GROUP. (1999; 130). Philip Burnham, 57 Tilford Road, Newstead Village, Nottingham, NG15 0BU. 01623 401980.

NOTTINGHAMSHIRE BIRDWATCHERS. (1935; 420). Ms Jenny Swindels, 21 Chaworth Road, West Bridgeford, Nottingham NG2 7AE. 0115 9812432; e-mail: j.swindells@btinternet.com www.nottmbirds.org.uk

WOLLATON NATURAL HISTORY SOCIETY. (1976; 99). Mrs P Price, 33 Coatsby Road, Hollycroft, Kimberley, Nottingham, NG16 2TH. 0115 938 4965.
Meetings: 7.30pm, 3rd Monday of the month, St mary's Church Hall, Wollaton until end December 2005, then: 7.30pm 3rd Wednesday of the month, St Leonards Church Hall, Wollaton Village from Jan 2006.

Ringing Groups
BIRKLANDS RG. A D Lowe, 12 Midhurst Way, Clifton Estate, Nottingham, NG11 8DY. e-mail: alowe@mansfield.gov.uk

NORTH NOTTS RG. Adrian Blackburn, Suleska, 1 Richmond Road, Retford, Notts, DN22 6SJ. 01777 706516; (M)07718 766873: e-mail: blackburns@suleska.freeserve.co.uk

SOUTH NOTTINGHAMSHIRE RG. K J Hemsley, 8 Grange Farm Close, Toton, Beeston, Notts, NG9 6EB. e-mail: k.hemsley@ntlworld.com

TRESWELL WOOD INTEGRATED POPULATION MONITORING GROUP. Chris du Feu, 66 High Street, Beckingham, Notts, DN10 4PF. e-mail: chris@beckingham0.demon.co.uk

RSPB Local Groups
MANSFIELD AND DISTRICT. (1986; 200). John Barlow, 240 Southwell Road West, Mansfield, Notts NG18 4LB.

01623 626647.
Meetings: 7.30pm, 1st Thursday of the month (Sep-Jun), Kay Hall, Ladybrook Lane.

NOTTINGHAM. (1974; 514). Andrew Griffin, Hawthorn Cottage, Thoroton, Notts, NG13 9DS. 01949 851426; e-mail: andrew@thoroton.f.sworld.co.uk www.notts-rspb.org.uk
Meetings: 7.30pm, 1st Wednesday of the month, Nottingham Mechanics, North Sherwood Street, Nottingham.

Wetland Bird Survey Organiser
Gary Hobson, 11 Sherwood Road, Harworth, Doncaster, DN11 8HY. 01302 743654.

Wildlife Trust
NOTTINGHAMSHIRE WILDLIFE TRUST. (1963; 4300). The Old Ragged School, Brook Street, Nottingham, NG1 1EA. 0115 958 8242; fax 0115 924 3175; e-mail: nottswt@cix.co.uk www.wildlifetrust.org.uk/nottinghamshire

OXFORDSHIRE

Bird Atlas/Avifauna
Birds of Oxfordshire by J W Brucker et al (Oxford, Pisces, 1992).

The New Birds of the Banbury Area by T G Easterbrook (Banbury Ornithological Society, 1995).

Bird Recorder
Ian Lewington, 119 Brasenose Road, Didcot, Oxon, OX11 7BP. 01235 819792; e-mail: ian@recorder.fsnet.co.uk

Bird Reports
BIRDS OF OXFORDSHIRE (1921-), from Roy Overall, 30 Hunsdon Road, Iffley, Oxford, OX4 4JE. 01865 775632.

BANBURY ORNITHOLOGICAL SOCIETY ANNUAL REPORT (1952-), from A Turner, 33 Newcombe Close, Milcombe, Nr Banbury, Oxon, OX15 4RN. 01295 720938.

BTO Regional Representatives & Regional Development Officer
NORTH. Frances Buckell, 15 Insall Road, Chipping Norton, Oxon, OX7 5LF. 01608 644425.

SOUTH RR & RDO. Mr John Melling, 17 Lime Grove, Southmoor, Nr Abingdon, Oxon OX13 5DN; e-mail: john@melling17lg.freeserve.co.uk

Clubs
BANBURY ORNITHOLOGICAL SOCIETY. (1952; 100). Frances Buckel, Witt's End, Radbone Hill, Overnorton, Chipping Norton, Oxfordshire OX7 5LF. 01608 644425; e-mail: fmarks@ukotcf.org

OXFORD ORNITHOLOGICAL SOCIETY. (1921; 330). Ivor Rhymes, 159 Rutten Lane, Yarnton, Kidlington, Oxon, OX5 1LT. 01865 372620; e-mail: secretary@oos.org.uk www.oos.org.uk
Meetings: Various dates, Stratford Brake, Kidlington.

Ringing Group
EDWARD GREY INSTITUTE. Dr A G Gosler, c/o Edward Grey Institute, Department of Zoology, South Parks Road, Oxford, OX1 3PS. 01865 271158.

RSPB Local Groups
OXFORD. (1977; 100). Ian Kilshaw, 6 Queens Court, Bicester, Oxon, OX26 6JX. Tel 01869 601901; (fax) 01869 600565; e-mail: ian.kilshaw@ntlworld.com www.rspb-oxford.org.uk
Meetings: Normally 1st Thursday of the month, Methodist Church Hall, New High Street, Headington.

VALE OF WHITE HORSE. (1977; 325). David Lovegrove, 17 Chiltern Crescent, Wallingford, Oxon, OX10 0PE. 01491 835692; e-mail: dflovegrove@hotmail.com.
Meetings: 7.30pm, 3rd Monday of the month (Sep-May). Didcot Civic Hall.

Wetland Bird Survey Organiser
OXFORDSHIRE (South). Catherine Ross, Duck End Cottage, 40 Sutton Lane, Witney, Oxfordshire, OX29 5RU. (H) 01865 881552; e-mail: catherine@duckend6332.freeserve.co.uk

Wildlife Trust
BBOWT. (1959; 11,000). The Lodge, 1 Armstrong Road, Littlemore, Oxford, OX4 4XT. 01865 775476; fax 01865 711301; e-mail: bbowt@cix.co.uk www.bbowt.org.uk

SHROPSHIRE

Bird Atlas/Avifauna
Atlas of the Breeding Birds of Shropshire (Shropshire Ornithological Society, 1995).

Bird Recorder
Geoff Holmes, 22 Tenbury Drive, Telford Estate, Shrewsbury, SY2 5YF. 01743 364621; e-mail: geoff.holmes4@btopenworld.com

Bird Report
SHROPSHIRE BIRD REPORT (1956-) Annual, from Helen Griffiths (Hon Secretary), 104 Noel Hill Road, Cross Houses, Shrewsbury SY5 6LD. 01743 761507; e-mail: helen.griffiths@english-nature.org.uk www.shropshirebirds.com

BTO Regional Representative
RR. Allan Dawes, Rosedale, Chapel Lane, Trefonen,

Oswestry, Shrops, SY10 9DX. 01691 654245; e-mail: allandawes@btinternet.com

Club
SHROPSHIRE ORNITHOLOGICAL SOCIETY. (1955; 800). Helen Griffiths, 104 Noel Hill Road, Cross Houses, Shrewsbury, SY5 6LD. 01743 761507. www.shropshirebirds.com
Meetings: 7.15pm, 1st Thursday of month (Oct-Apr), Shirehall, Shrewsbury.

RSPB Local Group
SHROPSHIRE. (1992; 320). Roger M Evans, 31 The Wheatlands, Bridgnorth, WV16 5BD.
Meetings: 3rd Thursday of the month (Sep-Apr), Council Chamber, Shirehall, Shrewsbury. Also field trip year round. 3rd Wednesday in the month (Oct-March) Secret Hills Centre Craven Arms.

Wetland Bird Survey Organiser
Bill Edwards, Hopton Villa, Maesbury Marsh, Oswestry, SY10 8JA. (H) 01691 656679.

Wildlife Trust
SHROPSHIRE WILDLIFE TRUST. (1962; 2000). 193 Abbey Foregate, Shrewsbury, Shropshire SY2 6AH. 01743 284280; fax 01743 284281; e-mail: enquiries@shropshirewildlifetrust.org.uk www.shropshirewildlifetrust.org.uk

SOMERSET & BRISTOL

Bird Atlas/Avifauna
The Birds of Exmoor and the Quantocks by DK Ballance and BD Gibbs. (Isabelline Books, 2 Highbury House, 8 Woodland Crescent, Falmouth TR11 4QS. 2003).

Atlas of Breeding Birds in Avon 1988-91 by R L Bland and John Tully (John Tully, 6 Falcondale Walk, Westbury-on-Trym, Bristol BS9 3JG, 1992).

Bird Recorders
Brian D Gibbs, 23 Lyngford Road, Taunton, Somerset, TA2 7EE. 01823 274887; e-mail: brian.gibbs@virgin.net

BATH, NE SOMERSET, BRISTOL, S GLOS. Harvey Rose, 12 Birbeck Road, Bristol, BS9 1BD. H:0117 968 1638; W:0117 331 1666; e-mail: h.e.rose@bris.ac.uk

Bird Reports
AVON BIRD REPORT (1977-), from Richard L Bland, 11 Percival Road, Bristol, BS8 3LN. Home/W:01179 734828; e-mail: richardbland@blueyonder.co.uk

EXMOOR NATURALIST (1974-), from Secretary, Exmoor Natural History Society.

SOMERSET BIRDS (1913-), from David Ballance, Flat 2, Dunboyne, Bratton Lane, Minehead, Somerset, TA24 8SQ. 01643 706820.

BTO Regional Representatives, Development Officer & Secretary
AVON RR. Richard L Bland, 11 Percival Road, Bristol, BS8 3LN. Home/W:01179 734828; e-mail: richardbland@blueyonder.co.uk

AVON ASSISTANT REGIONAL REPRESENTATIVE. John Tully, 6 Falcondale Walk, Westbury-on-Trym, Bristol, BS9 3JG. 0117 950 0992; e-mail: johntully4@aol.com

SOMERSET RR. Eve Tigwell, Hawthorne Cottage, 3 Friggle Street, Frome, Somerset, BA11 5LP. 01373 451630; e-mail: evetigwell@aol.com

Clubs
BRISTOL NATURALISTS' SOCIETY (Ornithological Section). (1862; 550). Dr Mary Hill, 15 Montrose Avenue, Redland, Bristol, BS6 6EH. 0117 942 2193; e-mail: mary@jhill15.fsnet.co.uk
www.bristolnats.org.uk
Meetings: 7.30pm, Wednesday or Friday (check for dates), Westmorland Hall, Westmorland Road, Bristol

BRISTOL ORNITHOLOGICAL CLUB. (1966; 660). Mrs Judy Copeland, 19 St George's Hill, Easton-in-Gordano, North Somerset, BS20 0PS. Tel/fax 01275 373554; judy.copeland@ukgateway.net
www.boc-bristol.org.uk
Meetings: 7.30pm, 3rd Thursday of the month, Newman Hall, Grange Court Road, Westbury-on-Trym.

CAM VALLEY WILDLIFE GROUP. (1994: 356). Helena Crouch, Bronwen, Farrington Road, Paulton, Bristol, BS39 7LP. 01761 410731.
e-mail: jim-helena@supanet.com
www.camvalleywildlifegroup.org.uk

EXMOOR NATURAL HISTORY SOCIETY. (1974; 480). Miss Caroline Giddens, 12 King George Road, Minehead, Somerset, TA24 5JD. 01643 707624; e-mail: carol.enhs@virgin.net; http://freespace.virgin.net/carol.enhs
Meetings: 7.30pm, 1st Wednesday of the month (Oct-Mar), Methodist Church Hall, The Avenue, Minehead.

SOMERSET ORNITHOLOGICAL SOCIETY. (1923; 350). Miss Sarah Beavis, The Old Surgery, 4 The Barton, Hatch Beauchamp, Taunton, Somerset, TA3 6SG. 01823 480948. www. somersetbirds.uko2.co.uk
Meetings: 7.30pm, various Thursdays (Oct-Apr), Ruishton Village Hall, Taunton.

Ringing Groups
CHEW VALLEY RS. W R White, Church View Cottage, Mead Lane, Blagdon, N Somerset, BS40 7UA. 01761 463157 (evgs); e-mail: warwickw@architen.com

GORDANO VALLEY RG. Lyndon Roberts, 20 Glebe Road, Long Ashton, Bristol, BS41 9LH. 01275 392722; e-mail: mail@lyndonroberts.com

RSPCA. S Powell, 1 Rosemill Cottage, Rosemill Lane, Ilminster, Somerset, TA19 5PR.

STEEP HOLM RS. A J Parsons, Barnfield, Tower Hill Road, Crewkerne, Somerset, TA18 8BJ. 01460 73640.

RSPB Local Groups
BATH AND DISTRICT. (1989; 220). Alan Barrett, 01225 310905; E-mail: alan_w_h_barrett@yahoo.co.uk
Meetings: 7.30pm, 3rd Wednesday of the month (Sep-Mar), Bath Society Meeting Room, Green Park Station, Bath

CREWKERNE & DISTRICT. (1979; 355). Denise Chamings, Daniels Farm, Lower Stratton, South Petherton, Somerset, TA13 5LP. 01460 240740; e-mail: rspb@crewkerne.fslife.co.uk
www.crewkerne.fslife.co.uk
Meetings: 7.30pm, 3rd Thursday of the month (Sep-Apr), The Day Centre, Crewkerne.

TAUNTON. (1975; 148). Eric Luxton, 33 Hoveland Lane, Taunton, Somerset, TA1 2EY. 01823 283033; e-mail: eric.luxton@btinternet.com

WESTON-SUPER-MARE (N SOMERSET). (1976; 215). Don Hurrell, Freeways, Star, Winscombe, BS25 1PS. 01934 842717.
Meetings: 7.30pm, 1st Thursday of the month (Sep-Apr), St Pauls Church Hall.

Wetland Bird Survey Organisers
SEVERN ESTUARY - SOMERSET & BRISTOL. Harvey Rose, 12 Birbek Road, Stoke Bishop, Bristol, BS9 1BD. (H)0117 9681638; (W)0117 9287992.

SOMERSET LEVELS. Steve Meen, RSPB West Sedgemoor, Dewlands Farm, Redhill, Curry Rivel, Langport, Somerset TA10 0PH. 01458 252805; e-mail: steve.meen@rspb.org.uk

Wildlife Trusts
AVON WILDLIFE TRUST. (1980; 4500). Wildlife Centre, 32 Jacobs Wells Road, Bristol, BS8 1DR. 0117 917 7270; fax 0117 929 7273;
e-mail: mail@avonwildlifetrust.org.uk
www.avonwildlifetrust.org.uk

SOMERSET WILDLIFE TRUST. (1964; 8000). Tonedale Mill, Tonedale, Wellington, Somerset TA21 0AW. 01823 652400; fax 01823 652411.
e-mail: enquiries@somersetwildlife.org
www.somersetwildlife.org

STAFFORDSHIRE

Bird Recorder
Mrs Gilly Jones, 4 The Poplars, Lichfield Road, Abbots

Bromley, Rugeley, Staffs, WS15 3AA;
e-mail: staffs-recorder@westmidlandbirdclub.com
www.westmidlandbirdclub.com

Bird Report See West Midlands

BTO Regional Representatives
NORTH EAST. Mrs Gilly Jones (address above), 01283
840555; e-mail: g.n.jones@wlv.ac.uk

SOUTH & EAST. Mrs Gilly Jones (address above),
01283 840555; e-mail: g.n.jones@wlv.ac.uk

WEST. Mrs Gilly Jones (address above), 01283
840555; e-mail: g.n.jones@wlv.ac.uk

Clubs
SOUTH PEAK RAPTOR STUDY
GROUP. (1998; 12). M E Taylor, 76
Hawksley Avenue, Newbold,
Chesterfield, Derbys, S40 4TL.
01246 277749.

WEST MIDLAND BIRD CLUB (STAFFORD BRANCH).
Gerald Ford, 01630 673409;
e-mail: gerald.ford@westmidlandbirdclub.com
www.westmidlandbirdclub.com/stafford
Meetings: 7.30pm, 3rd Friday of the month (Oct-
Mar), The Centre for The Blind, North Walls, Stafford.

WEST MIDLAND BIRD CLUB (TAMWORTH BRANCH).
(1992). Barbara Stubbs, 19 Alfred Street, Tamworth,
Staffs, B79 7RL. 01827 57865;
e-mail: tamworth@westmidlandbirdclub
www.westmidlandbirdclub.com/tamworth
Meetings: 7.30pm, 3rd Friday of the month (Sep-
Apr), Phil Dix Centre, Corporation Street, Tamworth.

RSPB Local Groups
BURTON-ON-TRENT. (1973; 50). Dave Lummis, 121
Wilmot Road, Swadlincote, Derbys, DE11 9BN. 01283
219902. www.basd-rspb.co.uk
Meetings: 7.30pm 1st Wednesday of the month, All
Saint's Church, Bronston Road, Burton.

LICHFIELD & DISTRICT. (1977; 1150). Ray Jennett,
12 St Margarets Road, Lichfield, Staffs, WS13 7RA.
01543 255195.
Meetings: 7.30pm, 2nd Tuesday of the month (Jan-
May, Sept-Dec), St Mary's Centre.

NORTH STAFFORDSHIRE. (1982; 187). John Booth, 32
St Margaret Drive, Sneyd Green, Stoke-on-Trent, ST1
6EW. 01782 262082; www.geocities.com/nsrspb
Meetings: 7.30pm, 3rd Wednesday of the month,
Medical Institute, Hartshill (no correspondence here
please).

SOUTH WEST STAFFORDSHIRE. (1972; 204). Mrs
Theresa Dorrance, 39 Wilkes Road, Codsall,
Wolverhampton, WV8 1RZ. 01902 847041;
e-mail: dorrancesteve@fsmail.net

Meetings: 8.00pm, 2nd Tuesday of the month,
Codsall Village Hall.

Wetland Bird Survey Organisers
Gilly Jones, 4 The Poplars, Lichfield Road, Abbots
Bromley, Rugeley, Staffordshire WS15 3AA.

Wildlife Hospitals
BRITISH WILDLIFE RESCUE CENTRE. Alfred Hardy,
Amerton Working Farm, Stowe-by-Chartley, Stafford,
ST18 0LA. 01889 271308. On A518 Stafford/Uttoxeter
road. All species, including imprints and permanently
injured. Hospital, large aviaries and caging. Open to
the public every day. Veterinary support.

GENTLESHAW BIRD OF PREY HOSPITAL. Jenny Smith,
5 Chestall Road, Cannock Wood, Rugeley, Staffs,
WS15 4RB. 01785 850379.
www.gentleshawwildlife.co.uk
Registered charity. All birds of prey (inc. owls).
Hospital cages and aviaries; release sites. Veterinary
support. Also GENTLESHAW BIRD OF PREY AND
WILDLIFE CENTRE, Fletchers Country Garden Centre,
Stone Road, Eccleshall, Stafford. 01785 850379
(1000-1700).

RAPTOR RESCUE, BIRD OF PREY REHABILITATION. J
M Cunningham, 8 Harvey Road, Handsacre, Rugeley,
Staffs, WS15 4HF. 01543 491712; (Nat. advice line)
0870 241 0609;
e-mail: mickcunningham@btinternet.com
www.raptorrescue.org.uk
Birds of prey only. Heated hospital units. Indoor
flights, secluded aviaries, hacking sites, rehabilitation
aviaries/flights. Falconry rehabilitation techniques,
foster birds for rearing young to avoid imprinting.
Veterinary support. Reg charity no. 283733

Wildlife Trust
STAFFORDSHIRE WILDLIFE TRUST. (1969; 4000). The
Wolseley Centre, Wolseley Bridge, Stafford, ST17 0YT.
01889 880100; fax 01889 880101;
e-mail: staffswt@cix.co.uk www.staffs-wildlife.org.uk

SUFFOLK

Bird Atlas/Avifauna
Birds of Suffolk by S H Piotrowski (February 2003).

Bird Recorders
NORTH EAST. Dave Thurlow,
e-mail: dave.thurlow@rspb.org.uk

SOUTH EAST (inc. coastal region from Slaughden Quay
southwards). Lee Woods,
e-mail: leejanwoods@ntlworld.com

WEST (whole of Suffolk W of Stowmarket, inc.
Breckland). Colin Jakes, 7 Maltward Avenue, Bury St
Edmunds, Suffolk, IP33 3XN. 01284 702215;
e-mail: cjjakes@supanet.com

ENGLAND

Bird Report
SUFFOLK BIRDS (inc Landguard Bird Observatory Report) (1950-), from Ipswich Museum, High Street, Ipswich, Suffolk.

BTO Regional Representative
Mick T Wright, 15 Avondale Road, Ipswich, IP3 9JT. 01473 710032; e-mail: micktwright@btinternet.com

Clubs
LAVENHAM BIRD CLUB. (1972; 54). Mike Lewis, 6 Grammar School Place, Sudbury, Suffolk CO10 2GE. 01787 324488.

SUFFOLK ORNITHOLOGISTS' GROUP. (1973; 650). Andrew M Gregory, 1 Holly Road, Ipswich, IP1 3QN. 01473 253816.
Meetings: Last Thursday of the month (Jan-Mar, Oct-Nov), Holiday Inn, Ipswich.

Ringing Groups
DINGLE BIRD CLUB. Dr D Pearson, 4 Lupin Close, Reydon, Southwold, Suffolk, IP18 6NW. 01502 722348.

LACKFORD RG. Dr Peter Lack, 11 Holden Road, Lackford, Bury St Edmunds, Suffolk, IP28 6HZ. e-mail: peter.diane@tinyworld.co.uk

LANDGUARD RG. Mr SH Piotrowski, 29 Churchfields Road, Long Stratton, Norfolk, NR15 2WH. 01508 531115.

MARKET WESTON RG. Dr R H W Langston, Walnut Tree Farm, Thorpe Street, Hinderclay, Diss, Norfolk, IP22 1HT. e-mail: rlangston@wntfarm.demon.co.uk

RSPB Local Groups
BURY ST EDMUNDS. (1982; 150). Trevor Hart, 7 Westgart Gardens, Bury St Edmunds, Suffolk, IP33 3LB. 01284 705165.
Meetings: 7.30pm, 3rd Tuesday of the month (Sep-May), County Upper School, Beetons Way, Bury St Edmunds.

IPSWICH. (1975; 250). BJ Cooper, Ipswich Group Leader, 115 Bucklesham Road, Ipswich, Suffolk IP3 8TX. 01473 431752; b.cooper9@ntlworld.com
Meetings: 7.30pm, 2nd Thursday of the month (Sep-Apr), Sidegate Primary School, Sidegate Lane, Ipswich.

LOWESTOFT & DISTRICT. (1976; 130). Mrs E Beaumont, 52 Squires Walk, Lowestoft, Suffolk, NR32 4LA. 01502 560126; e-mail: groupleader@lowestoft-rspb-group.org.uk
www.lowestoft-rspb-group.org.uk
Meetings: Friday 7.15pm 1st Monday in the month, St Marks Church Hall, Oulton Broad.

WOODBRIDGE. (1986; 400). Colin Coates, 42A Bredfield Road, Woodbridge, Suffolk, IP12 1JE. 01394 385209; e-mail: colinm.coates@virgin.net.
Meetings: 7.30pm, 1st Thursday of the month (Oct-Apr), Woodbridge Community Hall, and 1st Thursday in May, Wickham Market village hall.

Wetland Bird Survey Organisers
ALDE COMPLEX. Rodney West, Flint Cottage, Stone Common, Blaxhall, Woodbridge, IP12 2DP. (Office) 01728 689171; e-mail: rodwest@ndirect.co.uk

DEBEN ESTUARY. Nick Mason, Evening Hall, Hollesley, Nr Woodbridge, Ipswich, IP12 3QU. (H)01359 411150; e-mail: nick.mason@talk21.com

STOUR ESTUARY. Rick Vonk, RSPB, Unit 3 Court Farm, 3 Stutton Road, Brantham, Suffolk CO11 1PW. (D)01473 328006; e-mail:rick.vonk@rspb.org.uk

SUFFOLK (other sites). Alan Miller, Suffolk Wildlife Trust, 9 Valley Terrace, Valley Road, Leiston, IP16 4AP. (Day)01728 833405; e-mail: alanm@suffolkwildlife.cix.co.uk

Wildlife Trust
SUFFOLK WILDLIFE TRUST. (1961; 15,000). Brooke House, The Green, Ashbocking, Ipswich, IP6 9JY. 01473 890089; fax 01473 890165; e-mail: info@suffolkwildlife.cix.co.uk
www.wildlifetrust.org.uk/suffolk

SURREY

Bird Atlas/Avifauna
Birds of Surrey (avifauna). In preparation.

Bird Recorder (inc London S of Thames & E to Surrey Docks)
SURREY (includes Greater London south of the Thames and east to the Surrey Docks, excludes Spellhorne). Jeffery Wheatley, 9 Copse Edge, Elstead, Godalming, Surrey, GU8 6DJ. 01252 702450;(Fax) 01252 703650; e-mail: j.j.wheatley@btinternet.com

Bird Report
SURBITON AND DISTRICT BIRD WATCHING SOCIETY (1972-), from Thelma Caine, 21 More Lane, Esher, Surrey KT10 8AJ.

SURREY BIRD REPORT (1952-), from J Gates, 90 The Street, Wrecclesham, Farnham, Surrey, GU10 4QR. 01252 727683.

BTO Regional Representative
RR. Hugh Evans, 31 Crescent Road, Shepperton, Middx, TW17 8BL. 01932 227781; e-mail: hugh_w_evans@lineone.net

Clubs
SURBITON & DISTRICT BIRDWATCHING SOCIETY.

(1954; 200). Gary Caine, 21 More Lane, Esher, Surrey KT10 8AJ. 01372 468432;
e-mail: hockley@sdbws.ndo.co.uk
www.sdbws.ndo.co.uk
Meetings: 7.30pm, 3rd Tuesday of the month, Surbiton Library Annex.

SURREY BIRD CLUB. (1957; 400). Mrs Jill Cook, Moorings, Vale Wood Drive, Lower Bourne, Farnham, Surrey, GU10 3HW. 01252 792876;
e-mail: jilck@aol.com
www.surreybirdclub.org.uk

Ringing Groups
HERSHAM RG. A J Beasley, 29 Selbourne Avenue, New Haw, Weybridge, Surrey, KT15 3RB.
e-mail: abeasley00@hotmail.com

RUNNYMEDE RG. D G Harris, 22 Blossom Waye, Hounslow, TW5 9HD.
e-mail: daveharris@tinyonline.co.uk

RSPB Local Groups
DORKING & DISTRICT. (1982; 310). Alan Clark, 11 Maplehurst, Fetcham, Surrey, KT22 9NB. 01372 450607; e-mail: alanclark@tinyonline.co.uk
Meetings: Friday Night and Wednesday afternoons, Christian Centre, next to St Martin's Church, Dorking.

EAST SURREY. (1984; 150-200). Brian Hobley, 26 Alexandra Road, Warlingham, Surrey, CR6 9DU. 01883 625404; www.mouseline.demon.co.uk
Meetings: 8.00pm, 2nd Wednesday of the month (Sep-Jul), Whitehart Barn, Godstone.

EPSOM & EWELL. (1974; 168). Janet Gilbert, 78 Fairfax Avenue, Ewell, Epsom, Surrey, KT17 2QQ. 0208 394 0405; www.wsnhs.co.uk.
Meetings: 7.45pm, 2nd Tuesday of the month, Bourne Hall, Ewell. Venue changing December 05 to All Saints Church Hall, Fulford Road, West Ewell. 7.45pm 2nd Friday in month.

GUILDFORD AND DISTRICT. (1974; 550). Alan Bowen, Newlands, 13 Mountside, Guildford, Surrey, GU2 4JD. 01483 567041;
e-mail: alan.and.monica.bowen@care4free.net
Meetings: 2.15pm 2nd Tuesday and 7.45pm 4th Wednesday, Onslow Village Hall, Guildford.

NORTH WEST SURREY. (1973; 140). Ms Mary Braddock, 20 Meadway Drive, New Haw, Surrey, KT15 2DT. 01932 858692;
e-mail: mary@braddock3.wanadoo.co.uk
www.nwsurreyrspb.org.uk
Meetings: 7.45pm, 4th Wednesday of the month (not Dec, Jul, Aug), Sir William Perkins School, Chertsey.

Wetland Bird Survey Organiser
Jeffery Wheatley, 9 Copse Edge, Elstead, Godalming,

Surrey, GU8 6DJ. 01252 702540;
e-mail: j.j.wheatley@btinternet.com

Wildlife Hospitals
THE SWAN SANCTUARY. See National Directory

WILDLIFE AID. Simon Cowell, Randalls Farm House, Randalls Road, Leatherhead, Surrey, KT22 0AL. 01372 377332; 24-hr emergline 09061 800 132 (50p/min); fax 01372 375183; e-mail: wildlife@pncl.co.uk
www.wildlife-aid.org.uk/wildlife
Registered charity. Wildlife hospital and rehabilitation centre helping all native British species. Special housing for birds of prey. Membership scheme and fund raising activities. Veterinary support.

Wildlife Trust
SURREY WILDLIFE TRUST. (1959; 7500). School Lane, Pirbright, Woking, Surrey, GU24 0JN. 01483 795440; fax 01483 486505; e-mail: info@swtcs.co.uk
www.surreywildlifetrust.co.uk

SUSSEX

Bird Atlas/Avifauna
The Birds of Selsey Bill and the Selsey Peninsular (a checklist to year 2000). From: Mr O Mitchell, 21 Trundle View Close, Barnham, Bognor Regis, PO22 0JZ.

Birds of Sussex ed by Paul James (Sussex Ornithological Society, 1996).

Fifty Years of Birdwatching, a celebration of the acheivements of the Shoreham District OS from 1953 onwards. from Shoreham District Ornithological Society, 7 Berberis Court, Shoreham by Sea, West Sussex BN43 6JA. £15 plus £2.50 p&p.

Bird Recorder
Mr CW Melgar, 36 Victoria Road, Worthing, W Sussex, BN11 1XB. 01903 200064;
e-mail: cwmelgar@yahoo.com

Bird Reports
BIRDS OF RYE HARBOUR NR ANNUAL REPORT (1977-published every 5 years), from Dr Barry Yates, see Clubs.

PAGHAM HARBOUR LOCAL NATURE RESERVE ANNUAL REPORT, from Warden, see Reserves,

SHOREHAM DISTRICT ORNITHOLOGICAL SOCIETY ANNUAL REPORT (1952-) - back issues available, from The Secretary, 01273 452 497.

SUSSEX BIRD REPORT (1963-), from J E Trowell, Lorrimer, Main Road, Icklesham, Winchelsea, E Sussex, TN36 4BS. e-mail: membership@susos.org.uk
www.susos.org.uk

COUNTY DIRECTORY

BTO Regional Representative
Dr A Barrie Watson, 83 Buckingham Road, Shoreham-by-Sea, W Sussex, BN43 5UD. 01273 452472; e-mail: barrie@watsonbarrie.wanadoo.co.uk

Clubs
FRIENDS OF RYE HARBOUR NATURE RESERVE. (1973; 1600). Dr Barry Yates, 2 Watch Cottages, Nook Beach, Winchelsea, E Sussex, TN36 4LU. 01797 223862; e-mail: yates@clara.net
www.naturereserve.ryeharbour.org
Meetings: Monthly talks in winter, monthly walks all year.

HENFIELD BIRDWATCH. (1999; 110). Mike Russell, 31 Downsview, Small Dole, Henfield, West Sussex, BN5 9YB. 01273 494311;
e-mail: mikerussell@sussexwt.org.uk

S
D SHOREHAM DISTRICT
O ORNITHOLOGICAL SOCIETY.
S (1953; 120). SDOS Membership administrator, 7 Berberis Court, Shoreham By Sea, West Sussex BN43 6JA.
www.sdos.org
Meetings: 7.30pm, 1st Tuesday of the month, St Peter's Church Hall, Shoreham-by-Sea.

SUSSEX ORNITHOLOGICAL SOCIETY. (1962; 1612). Mr Richard Cowser, Beavers Brook, The Thatchway, Angmering, BN16 4HJ. 01903 770259;
e-mail:secretary@sos.org.uk
www.sos.org.uk

Ringing Groups
BEACHY HEAD RS. R D M Edgar, 32 Hartfield Road, Seaford, E Sussex BN25 4PW.

CUCKMERE RG. Tim Parmenter, 22 The Kiln, Burgess Hill, W Sussex, RH15 0LU. 01444 236526.

RYE BAY RG. P Jones, Elms Farm, Pett Lane, Icklesham, Winchelsea, E Sussex, TN36 4AH. 01797 226374; e-mail: phil@wetlandtrust.org

STEYNING RINGING GROUP. B R Clay, 30 The Drive, Worthing, W Sussex, BN11 5LL. e-mail:
brian.clay@ntlworld.com

RSPB Local Groups
BATTLE. (1973; 100). Miss Lynn Jenkins, 61 Austen Way, Guestling, Hastings, E Sussex, TN35 4JH. 01424 432076; e-mail: battlerspb@freewire.co.uk
www.battlerspb.freewire.co.uk
Meetings: 7.45pm, last Tuesday of the month, St Mary's Church Hall, Battle.

BRIGHTON & DISTRICT. (1974; 350). Marion Couldery, 81 Hove Park Road, Hove, East Sussex BN3 6LN. e-mail: MarionCouldery@aol.com
www.rspb.port5.com

Meetings: 7.30pm, 4th Thursday of the month, All Saints Church Hall, Eaton Road, Hove.

CHICHESTER & SW SUSSEX. (1979; 245). Mr R Storkey, 216 Goring Road, Goring by Sea, Worthing, W Sussex BN12 2PQ.

CRAWLEY & HORSHAM. (1978; 148). Andrea Saxton, 104 Heath Way, Horsham, W Sussex, RH12 5XS. 01403 242218.
Meetings: 8.00pm, 3rd Wednesday of the month (Sept-Apr), The Friary Hall, Crawley.

EAST GRINSTEAD. (1998; 185). Nick Walker, 14 York Avenue, East Grinstead, W Sussex, RH19 4TL. 01342 315825.
Meetings: 8.00pm, last Wednesday of the month, Larec Parish Hall, De La Ware Road, East Grinstead.

EASTBOURNE & DISTRICT. (1993; 520). David Jode, 01323 422368; e-mail: david@parkmove.com

HASTINGS & ST LEONARDS. (1983; 110). Richard Prebble, 1 Wayside, 490 Sedlescombe Road North, St Leonards-on-Sea, E Sussex, TN37 7PH. 01424 751790.
Meetings: 7.30pm, 3rd Friday of the month, Taplin Centre, Upper Maze Hill.

Wildlife Hospital
BRENT LODGE BIRD & WILDLIFE TRUST. Penny Cooper, Brent Lodge, Cow Lane, Sidlesham, Chichester, West Sussex, PO20 7LN. 01243 641672. All species of wild birds and small mammals. Full surgical and medical facilities (inc. X-ray). Purpose-built oiled bird washing unit. Veterinary support.

Wildlife Trust
SUSSEX WILDLIFE TRUST. (1961; 11000). Woods Mill, Shoreham Road, Henfield, W Sussex, BN5 9SD. 01273 492630; fax 01273 494500;
e-mail: enquiries@sussexwt.org.uk
www.sussexwt.org.uk

TYNE & WEAR

Bird Recorders
See Durham; Northumberland.

Bird Report See Durham; Northumberland.

Clubs
NATURAL HISTORY SOCIETY OF NORTHUMBRIA. (1829; 900). David C Noble-Rollin, Hancock Museum, Barras Bridge, Newcastle upon Tyne, NE2 4PT. 0191 232 6386; e-mail: nhsn@ncl.ac.uk

NORTHUMBERLAND & TYNESIDE BIRD CLUB. (1958; 270). Sarah Barratt, 3 Haydon Close, Red House Farm, Gosforth, Newcastle upon Tyne, NE3 2BY. 0191 213 6665.

ENGLAND

RSPB Local Groups
NEWCASTLE UPON TYNE. (1969; 250). Brian
Moorhead, 07903 387429;
e-mail: NcastleRSPBgroup@btinternet.com
Meetings: Meetings held at Quaker Meeting House, 1
Archbold Terrace, Newcastle upon tyne.

Wetland Bird Survey Organiser
NORTHUMBERLAND COAST. Roger Norman, 1
Prestwick Gardens, Kenton, Newcastle-upon-Tyne,
NE3 3DN. (H)01912 858314;
e-mail: r.norman@clara.net

WARWICKSHIRE

Bird Recorder
Jonathan Bowley, 17 Meadow Way, Fenny Compton,
Southam, Warks, CV47 2WD. 01295 770069;
e-mail: warwks-recorder@westmidlandbirdclub.com

Bird Report See West Midlands.

BTO Regional Representatives
WARWICKSHIRE. Mark Smith, Tel: 01926 735 398;
e-mail: mark.smith36@ntlworld.com

RUGBY. Barrington Jackson, 5 Harris Drive, Rugby,
Warks, CV22 6DX. 01788 814466;
e-mail: jacksonbj2@aol.com

Clubs
NUNEATON & DISTRICT
BIRDWATCHERS' CLUB. (1950;
87). Alvin K Burton, 23 Redruth
Close, Horeston Grange,
Nuneaton, Warwicks, CV11 6FG.
024 7664 1591.
Meetings: 7.30pm, 3rd Thursday
of the month (Sep-May), Hatters
Space Community Centre, Upper Abbey Street,
Nuneaton.

Ringing Groups
ARDEN RG. Roger J Juckes, 24 Croft Lane, Temple
Grafton, Alcester, Warks B49 6PA. 01789 778748.

BRANDON RG. David Stone, Overbury, Wolverton,
Stratford-on-Avon, Warks, CV37 0HG. 01789 731488.

RSPB Local Group
See West Midlands.

Wildlife Trust
WARWICKSHIRE WILDLIFE TRUST. (1970; 7000).
Brandon Marsh Nature Centre, Brandon Lane,
Coventry, CV3 3GW. 024 7630 2912; fax 024 7663
9556; e-mail: enquiries@wkwt.org.uk
www.warwickshire-wildlife-trust.org.uk

WEST MIDLANDS

Bird Atlas/Avifauna
The New Birds of the West Midlands edited by
Graham Harrison et al (West Midland Bird Club, 2005).

Bird Recorder
Tim Hextell, 39 Windermere Road, Handsworth,
Birmingham, B21 9RQ. 0121 551 9997;
e-mail: west-mids-recorder@westmidlandbirdclub.com
www.westmidlandbirdclub.com

Bird Reports
THE BIRDS OF SMESTOW VALLEY AND DUNSTALL
PARK (1988-), from Secretary, Smestow Valley Bird
Group.

WEST MIDLAND BIRD REPORT (inc Staffs, Warks,
Worcs and W Midlands) (1934-), from Mr J Reeves, 9
Hintons Coppice, Knowle, Solihull, B93 9RF. .

BTO Regional Representative
BIRMINGHAM & WEST MIDLANDS. Position vacant.

Clubs
SMESTOW VALLEY BIRD GROUP. (1988; 56). Frank
Dickson, 11 Bow Street, Bilston, Wolverhampton,
WV14 7NB. 01902 493733.

WEST MIDLAND BIRD CLUB. (1929; 2000). Mr MJ
West, 6 Woodend Road, Walsall, WS5 3BG. 01922
639931; e-mail: secretary@westmidlandbirdclub.com
www.westmidlandbirdclub.com

WEST MIDLAND BIRD CLUB (BIRMINGHAM BRANCH).
(1995; 800). Martin Kenrick;
e-mail: birmingham@westmidlandbirdclub.com
www.westmidlandbirdclub.com/birmingham
Meetings: Usually last Tuesday of the month (Oct-
Apr), Birmingham Medical Institute, Harborne Road,
Edgbaston, near Five Ways

WEST MIDLAND BIRD CLUB (SOLIHULL BRANCH). Jim
Winsper, 32 Links Road, Hollywood, Birmingham, B14
4TP. e-mail: solihull@westmidlandbirdclub
www.westmidlandbirdclub.com/solihull

Ringing Groups
MERCIAN RG (Sutton Coldfield). R L Castle, 91 Maney
Hill Road, Sutton Coldfield, West Midlands, B72 1JT.
0121 686 7568.

RSPB Local Groups
BIRMINGHAM. (1975; 100). John Bailey, 52 Gresham
Road, Hall Green, Birmingham, B28 0HY. 0121 777
4389. www.rspb-birmingham.org.uk
Meetings: 7.30pm, 3rd Thursday of the month (Sep-
Jun), Salvation Army Citadel, St Chads, Queensway,
Birmingham.

ENGLAND

COVENTRY & WARWICKSHIRE. (1969; 130). Alan King, 69 Westmorland Road, Coventry, CV2 5BO. 024 7672 7348.

SOLIHULL. (1983; 2600). John Gibney, 01564 775 878; e-mail: john.gibney@virgin.net
Meetings: Thurday afternoons, Solihull Methodist Church, Blossomfield Road, Solihull.

STOURBRIDGE. (1978; 150). Paul Banks, 4 Sandpiper Close, Wollescote, Stourbridge, DY9 8TD. 01384 898948; e-mail: picapica@tinyworld.co.uk
Meetings: 7.30pm, 2nd Wednesday of the month (Sep-May), Stourbridge Town Hall, Crown Centre, Stourbridge.

SUTTON COLDFIELD. (1986; 250). Joanna Bazen, 66 Station Road, Wylde Green, Sutton Coldfield, B73 5LA. 0121 354 5626; e-mail: jobazen@yahoo.co.uk
Meetings: 7.30pm, 1st Monday of the month, Bishop Vesey's Grammer School.

WALSALL. (1970). Mike Pittaway, 2 Kedleston Close, Bloxwich, Walsall WS3 3TW. 01922 710568; e-mail: chair@rspb-walsall.org.uk
www.rspb-walsall.org.uk
Meetings: 7.30pm, 3rd Wednesday of the month, St Marys School, Jesson Road, Walsall.

WOLVERHAMPTON. (1974; 110). Ian Wiltshire, 25 Oakridge Drive, Willenhall, WV12 4EN. 01902 630418.
Meetings: 7.30pm, 2nd Wednesday of the month (Sept-Apr), The Newman Centre, Haywood Drive, Tettenhall, Wolverhampton.

Wildlife Hospitals
KIDD, D J. 20 Parry Road, Ashmore Park, Wednesfield, Wolverhampton, WV11 2PS. 01902 863971. All birds of prey, esp. owls. Aviaries, isolation pens. Veterinary support.

WEDNESFIELD ANIMAL SANCTUARY. Jimmy Wick, 92 Vicarage Road, Nordley, Wednesfield, Wolverhampton, WV11 1SF. 01902 823064. Birds of prey, softbills, seed-eaters. Brooders, incubators, outdoor aviaries, heated accommodation. Telephone first. Veterinary support.

Wildlife Trust
THE BIRMINGHAM AND BLACK COUNTRY WILDLIFE TRUST. (1980; 900). 28 Harborne Road, Edgbaston, Birmingham, B15 3AA. 0121 454 1199; fax 0121 454 6556; e-mail: info@bbcwildlife.org.uk
www.bbcwildlife.org.uk

WILTSHIRE

Bird Atlas/Avifauna
Birds of Wiltshire. Due early 2005.

Bird Recorder
Rob Turner, 14 Ethendun, Bratton, Westbury, Wilts, BA13 4RX. 01380 830862; e-mail: robt14@btopenworld.com

Bird Report
Published in Hobby (journal of the Wiltshire OS) (1975-), from John Osborne, 4 Fairdown Avenue, Westbury, Wiltshire BA13 3HS. 01373 864598

BTO Regional Representatives
NORTH. Position vacant.

SOUTH. Andrew Carter, Standlynch Farm, Downton, Salisbury, SP5 3QR. 01722 710382; e-mail: standlynch@aol.com

Clubs
SALISBURY & DISTRICT NATURAL HISTORY SOCIETY. (1952; 175). J Pitman, 10 The Hardings, Devizes Road, Salisbury, SP2 9LZ. 01722 327395.
Meetings: 7.30pm, 3rd Thursday of the month (Sept-Apr), Lecture Hall, Salisbury Museum, King House, The Close, Salisbury.

WILTSHIRE ORNITHOLOGICAL SOCIETY. (1974; 480). Phil Deacon, 12 Rawston Close, Nythe, Swindon, Wilts SN3 3PW. 01793 528930; e-mail: phil.deacon@tinyworld.co.uk

Ringing Group
COTSWOLD WATER PARK RG. John Wells, 25 Pipers Grove, Highnam, Glos, GL2 8NJ. e-mail: john.wells2@btinternet.com

WEST WILTSHIRE RG. Mr M.J. Hamzij, 13 Halfway Close , Trowbridge, Wilts, BA14 7HQ. e-mail: mikehamzij@halfway11.freeserve.co.uk

RSPB Local Groups
NORTH WILTSHIRE. (1973; 104). Derek Lyford, 9 Devon Road, Swindon, SN2 1PQ. 01793 520997; e-mail: derek.lyford@virgin.net
www.rspb.org.uk/groups/northwiltshire
Meetings: 7.30pm, 1st Tuesday of the month (Sep-Jun), Even Swindon Community Centre, Jennings St, Swindon SU 137 849.

SOUTH WILTSHIRE. (1986; 800). Tony Goddard, Clovelly, Lower Road, Charlton All Saints, Salisbury, SP5 4HQ. 01725 510309.
Meetings: 7.30pm, Tuesday evenings (monthly), City Hall, Salisbury.

Wildlife Hospital
CALNE WILD BIRD AND ANIMAL RESCUE CENTRE. Tom and Caroline Baker, 2 North Cote, Calne, Wilts, SN11 9DL. 01249 817893. All species of birds. Large natural aviaries (all with ponds), release areas, incubators, heated cages. Day and night collection. Veterinary support.

Wildlife Trust

WILTSHIRE WILDLIFE TRUST. (1962; 10,000). Elm Tree Court, Long Street, Devizes, Wilts, SN10 1NJ. 01380 725670; fax 01380 729017; e-mail: info@wiltshirewildlife.org www.wiltshirewildlife.org

WORCESTERSHIRE

Bird Recorder

Andy Warr, 14 Bromsgrove Street, Worcester WR3 8AR. 01905 28281; e-mail: worcs-recorder@westmidlandbirdclub.com

Bird Report See West Midlands.

BTO Regional Representative

G Harry Green MBE, Windy Ridge, Pershore Road, Little Comberton, Pershore, Worcs, WR10 3EW. 01386 710377; e-mail: harrygreen@britishlibrary.net

Ringing Group

WYCHAVON RG. J R Hodson, 15 High Green, Severn Stoke, Worcester, WR8 9JS. 01905 754919(day), 01905 371333(eve); e-mail: john.hodson@tesco.net

Club

WEST MIDLAND BIRD CLUB (KIDDERMINSTER BRANCH). Celia Barton, 28A Albert Street, Wall Heath, Kingswinford DY6 0NA. 01384 839838; e-mail: kidderminster@ westmidlandbirdclub.com **Meetings:** 7.30pm, 4th Wednesday of the month (Sep-Apr), St Oswalds Church Centre, Broadwaters, Kidderminster.

RSPB Local Group

WORCESTER & MALVERN. (1980; 300). Garth Lowe, Sunnymead, Old Storridge, Alfrick, Worcester, WR6 5HT. 01886 833362. **Meetings:** 7.30pm, 2nd Wednesday in month (Sept-May), Powick Village Hall.

Wildlife Trust

WORCESTERSHIRE WILDLIFE TRUST. (1968; 8000). Lower Smite Farm, Smite Hill, Hindlip, Worcester, WR3 8SZ. 01905 754919; fax 01905 755868; e-mail: worcswt@cix.co.uk www.worcswildlifetrust.co.uk

YORKSHIRE

Bird Atlas/Avifauna

Atlas of Breeding Birds in the Leeds Area 1987-1991

by Richard Fuller et al (Leeds Birdwatchers' Club, 1994).

The Birds of Halifax by Nick Dawtrey (only 20 left), 14 Moorend Gardens, Pellon, Halifax, W Yorks, HX2 0SD.

The Birds of Yorkshire by John Mather (Croom Helm, 1986).

*An Atlas of the Breeding Birds of the Huddersfield Area, 1987-1992.*by Brian Armitage et al (2000) - very few copies left.

Birds of Barnsley by Nick Addey (Pub by author, 114 Everill Gate Lane, Broomhill, Barnsley S73 0YJ, 1998).

Bird Recorders

VC61 (East Yorkshire) AND EDITOR. Geoff Dobbs, 12 Park Avenue, Hull, HU5 3ER. 01482 341524; e-mail: geoffdobbs@aol.com

VC62 (North Yorkshire East). Russell Slack, 18 Ruffhams Close, Wheldrake, York, YO19 6TD. 01904 449098; e-mail: russ@birdguides.com

VC63 (South & West Yorkshire). Covering the following groups - Barnsley Bird Study, Blacktoft Sands RSPB, Doncaster and District OS, Rotherham and District OS, Sheffield Bird Study and SK58 Birders. John Wint, 9 Yew Tree Park, Whitley, Goole, DN14 0NZ. 01977 662826; e-mail: john.wint@tesco.net

VC64 (West Yorkshire)/HARROGATE & CRAVEN. Jim Pewtress, 31 Piercy End, Kirbymoorside, York, YO62 6DQ. 01751 431001; e-mail: trivialis@operamail.com

VC65 (North Yorkshire West). Steve Worwood, 18 Coltsgate Hill, Ripon HG4 2AB. 01765 602518; e-mail: s.worwood@bronco.com

Bird Reports

BARNSLEY & DISTRICT BIRD STUDY GROUP REPORT (1971-), from Secretary.

BRADFORD NATURALISTS' SOCIETY ANNUAL REPORT, from Mr I Hogg, 23 St Matthews Road, Bankfoot, Bradford, BD5 9AB. 01274 727902.

BRADFORD ORNITHOLOGICAL GROUP REPORT (1987-), from Jenny Barker, 4 Chapel Fold, Slack Lane, Oakworth, Keighley, BD22 0RQ.

DONCASTER BIRD REPORT (1955-), from Mr M Roberts, 30 St Cecillia's Road, Belle Vue, Doncaster, DN4 5EG. 01302 361731.

FILEY BRIGG BIRD REPORT (1976-), from Mr C Court, 12 Pinewood Avenue, Filey, YO14 9NS.

HALIFAX BIRDWATCHERS' CLUB ANNUAL REPORT (1991-), from Nick C Dawtrey, 14 Moorend Gardens, Pellon, Halifax, W Yorks, HX2 0SD. 01422 364228.

*HARROGATE & DISTRICT NATURALISTS'
ORNITHOLOGY REPORT (1996-)*, from Secretary.

HULL VALLEY WILDLIFE GROUP REPORT (2000-)
incorporating Tophill Low recording area). from Geoff
Dobbs, 12 Park Avenue, Hull, HU5 3ER. 01482
341524; e-mail: geoffdobbs@aol.com

BIRDS IN HUDDERSFIELD (1966-), from Mr Brian
Armitage, 106 Forest Road, Dalton, Huddersfield HD5
8ET.01484 305054;
e-mail: brian.armitage@ntlworld.com

*LEEDS BIRDWATCHERS' CLUB ANNUAL REPORT
(1949-)*, from Secretary.

BIRDS OF ROTHERHAM (1975-), from Secretary,
Rotherham Orn Soc, www.rotherhambirds.co.uk
(check website for current publication details).

BIRDS IN THE SHEFFIELD AREA (1973-), from Tony
Morris, 4A Raven Road, Sheffield, S7 1SB. e-mail:
ajmo@blueyonder.co.uk
www.sbsg.org

THE BIRDS OF SK58 (1993-), from Secretary, SK58
Birders.

SPURN BIRD OBSERVATORY ANNUAL REPORT, from
Warden, see Reserves.

WINTERSETT AREA ANNUAL REPORT (1988-), from
Steve Denny, 13 Rutland Drive, Crofton, Wakefield,
WF4 1SA.01924 864487.

*YORK ORNITHOLOGICAL CLUB ANNUAL REPORT
(1970-)*, from Peter Watson, 1 Oak Villa, Hodgson
Lane, Upper Poppleton, York YO26 6EA. 01904
795063; e-mail: peterewwatson@aol.com.

*YORKSHIRE NATURALISTS' UNION: BIRD REPORT
(1940-)*, from John A Newbould, Stonecroft, 3
Brookmead Close, Sutton Poyntz, Wemouth, Dorset,
DT3 6RS.

**BTO Regional Representatives & Regional
Development Officers**
NORTH-EAST RR. Michael Carroll, 01751 476550.

NORTH-WEST RR. Gerald Light, 01756 753720.

SOUTH-EAST AND SOUTH-WEST RR. Chris Falshaw, 6
Den Bank Crescent, Sheffield, S10 5PD. 0114 230
3857; e-mail: chris@falshaw.f9.co.uk

EAST RR. Position vacant.

BRADFORD RR & RDO. Mike L Denton, 77 Hawthorne
Terrace, Crosland Moor, Huddersfield, HD4 5RP. 01484
646990.

HARROGATE RR. Mike Brown, 48 Pannal Ash Drive,

Harrogate, N Yorks, HG2 0HU. H:01423 567382;
W:01423 507237;
e-mail: mike@thebrownsathome.plus.com

HULL RR. Martin Chadwick, 01482 653 391;
e-mail: martin_chadwick@hotmail.com.

RICHMOND RR. John Edwards, 7 Church Garth, Great
Smeaton, Northallerton, N Yorks, DL6 2HW. H:01609
881476; W:01609 780780 extn 2452;
e-mail: john@garthwards.fsnet.co.uk

YORK RR. Rob Chapman, 12 Moorland Road, York,
YO10 4HF. 01904 633558;
e-mail: robert.chapman@tinyworld.co.uk

Clubs
BARNSLEY BIRD STUDY GROUP. (1970; 35). Christine
Carr, 300 Higham Common Road, Higham, Barnsley,
South Yorkshire S75 1PS. 01226 384694.
Meetings: 7.15pm, 1st Thursday in the month (Nov-
Mar), Old Moor Wetland Centre, Wombwell.

BRADFORD NATURALISTS' SOCIETY. (1875; 30). D R
Grant, 19 The Wheatings, Ossett, W Yorks, WF5 0QQ.
01924 273628.
Meetings: 7.30pm, Mondays, Richmond Building,
University of Bradford.

BRADFORD ORNITHOLOGICAL GROUP. (1987; 180).
Shaun Radcliffe, 8 Longwood Avenue, Bingley, W
Yorks, BD16 2RX. 01274 770960;
www.bradfordbirding.org

CASTLEFORD & DISTRICT NATURALISTS' SOCIETY.
(1956; 20). Michael J Warrington, 31 Mount Avenue,
Hemsworth, Pontefract, W Yorks, WF9 4QE. 01977
614954; e-mail:
michael@warrington31mount.freeserve.co.uk.
Meetings: 7.30pm, Tuesdays monthly (Sep-Mar),
Whitwood College, Castleford (check with above for
dates).

DONCASTER & DISTRICT ORNITHOLOGICAL SOCIETY.
(1955; 40). D Hazard, 01302 788044;
e-mail: secretary@birdingdoncaster.org.uk

FILEY BRIGG ORNITHOLOGICAL
GROUP. (1977; 70). Jack
Whitehead, 15 The Beach,
Filey, N Yorkshire, YO14 9LA.
01723 514565.

HALIFAX BIRDWATCHERS' CLUB. (1992). Nick C
Dawtrey, 14 Moorend Gardens, Pellon, Halifax, W
Yorks, HX2 0SD. 01422 364228.

HARROGATE & DISTRICT NATURALISTS' SOCIETY.
(1947; 400). Mrs J McClean, 6 Rossett Park Road,
Harrogate, N Yorks, HG2 9NP. 01423 879095;

e-mail: joan_mcclean@hotmail.com
Meetings: 7.45pm, Wednesday fornightly from 12/
10/05, St Robert's Centre, Harrogate.

HORNSEA BIRD CLUB. (1967; 35). John Eldret, 44
Rolston Road, Hornsea, HU18 1UH. 01964 532854.
Meetings: 7.30pm, 3rd Friday of the month (Sep-
Mar), Hornsea Library.

HUDDERSFIELD BIRDWATCHERS' CLUB. (1966; 80).
Chris Abell, 57 Butterley Lane, New Mill, Holmfirth, HD9
7EZ. 01484 862006; e-mail: chrisabell57@aol.com
www.HuddersfieldBirdwatchersClub@groups.msn.com
Meetings: 7.30pm, Tuesday's fortnightly, Children's
Library (section), Huddersfield Library and Art Gallery,
Princess Alexandra Walk, Huddersfield.

HULL VALLEY WILDLIFE
GROUP. (1997; 175). The
Secretary, Roy Lyon, 670
Hotham Road South, Hull
HU5 5LE. www.hvwg.co.uk

LEEDS BIRDWATCHERS' CLUB. (1949; 60). Mrs Shirley
Carson, 2 Woodhall Park Gardens, Stanningley,
Pudsey, W Yorks, LS28 7XQ. 0113 255 2145; e-mail:
shirley.carson@care4free.net
Meetings: 7.15pm Monday fortnightly, Quaker
Meeting House, Wordhouse Lane, Leeds.

NEW SWILLINGTON INGS BIRD GROUP. (1989; 36).
Nick Smith, 40 Holmsley Lane, Woodlesford, Leeds,
LS26 8RN. 0113 282 6154;
e-mail: nick@woodlesford.fsbusiness.co.uk.
Meetings: 7.30pm, 1st Thursday of even months,
Two Pointers Inn, Woodlesford, Leeds.

ROTHERHAM & DISTRICT ORNITHOLOGICAL
SOCIETY. (1974; 90). Malcolm Taylor, 18 Maple Place,
Chapeltown, Sheffield, S35 1QW. 0114 246 1848.
http://members.lycos.co.uk/RDOS8
Meetings: 7.30pm, 2nd Friday of the month.

SCALBY NABS ORNITHOLOGICAL GROUP. (1993; 15).
R.N.Hopper (Membership Secretary), 10A Ramshill
Road, Scarborough, N Yorkshire, YO11 2QE. 01723
369537. www.scarborough-birding.org.uk

SHEFFIELD BIRD STUDY GROUP. (1972; 160). Richard
Dale, 109 Main Road, Wharncliffe Side, Sheffield S35
0DP. 0114 286 2513; e-mail: richarddale9@hotmail.com
Meetings: 7.15pm, 2nd Wednesday of the month
(Sep-Jun), Lecture Theatre 5, Sheffield University Arts
Tower.

SK58 BIRDERS. (1993; 60). Andy Hirst, 15 Hunters
Drive, Dinnington, Sheffield, S25 2TG. 07947 068125;
e-mail: sk58birders@sk58.freeserve.co.uk
www.sk58.freeserve.co.uk

Chair: Mick Clay, 2 High St, S.Anston, Sheffield. 01909
566000.
Meetings: 7.30pm, last Wednesday of the month
(except Aug), Upstairs Room, Loyal Trooper pub,
South Anston.

SORBY NHS (ORNITHOLOGICAL SECTION). (1918;
40). Mr R Butterfield, General Secretary, 159 Bell Hagg
Road, Sheffield S6 5DA.
e-mail: secretary@sorby.org.uk
www.sorby.org.uk

WAKEFIELD NATURALISTS' SOCIETY. (1851; 40). Philip
Harrison, 392 Dewsbury Road, Wakefield, W Yorks,
WF2 9DS. 01924 373604.
Meetings: 7.30pm, 2nd Tuesday of the month (Sep-
Apr), Friends Meeting House, Thornhill Street,
Wakefield.

YORK ORNITHOLOGICAL CLUB. (1967; 80). Ian
Traynor, The Owl House, 137 Osbaldwick Lane, York,
YO10 3AY. e-mail: info@yorkbirding.org.uk
www.yorkbirding.org.uk
Meetings: 7.30pm, 1st Tuesday of the month,
Friends' Meeting House, Friargate, York (see website).

YORKSHIRE NATURALISTS' UNION (Ornithological
Section). (1940; 500). W F Curtis, Farm Cottage,
Atwick, Driffield, YO25 8DH. 01964 532477.
www.ynu.org.uk

Ringing Groups
BARNSLEY RG. M C Wells, 715 Manchester Road,
Stocksbridge, Sheffield, S36 1DQ. 0114 288 4211.

DONCASTER RG. D Hazard, 41 Jossey Lane,
Scawthorpe, Doncaster, S Yorks, DN5 9DB. 01302
788044; e-mail: davehazard@btopenworld.com

EAST DALES RG. S P Worwood, 18 Coltsgate Hill,
Ripon, N Yorks, HG4 2AB.

EAST YORKS RG. Peter J Dunn, 43 West Garth
Gardens, Cayton, Scarborough, N Yorks, YO11 3SF.
01723 583149; e-mail: pjd@fbog.co.uk

SORBY-BRECK RG. Geoff P Mawson, Moonpenny Farm,
Farwater Lane, Dronfield, Sheffield, S18 1RA. 01246
415097; e-mail: gpmawson@hotmail.com

SOUTH CLEVELAND RG. W Norman, 2 Station
Cottages, Grosmont, Whitby, N Yorks, YO22 5PB.
01947 895226; e-mail: wilfgros@lineone.net

SPURN BIRD OBSERVATORY. I D Walker, 31 Walton
Park, Pannal, Harrogate, N Yorks, HG3 1EJ. 01423
879408.

TEES RG. E Wood, Southfields, 16 Marton Moor Road,
Nunthorpe, Middlesbrough, Cleveland, TS7 0BH.
01642 323563; e-mail: redshank@ntlworld.co.uk

ENGLAND

WINTERSETT RG. P Smith, 16 Templar Street, Wakefield, W Yorks, WF1 5HB. 01924 375082.

RSPB Local Groups

AIREDALE AND BRADFORD. (1972; 3500 in catchement area). Ruth Porter, 01484 868415; e-mail: Kirsten.whittaker@RSPB.org.uk
Meetings: 7.30pm, monthly on Fridays, Room 3, Shipley Library.

CLEVELAND. (1974; 200). Mark Stokeld, 38 Ash Grove, Kirklevington, Cleveland, TS15 9NQ. 01642 783819; e-mail: mark@stokeld.demon.co.uk
www.stokeld.demon.co.uk

CRAVEN & PENDLE. (1986; 250). Ian Cresswell, Dove House, Skyreholme, Skipton, N Yorks, BD23 6DE. 01756 720355; e-mail: ian@cravenandpendlerspb.org
www.cravenandpendlerspb.org

DONCASTER. (1984; 115). Sue Clifton, West Lodge, Wadworth Hall Lane, Wadworth, Doncaster, DN11 9BH. Tel/fax 01302 854956; e-mail: sue@wlwad.wanadoo.co.uk
Meetings: 7.30pm 2nd Wednesday of the month (Sept-May), contact Sue Clifton for venue.

EAST YORKSHIRE. (1986;140). Trevor Malkin, 49 Taylors Field, Driffield, E Yorks, YO25 6FQ. 01377 257325. www.eymg.freeserve.co.uk

HUDDERSFIELD & HALIFAX. (1981; 200). David Hemingway, 267 Long Lane, Dalton, Huddersfield, HD5 9SH. 01484 301920; e-mail: d.hemingway@ntlworld.com

LEEDS. (1974; 450). Linda Jenkinson, 112 Eden Crescent, Burley, Leeds, LS4 2TR. 0113 230 4595 www.rspb-leeds.ndo.co.uk
Meetings: 7.30pm, 3rd Wednesday of the month (Sep-Apr), Lecture Theatre B, School of Mechanical Engineering, University of Leeds.

SHEFFIELD. (1981; 500). John Badger, 24 Athersley Gardens, Owlthorpe, Sheffield, S20 6RW. 0114 247 6622; www.rspb-sheffield.org.uk

Meetings: 7.30pm 1st Thursday of the month (Sept-May), Central United Reformed Church, Norfolk St, Sheffield.

WAKEFIELD. (1987; 150). Bob Coursey, 21 Greenside, Walton, Wakefield, West Yorkshire WF2 6NN. 01924 256289.
Meetings: 7.30pm, 4th Thursday of the month (Sep-Apr), Ossett War Memorial community Centre, Prospect Road, Ossett.

WHITBY. (1977; 120). John Woolley, 01947 604505. Meetings: St. Johns Ambulance Hall, St. Hildas Terrace, Whitby.

YORK. (1973; 600). Chris Lloyd, 7 School Lane, Upper Poppleton, York, YO26 6JS. 01904 794865; e-mail: chris.a.lloyd@care4free.net
www.yorkrspb.org.uk
Meetings: 7.30pm, Tues, Wed or Thurs, Temple Hall, York St John College, Lord Mayors Walk, York.

Wldlife Hospital

ANIMAL HOUSE WILDLIFE WELFARE. Mrs C Buckroyd, 14 Victoria Street, Scarborough, YO12 7SS. 01723 371256; shop 01723 375162. All species of wild birds. Oiled birds given treatment before forwarding to cleaning stations. Incubators, hospital cages, heat pads, release sites. Birds ringed before release. Prior telephone call requested. Collection if required. Veterinary support. Charity shop at 127 Victoria Road.

Wildlife Trusts

TEES VALLEY WILDLIFE TRUST. (1979; 4000). Bellamy Pavilion, Kirkleatham Old Hall, Kirkleatham, Redcar, Cleveland, TS10 5NW. 01642 759900; fax 01642 480401; e-mail: teesvalleywt@cix.co.uk
www.wildlifetrust.org.uk/teesvalley

SHEFFIELD WILDLIFE TRUST. (1985; 250). 37 Stafford Road, Sheffield, S2 2SF. 0114 263 4335 ; fax 0114 263 4345; e-mail: mail@wildsheffield.com
www.wildsheffield.com

YORKSHIRE WILDLIFE TRUST. (1946; 8000). 10 Toft Green, York, YO1 6JT. 01904 659570; fax 01904 613467; e-mail: yorkshirewt@cix.co.uk
www.yorkshire-wildlife-trust.org.uk

SCOTLAND

For this section we are following the arrangement of the Scottish recording areas as set out by the Scottish Ornithologists' Club.

Bird Report
See Scottish Ornithologists' Club in National Directory

Club
See Scottish Ornithologists' Club in National Directory.

ANGUS & DUNDEE

Bird Recorder
ANGUS & DUNDEE. Graham Christer, 1 Balzeordie Cottages, Menmuir, by Brechin DD9 7RQ.

Bird Report
ANGUS & DUNDEE BIRD REPORT (1974-), From Secretary, SOC Tayside Branch.

BTO Regional Representatives & Regional Development Officer
ANGUS RR & RDO. Ken Slater, Braedownie Farmhouse, Glen Clova, Kirriemuir, Angus, DD8 4RD. 01575 550233

Clubs
ANGUS & DUNDEE BIRD CLUB. (1997; 185). Bob McCurley, 22 Kinnordy Terrace, Dundee,DD4 7NW. 01382 462944;
e-mail: lunanbay@onetel.com
www.angusbirding.homestead.com
Meetings: 7.30pm, Tuesdays, Montrose Basin Wildlife Centre.

SOC TAYSIDE BRANCH. (145). James Whitelaw, 36 Burn Street, Dundee,DD3 0LB. 01382 819391.

Ringing Group
TAY RG. Ms S Millar, Edenvale Cottage, 1 Lydox Cottages, Dairsie, Fife, KY15 4RN.
e-mail: les@lydox.fsnet.co.uk

RSPB Members' Groups
DUNDEE. (1972;110). Ron Downing, 3 Lynnewood Place, Dundee,DD4 7HB. 01382 451987

Wetland Bird Survey Organisers
ANGUS INLAND (Excluding Montrose Basin). Graham Christer, The Ivy, 8 West Hemming Street, Letham, Forfar, Angus DD8 2PU.

MONTROSE BASIN. The Warden, SWT, Montrose Basin Wildlife Centre, Rossie Braes, Montrose, DD10 9JT. 01674 676336; e-mail montrosebasin@swt.org.uk

ARGYLL

Bird Recorder
ARGYLL. Paul Daw, Tigh-na-Tulloch, Tullochgorm, Minard, Argyll, PA32 8YQ. 01546 886260;
e-mail: monedula@globalnet.co.uk

Bird Reports
ARGYLL BIRD REPORT (1984-), FromDr Bob Furness, The Cnoc, Tarbet, Dunbartonshire G83 7DG.01301 702603; e-mail: r.furness@bio.gla.ac.uk

MACHRIHANISH SEABIRD OBSERVATORY REPORT (1992-), From Observatory, see Reserves & Observatories.

BTO Regional Representatives
ARGYLL (MULL, COLL, TIREE AND MORVERN). Sue Dewar, 01680 812594;
e-mail: sue@wingsovermull.fsnet.co.uk

ARGYLL MAINLAND, BUTE, GIGHA AND ARRAN. Richard Allen, E-MAIL: richardallan@compuserve.com

ISLAY, JURA, COLONSAY RR. Dr Malcolm Ogilvie, Glencairn, Bruichladdich, Isle of Islay, PA49 7UN. 01496 850218; e-mail: maogilvie@indaal.demon.co.uk

Club
ARGYLL BIRD CLUB. (1983;270). Sue Furness, The Cnoc, Tarbet, Argyll, G83 7DG.01301 702603;
e-mail: r.furness@bio.gla.ac.uk
www.argyllbirdclub.org

ISLE OF MULL BIRD CLUB. (2001;142), Len White, Ard Dochas, Lochdon, Isle of mull, ArgyllPA64 6AP.01680 812335;
e-mail: arddochas@aol.com
www.mullbirds.com
Meetings: 7.30pm, Craignure Village Hall.

Ringing Group
TRESHNISH AUK RG. S W Walker, Snipe Cottage, Hamsterley, Bishop Auckland, Co Durham, DL13 3NX.
e-mail: snipe@snipe.screaming.net

HELENSBURGH. (1975; 62). Steve Chadwin, 01436 670158.

AYRSHIRE

Bird Recorder
AYRSHIRE. Angus Hogg, 11 Kirkmichael Road, Crosshill, Maybole, Ayrshire, KA19 7RJ.
e-mail: recorder@ayrshire-birding.org.uk

SCOTLAND

Bird Reports
AYRSHIRE BIRD REPORT (1976-), From Recorder or Dr RG Vernon, 29 Knoll Park, Ayr KA7 4RH.

BTO Regional Representatives
AYRSHIRE RR. Brian Broadley, 01290 424 241;
e-mail: maggie_broadley@hotmail.com

Club
SOC AYRSHIRE. (1962; 154). Henry Martin, 9 Shawfield Avenue, Ayr, KA7 4RE. 01292 442086; www.ayrshire-birding.org.uk
Meetings: 7.30pm, Tuesdays monthly, Monkton Community Church, Monkton by Prestwick.

RSPB Members' Groups
CENTRAL AYRSHIRE. (1978; 50). Tony Scott, 01292 281085; E-mail: dascott@zoom.co.uk
www.ayrshire-birding.org.uk

NORTH AYRSHIRE. (1976; 180). Duncan Watt, 28 Greenbank, Dalry, Ayrshire, KA24 5AY.
www.narspb.org.uk
Meetings: 7.30pm, various Fridays (Aug-Apr), Ardrossan Civic Centre, open to all. Full list available.

Wetland Bird Survey Organiser
Mr David Grant, 16 Thorn Avenue, Coylton, Ayr KA6 6NL. (H)01292 570491; e-mail: d.grant@au.sac.as.uk

Wildlife Hospital
HESSILHEAD WILDLIFE RESCUE CENTRE. Gay & Andy Christie, Gateside, Beith, Ayrshire, KA15 1HT. 01505 502415; e-mail: info@hessilhead.org.uk
www.hessilhead.org.uk
All species. Releasing aviaries. Veterinary support. Visits only on open days please.

BORDERS

Bird Atlas/Avifauna
The Breeding Birds of South-east Scotland, a tetrad atlas 1988-1994 by R D Murray et al. (Scottish Ornithologists' Club, 1998).

Bird Recorder
Ray Murray, 4 Bellfield Crescent, Eddleston, Peebles, EH45 8RQ. 01721 730677;
e-mail: ray.d.murray@ukgateway.net

Bird Report
BORDERS BIRD REPORT (1979-), From Malcolm Ross, The Tubs, Dingleton Road, Melrose, Borders TD6 9QP. 01896 822132; e-mail: elise.ross@virgin.net

BTO Regional Representative & Regional Development Officer
RR. Alex Copland; e-mail: crex@eircom.net

Club
SOC BORDERS BRANCH. (90). Vicky McLellan, 18 Glen Crescent, Peebles, EH45 9BS. 01721 724580.
Meetings: 7.30pm usually 2nd Monday of the month, Scott Room, George and Abbotsford Hotel, Melrose.

Ringing Group
BORDERS RG. (1991; 10)Dr T W Dougall, 38 Leamington Terrace, Edinburgh,EH10 4JL. Office tel 0131 469 5557; Office fax 0131 469 5599

RSPB Members' Group
BORDERS. (1995; 94). Jim Stillie, 01750 20660.
Meetings: The Corn Exchange, Market Square, Melrose.

CAITHNESS

Bird Recorders
CAITHNESS. Julian Smith (Report Editor), St John, Broch, Dunnet, Caithness;
e-mail: designsmith@madasafish.com

Bird Reports
CAITHNESS BIRD REPORT (1983-97), From Julian Smith, St John's, Brough, Dunnet, Caithness; e-mail: designsmith@madasafish.com

BTO Regional Representatives & Regional Development Officers
CAITHNESS. Position vacant.

Clubs
SOC CAITHNESS BRANCH. (51). Stan Laybourne, Old Schoolhouse, Harpsdale, Halkirk, Caithness, KW12 6UN. 01847 841244;
e-mail:stanlaybourne@talk21.com

CLYDE

Bird Atlas/Avifauna
*A Guide to Birdwatching in the Clyde Area (2001)*by Cliff Baister and Marin Osler (Scottish Ornithologists' Club, Clyde branch).

Clyde Breeding Bird Atlas (working title). In preparation.

Bird Recorder
CLYDE. Iain P Gibson, 8 Kenmure View, Howwood, Johnstone, Renfrewshire, PA9 1DR. 01505 705874; e-mail:iain.gibson@land.glasgow.gov.uk

CLYDE ISLANDS. Bernard Zonfrillo, 28 Brodie Road, Glasgow,G21 3SB. 0141 557 079;
e-mail:b.zonfrillo@bio.gla.ac.uk

Bird Reports
CLYDE BIRDS (1973-), From Jim & Valerie Wilson, 76 Laigh Road, Newton Mearns, Glasgow, G77 5EQ.
e-mail: jim.val@btinternet.com

BTO Regional Representatives
LANARK, RENFREW, DUMBARTON. John Knowler, 0141 584 9117; e-mail: john.knowler@ntlworld.com

Club
SOC CLYDE BRANCH. (300). Sandra Hutchinson, 52 Station Road, Bearsden, Glasgow, G61 4AL. 0141 943 1816; e-mail: hutchinson_80@hotmail.com

Ringing Groups

CLYDE RG. (1979; 18)I Livingstone, 57 Strathview Road, Bellshill, Lanarkshire, ML4 2UY.01698 749844; e-mail: iainlivcrg@blueyonder.co.uk

RSPB Members' Groups

GLASGOW. (1972;141). Jim Coyle, 6 Westerlands, Anniesland, Glasgow, G12 0FB. 0141 579 7565; e-mail: j.coyle13@ntlworld.com
Meetings: 7.30pm, generally 1st Wednesday of the month (Sep-Apr), Woodside Halls or Fotheringay Centre.

HAMILTON. (1976;90). Mr Niall Whyte, Secretary, 12 Balmoral Place, West Mains, East Kilbride G74 1EP. 01355 900099;
e-mail: niall.whyte@blueyonder.co.uk
www.baronshaugh.co.uk
Meetings: 7.30pm, 3rd Thursday of the month, Strathclyde Water Centre, Strathclyde Country Park, Motherwell.

RENFREWSHIRE. (1986; 200). Jim Sutherland, 0141 6397028; e-mail: sutherland.jim@btopenworld.com
Meetings: The McMaster Centre, 2a Donaldson Drive, Renfrew.

Wetland Bird Survey Organisers

ARGYLL & ISLANDS. Malcolm Ogilvie, Glencairn, Bruichladdich, Isle of Islay, PA49 7UN. 01496 850218; e-mail: maogilvie@indaal.demon.co.uk

ARRAN. Audrey Walters, Sula, Margnaheglish Road, Lamlash, Isle of Arran KA27 8LE.

CLYDE ESTUARY. Jim & Valerie Wilson, 76 Laigh Road, Newton Mearns, Glasgow G77 5EQ. (H)0141 639 2516; e-mail: Jim.Val@btinternet.com

GLASGOW/RENFREWSHIRE/LANARKSHIRE. Jim & Valerie Wilson, 76 Laigh Road, Newton Mearns, Glasgow G77 5EQ. (H)0141 639 2516; e-mail: Jim.Val@btinternet.com

DUMFRIES & GALLOWAY

Bird Recorders

Paul N Collin, Gairland, Old Edinburgh Road, Minnigaff, Newton Stewart, Wigtownshire, DG8 6PL. 01671 402861; e-mail: paul.collin@rspb.org.uk

Steve Cooper, Wildfowl & Wetlands Trust, Eastpark Farm, Caerlaverock, Dumfries DG1 4RS. 01387 770200; e-mail: steve.cooper@wwt.org.uk

Bird Report

DUMFRIES & GALLOWAY REGION BIRD REPORT (1985-), From Peter Norman, Low Boreland, Tongland Road, Kirkcudbright, DG6 4UU. 01557 331429.

BTO Regional Representatives & Regional Development Officer

DUMFRIES RR. Postition vacant.

KIRKCUDBRIGHT RR. Andrew Bielinski, 41 Main Street, St Johns Town of Dalry, Castle Douglas, Kirkcudbright, DG7 3UP. 01644 430418 (evening);
e-mail: andrewb@bielinski.fsnet.co.uk

WIGTOWN RR. Geoff Sheppard, The Roddens, Leswalt, Stranraer, Wigtownshire, DG9 0QR. 01776 870 685; e-mail: geoff.sheppard@tesco.net

Clubs

SOC DUMFRIES BRANCH. (1961; 105). Brian Smith, Rockiemount, Colvend, Dalbeattie, Dumfries, DG5 4QW. 01556 620617
Meetings: 7.30pm, 2nd Wednesday of the month (Sept-Apr), Cumberland St Day Centre.

SOC STEWARTRY BRANCH. (1976; 85). Miss Joan Howie, 60 Main Street, St Johns Town of Dalry, Castle Douglas, Kirkcudbrightshire, DG7 3UW. 01644 430226
Meetings: 7.30pm, usually 2nd Thursday of the month (Sep-Apr), Kells School, New Galloway.

SOC WEST GALLOWAY BRANCH. (1975; 50). Geoff Sheppard, The Roddens, Leswalt, Stranraer, Wigtownshire, DG9 0QR.
e-mail: geoff.sheppard@tesco.net
Meetings: 7.30pm, 2nd Tuesday of the month (Oct-Mar), Stranraer Library.

Ringing Group

NORTH SOLWAY RG. Geoff Sheppard, The Roddens, Leswalt, Stranraer, Wigtownshire, DG9 0QR. e-mail: geoff.sheppard@tesco.net

RSPB Members' Group

GALLOWAY. (1985;170). Robert M Greenshields, Nether Linkins, Gelston, Castle Douglas, DG7 1SU. 01556 680217; website: www.gallowayrspb-localgroup.org.uk
Meetings: 7.30pm 3rd Tuesday in the month, Castle Douglas High School.

Wetland Bird Survey Organisers

AUCHENCAIRN. Euan MacAlpine, Auchenshore, Auchencairn, Castle Douglas, Galloway DG7 1QZ .

DUMFRIES & GALLOWAY (OTHER SITES). Steve Cooper, Wildfowl & Wetlands Trust, Eastpark Farm, Caerlaverock, Dumfries DG1 4RS. 01387 770200; e-mail: steve.cooper@wwt.org.uk

LOCH RYAN. Geoff Shepherd, The Roddens, Leswalt, Stranraer, Wigtownshire DG9 0QR. e-mail: geoff.sheppard@tesco.net

ROUGH FIRTH. Judy Baxter, Saltflats Cottage, Rockcliffe, Dalbeattie, DG5 4QQ. 01556 630262; e-mail: Jbaxter@nts.org.uk

SOLWAY ESTUARY (NORTH). Steve Cooper, Wildfowl & Wetlands Trust, Eastpark Farm, Caerlaverock, Dumfries DG1 4RS. 01387 770200; e-mail: steve.cooper@wwt.org.uk

WIGTOWN. Paul Collin, Gairland, Old Edinburgh Road, Minnigaff, Newton Stewart, DG8 6PL. 01671 402861.

FIFE

Bird Atlas/Avifauna
The Fife Bird Atlas 2003 by Norman Elkins, Jim Reid, Allan Brown, Derek Robertson & Anne-Marie Smout. Available from Allan W. Brown (FOAG), 61 Watts Gardens, Cupar, Fife KY15 4UG, Tel. 01334 656804, email: swans@allanwbrown.co.uk

Bird Recorders
FIFE REGION INC OFFSHORE ISLANDS (NORTH FORTH). Rab Shand, 33 Liddle Drive, Bo'ness, West Lothian EH51 0PA. 01506 825101 (H); 07799 532954 (M); e-mail:rabshand@blueyonder.co.uk

ISLE OF MAY BIRD OBSERVATORY. Iain English, 19 Nethan Gate, Hamilton, S Lanarks, ML3 8NH. e-mail: i.english@talk21.com

Bird Reports
FIFE BIRD REPORT (1988-) (FIFE & KINROSS BR 1980-87), From Willie McBay, 41 Shamrock Street, Dunfermline, Fife, KY12 0JQ.01383 723464; e-mail: wmcbay@aol.com

ISLE OF MAY BIRD OBSERVATORY REPORT (1985-), From David Thorne, Craigurd House, Blyth Bridge, West Linton, Peeblesshire, EH46 7AH.

BTO Regional Representative
FIFE & KINROSS RR. Norman Elkins, 18 Scotstarvit View, Cupar, Fife, KY15 5DX. 01334 654348; e-mail: jandnelkins@btinternet.com

Clubs
FIFE BIRD CLUB. (1985; 250). Willie McBay, 41 Shamrock Street, Dunfermline, Fife, KY12 0JQ. 01383 723464, www.fifebirdclub.org.uk
Meetings: 7.30pm, (various evenings), Dean Park Hotel, Chapel Level, Kirkcaldy.

LOTHIANS AND FIFE MUTE SWAN STUDY GROUP. (1978)Allan & Lyndesay Brown, 61 Watts Gardens, Cupar, Fife, KY15 4UG.
e-mail: swans@allanwbrown.co.uk
www.swanscot.org.uk

SOC FIFE BRANCH. (1956;170). Howard Chapman, 5 Woodville Park, Dairsie, Cupar, Fife KY15 4TE. 01334 870768.
Meetings: 7.30pm, 2nd Wednesday of the month (Sep-Apr), St Andrews Town Hall.

Ringing Groups
ISLE OF MAY BIRD OBSERVATORY. Margaret Thorne, Craigurd House, West Linton, Peebles EH46 7AH.01721 752612.

TAY RG. Ms S Millar, Edenvale Cottage, 1 Lydox Cottages, Dairsie, Fife, KY15 4RN.

Wetland Bird Survey Organisers
FIFE (excluding estuaries). Allan Brown, 61 Watts Gardens, Cupar, Fife KY15 4UG;
e-mail: swans@allanwbrown.co.uk

EDEN ESTUARY. Les Hatton, Fife Ranger Service, Silverdaleburn House, Largo Road, By Leven, KY8 5PU. (Day)01333 429785.

TAY ESTUARY. Norman Elkins, 18 Scotstarvit View, Cupar, Fife KY15 5DX. 01334 654348;
e-mail: jandnelkins@rapidial.co.uk

Wildlife Hospital
SCOTTISH SPCA WILD LIFE REHABILITATION CENTRE. Middlebank Farm, Masterton Road, Dunfermline, Fife, KY11 8QN. 01383 412520. All species. Open to visitors, groups and school parties. Illustrated talk on oiled bird cleaning and other aspects of wildlife rehabilitation available. Veterinary support.

FORTH

Bird Recorder
UPPER FORTH (Does not include parts of Stirling in Loch Lomonside/Clyde Basin). Dr C J Henty, Edgehill East, 7b Coneyhill Road, Bridge of Allan, Stirling, FK9 4EL. 01786 832166

Bird Report
FORTH AREA BIRD REPORT (1975-) - enlarged report published annually in The Forth Naturalist and Historian, University of Stirling, From Dr Henty (see recorder) or Hon Sec. Lindsay Corbett, University of Stirling, Stirling FK9 4LA. 01259 215091.

BTO Regional Representative
CENTRAL RR. Neil Bielby, 56 Ochiltree, Dunblane, Perthshire, FK15 0DF. 01786 823830; e -mail: neil.bielby@tiscali.co.uk

Club
SOC CENTRAL SCOTLAND BRANCH. (1968; 101). Mr RL Gooch, The Red House, Dollarfield, Dollar, Clacks FK14 7LX.01259 742326.
Meetings: 7.30pm, 1st Thursday of the month (Sep-Apr), The Smith Art Gallery and Museum, Dumbarton Road, Stirling.

RSPB Members' Group
FORTH VALLEY. (1996; 150). David Redwood, 8 Strathmore Avenue, Dunblane, Perthshire FK15 9HX. 01786 825493; e-mail: d.redwood@tesco.net
http://forthrspb.p5.org.uk
Meetings: 7.30pm, 3rd Wednesday of the month (Sept-Apr), Cowane Centre, Stirling.

Wetland Bird Survey Organiser
CENTRAL (excl Forth Estuary. Neil Bielby, 56 Ochiltree, Dunblane, Perthshire FK15 0DF. (H)01786 823830; e-mail: neil.bielby@tiscali.co.uk

HIGHLAND

Bird Atlas/Avifauna
The Birds of Sutherland by Alan Vittery (Colin Baxter Photography Ltd, 1997).
Birds of Skye by Andrew Currie. In preparation.

Bird Recorders
ROSS-SHIRE, INVERNESS-SHIRE, SUTHERLAND. Alastair McNee, Liathach, 4 Balnafettack Place, Inverness IV3 8TQ. 01463 220493; (M)07763 927814; e-mail: aj.mcnee@care4free.net

Bird Reports
HIGHLAND BIRD REPORT (1991-), From Recorder. 2003 edition £7.50 including p&p.

SUTHERLAND BIRD REVIEW (2002-), From Recorder. (sold out).

BTO Regional Representatives & Regional Development Officers
INVERNESS & SPEYSIDE RR & RDO. Hugh Insley, 1 Drummond Place, Inverness,IV2 4JT. 01463 230652; e-mail: hugh.insley@gmail.com

RUM, EIGG, CANNA & MUCK RR & RDO. Bob Swann, 14 St Vincent Road, Tain, Ross-shire, IV19 1JR. 01862 894329; e-mail: bob.swann@freeuk.com

ROSS-SHIRE RR. Simon Cohen; e-mail: saraandsimon@hotmail.com

SUTHERLAND. David Devonport, e-mail: dave.forbirds@fsmail.net

SKYE. Robert McMillan, 01471 866305; e-mail: Bob@Skye-birds.com

Clubs
EAST SUTHERLAND BIRD GROUP. (1976; 80). Tony Mainwood, 13 Ben Bhraggie Drive, Golspie, Sutherland KW10 6SX. 01408 633247; e-mail: tony.mainwood@which.net
Meetings: 7.30pm, Last Monday of the month (Oct, Nov, Jan, Feb, Mar), Golspie Community Centre.

SOC HIGHLAND BRANCH. (1955; 151). Ann Sime, Drumrunie House, Myrtlefield Lane, Westhill, Inverness, IV2 5UE, 01463 790249; e-mail: ann@drumrunie.fsnet.co.uk
Meetings: 7.45pm, 1st Tuesday of the month, Inverness Marriott Hotel.

Ringing Groups
HIGHLAND RG. Bob Swann, 14 St Vincent Road, Tain, Ross-shire, IV19 1JR. e-mail: bob.swann@freeuk.com

RSPB Members' Group
HIGHLAND. (1987; 198). Richard Prentice, Lingay, Lewiston, Drumnadrochit, Inverness, IV63 6UW. 01456 450526;

e-mail: richard@rprentice.wanadoo.co.uk
Meetings: 7.30pm, Inverness Marriot Hotel, Culcabock Road, Inverness.

Wetland Bird Survey Organisers
MORAY BASIN COAST. Bob Swann, 14 St Vincent road, Tain, Ross-shire IV19 1JR. 01862 894329; e-mail: bob.swann@hcs.uhi.ac.uk

MORAY & NAIRN (Inland). Martin Cook, Rowanbrae, Clochan, Buckie, Banffshire AB56 5EQ. (H)01542 850296.

SKYE & LOCHALSH. Bob McMillan, 10/11 Elgol, Nr Broadford, Isle of Skye IV49 9BL. 01471 866305; e-mail: bob@skye-birds.com

LOTHIAN

Bird Atlas/Avifauna
The Breeding Birds of South-east Scotland, a tetrad atlas 1988-1994 by R D Murray et al. (Scottish Ornithologists' Club, 1998).

Bird Recorder
David J Kelly, 20 Market View, Tranent, East Lothian, EH32 9AX. e-mail: dj_kelly@btinternet.com

Bird Reports
LOTHIAN BIRD REPORT (1979-), From Lothian SOC Branch Secretary.

WEST LOTHIAN BIRD CLUB REPORT (1991-), From Secretary, West Lothian Bird Club.

BTO Regional Representative
RR. Alan Heavisides, 9 Addiston Crescent, Balerno, Edinburgh, EH14 7DB. 0131 449 3816; e-mail: a.heavisides@napier.ac.uk

Clubs
EDINBURGH NATURAL HISTORY SOCIETY. (1869; 200). Miss Joan Fairlie, 14 Regulas Road, Edinburgh EH9 2ND. 0131 6691470; e-mail: enquiries@edinburghnaturalhistorysociety.org.uk www.edinburghnaturalhistorysociety.org.uk

FOULSHIELS BIRD GROUP. (1991; 5). Frazer Henderson, 2 Elizabeth Gardens, Stoneyburn, W Lothian, EH47 8BP. 01501 762972

LOTHIANS AND FIFE MUTE SWAN STUDY GROUP. (1978; 12) Allan & Lyndesay Brown, 61 Watts Gardens, Cupar, Fife, KY15 4UG. e-mail: swans@allanwbrown.co.uk

LOTHIAN SOC. (1936; 370). John Hamilton, 30 Swanston Gardens, Edinburgh, EH10 7DL. 0131 445 5317; e-mail: john.r.hamilton31@btopenworld.com www.lsoc.btinternet.co.uk
Meetings: 7.30pm, 2nd Tuesday (Sep-Apr), Lounge 2, Meadowbank Sports Stadium.

WEST LOTHIAN BIRD CLUB.
(1990; 20). Alan Paterson,
17 Main Street, Winchburgh,
Broxburn, W Lothian.

Ringing Group
LOTHIAN RG. Mr M Cubitt,
12 Burgh Mills Lane,
Linlithgow,West Lothian
EH49 7TA.

RSPB Members' Group
EDINBURGH. (1974;480). Hugh Conner, 22 Tippet
Knowes Court, Winchburgh,West Lothian, EH52 6UW.
e-mail: h.m.conner@blueyonder.co.uk
http://rspb-edin.pwp.blueyonder.co.uk
Meetings: 7.30pm, 3rd Tuesday or Wednesday of
the month (Sep-Apr), Napier University, Craiglockhart
Campus, Edinburgh.

Wetland Bird Survey Organisers
FORTH ESTUARY (North). Alastair Inglis, 5 Crowhill
Road, Dalgety Bay, Fife KY11 5LJ.

FORTH ESTUARY (Outer South). Duncan Priddle, c/o
City of Edinburgh Countryside Ranger Service,
Hermitage House, 69a Braid Road, Edinburgh EH10
6JF. (Day) 0131 4477145; e-mail:
duncan@cecrangerservice.demon.uk   day 0131
4477145

LOTHIAN (excl estuaries). Joan Wilcox, 18 Howdenhall
Gardens, Edinburgh, Midlothian EH16 6UN. (H)0131
6648893

TYNINGHAME ESTUARY. John Muir Country Park, Town
House, Dunbar, East Lothian EH42 1ER. (W)01368
863886; e-mail: randerson@eastlothian.gov.uk

MORAY & NAIRN

Bird Atlas/Avifauna *The Birds of Moray and Nairn* by
Martin Cook (Mercat Press, 1992.

Bird Recorder
NAIRN. Martin J H Cook, Rowanbrae, Clochan, Buckie,
Banffshire, AB56 5EQ. 01542 850296;
e-mail: martin.cook9@virgin.net

MORAY. Martin J H Cook, Rowanbrae, Clochan,
Buckie, Banffshire, AB56 5EQ. 01542 850296;
e-mail: martin.cook9@virgin.net

Bird Reports
BIRDS IN MORAY AND NAIRN (1999-), From Moray
Recorder, 01542 850296;
e-mail: martin.cook9@virgin.net

MORAY & NAIRN BIRD REPORT (1985-1998), From
Moray Recorder, 01542 850296;
e-mail: martin.cook9@virgin.net

**BTO Regional Representatives & Regional
Development Officer**
MORAY AND NAIRN RR. Bob Proctor, 78 Marleon Field,
Elgin, Moray, IV30 4GE. 01343 548395; (W)01479
821409; e-mail: bob.proctor@rspb.org.uk

Wetland Bird Survey Organisers
LOSSIE ESTUARY. Bob Proctor, 78 Marleon Field,
Silvercrest, Bishopmill, Elgin, IV30 4GE;
e-mail: bob.proctor@rspb.org.uk

MORAY & NAIRN (Inland). Martin Cook, Rowanbrae,
Clochan, Buckie, Banffshire, AB56 5EQ. 01542
850296.

NORTH EAST SCOTLAND

Bird Atlas/Avifauna
The Birds of North East Scotland by S T Buckland, M V
Bell & N Picozzi (North East Scotland Bird Club, 1990).

Bird Recorder
NORTH-EAST SCOTLAND. Andrew Thorpe, 30
Monearn Gardens, Milltimber, Aberdeen, AB13 0EA.
e-mail: andrewthorpe4@aol.com

Bird Reports
NORTH-EAST SCOTLAND BIRD REPORT (1974-), From
Dave Gill, Drakemyre Croft, Cairnorrie, Methlick,
Aberdeenshire, AB41 7JN. 01651 806252;
e-mail: dave@drakemyre.freeserve.co.uk

NORTH SEA BIRD CLUB ANNUAL REPORT (1979-),
From NSBC Recorder, see below, 01224 274428;
e-mail: nsbc@abdn.ac.uk

**BTO Regional Representatives & Regional
Development Officer**
ABERDEEN. John Littlejohn,
e-mail: j.w.littlejohn@talk21.com

KINCARDINE & DEESIDE. Graham Cooper, Westbank,
Beltie Road, Torphins, Banchory, Aberdeen, AB31 4JT.
H:01339 882706; e-mail: grahamwcooper@beeb.net

Clubs
SOC GRAMPIAN BRANCH. (1956; 110). John Wills,
Bilbo, Monymusk, Inverurie, Aberdeenshire, AB51
7HA. 01467 651296;
e-mail: bilbo@monymusk.freeserve.co.uk
Meetings: 7.30pm, 1st Monday of the month (Sep-
Apr), Sportsman's Club, 11 Queens Road, Aberdeen.

Ringing Groups
ABERDEEN UNIVERSITY RG. Andrew Thorpe, Ocean
Laboratory and Centre for Ecology, Aberdeen
University, Newburgh,   Ellon, Aberdeenshire, AB41
6AA; e-mail: a.thorpe@abdn.ac.uk

GRAMPIAN RG. R Duncan, 86 Broadfold Drive, Bridge
of Don, Aberdeen, AB23 8PP.
e-mail: Raymond@waxwing.fsnet.co.uk

RSPB Members' Group
ABERDEEN & DISTRICT. (1975; 190). Robert Payne,

c/o The RSPB - East Scotland Regional Office, 10 Albyn Terrace, Aberdeen AB10 1YP;
e-mail: Rodney@rspb-abdn-mbrsgp.org.uk
www.rspb-abdn-mbrsgp.org.uk
Meetings: 7.30pm, monthly in the winter, Lecture Theatre, Zoology Dept, Tillydrone Av, Aberdeen. Two birding trips monthly throughout the year.

Wildlife Hospital
GRAMPIAN WILDLIFE REHABILITATION TRUST. 40 High Street, New Deer, Turriff, Aberdeenshire, AB53 6SX. 01771 644489. Veterinary surgeon. Access to full practice facilities. Will care for all species of birds.

ORKNEY

Bird Atlas/Avifauna
The Birds of Orkney by CJ Booth et al (The Orkney Press, 1984).

Bird Recorder
Mr EJ Williams, Fairholm, Finstown, Orkney, KW17 2EQ; e-mail: jim@geniefea.freeserve.co.uk

Bird Report
ORKNEY BIRD REPORT (inc North Ronaldsay Bird Report) (1974-), From Mr EJ Williams, Fairholm, Finstown, Orkney, KW17 2EQ.
e-mail: jim@geniefea.freeserve.co.uk

BTO Regional Representative & Regional Development Officer
Colin Corse, Garrisdale, Lynn Park, Kirkwall, Orkney, KW15 1SL. H:01856 874484; W:01856 884156; e-mail: ccorse@garrisdale1.fstnet.co.uk

Club
SOC ORKNEY BRANCH. (1993; 15). Colin Corse, Garrisdale, Lynn Park, Kirkwall, Orkney, KW15 1SL. H:01856 874484; W:01856 884156; e-mail: ccorse@garrisdale1.fstnet.co.uk

Ringing Groups
NORTH RONALDSAY BIRD OBSERVATORY. Ms A E Duncan, Twingness, North Ronaldsay, Orkney, KW17 2BE. e-mail: alison@nrbo.prestel.co.uk
www.nrbo.f2s.com

ORKNEY RG. Colin J Corse, Garrisdale, Lynn Park, Kirkwall, Orkney, KW15 1SL. H:01856 874484; W:01856 884156; e-mail: ccorse@garrisdale1.fstnet.co.uk

SULE SKERRY RG. Dave Budworth, 121 Wood Lane, Newhall, Swadlincote, Derbys, DE11 0LX. 0121 6953384.

RSPB Members' Group
ORKNEY. (1985;300 in catchment area). Mrs Pauline Wilson, Sunny Bank, Deerness, Orkney KW17 2QQ. 01856 741382; e-mail: pwilso@tiscali.co.uk
Meetings: Meetings advertised in newsletter and local press, held at Kirkwall Community Centre.

Wetland Bird Survey Organiser
ORKNEY (other sites). Eric Meek, RSPB, 12/14 North End Road, Stromness, Orkney KW16 3AG. 01856 850176.

OUTER HEBRIDES

Bird Recorder
OUTER HEBRIDES AND WESTERN ISLES. Andrew Stevenson, The Old Stores, Bornish, Isle of South Uist Western Isles HS8 5SA. 01878 710372; e-mail: andrew@bornish.fsnet.co.uk

Bird Report
OUTER HEBRIDES BIRD REPORT (1989-), From Recorder.

BTO Regional Representatives & Regional Development Officer
BENBECULA & THE UISTS RR & RDO. Brian Rabbitts, 01876 580328; e-mail: brian.rabbitts@virgin.net
e-mail: ellerpendle@madasafish.com

LEWIS & HARRIS RR. Chris Reynolds, 11 Reef, Isle of Lewis, HS2 9HU. 01851 672376; e-mail: emreynolds@btinternet.com

Ringing Group
SHIANTS AUK RG. David Steventon, Welland House, 207 Hurdsfield Road, Macclesfield, Cheshire, SK10 2PX. 01625 421936

UISTS AND BENBECULA. Brian Rabbitts, 01876 580328 e-mail: brian.rabbits@virgin.net

PERTH & KINROSS

Bird Recorder
PERTH & KINROSS. Ron Youngman, Blairchroisk Cottage, Ballinluig, Pitlochry, Perthshire, PH9 0NE. 01796 482324; e-mail: blairchroisk@aol.com

Bird Report
PERTH & KINROSS BIRD REPORT (1974-), From Recorder. Now published on-line and will be e-mailed.

BTO Regional Representatives & Regional Development Officer
PERTHSHIRE RR. Position vacant.

Clubs
PERTHSHIRE SOCIETY OF NATURAL SCIENCE (Ornithological Section). (1964; 48). Miss Esther Taylor, 23 Verena Terrace, Perth,PH2 0BZ. 01738 621986.
Meetings: 7.30pm, Wednesdays monthly (Oct-Mar), Perth Museum.

Wetland Bird Survey Organiser
TAY ESTUARY. Norman Elkins, 18 Scotstarvit View, Cupar, Fife KY15 5DX. 01334 654348; e-mail: jandnelkins@rapidial.co.uk

SHETLAND

Bird Recorders
FAIR ISLE. Deryk Shaw, Bird Observatory, Fair Isle, Shetland, ZE2 9JU.
e-mail: fairisle.birdobs@zetnet.co.uk

SHETLAND. Micky Maher, Hamarsgarth, Haroldswick, Shetland, ZE2 9ED. (01595) 711677 or 711528; e-mail: recorder@birdclub.shetland.co.uk

Bird Reports
FAIR ISLE BIRD OBSERVATORY REPORT (1949-), From Scottish Ornithologists' Club, 21 Regent Terrace, Edinburgh, EH7 5BT. 0131 556 6042

SHETLAND BIRD REPORT (1969-) no pre 1973 available, From Martin Heubeck, East House, Sumburgh Lighthouse, Virkie, Shetland, ZE3 9JN.e-mail: martinheubeck@btinternet.com

BTO Regional Representative
RR and RDO. Dave Okill, Heilinabretta, Cauldhame, Trondra, Shetland, ZE1 0XL. 01595 880450.

Club
SHETLAND BIRD CLUB. (1973; 200). Reinoud Norde, Lindale, Ireland, Bigton, Shetland, ZE2 9JA. 01950 422467: e-mail: reinoud.norde@lineone.net

Ringing Groups
FAIR ISLE BIRD OBSERVATORY. Deryk Shaw, Bird Observatory, Fair Isle, Shetland, ZE2 9JU. e-mail: fairisle.birdobs@zetnet.co.uk

SHETLAND RG. Dave Okill, Heilinabretta, Cauldhame, Trondra, Shetland, ZE1 0XL. H:01595 880450; W:01595 696926

Wetland Bird Survey Organiser
Paul Harvey, Shetland Biological Records Centre, Shetland Amenity Trust, 22-24 North Road, Lerwick, Shetland, ZE1 3NG. (Day)01595 694688; e-mail: sbrc@zetnet.co.uk

WALES

Bird Report
See Welsh Ornithological Society in National Directory

BTO Honorary Wales Officer
BTO WALES OFFICER. John Lloyd, Cynghordy, Llandovery, SA20 0LN.
e-mail: thelloyds@dial.pipex.com

Club
See Welsh Ornithological Society in National Directory.

EAST WALES

Bird Atlas/Avifauna

Bird Recorders
BRECONSHIRE. Andrew King, Heddfan, Pennorth, Brecon, Powys LD3 7EX. 01874 658351.

GWENT. Chris Jones, 22 Walnut Drive, Caerleon, Newport, Gwent, NP6 1SB. 01633 423439; e-mail: countyrecorder@gwentbirds.org.uk

MONTGOMERYSHIRE. Brayton Holt, Scops Cottage, Pentrebeirdd, Welshpool, Powys, SY21 9DL. 01938 500266.

RADNORSHIRE. Pete Jennings, Penbont House, Elan Valley, Rhayader, Powys, LD6 5HS. H:01597 811522; W:01597 810880;
e-mail: petejelanvalley@hotmail.com

Bird Reports
BRECONSHIRE BIRDS (1962-), from Brecknock Wildlife Trust.

GWENT BIRD REPORT (1964-), from Jerry Lewis, Y Bwthyn Gwyn, Coldbrook, Abergavenny, Monmouthshire, NP7 9TD. (H)01873 855091; (W)01633 644856
MONTGOMERYSHIRE BIRD REPORT (1981-82-), from Montgomeryshire Wildlife Trust.

RADNOR BIRDS (1987/92-), from Radnorshire Recorder.

BTO Regional Representatives
BRECKNOCK RR. John Lloyd, Cynghordy, Llandovery, Carms, SA20 0LN. e-mail; thelloyds@dial.pipex.com

GWENT RR. Jerry Lewis, Y Bwthyn Gwyn, Coldbrook, Abergavenny, Monmouthshire, NP7 9TD. H:01873 855091; W:01633 644856

WALES

MONTGOMERY RR. Jane Kelsall, Holly Bank, Moel y Garth, Welshpool, Powys SY21 9JA. 01938 556438; e-mail: jane@melodeons.com

RADNORSHIRE RR. Brian Jones;
e-mail: jones.brn10@virgin.net

Clubs
THE GWENT ORNITHOLOGICAL SOCIETY. (1964; 420). T J Russell, The Pines, Highfield Road, Monmouth, Gwent, NP25 3HR. 01600 716266;
e-mail: secretary@GwentBirds.org.uk
www.gwentbirds.org.uk.
Meetings: 7.30pm, alternate Saturdays (Sept-Apr), Goytre Village Hall.

MONTGOMERYSHIRE FIELD SOCIETY. (1946; 190). Maureen Preen, Ivy House, Deep Cutting, Pool Quay, Welshpool, Powys, SY21 9LJ.Tel: Mary Oliver, 01686 413518.
Meetings: 3rd Saturday of the month (Nov, Jan, Feb, Mar), Methordist Church Hall, Welshpool. Field trips (Apr-Oct)

MONTGOMERYSHIRE WILDLIFE TRUST BIRD GROUP. (1997; 110). A M Puzey, Four Seasons, Arddleen, Llanymynech, Powys, SY22 6RU. 01938 590578.

RADNOR BIRD GROUP. (1986; 60). Pete Jennings, Penbont House, Elan Valley, Rhayader, Powys, LD6 5HS. 01597 811522; e-mail:
petejelanvalley@hotmail.com

Ringing Groups
GOLDCLIFF RG. Vaughan Thomas, Gilgal Cottage, Gilfach, Llanvaches, S Wales, NP26 3AZ. 01633 817161.

LLANGORSE RG. Jerry Lewis, Y Bwthyn Gwyn, Coldbrook, Abergavenny, Monmouthshire, NP7 9TD. H:01873 855091; W:01633 644856

Wetland Bird Survey Organisers
RADNORSHIRE. Peter Jennings, Pentbont House, Elan Valley, Rhayader, Powys, LD6 5HS. (H)01597 811522; (Day)01597 810880

BRECONSHIRE. Andrew King, Heddfan, Pennorth, Brecon LD3 7EX; e-mail: heddfan25@hotmail.com

Wildlife Trusts
BRECKNOCK WILDLIFE TRUST. (1963; 893). Lion House, Bethel Square, Brecon, Powys, LD3 7AY. 01874 625 708; fax 01874 625 708; e-mail:
brecknockwt@cix.co.uk
www.wildlifetrust.org.uk/brecknock

GWENT WILDLIFE TRUST. (1963;1200). Seddon House, Dingestow, Monmouth NP25 4DY. 01600 740358; fax 01600 740299; e-mail: gwentwildlife@cix.co.uk
www.wildlifetrust.org.uk/gwent

MONTGOMERYSHIRE WILDLIFE TRUST. (1982; 1000). Collot House, 20 Severn Street, Welshpool, Powys, SY21 7AD. 01938 555654; fax 01938 556161; e-mail: montwt@cix.co.uk
www.wildlifetrust.org.uk/montgomeryshire

RADNORSHIRE WILDLIFE TRUST. (1987; 789). Warwick House, High Street, Llandrindod Wells, Powys, LD1 6AG. 01597 823298; fax 01597 823274; e-mail: info@ radnorshirewildlifetrust.org.uk
www.radnorshirewildlifetrust.org.uk

NORTH WALES

Bird Recorders
ANGLESEY. Stephen Culley, Millhouse, Penmynydd Road, Menai Bridge, Anglesey, LL59 5RT; e-mail: SteCul10@aol.com

CAERNARFON. John Barnes, Fach Goch, Waunfawr, Caernarfon, LL55 4YS. 01286 650362.

DENBIGHSHIRE & FLINTSHIRE. Ian Spence, 43 Blackbrook, Sychdyn, Mold, Flintshire,CH7 6LT. Tel/fax 01352 750118; e-mail: ianspence.cr@imsab.idps.co.uk

MEIRIONNYDD. D L Smith, 3 Smithfield Lane, Dolgellau, Gwynedd, LL40 1BU. 01341 421064;
e-mail: d.smith@ccw.gov.uk

Bird Reports
BARDSEY BIRD OBSERVATORY ANNUAL REPORT, from Warden, see Reserves.

CAMBRIAN BIRD REPORT (sometime Gwynedd Bird Report) (1953-), fromMr Rhion Pritchard, Pant Afonig, Hafod Lane, Bangor, Gwynedd, LL57 4BU
e-mail: rhion@pritchardr.freeserve.co.uk

CLWYD BIRD REPORT (2002-), fromDr Anne Brenchley, Ty'r Fawnog, 43 Black Brook, Sychdyn, Mold, Flints, CH7 6LT. 01352 750118.

MEIRIONNYDD BIRD REPORT Published in Cambrian Bird Report (above).

WREXHAM BIRDWATCHERS' SOCIETY ANNUAL REPORT (1982-), from Secretary, Wrexham Birdwatchers' Society.

BTO Regional Representatives & Regional Development Officer
ANGLESEY RR. Tony White, 01407 710 137;
e-mail: wylfor@treg5360.freeserve.co.uk

CAERNARFON RR. Geoff Gibbs, 01248 681 936;
e-mail: geoffkate.gibbs@care4free.net

CLWYD EAST RR. Anne Brenchley, Ty'r Fawnog, 43 Black Brook, Sychdyn, Mold, CH7 6LT.
e-mail: anne.brenchley@cbrg1.idps.co.uk

CLWYD WEST RR. Mel ab Owain, 31 Coed Bedw, Abergele, Conwy, LL22 7EH. 01745 826528;
e-mail: melabowain@cix.co.uk

MEIRIONNYDD RR. Peter Haveland, Ty Manceinion, Penmachno, Betws-y-Coed, Sir Gonwy, LL24 0UD. 01690 760337; e-mail: peter.haveland@tesco.net

Clubs
CAMBRIAN ORNITHOLOGICAL SOCIETY. (1952; 190). Mr Rhion Pritchard, Pant Afonig, Hafod Lane, Bangor,

Gwynedd, LL57 4BU. 01248 671301; http://mysite.freeserve.com/cambrianos
Meetings: 7.30pm, 1st Friday of the month, Pensychnant Centre, Sychnant Pass.

CLWYD BIRD RECORDING GROUP. Anne Brenchley, Ty'r Fawnog, 43 Black Brook, Sychdyn, Mold, CH7 6LT. e-mail: anne.brenchley@cbrg1.idps.co.uk

DEE ESTUARY CONSERVATION GROUP. (1973; 25 grps). Tony Perry, 10 Ridgeway Close, Connah's Quay, Deeside, CH5 4LZ.01244 831725.

DEESIDE NATURALISTS' SOCIETY. (1973; 500). Secretary, Deeside Naturalists' Society, 38 Kelsterton Road, Connah's Quay, Flintshire CH5 4BJ; e-mail: richard@deeestuary.co.uk www.deeestuary.co.uk

WREXHAM BIRDWATCHERS' SOCIETY. (1974; 90). Miss Marian Williams, 10 Lake View, Gresford, Wrexham, Clwyd, LL12 8PU. 01978 854633.
Meetings: 7.30pm, 1st Friday of the month (Sep-Apr), Gresford Memorial Hall, Gresford.

Ringing Groups
BARDSEY BIRD OBSERVATORY. Steven Stansfield, Bardsey Island, off Aberdaron, Pwllheli, Gwynedd, LL53 8DE. 07855 264151; e-mail: warden@bbfo.org.uk

MERSEYSIDE RG. P Slater, 45 Greenway Road, Speke, Liverpool, L24 7RY.

SCAN RG. D J Stanyard, Court Farm, Groeslon, Caernarfon, Gwynedd, LL54 7UE.01286 881 669.

RSPB Local Group
NORTH WALES. (1986; 130). Maureen Douglas, 01492 547768.
Meetings: St Davids Church Hall, Penrhyn Bay, Lladudno.

Wetland Bird Survey Organisers
CONWY ESTUARY. Alan Davies, RSPB Conwy Reserve, Llandudno Junction, LL31 9XZ. 01492 584091: e-mail: alan.davies@rspb.org.uk

ANGLESEY (other sites). Ian Sims, Plas Nico, South Stack, Holyhead, LL65 1TH.

CAERNARFONSHIRE (excl Traeth Lafan). Rhion Pritchard, Pant Afonig, Hafod Lane, Bangor, Gwynedd LL57 4BU. (H)01248 671301; e-mail: RhionPritchard@gwynedd.gov.uk

CLWYD (Coastal). Mr Peter Wellington, 4 Cheltenham Avenue, Rhyl, Clwyd LL18 4DN. (H)01745 354232.

MERIONETH (other sites). Mr Trefor Owen, Crochendy Twrog, Maentwrog, Blaenau Ffestiniog, LL41 3YU. (H)01766 590302.

TRAETH LAFAN. Mr Rhion Pritchard, Pant Afonig, Hafod Lane, Bangor, Gwynedd LL57 4BU. (Day)01286 679462; e-mail: RhionPritchard@gwynedd.gov.uk

Wildlife Trust
NORTH WALES WILDLIFE TRUST. (1963; 2400). 376 High Street, Bangor, Gwynedd, LL57 1YE. 01248 35154; fax 01248 353192; e-mail: nwwt@cix.co.uk www.wildlifetrust.org.uk/northwales

SOUTH WALES

Bird Recorders
GLAMORGAN (EAST). Steve Moon, 36 Rest Bay Close, Porthcawl, Bridgend, CF36 3UN. e-mail: sjmbirds@aol.com

GOWER (WEST GLAMORGAN). Robert Taylor, 285 Llangyfelach Road, Brynhyfryd, Swansea, SA5 9LB. 01792 464780; (M)07970 567007; e-mail: rob@birding.freeserve.co.uk

Bird Reports
AST GLAMORGAN BIRD REPORT (title varies 1963-95) 1996-2003, from Richard G Smith, 35 Manor Chase, Gwaun Miskin, Pontypridd, Rhondda Cynon Taff, S Wales, CF38 2JD. e-mail: rgsmith@birdpix.freeserve.co.uk

GOWER BIRDS (1965-), from Audrey Jones, 24 Hazel Road, Uplands, Swansea, SA2 0LX. 01792 298859.

BTO Regional Representatives & Regional Development Officer
EAST GLAMORGAN (former Mid & South Glam) RR. Rob Nottage, 32 Village Farm, Bonvilston, Cardiff, CF5 6TY. e-mail: rob@nottages.freeserve.co.uk

WEST RR. Bob Howells, Ynys Enlli, 14 Dolgoy Close, West Cross, Swansea, SA3 5LT. e-mail: bobhowells31@hotmail.com

Clubs
CARDIFF NATURALISTS' SOCIETY. (1867; 225). Stephen R Howe, Department of Geology, National Museum of Wales, Cardiff, CF10 3NP. e-mail: steve.howe@nmgw.ac.uk
Meetings: 7.30pm, various evenings, Lecture Theatre EO.02, Llandaff Campus Unic, Western Avenue, Cardiff.

GLAMORGAN BIRD CLUB. (1990; 300+) John Wilson (Chairman), 122 Westbourne Road, Penarth CF64 3HH. 029 20339424; (M)07742 728069; e-mail: john.wilson@glamorganbirds.org.uk www.glamorganbirds.org.uk
Meetings: 8pm, winter months, Kenfig Reserve Centre.

GOWER ORNITHOLOGICAL SOCIETY. (1956; 120). Audrey Jones, 24 Hazel Road, Uplands, Swansea, SA2 0LX. 01792 298859. www.glamorganbirds.org.uk
Meetings: 7.15pm, last Friday of the month (Sep-Mar), Enviroment Centre, Swansea.

Ringing Groups
FLAT HOLM RG. Brian Bailey, Tamarisk House, Wards

WALES

Court, Frampton-on-Severn, Glos, GL2 7DY.
e-mail: brian@sandbservices.fsnet.co.uk

KENFIG RG. Mr D.G. Carrington, 25 Bryneglwys
Gardens , Porthcawl, Bridgend, Mid Glamorgan, CF36
5PR.

RSPB Local Groups
CARDIFF & DISTRICT. -1973. Joy Lyman, 5 Dros-Y-
Morfa, Rumney, Cardiff, CF3 3BL. 029 2077 0031;
e-mail: joy@lyman.plus.com.
Meetings: 7.30pm, various Fridays (Sept-May), UWIC,
Cyncoed Road, Cardiff.

WEST GLAMORGAN. (1985; 421). Mr Russel Evans,
07800 740399 after 5 pm or 01792 208038 after 9 pm.
Meetings: 7.30pm, Environment Centre, Pier Street,
Swansea.

Wetland Bird Survey Organisers
WEST GLAMORGAN. Bob Howells, Ynys Enlli, 14 Dolgoy
Close, West Cross, Swansea, SA3 5LT. (H)01792
405363; e-mail: bobhowells31@hotmail.com

EAST GLAMORGAN. Rob Nottage, 32 Village Farm,
Bonvilston, Cardiff, CF5 6TY;
e-mail: rob@nottages.freeserve.co.uk

SEVERN ESTUARY. Niall Burton, c/o The BTO, The
Nunnery, Thetford, Norfolk IP27 2PU. 01842 750050;
e-mail: niall.burton@bto.org

Wildlife Hospitals
GOWER BIRD HOSPITAL. Karen Kingsnorth and Simon
Allen, Valetta, Sandy Lane, Pennard, Swansea, SA3
2EW. 01792 371630;
e-mail: info@gowerbirdhospital.org.uk
All species of wild birds, also hedgehogs and small
mammals. Prior phone call essential. Gower Bird Hospital
cares for sick, injured and orphaned wild birds and
animals with the sole intention of returning them to the
wild. Post release radio tracking projects, ringing
scheme. Contact us for more information.

Wildlife Trust
THE WILDLIFE TRUST OF SOUTH AND WEST WALES.
Fountain Road, Tondu, Bridgend, CF32 0EH. 01656
724100; fax 01656 729880;
e-mail: tondu@wildlifetrustswales.org

WEST WALES

Bird Recorders
CARMARTHENSHIRE. Tony Forster, Ffosddu, Salem,
Llandeilo, Carmarthenshire, SA19 7NS. 01558 824237;
e-mail: tony-forster@supanet.com

CEREDIGION. Hywel Roderick, 32 Prospect Street,
Aberystwyth, Ceredigion, SY23 1JJ. 01970 617681, e-
mail: hywel@adar.freeserve.co.uk

PEMBROKESHIRE. Graham Rees, 22 Priory Avenue,
Haverfordwest, Pembrokeshire, SA61 1SQ. 01437
762877.

Bird Reports
CARMARTHENSHIRE BIRD REPORT (1982-), from
Carmarthenshire Recorder.

CEREDIGION BIRD REPORT (biennial 1982-87; annual
1988-), from Wildlife Trust West Wales.

PEMBROKESHIRE BIRD REPORT (1981-), from TJ
Price, 2 Wordsworth Ave, Haverfordwest,
Pembrokeshire, SA61 1SN.

BTO Regional Representatives & Regional Development Officer
CARDIGAN RR. Moira Convery, 41 Danycoed,
Aberystwyth,SY23 2HD. 01970 612998.

CARMARTHEN RR. Colin Jones, 01554 821 632;
e-mail: colinjones25@yahoo.co.uk

PEMBROKE RR. Annie Haycock and Bob Haycock, 1
Rushmoor, Martletwy, Pembrokeshire, SA67 8BB.
e-mail: rushmoor1@tiscali.co.uk

Clubs
CARMARTHENSHIRE BIRD CLUB.(2003; 80).Ian
Hainsworth, 23 Rhyd y Defaid Drive, Swansea, SA2
8AJ. 01792 205693; e-mail: ian.hains@ntlworld.com
www.carmarthenshirebirds.co.uk

LLANELLI NATURALISTS. (1971; 100). Richard Pryce,
Trevethin, School Road, Pwll, Llanelli,
Carmarthenshire, SA15 4AL; e-mail:
pryceeco@aol.com
Meetings: 1st Thursday of the month, YWCA Llanelli
(see programme in local libraries).

PEMBROKESHIRE BIRD GROUP. (1993; 60). T J Price, 2
Wordsworth Ave, Haverfordwest, Pembs, SA61 1SN.
01437 779667.
Meetings: 7.30pm, 2nd Monday of the month (Oct-
Apr), The Patch, Furzy Park, Haverfordwest.

Ringing Group
PEMBROKESHIRE RG. J Hayes, 3 Wades Close,
Holyland Road, Pembroke, SA71 4BN. 01646 687036.

Wetland Bird Survey Organiser
BURRY INLET (North). Graham Rutt, 13 St James
Gardens, Uplands, Swansea, A1 6DY. (Day) 01792
325603.

DYFI/DYSYNNI ESTUARIES. Dick Squires, Cae'r Berllan,
Eglwys-Fach, Machynlleth, SY20 8TA. 01654 781265;
e-mail: dick.squires@rspb.org.uk

CARDIGAN (excl Dyfi Estuary). Dick Squires, Cae'r
Berllan, Eglwys-Fach, Machynlleth, SY20 8TA. 01654
781265; e-mail: dick.squires@rspb.org.uk

Wildlife Hospitals
NEW QUAY BIRD HOSPITAL. Jean Bryant, Penfoel,
Cross Inn, Llandysul, Ceredigion, SA44 6NR. 01545
560462. All species of birds. Fully equipped for
cleansing oiled seabirds. Veterinary support.

WEST WILLIAMSTON OILED BIRD CENTRE. Mrs J
Hains, Lower House Farm, West Williamston, Kilgetty,

Pembs, SA68 0TL. 01646 651236.
Facilities for holding up to 200 Guillemots, etc. for short periods. Initial treatment is given prior to despatch to other washing centres during very large oil spills; otherwise birds are washed at the Centre with intensive care and rehabilitation facilities. Also other species. Veterinary support.

Wildlife Trust
THE WILDLIFE TRUST OF SOUTH AND WEST WALES. Fountain Road, Tondu, Bridgend, CF32 0EH. 01656 724100; fax 01656 729880; e-mail: tondu@wildlifetrustswales.org

ISLE OF MAN

Bird Atlas/Avifauna
Manx Bird Atlas.

Bird Recorder
Dr Pat Cullen,
Troutbeck,
Cronkbourne, Braddan,
Isle of Man, IM4 4QA.
Home: 01624 623308;
Work 01624 676774; e-mail: bridgeen@mcb.net

Bird Reports
MANX BIRD REPORT (1947-), published in Peregrine. From G D Craine, 8 Kissack Road, Castletown, Isle of Man, IM9 1NP.e-mail: g.craine@advsys.co.uk

CALF OF MAN BIRD OBSERVATORY ANNUAL REPORT, from Secretary, Manx National Heritage, Manx Museum, Douglas, Isle of Man, IM1 3LY.

BTO Regional Representative & Regional Development Officer
RR. Dr Pat Cullen, as above, 01624 623308

Club
MANX ORNITHOLOGICAL SOCIETY. (1967; 150). Mrs A C Kaye, Cronk Ny Ollee, Glen Chass, Port St Mary, Isle of Man, IM9 5PL. 01624 834015.
Meetings:1st Tues in month, 7.30pm, Union Mills Hall

Ringing Group
CALF OF MAN BIRD OBSERVATORY. Tim Bagworth, Calf of Man, c/o Kionsleau, Plantation Road, Port St Mary, Isle of Man, IM9 5AY. Mobile 07624 462858
MANX RINGING GROUP. Chris Sharpe, 33 Mines Road, Laxey, Isle of Man, IM4 7NH. 01624 861130; e-mail: chris@manxbirdatlas.org.uk

Wetland Bird Survey Organiser Pat Cullen, Troutbeck, Cronkbourne, Braddan, Isle of Man, IM4 4QA. (H)01624 623308; (W)01624 676774; e-mail: bridgeen@mcb.net

Wildlife Trust
MANX WILDLIFE TRUST. (1973; 900). The Courtyard, Tynwald Mills, St Johns, Isle of Man IM4 3AE. 01624 801985; fax 01624 801022; e-mail: manxwt@cix.co.uk www.wildlifetrust.org.uk/manxwt/

CHANNEL ISLANDS

Ringing Group
The Channel Islands ringing scheme is run by the Société Jersiaise.

ALDERNEY

Bird Recorder
Mark Atkinson, 1 Birdcage Terrace, Birdcage Row, Alderney GY9 3XR; 01481 823286; (M)07781 421215. e-mail: atkinson@cwgsy.net

Bird Report
ALDERNEY SOCIETY ORNITHOLOGY REPORT (1992-), from Recorder.

BTO Regional Representative
Jamie Hooper, 1 Trinity Cottages, Torteval, Guernsey, GY8 0QD. Tel/fax 01481 266924; e-mail: jamie.hooper@cwgsy.net

Wildlife Trust
ALDERNEY WILDLIFE TRUST
Alderney Information Centre, Victoria Street, St Anne, Alderney GY9 3AA. 01481 822935; (Fax)01481 822935; e-mail: info@alderneywildlife.org
www.alderneywildlife.org

GUERNSEY

Bird Atlas/Avifauna
Birds of the Bailiwick of Guernsey

Bird Recorder
Mark Lawlor, Pentland, 15 Clos des Pecqueries, La Passee, St Sampson's, Guernsey, GY2 4TU. 01481 258168. e-mail: mplawlor@cwgsy.net

Bird Report
REPORT & TRANSACTIONS OF LA SOCIÉTIÉ GUERNESIAISE (1882-), from Recorder.
Jamie Hooper, 1 Trinity Cottages, Torteval, Guernsey, GY8 0QD. Tel/fax 01481 266924; e-mail: jamie.hooper@cwgsy.net

BTO Regional Representative
Jamie Hooper, 1 Trinity Cottages, Torteval, Guernsey, GY8 0QD. Tel/fax 01481 266924; e-mail: jamie.hooper@cwgsy.net

Clubs
LA SOCIÉTIÉ GUERNESIAISE (Ornithological Section). (1882; 30). MP Lawlor, St Etienne, Les Effards, Guernsey GY2 4YN. 01481 241336.
www.societe.org.gg

Meetings: 8pm, 1st Thurs of the month, Candie Gardens lecture theatre.

RSPB Local Group
GUERNSEY. (1975; 350+). Michael Bairds, Les Quatre Vents, La Passee, St Sampsons, Guernsey, GY2 4TS. 01481 255524; e-mail: mikebairds@cwgsy.net
www.rspbguernsey.co.uk
Meetings: Candie Museum, Candie Road, St. Peter Port.

Wetland Bird Survey Organiser
GUERNSEY COAST. Wayne Turner, Rooster's View, Rue de La Boullerie, St Andrews, Guernsey, GY6 8XQ. 01481 239832; e-mail: roosters@cwgsy.net

Wildlife Hospital
GUERNSEY. GSPCA ANIMAL SHELTER. Mrs Jayne Le Cras, Rue des Truchots, Les Fiers Moutons, St Andrews, Guernsey, Channel Islands, GY6 8UD. 01481 257261; e-mail: jaynelecras@gspca.org.gg
All species. Modern cleansing unit for oiled seabirds. 24-hour emergency service. Veterinary support.

JERSEY

Bird Recorder
Tony Paintin, 16 Quennevais Gardens, St Brelade, Jersey, Channel Islands, JE3 8LH. 01534 741928; e-mail: cavokjersey@hotmail.com

Bird Report
JERSEY BIRD REPORT, from Secretary (Publications), Société Jersiaise.

BTO Regional Representative
Tony Paintin, 16 Quennevais Gardens, St Brelade, Jersey, Channel Islands, JE3 8FQ; e-mail: cavokjersey@hotmail.com

Club
SOCIÉTIÉ JERSIAISE (Ornithological Section). (1948; 40). Roger Noel, 7 Pier Road, St Helier, Jersey, JE2 4XW. 01534 758314.
www.societe-jersiaise.org
Meetings: 8.00pm, alternate Thursdays throughout the year, Museum in St.Helier.

Wildlife Hospital
JERSEY. JSPCA ANIMALS' SHELTER. Pru Bannier, 89 St Saviour's Road, St Helier, Jersey, JE2 4GJ. 01534 724331; fax 01534 871797; e-mail: info@jspca.org.je. All species. Expert outside support for owls and raptors. Oiled seabird unit. Veterinary surgeon on site. Educational Centre.

NORTHERN IRELAND

Bird Recorder
George Gordon, 2 Brooklyn Avenue, Bangor, Co Down, BT20 5RB. 028 9145 5763;
e-mail: gordon@ballyholme2.freeserve.co.uk

Bird Reports
NORTHERN IRELAND BIRD REPORT, from Secretary, Northern Ireland, Birdwatchers' Association (see National Directory).

IRISH BIRD REPORT, Included in Irish Birds, BirdWatch Ireland in National Directory.

COPELAND BIRD OBSERVATORY REPORT, from see Reserves.

BTO Regional Representatives
ANTRIM & BELFAST. Position vacant.

ARMAGH. David W A Knight, 20 Mandeville Drive, Tandragee, Craigavon, Co Armagh, BT62 2DQ. 01762 840658 or 028 38 840658;
e-mail: david.knight@waterni.gov.uk

CO DOWN. Position vacant.

LONDONDERRY. Charles Stewart, Bravallen, 18 Duncrun Road, Bellarena, Limavady, Co Londonderry, BT49 0JD. 028 77 750468.

TYRONE SOUTH. Declan Coney, 028 867 35868;
e-mail: declanconey@utvinternet.com

TYRONE NORTH. Mary Mooney, 20 Leckpatrick Road, Ballymagorry, Strabane, Co Tyrone, BT82 0AL. 028 7188 2442;
e-mail: memooney@foxlodge.healthnet.co.uk

Clubs
NORTHERN IRELAND BIRDWATCHERS' ASSOCIATION See National Directory.

NORTHERN IRELAND ORNITHOLOGISTS' CLUB See National Directory.

CASTLE ESPIE BIRDWATCHING CLUB. (1995; 60). Dot Blakely, 8 Rosemary Park, Bangor, Co Down, BT20 3EX. 028 9145 0784

Ringing Groups
ANTRIM & ARDS RG. M McNeely, 35 Balleyvalley Heights, Banbridge, Co Down, BT32 4AQ. 028 406 29823

COPELAND BIRD OBSERVATORY. C W Acheson, 28 Church Avenue, Dunmurry, Belfast, BT17 9RS.

NORTH DOWN RINGING GROUP. Hugh Thurgate, 16 Inishmore, Killyleagh, Downpatrick, Co Down, BT30 9TP.

RSPB Local Groups
ANTRIM. (1977; 23). Brenda Campbell . 028 9332 3657; email: brender49@hotmail.com
Meetings: 8pm, 2nd Monday of the month. College

of Agriculture Food & Rural Enterprise, 22 Greenmount Road, Antrim.

BANGOR. (1973; 25). Michael Richardson, 028 9185 7041; e-mail: m_t.richardson@lineone.net
Meetings: Trinity Presbyterian Church Hall, Main Street, Bangor.

BELFAST. (1970; 130). Ron Houston, 028 9076 8887.
Meetings: Cooke Centenary Church Hall, Cooke Centenary Church Hall, Ormeau Rd, Belfast.

COLERAINE. (1978; 45). Peter Robinson, 34 Blackthorn Court, Coleraine, Co Londonderry, BT52 2EX. 028 7034 4361; (mob)0780 3529472;
e-mail: peter-g@pgrobinson.freeserve.co.uk
Meetings: 7.30pm, third Monday of the month (Sept-Apr), Alliance Youth Works, Atillery Road, Coleraine.

FERMANAGH. (1977; 28). Barbara Johnston, 028 6634 1708;
e-mail: barbara.johnston@btopenworld.com
Meetings: 7.30pm, 4th Tuesday of the month, St Macartans Church Hall.

LARNE. (1974; 35). Jimmy Christie, 314 Coast Road, Ballygally, Co Antrim, BT40 2QZ. 028 2858 3223
Meetings: 7.30pm, 1st Wednesday of the month, Larne Grammar School.

LISBURN. (1978; 30). David McCreedy, 10 Downside Avenue, Banbridge, Co Down, BT32 4BP. 028 4062 6125; e-mail: david@dmccreedy@wanadoo.co.uk
Meetings: Friends Meeting House, 4 Magheralave Road, Lisburn.

Wetland Bird Survey Organisers
ANTRIM, BELFAST LOUGH. John O'Boyle, Environment & Heritage Service, Commonwealth House, 35 Castle Street, Belfast BT1 1GU. 028 9054 6521;
e-mail: john.oboyle@doeni.gov.uk

ANTRIM, LOUGHS NEAGH & BEG. Steve Foster, Peatlands Park, 33 Derryhubbert Road, Verner's Bridge, Dungannon BT71 6NW. (W)028 3832 2398;
e-mail: Stephen.foster@doeni.gov.uk

ARMAGH, LOUGHS NEAGH & BEG. Steve Foster, Peatlands Park, 33 Derryhubbert Road, Verner's Bridge, Dungannon, BT71 6NW. (W)028 3832 2398;
e-mail: Stephen.foster@doeni.gov.uk

DOWN, BELFAST LOUGH. John O'Boyle, Environment & Heritage Service, Commonwealth House, 35 Castle Street, Belfast BT1 1GU. 028 9054 6521;
e-mail: john.oboyle@doeni.gov.uk

DOWN, CARLINGFORD LOUGH. Frank Carroll, 292 Barcroft Park, Newry, Co. Down BT35 8ET. (H) 01693 68015

DOWN, LOUGHS NEAGH & BEG. Steve Foster, Peatlands Park, 33 Derryhubbert Road, Verner's

Bridge, Dungannon, BT71 6NW. (W)028 3832 2398;
e-mail: Stephen.foster@doeni.gov.uk

DOWN, OUTER ARDS. Neil McCulloch, Environment &
Heritage Service, Commonwealth House, 35 Castle
Street, Belfast BT1 1GU. 01232 251477;
e-mail: neil.mcCulloch@doeni.gov.uk

DOWN, STRANGFORD LOUGH. Paddy Mackie, Mahee
island, Comber, Newtonards, Co. Down, BT23 6EP.
(Tel/fax)028 9754 1420

FERMANAGH. Neil McCulloch, Environment & Heritage
Service, Commonwealth House, 35 Castle Street,
Belfast BT1 1GU. 01232 251477;
e-mail: neil.mcCulloch@doeni.gov.uk

LONDONDERRY, BANN ESTUARY. Hill Dick, 33 Hopefield
Avenue, Portrush, Co. Antrim BT56 8HB.

LONDONDERRY, LOUGHS NEAGH & BEG. . Steve
Foster, Peatlands Park, 33 Derryhubbert Road,

Verner's Bridge, Dungannon, BT71 6NW. (W)028 3832
2398; e-mail: Stephen.foster@doeni.gov.uk

TYRONE, LOUGHS NEAGH & BEG. Steve Foster,
Peatlands Park, 33 Derryhubbert Road, Verner's
Bridge, Dungannon, BT71 6NW. (W)028 3832 2398;
e-mail: Stephen.foster@doeni.gov.uk

Wildlife Hospital
TACT WILDLIFE CENTRE. Mrs Patricia Nevines, 2
Crumlin Road, Crumlin, Co Antrim, BT29 4AD. Tel/fax
028 944 22900; e-mail: t.a.c.t@care4free.net
All categories of birds treated and rehabilitated;
released where practicable, otherwise given a home.
Visitors (inc. school groups and organisations) welcome
by prior arrangement. Veterinary support.

Wildlife Trust
ULSTER WILDLIFE TRUST. (1978; 2100). 3 New Line,
Crossgar, Co Down, BT30 9EP. 028 4483 0282; (fax)
028 4483 0888; e-mail: info@ulsterwildlifetrust.org
www.ulsterwildlifetrust.org

REPUBLIC OF IRELAND

Bird Recorders
Non rarities. BirdWatch Ireland, Rockingham House,
Newcastle, Co. Wicklow, Ireland. +353 (0)1 2819878;
e-mail: info@birdwatchireland.org

Rarities. Paul Milne, 100 Dublin Road, Sutton, Dublin
13, +353 (0)1 8325653;
e-mail: paul.milne@oceanfree.net

CLARE, John Murphy, Applewood, Ballycarr,
Newmarket-on-Fergus

CORK, Mark Shorten, 106 Sunday's Well Road,
Sunday's Well

DONEGAL, Ralph Sheppard, Carnowen House, Lifford

EAST COAST, Dick Coombes, C/o BirdWatch Ireland,
Rockingham House, Newcastle

GALWAY, Tim Griffin, 74 Monalee Heights,
Knocknacarra

KERRY, Edward Carty, 3 The Orchard, Ballyrickard,
Tralee

MAYO, Tony Murray, National Parks and Wildlife,
Lagduff More, Ballycroy, Westport

MID-SHANNON, Stephen Heery, Laurencetown,
Ballinasloe

WATERFORD, Paul Walsh, 16 Castlepoint,
Crosshaven, Co. Cork;

e-mail: pmwalsh@waterfordbirds.com

WEXFORD, Chris Wilson, Wexford Wildfowl Reserve,
North Slob

Bird Reports
IRISH BIRD REPORT, Warden, BirdWatch Ireland in
National Directory).

CAPE CLEAR BIRD OBSERVATORY ANNUAL REPORT,

CORK BIRD REPORT (1963-71; 1976-), contact
BirdWatch Ireland.

EAST COAST BIRD REPORT (1980-), contact
BirdWatch Ireland.

BirdWatch Ireland Branches
BirdWatch Ireland has branches in most of the
counties in The Republic of Ireland. Contact Birdwatch
Ireland HQ for details (See National Directory)

Ringing Groups
CAPE CLEAR BIRD OBSERVATORY. S Wing, 30 Irsher
Street, Appledore, Devon, EX39 1RZ.

GREAT SALTEE RS. O J Merne, 20 Cuala Road, Bray,
Co Wicklow,

MUNSTER RG. K P Collins, Ballygambon , Lisronagh,
Clonmel, Co Tipperary. e-mail: kevcoll@indigo.ie

SHANNON WADER RG. P A Brennan, The Crag,
Stonehall, Newmarket-on-Fergus, Co Clare,

ARTICLES IN BIRD REPORTS

Ayrshire Bird and Butterfly Report 2004
The Lesser Whitethroat in Ayrshire by Tom Byars.
The Kestrel In Ayrshire 2004 by Gordon Riddle.
Sparrowhawk - Ayrshire Breeding Details 2004 by Ian Todd.
Wood Nuthatch Nesting in Ayrshire by Roger and Angela Hissett.

The Birds of Berkshire 2002
Scarce Gulls in Berkshire by Chris Heard.
Berkshire Bird Index 2002 by Chris Robinson.
Breeding Waders of Wet Meadows by Chris Robinson.
Nestbox Survey at Moor Green Lakes by Kevin Briggs.
Jubilee River and Dorney Wetlands in 2002 by Brian Clews.

Breconshire Bird Report 2004
Breconshire Birds – Systematic List 2004.
Paddyfield Warbler (*Acrocephalus agricola*) – A New Species for Breconshire and Wales by Jerry Lewis.
Aquatic Warblers (*Acrocephalus paludicola*) at Llangorse Lake by Jerry Lewis.
Llangorse Lake Ringing Group Report.

Buckinghamshire Bird Report 2003
The two-barred Crossbill at Hillmotts Wood, Hedgerley by Mark and Paul McManus.
Black Kite at Little Marlow by Alan Stevens, Mick McQuand, Chris Bullock.
Black Kite at Ivinghoe Beacon by Mike Walker.
Least Sandpiper at Startops Reservoir by Mike Campbell.
A New County Record by Simon Nichols and Ashley Beolens.
Colour-ringed gulls at Little

Marlow in 2003 by Chris Goodlie.
A Site Guide to Little Linford Wood by Robert Hill.

Cambridgeshire Bird Report 2003
Caspian Gull Identification by Dick Newell.
Baseline Bird Populations on the New Coton Farm Reserve near Cambridge and Comparison with 1967-81 Populations by Louise Bacon and Vince Lea.
Red Data Woodland Species – Willow tit, Marsh Tit and Lesser Spotted Woodpecker, A Survey in Summer 2003 by Louise Bacon and Bill Jordan.
Bird Biodiversity Action Plans by Vince Lea and Catherine Weightman.
The Flooding of Holt Fen in 2003 by Bruce Martin.

Ceredigion Bird Report 2003
Squacco Heron: first county record by Roy Bamford.
Cory's Shearwater: first county record by Lynden Lomax.
The Heron in Ceredigion by Peter Davis.
A Century of Breeding Birds In Ceredigion by A Williams.

Cheshire and Wirral Bird Report 2004
Articles will cover the following topics:
Cormorants – their population, breeding and taxonomic status in the county.
Mediterranean Gull – a new breeding species.
Iberian Chiffchaffs – a 'possible' being dealt with by BBRC.
Spotted Sandpiper – a county second.

County Durham Bird Report 2004 (in preparation)
Long-eared Owls in Durham by Steve Evans.
The Reintroduction of Red Kites to County Durham by Keith Bowey.

Derbyshire Bird Report 2003
Bittern (*Botarus stellaris*) at Drakelow Wildfowl Reserve 2002 and 2003 by T Cockburn.
Great White Egret (*Ardea alba*) at Drakelow Wildfowl Reserve, 12th August to 12th October 2003 by T Cockburn.
Tawny Pipit (*Anthus campestris*) at Aston-on-Trent Gravel Pits: a new species for Derbyshire by R W Key.
A Short History of the Derbyshire Ornithological Society – The first 50 years by Steve Shaw.
Species Added to the County List Since the Formation of the Society by Richard James.

Devon Bird Report 2003
Roosting White and Pied Wagtails at Slapton Ley by Dennis Elphick.
The Status of Cirl Buntings in Devon in 2003 by Simon Wottan (RSPB Sandy).
Glossy Ibis on The Exeter Estuary 2002-2004 by Malcolm Davies (RSPB Exeter).
The Wintering Dusky Warbler in Paignton by Mike Langman.
A Parish Winter Bird Survey by Rod Bone and Peter Reay.
The 2003 Devon Barn Owl Survey – A Summary by David Ramsden and John Howells.
A Study of House Sparrows in Kingsteignton by Jon Avon.
Swifts In A Plymouth Tower 2003 by Robert, Leonard and Anne Hurrell.
Looking Back on the 1953 and 1978 Reports by Dave Jenks.
Devon Ringing Report for 2003 by Roger Dusinfen.
Staddon Point – Review of Visible Migration in a County Context, Autumn 2003 by Simon Geary.

Hampshire Bird Report 2003
Hampshire Ringing Report 2003 by DA Bell.
Mongolian Plover – a new

ARTICLES IN BIRD REPORTS

species for Hampshire by L Chappell.
Changes In Breeding Populations on a Mixed Farm 1977-2003 by T Gutteridge.
The Change in Status and Distribution of Little Egret in Hampshire 1993-2003.
Low Tide Counts in Portsmouth Harbour 2003/3 by DJ Unsworth.
The Breeding Birds Survey 1994-2003 by B Sharkey.
Hampshire River Valley Survey 2002 by GC Evans and J Eyre.
Discovery of Winter Roost of Shags in the Solent by J Crook.

The Hertfordshire Bird Report 2002 (published April 2005)
WeBS Counts for Herts in 2002 by Chris Dee.
First and Last dates of Migrants Bird Ringing Report 2002 by Chris Dee.
BTO Breeding Bird Survey results for 2002 by Chris Dee.
Black Stork at Amwell CP by Barry Reed.
Mute Swan in Herts 2002 by Jim Terry.
Solitary Sandpiper at Rye Meads by Paul Roper.
Notes on the Status of Barn Owl in Herts by Peter Wilkinson.
Lesser Yellowlegs at Amwell GP by Barry Reed.
Herts Red Data List for Birds 2003; Species text by Herts Scientific Committee.

Highland Bird Report 2003
Highland Ringing Group Report 2003 by Bob Swann and Dave Butterfield.
Two-barred Crossbills in Easter Ross by Bob Dawson.
The Status of Magpies in the Highlands by Dave Butterfield.
The Storm Petrel Colonies At Eilean Hoan and Priest Island by

Steph Elliot and Kenny Graham.

Kent Bird Report 2002 (July 2005)
A Century of Dartford Warblers by John Van der Dol.
Dartford Warbler survey 2002 by Norman McCanch.
Diet of Long-eared Owls at Sheppey by Peter Oliver.
Audouin's Gull at Dungeness by David Walker.
Paddyfield Warbler at Dungeness by David Walker.
Colonisation of Little Egrets in Kent by Gordon Allison.

Lancashire Bird Report 2003
Status of Greylag Geese in North Lancashire by John Wilson.
Water Pipits on the North Ribble Marshes by Stephen Dunston.
Status of Quail by Chris Kehoe.

Leicestershire and Rutland Bird Report 2004
Leicestershire and Rutland Ringing Report by Nigel Judson.
The Great Snipe in Leicestershire and Rutland by Andrew Harrop.
Two Interesting Shrike Specimens from Leicestershire by Andrew Harrop.
Memories of the Formation of the Leicestershire and Rutland Ornithological Society by Eric Duffey.

London Bird Report 2001
Ringing Report by R Taylor
Breeding Bird Survey by DA Coleman.
The House Martin in North London by NJ Ellis and C Herbet.
House Sparrow Monitoring in London by H Baker.
Radde's Warbler: New to the London Area by AA Bell.

Norfolk Bird and Mammal Report 2004 (to be

published autumn 2005)
Berney Marshes Reserve by Peter Allard.
Little Egret – Colonisation of Norfolk by Ron Harold.
Norfolk Bird Names by Richard Richardson, Introduction by Richard Fitter.
Bearded Tit Populations, Hickling Broad 1980-2004 by James Cadbury.
Grey-Cheeked Thrush – 1st for Norfolk by David Leech.
Alpine Accentor – 2nd for Norfolk by Simon Chidwick and Ben Murphy.
Pine Bunting – 2nd for Norfolk by Ashley Saunders.

North East Scotland Bird Report 2004
Black Duck first for North East Scotland by J Wills.
Pallid Swift first for North East Scotland by P Baxter and P&S Morrison.
The early-autumn fall of migrants by P Crockett.
North East Scotland Breeding Bird Atlas 2002-2004 by I Francis.

Northern Ireland Bird Report 2001
Red-throated Pipit – A first for Northern Ireland by Neville McUse.
Stilt Sandpiper at Lough Beg – A first for Northern Ireland by Davy Hunter.
The Demise of a Hen Harrier Roost Site by Don Scott and Philip McHaffie.
Breeding Seabirds on The Isle of Muck (Antrim) by Andrew Upton.

Wiltshire Bird Report 2004 (Hobby No.30)
Roof Nesting Gulls in Wiltshire by Peter Rock.
Late nesting Swifts using nestbox by Jonathon Pomroy.

NATIONAL
DIRECTORY

Snipe by Nick Williams

303

ARMY ORNITHOLOGICAL SOCIETY (1960; 250).

Open to MOD employees and civilians who have an interest in their local MOD estate. Activities include field meetings, expeditions, the preparation of checklists of birds on Ministry of Defence property, conservation advice and an annual bird count. Annual journal *The Osprey*, published with the RNBWS and RAFOS from easter 2001. Bulletins/newsletters twice a year.
Contact: Secretary AOS, GSV IPT, DLO Andover, Monxton Road, Andover, SP11 8HT. 01264 382910;
e-mail: secretary@aos.org.uk www.aos.org.uk

ASSOCIATION FOR THE PROTECTION OF RURAL SCOTLAND (1926).
Works to protect Scotland's countryside from unnecessary or inappropriate development, recognising the needs of those who live and work there and the necessity of reconciling these with the sometimes competing requirements of recreational use.
Contact: Director, Mr Bill Wright, Gladstone's Land, 3rd Floor, 483 Lawnmarket, Edinburgh EH1 2NT. 0131 225 7012; (Fax)0131 225 6592;
e-mail:info@ruralscotland.org
www.ruralscotland.org

ASSOCIATION OF COUNTY RECORDERS AND EDITORS (1993; 120).
The basic aim of ACRE is to promote best practice in the business of producing county bird reports, in the work of Recorders and in problems arising in managing record systems and archives. Organises periodic conferences and publishes *newsACRE*.
Contact: Secretary, M J Rogers, 2 Churchtown Cottages, Towednack, St Ives, Cornwall TR26 3AZ. 01736 796223;
e-mail: judith@gmbirds.freeserve.co.uk

BARN OWL TRUST(1988)
Registered charity. Aims to conserve the Barn Owl and its environment through conservation, education, research and information. Free leaflets on all aspects of Barn Owl conservation. Educational material inc. video and resource pack. Book 'Barn Owls on Site', a guide for planners and developers (priced). Works with and advises landowners, farmers, planners, countryside bodies and others to promote a brighter future for Britain's Barn Owls. Currently pursuing proactive conservation schemes in SW England to secure breeding sites and form a stable basis for population expansion. Open to phone calls Mon-Fri (9.30-5.30). Send SAE for information.
Contact: Secretary, Barn Owl Trust, Waterleat, Ashburton, Devon TQ13 7HU. 01364 653026;
e-mail: info@barnowltrust.org.uk
www.barnowltrust.org.uk

BIRD OBSERVATORIES COUNCIL (1970).
Objectives are to provide a forum for establishing closer links and co-operation between individual autonomous observatories and to help co-ordinate the work carried out by them. All accredited bird observatories affiliated to the Council undertake a ringing programme and provide ringing experience to those interested, most also provide accommodation for visiting birdwatchers.
Contact: Secretary, Peter Howlett, c/o Dept of Biodiversity, National Museums & Galleries, Cardiff CF10 3NP. 0292 057 3233; (Fax)0292 023 9009;
e-mail: peter.howlett@nmgw.ac.uk
www.birdobscouncil.org.uk

BIRD STAMP SOCIETY (1986; 250).
Quarterly journal *Flight* contains philatelic and ornithological articles. Lists all new issues and identifies species. Runs a quarterly Postal Auction; number of lots range from 400 to 800 per auction. UK subs £14 per annum from 1st August.
Contact: Mrs R Bradley, 31 Park View, Crossway Green, Chepstow NP16 5NA. 01291 625412;
e-mail: bradley666@lycos.co.uk
www.bird-stamps.org

BIRDWATCH IRELAND (1968; 10,000).
The trading name of the Irish Wildbird Conservancy, a voluntary body founded in 1968 by the amalgamation of the Irish Society for the Protection of Birds, the Irish Wildfowl Conservancy and the Irish Ornithologists' Club. Now the BirdLife International partner in Ireland with 21 voluntary branches. Conservation policy is based on formal research and surveys of birds and their habitats. Owns or manages an increasing number of reserves to protect threatened species and habitats. Publishes *Wings* quarterly and *Irish Birds* annually, in addition to annual project reports and survey results.
Contact: BirdWatch Ireland, Rockingham House, Newcastle, Co. Wicklow, Ireland. +353 (0)1 2819878; (Fax)+353 (0)1 2819763;
e-mail: info@birdwatchireland.org
www.birdwatchireland.ie

Figures appearing in brackets following the names of organisations indicate the date of formation and, if relevant, the current membership.

BRITISH BIRDS RARITIES COMMITTEE (1959).

The Committee adjudicates records of species of rare occurrence in Britain (marked 'R' in the Log Charts). Its annual report, which is published in *British Birds*. The BBRC also assesses records from the Channel Islands. In the case of rarities trapped for ringing, records should be sent to the Ringing Office of the British Trust for Ornithology, who will in turn forward them to the BBRC.
Contact: Hon Secretary, M J Rogers, 2 Churchtown Cottages, Towednack, St Ives, Cornwall TR26 3AZ. 01736 796223; e-mail: secretary@bbrc.org.uk
www.bbrc.org.uk

BRITISH DRAGONFLY SOCIETY 1983; 1604.

The BDS aims to promote the conservation and study of dragonflies. Members receive two issues of *Dragonfly News* and *BDS Journal* each year in spring and autumn. There are countrywide field trips, an annual members day and training is available on aspects of dragonfly ecology.
Contact: Hon Secretary, Dr WH Wain, The Haywain, Hollywater Road, Bordon, Hants GU35 0AD; e-mail: thewains@ukonline.co.uk
www.dragonflysoc.org.uk

BRITISH FALCONERS' CLUB (1927; 1200).

Largest falconry club in Europe, with regional branches. Its aim is to encourage responsible falconers and conserve birds of prey by breeding, holding educational meetings and providing facilities, guidance and advice to those wishing to take up the sport. Publishes *The Falconer* annually and newsletter twice yearly.
Contact: Director, John Callaghan, Home Farm, Hints, Tamworth, Staffs B78 2DW. Tel/(Fax)01543 481737;
e-mail: admin@ britishfalconersclub.co.uk
www.britishfalconersclub.co.uk

BRITISH MUSEUM (NAT HIST) see Walter Rothschild Zoological Museum

BRITISH ORNITHOLOGISTS' CLUB (1892; 600).

A registered charity, the Club's objects are 'the promotion of scientific discussion between members of the BOU, and others interested in ornithology, and to facilitate the publication of scientific information in connection with ornithology'. The Club maintains a special interest in avian systematics, taxonomy and distribution. About eight dinner meetings are held each year. Publishes the *Bulletin of the British Ornithologists' Club* quarterly, also (since

1992) a continuing series of occasional publications.
Contact: BOC Office, PO Box 417, Peterborough PE7 3FX. 01733 844 820; e-mail: boc.admin@bou.org.uk
www.boc-online.org

BRITISH ORNITHOLOGISTS' UNION (1858; 2,000).

Founded by Professor Alfred Newton FRS and one of the world's oldest and most respected ornithological societies. It aims to promote ornithology within the scientific and birdwatching communities, both in Britain and around the world. This is largely achieved by the publication of its quarterly international journal, *Ibis* (1859-), featuring work at the cutting edge of our understanding of the world's birdlife. An active programme of meetings, seminars and conferences inform birdwatchers and ornithologists about the work being undertaken around the world. This often includes research projects that have received financial assistance from the BOU's ongoing programme of Ornithological Research Grants, which includes student sponsorship. The BOU also runs the Bird Action Grant scheme to assist projects aimed at conserving or researching species on the UK's Biodiversity Action Plan (BAP) list. Part of the BOU Library is housed at the Linnean Society, whilst copies of exchange journals, books reviewed in *Ibis* and offprints are held as part of the Alexander Library in the Zoology Department of the University of Oxford (see Edward Grey Institute). The BOU Records Committee maintains the official British List) see below.
Contact: Steve Dudley, Dept of Zoology, University of Oxford, South Parks Road, Oxford OX1 3PS. (Tel/fax)01865 281 842; e-mail: bou@bou.org.uk
www.bou.org.uk   www.ibis.ac.uk

BRITISH ORNITHOLOGISTS' UNION RECORDS COMMITTEE

The BOURC is a standing committee of the British Ornithologists' Union. Its function is to maintain the British List, the official list of birds recorded in Great Britain. The up-to-date list can be viewd on the BOU website. Where vagrants are involved it is concerned only with those which relate to potential additions to the British List (ie first records). In this it differs from the British Birds Rarities Committee (qv). In maintaining the British List, it also differs from the BBRC in that it examines, where necessary, important pre-1950 records, monitors introduced species for possible admission to or deletion from the List, and reviews taxonomy and nomenclature generally. BOURC reports are published in Ibis. Decisions contained in these reports which affect the List are also announced via the popular birdwatching press. **Contact:** Steve Dudley, Dept of Zoology, University of Oxford, South Parks Road, Oxford OX1 3PS. (Tel/fax)01865 281 842; e-mail: bourc@bou.org.uk www.bou.org.uk

BRITISH TRUST FOR ORNITHOLOGY (1933; 12,500).

A registered charity governed by an elected Council, it has a rapidly growing membership and enjoys the support of a large number of county and local birdwatching clubs and societies through the BTO/Bird Clubs Partnership. Its aims are: 'To promote and encourage the wider understanding, appreciation and conservation of birds through scientific studies using the combined skills and enthusiasm of its members, other birdwatchers and staff.' Through the fieldwork of its members and other birdwatchers, the BTO is responsible for the majority of the monitoring of British birds, British bird population and their habitats. BTO surveys include the National Ringing Scheme, the Nest Record Scheme, the Breeding Bird Survey (in collaboration with JNCC and RSPB), and the Waterways Breeding Bird Survey - all contributing to an integrated programme of population monitoring. The BTO also runs projects on the birds of farmland and woodland, also (in collaboration with WWT, RSPB and JNCC) the Wetland Bird Survey, in particular Low Tide Counts. Garden BirdWatch, which started in 1995, now has more than 14,000 participants. The Trust has 140 voluntary regional representatives (see County Directory) who organise fieldworkers for the BTO's programme of national surveys in which members participate. The results of these co-operative efforts are communicated to government departments, local authorities, industry and conservation bodies for effective action. For details of current activities see National Projects. Members receive BTO News six times a year and have the option of subscribing to the thrice-yearly journal, Bird Study and twice yearly Ringing & Migration. Local meetings are held in conjunction with bird clubs and societies; there are regional and national birdwatchers' conferences, and specialist courses in bird identification and modern censusing techniques. Grants are made for research, and members have the use of a lending and reference library at Thetford and the Alexander Library at the Edward Grey Institute of Field Ornithology (qv). **Contact:** Director, Professor Jeremy J D Greenwood, British Trust for Ornithology, The Nunnery, Thetford, Norfolk IP24 2PU. 01842 750050; (Fax)01842 750030; e-mail: info@bto.org www.bto.org

BRITISH WATERFOWL ASSOCIATION

The BWA is an association of enthusiasts interested in keeping, breeding and conserving all types of waterfowl, including wildfowl and domestic ducks and geese. It is a registered charity, without trade affiliations, dedicated to educating the public about waterfowl and the need for conservation as well as to raising the standards of keeping and breeding ducks, geese and swans in captivity. **Contact:** Mrs Sue Schubert, PO Box 163, OxtedRH8 0WP. 01892 740212; e-mail: info@waterfowl.org.uk www.waterfowl.org.uk

BRITISH WILDLIFE REHABILITATION COUNCIL (1987).

Its aim is to promote the care and rehabilitation of wildlife casualties through the exchange of information between people such as rehabilitators, zoologists and veterinary surgeons who are active in this field. Organises an annual symposium or workshop. Publishes a regular newsletter. Supported by many national bodies including the Zoological Society of London, the British Veterinary Zoological Society, the RSPCA, the SSPCA, and the Vincent Wildlife Trust. **Contact:** Secretary, Tim Thomas, Wildlife Department, RSPCA, Wilberforce Way, Southwater, Horsham, W Sussex RH13 9RS. ; e-mail: tim@bwrc.org.uk www.bwrc.org.uk

BTCV (formerly British Trust for Conservation Volunteers) (1959).

It's mission is to create a more sustainable future by inspiring people and improving places. Between 2004 and 2008 it aims to enrich the lives of one million people, through involvement with BTCV, through volunteering opportunities, employment, improved health, and life skills

NATIONAL ORGANISATIONS

development; to Improve the biodiversity and local environment of 20,000 places and to support active citizenship in 5,000 community based groups. BTCV is governed by a board of 6 volunteer trustees elected by the charity membership and currently supports 140,000 volunteers to take practical action to improve their urban and rural environments. Publishes a quarterly newsletter, *The Conserver*, a series of practical handbooks and a wide range of other publications. Further information and a list of local offices is available from the above address.
Contact: Conservation Centre, 163 Balby Road, Doncaster, South Yorkshire DN4 0RH. 01302 572 244; (Fax) 01302 310 167. 01302 572 244; (Fax) 01302 310 167; e-mail: Information@btcv.org.uk
www.btcv.org.uk

BTCV SCOTLAND
Runs 7-14 day 'Action Breaks' in Scotland during which participants undertake conservation projects; weekend training courses in environmental skills; midweek projects in Edinburgh, Glasgow, Aberdeen, Stirling and Inverness.
Contact: Balallan House, 24 Allan Park, Stirling FK8 2QG. 01786 479697; (Fax)01786 465359; e-mail: scotland@btcv.org.uk www.btcv.org.uk

CAMPAIGN FOR THE PROTECTION OF RURAL WALES (1928; 3,500)
Its aims are to help the conservation and enhancement of the landscape, environment and amenities of the countryside, towns and villages of rural Wales and to form and educate opinion to ensure the promotion of its objectives. It recognises the importance of the indigenous cultures of rural Wales and gives advice and information upon matters affecting protection, conservation and improvement of the visual environment.
Contact: Director, Peter Ogden, Ty Gwyn, 31 High Street, Welshpool, Powys SY21 7YD. 01938 552525/556212; (Fax)552741; e-mail: info@cprwmail.org.uk
www.cprw.org.uk

CENTRE FOR ECOLOGY & HYDROLOGY (CEH)
The work of the CEH, a component body of the Natural Environment Research Council, includes a range of ornithological research, covering population studies, habitat management and work on the effects of pollution. The CEH has a long-term programme to monitor pesticide and pollutant residues in the corpses of predatory birds sent in by birdwatchers, and carries out detailed studies on affected species. The Biological Records Centre (BRC), which is part of the CEH, is responsible for the national biological data bank on plant and animal distributions (except birds).
Contact: Director, Prof Pat Nuttall, Centre for Ecology and Hydrology, Polaris House, North Star ,SwindonSN2 1EU. 01793 442516; (Fax) 01793 442528.
e-mail: director@ceh.ac.uk
www.ceh.ac.uk

COUNTRY LAND AND BUSINESS ASSOCIATION (1907; 50,000).
The CLA is at the heart of rural life and is the voice of the countryside for England and Wales, campaigning on issues which directly affect those who live and work in rural communities. Its members together manage 60% of the countryside. CLA members range from some of the largest landowners, with interests in forest, moorland, water and agriculture, to some of the smallest with little more than a paddock or garden.
Contact: Secretary, 16 Belgrave Square, London, SW1X 8PQ. 020 7235 0511;(Fax) 020 7235 4696;
e-mail: mail@cla.org.uk www.cla.org.uk

COUNTRYSIDE AGENCY
Is the statutory body working to make life better for people in the countryside and improve the quality of the countryside for everyone. The Countryside Agency will help to achieve the following; empowered, active and inclusive communities, high standards of rural services, vibrant local economies, all countryside managed sustainably, recreation opportunities for all, realising the potential of the urban fringe. The Countryside Agency is funded by Defra who is a major customer for their work.
Offices:
North East Region. Cross House, Westgate Road, Newcastle upon Tyne NE1 4XX, 0191 269 1600; (Fax)0191 269 1601
North West Region. 7th Floor, Bridgewater House, Whitworth Street, Manchester M1 6LT. 0161 237 1061;(fax)0161 237 1062
Haweswater Road, Penrith, Cumbria CA11 7EH. 01768 865752;(fax)01768 890414.
South West Region. Bridge House, Sion Place, Clifton Down, Bristol BS8 4AS. 0117 973 9966;(fax)0117 923 8086.
Second Floor, 11-15 Dix's Field, Exeter EX1

NATIONAL ORGANISATIONS

1QA. 01392 477150;(fax)01392 477151.
Yorkshire & The Humber Region. 4th Floor
Victoria Wharf, No 4 The Embankment,
Sovereign Street, Leeds LS1 4BA. 0113 246
9222;(fax)0113 246 0353
East Midlands Region. 18, Market Place,
Bingham, Nottingham NG13 9AP. 01949
876200;(fax)01949 876222
West Midlands Region. 1st Floor, Vincent House,
Tindal Bridge, 92-93 Edward Street, Birmingham
B1 2RA. 0121 233 9399:(fax)0121 233 9286
Eastern Region. 2nd Floor, City House, 126-128
Hills Road, Cambridge CB2 1PT. 01223
354462;(fax)01223 273550.
South East Region. Dacre House, Dacre Street,
London SW1H 0DH, 020 7340 2900;(fax)020
7340 2911.
Sterling House, 7 Ashford Road, Maidstone
ME14 5BJ. 01622 765222;(fax)01622 662102.
Contact: Acting Chief Executive, Margaret
Clark, John Dower House, Crescent Place,
Cheltenham, Glos GL50 3RA. 01242 521381;
(Fax)01242 584270;
e-mail: info@countryside.gov.uk
www.countryside.gov.uk

COUNTRYSIDE COUNCIL FOR WALES

The Government's statutory adviser on wildlife,
countryside and maritime conservation matters
in Wales. It is the executive authority for the
conservation of habitats and wildlife. Through
partners, CCW promotes protection of
landscape, opportunities for enjoyment, and
support of those who live, work in, and manage
the countryside. It enables thesepartners,
including local authorities, voluntary
organisations and interested individuals, to
pursue countryside management projects
through grant aid. CCW is accountable to the
National Assembly for Wales which appoints its
Council members and provides its annual grant-
in-aid.

Area Offices

West Area. Plas Gogerddan, Aberystwyth,
Ceredigian, SY23 3EE. 01970 821100;
e-mail: westarea@ccw.gov.uk
North West Area. Llys y bont, Ffordd y Parc,
Parc Menai, Bangor, Gwynedd, LL57 4BH.
01248 672500;
e-mail: northwestarea@ccw.gov.uk
South and East Area. Unit 7, Castleton Court,
Fortran Road, St Mellons, Cardiff CF3 0LT. 02920
772400.
South and East Area. Eden House, Ithon Road,
Llandrindod, Powys, LD1 6AS, 01597 827400;
email: llandrindod@ccw.gov.uk
North East Area. Victoria House, Grosvenor
Street, Mold CH7 1EJ. 01352 706600

Contact: Maes-y-Ffynnon, Penrhosgarnedd,
Bangor, Gwynedd. 01248 385500; (Fax)01248
355782; (Enquiry unit) 0845 1306229.
e-mail: enquiries@ccw.gov.uk
www.ccw.gov.uk

CPRE (formerly Council for the Protection of Rural England) (1926; 60,000).

Patron HM The Queen. CPRE now has 43
county branches and 200 local groups. We are
people who care passionately about our
countryside and campaign for it to be
protected and enhanced for the benefit of
everyone. Membership open to all.
Contact: CPRE National Office, 128 Southwark
Street, London SE1 0SW. 020 7981 2800;
(fax)020 7981 2899; e-mail: info@cpre.org.uk
www.cpre.org.uk

DEPARTMENT OF THE ENVIRONMENT FOR NORTHERN IRELAND

Responsible for the
declaration and management
of National Nature Reserves,
the declaration of Areas of
Special Scientific Interest,
the administration of Wildlife
Refuges, the classification of
Special Protection Areas under the EC Birds
Directive, the designation of Special Areas of
Conservation under the EC Habitats Directive
and the designation of Ramsar sites under the
Ramsar Convention. It administers the Nature
Conservation and Amenity Lands (Northern
Ireland) Order 1985, the Wildlife (Northern
Ireland) Order 1985, the Game Acts and the
Conservation (Natural Habitats, etc) Regulations
(NI) 1995 and the Environment (Northern
Ireland) Order 2002.
Contact: Environment and Heritage Service,
Commonwealth House, 35 Castle Street, Belfast
BT1 1GU. 028 9054 6521; (polution hotline;
0800 807 060); e-mail:
bob.bleakley@doeni.gov.uk
www.ehsni.gov.uk

DISABLED BIRDER'S ASSOCIATION (2000;500).

The DBA is a registered charity and international
movement, which aims to promote access to
reserves and other birding places and to a
range of services, so that people with different
needs can follow the birding obsession as freely
as able-bodied people. Membership is currently
free and open to all, either disabled or able-
bodied. We are keen for new members to help
give a strong voice to get our message across
to those who own and manage nature reserves
to ensure that they think access when they are

planning and improving their facilities. We are also seeking to influence those who provide birdwatching services and equipment. The DBA runs overseas trips.

Contact: DBA, 18 St Mildreds Road, Cliftonville, Margate, Kent CT9 2LT. e-mail: bo@fatbirder.com
www.disabledbirdersassociation.org.uk

EDWARD GREY INSTITUTE OF FIELD ORNITHOLOGY (1938).

The EGI takes its name from Edward Grey, first Viscount Grey of Fallodon, a life-long lover of birds and former Chancellor of the University of Oxford, who gave his support to an appeal for its foundation capital. The Institute now has a permanent research staff; it usually houses some 12-15 research students, two or three senior visitors and post-doctoral research workers. The EGI also houses Prof Sir John Krebs's Ecology & Behaviour Group, which studies the ecology, demography and conservation of declining farmland birds. Field research is carried out mainly in Wytham Woods near Oxford and on the island of Skomer in West Wales. In addition there are laboratory facilities and aviary space for experimental work. The Institute houses the Alexander Library, one of the largest collections of 20th century material on birds in the world. The library is supported by the British Ornithologists Union who provides much of the material. Included in its manuscript collections are diaries, notebooks and papers of ornithologists. It also houses the British Falconers Club library. The Library is open to members of the BOU and the Oxford Ornithological Society; other bona fide ornithologists may use the library by prior arrangement.

Contact: Clare Rowsell, PA to Prof. Sheldon, Department of Zoology, South Parks Road, Oxford OX1 3PS. 01865 271274, Alexander Library 01865 271143; e-mail: lynne.bradley@zoology.oxford.ac.uk; e-mail:clare.rowsell@zoo.ox.ac.ukweb-site, EGI: http://egizoosrv.zoo.ox.ac.uk/EGI/EGIhome.htm
web-site library http://users.ox.ac.uk/~zoolib/

ENGLISH NATURE

Advises Government on nature conservation in England. It promotes, directly and through others, the conservation of England's wildlife and geology within the wider setting of the UK and its international responsibilities. It selects, establishes and manages National Nature Reserves (many of which are described in Part 9: Reserves and Observatories), and identifies and notifies Sites of Special Scientific Interest.

ENGLISH NATURE

It provides advice and information about nature conservation and supports and conducts research relevant to these functions. Through the Joint Nature Conservation Committee (qv), English Nature works with sister organisations in Scotland and Wales on UK and international nature conservation issues.

Local Teams:

Bedfordshire and Cambridgeshire. Ham Lane House, Ham Lane, Nene Park, Orton Waterville, Peterborough PE2 5UR, 01733 405850; fax 01733 394093; e-mail beds.cambs.nhants@english-nature.org.uk.

Cheshire and Lancashire. Pier House, 1st Floor, Wallgate, Wigan WN3 4AL. 01942 820342; fax 01942 820364; e-mail northwest@english-nature.org.uk.

Cornwall & Isles of Scilly. Trevint House, Strangways Villas, Truro TR1 2PA. 01872 265710; fax 01872 262551; e-mail cornwall@english-nature.org.uk.

Cumbria. Juniper House, Murley Moss, Oxenholme Road, Kendal LA9 7RL. 01539 792800; fax 01539 792830; e-mail; cumbria@english-nature.org.uk.

Devon. Level 2, Rensdale House, Bonhay Rd, Exeter, EX4 3AW, 01392 889770; fax 01392 437999; e-mail devon@english-nature.org.uk.

Dorset. Slepe Farm, Arne, Wareham, Dorset BH20 5BN. 01929 557450; fax 01929 554752; e-mail dorset@english-nature.org.uk.

Eastern Area. The Maltings, Wharf Road, Grantham, Lincs NG31 6BH. 01476 584800; fax 01476 570927; e-mail eastmidlands@english-nature.org.uk

Essex, London and Hertfordshire. Harbour House, Hythe Quay, Colchester CO2 8JF. 01206 796666; fax 01206 794466; e-mail; essex.herts@english-nature.org.uk.

Hampshire and Isle of Wight. 1 Southampton Road, Lyndhurst, Hants SO43 7BU. 02380 283944; fax 02380 283834; e-mail; hants.iwight@english-nature.org.uk.

Hereford and Worcester. Bronsil House, Eastnor, Nr Ledbury HR8 1EP. 01531 638500; fax 01531 638501; e-mail Hereford and Worcester @english-nature.org.uk.

Humber to Pennines. Bull Ring House, Northgate, Wakefield, W Yorks WF1 1HD. 01924 334500; fax 01924 201507;

e-mail; humber.pennines@english-nature.org.uk.
Kent. The Countryside Management Centre,
Coldharbour Farm, Wye, Ashford, Kent TN25
5DB. 01233 812525; fax 01233 812520;
e-mail; kent@english-nature.org.uk.
Norfolk. 60 Bracondale, Norwich NR1 2BE.
01603 598400; fax 01603 762552;
e-mail norfolk@english-nature.org.uk.
North and East Yorkshire. Genesis 1, University
Road, Heslington, York YO10 5ZQ. 01904
435500; fax 01904 435520; e-mail
york@english-nature.org.uk.
North Mercia (Shrops, Staffs, Warks, W Mid).
Attingham Park, Shrewsbury SY4 4TW. 01743
709611; fax 01743 709303;
e-mail North Mercia @english-nature.org.uk.
Northumbria. Stocksfield Hall, Stocksfield,
Northumberland NE4 7TN. 01661 845500; fax
01661 845501; e-mail northumbria@english-nature.org.uk.
Peak District and Derbyshire. Endcliffe,
Deepdale Business Park, Ashford Road, Bakewell
DE45 1GT. 01629 816640; fax 01629 816679.
e-mail peak.derbys@english-nature.org.uk.
Somerset and Gloucester. Roughmoor, Bishop's
Hull, Taunton, Somerset TA1 5AA. 01823
283211;
fax 1823 272978;
e-mail somerset@english-nature.org.uk.
Suffolk. Regent House, 110 Northgate Street,
Bury St Edmunds, Suffolk IP33 1HP. 01284
762218; fax 01284 764318; e-mail
suffolk@english-nature.org.uk.
Sussex and Surrey. Phoenix House, 32-33 North
Street, Lewes, E Sussex BN7 2PH. 01273
476595;
fax 01273 483063;
e-mail sussex.surrey@english-nature.org.uk.
Thames and Chilterns. Foxhold House,
Thornford Road, Crookham Common,
Thatcham, Berks RG19 8EL. 01635 268881; fax
01635 267027;
e-mail thames.chilterns@english-nature.org.uk.
Wiltshire. Prince Maurice Court, Hambleton
Avenue, Devizes, Wilts SN10 2RT. 01380
726344; fax 01380 721411; e-mail
wiltshire@english-nature.org.uk.
Contact: Northminster House, Peterborough,
PE1 1UA. 01733 455100; (Fax)01733 455103;
e-mail: enquiries@english-nature.org.uk
www.english-nature.org.uk

ENVIRONMENT AGENCY (THE)

A non-departmental body that aims to protect
and improve the environment and to contribute
towards the delivery of sustainable
developmentthrough the integrated
management of air, land and water. Functions
include pollution prevention and control, waste
minimisation,management of water resources,
flood defence, improvement of salmon and
freshwater fisheries, conservation of aquatic
species,navigation and use of inland and coastal
waters for recreation. Sponsored by the
Department of the Environment, Transport and
the Regions, MAFF and the Welsh Office.

Regional Offices
Anglian. Kingfisher House, Goldhay Way, Orton
Goldhay, Peterborough PE2 5ZR. 01733
371811; fax 01733 231840.
North East. Rivers House, 21 Park Square
South, Leeds LS1 2QG. 0113 244 0191; fax
0113 246 1889.
North West. Richard Fairclough House,
Knutsford Road, Warrington WA4 1HG. 01925
653999; fax 01925 415961.
Midlands. Sapphire East, 550 Streetsbrook
Road, Solihull B91 1QT. 0121 711 2324; fax
0121 711 5824.
Southern. Guildbourne House, Chatsworth
Road, Worthing, W Sussex BN11 1LD. 01903
832000; fax 01903 821832.
South West. Manley House, Kestrel Way,
Exeter EX2 7LQ. 01392 444000; fax 01392
444238.
Thames. Kings Meadow House, Kings Meadow
Road, Reading RG1 8DQ. 0118 953 5000; fax
0118 950 0388.
Wales. Rivers House, St Mellons Business Park,
St Mellons, Cardiff CF3 0EY. 029 2077 0088; fax
029 2079 8555.
Contact: Rio House, Waterside Drive, Aztec
West, Almondsbury, Bristol BS32 4UD. 01454
624400; (Fax)01454 624409;
www.environment-agency.gov.uk

FARMING AND WILDLIFE ADVISORY GROUP (FWAG) (1969).

An independent UK registered charity led by
farmers and supported by government and
leading countryside organisations. Its aim is to
unite farming and forestry with wildlife and
landscape conservation. Active in most UK
counties. There are 120 Farm Conservation
Advisers who give practical advice to farmers
and landowners to
help them integrate
environmental
objectives with
commercial farming
practices.
Contact:
English Head Office,
FWAG, National
Agricultural Centre,
Stoneleigh,

Kenilworth, Warwickshire CV8 2RX. 02476 696 699;(Fax) 02476 696 699;
 e-mail: info@fwag.org.uk   www.fwag.org.uk
Northern Ireland, FWAG, National Agricultural Centre, 46b Rainey Street, Magherafelt, Co. Derry BT45 5AH. 028 7930 0606; (fax)028 7930 0599; e-mail: n.ireland@fwag.org.uk
Scottish Head Office, FWAG Scotland, Algo Business Centre, Glenearn Road, Perth PH2 ONJ. 01738 450500; (Fax) 01738 450495; e-mail: steven.hunt@fwag.org.uk
Wales Head Office. FWAG Cymru, Ffordd Arran, Dolgellau, Gwynedd LL40 1LW. 01341 421456; (Fax) 01341 422757;
 e-mail: cymru@fwag.org.uk

FIELD STUDIES COUNCIL (1943).

Manages Centres where students from schools, universities and colleges of education, as well as individuals of all ages, could stay and study various aspects of the environment under expert guidance. The courses include many for birdwatchers, providing opportunities to study birdlife on coasts, estuaries, mountains and islands. There are some courses demonstrating bird ringing and others for members of the BTO. The length of the courses varies: from a weekend up to seven days' duration. Research workers and naturalists wishing to use the records and resources are welcome.

Centres:
Blencathra Field Centre, Threlkeld, Keswick, Cumbria CA12 4SG, 017687 79601;
e-mail enquiries.bl@field-studies-council.org
Castle Head Field Centre, Grange-over-Sands, Cumbria LA11 6QT, 015395 38120,
e-mail enquiries.ch@field-studies-council.org
Dale Fort Field Centre, Haverfordwest, Pembs SA62 3RD, 01646 636205,
e-mail enquiries.df@field-studies-council.org
Epping Forest Field Centre, High Beach, Loughton, Essex, IG10 4AF, 020 8502 8500,
e-mail enquiries.ef@field-studies-council.org
Flatford Mill Field Centre, East Bergholt, Suffolk, CO7 6UL, 01206 297110,
e-mail enquiries.fm@field-studies-council.org
Derrygonnelly Field Centre, Tir Navar, Creamery St, Derrygonnelly, Co Fermanagh, BT93 6HW. 028 686 41673, e-mail: enquiries.dg@field-studies-council.org
Juniper Hall Field Centre, Dorking, Surrey, RH5 6DA, 0845 458 3507,
e-mail enquiries.jh@field-studies-council.org
Kindrogan Field Centre, Enochdhu, Blairgowrie, Perthshire PH10 7PG. 01250 870150,

e-mail: admin.kd@field-studies-council.org
Margam Park Field Centre, Port Talbot SA13 2TJ. 01639 895636,
e-mail:
margam_sustainable_centre@hotmail.com
Malham Tarn Field Centre, Settle, N Yorks, BD24 9PU, 01729 830331,
e-mail fsc.malham@ukonline.co.uk
Nettlecombe Court, The Leonard Wills Field Centre, Williton, Taunton, Somerset, TA4 4HT, 01984 640320,
e-mail enquiries.nc@field-studies-council.org
Orielton Field Centre, Pembroke, Pembs,SA71 5EZ, 01646 623920,
e-mail enquiries.or@field-studies-council.org
Preston Montford Field Centre, Montford Bridge, Shrewsbury, SY4 1DX, 01743 852040,
e-mail enquiries.pm@field-studies-council.org
Rhyd-y-creuau, the Drapers' Field Centre Betws-y-coed, Conwy, LL24 0HB, 01690 710494,
e-mail enquiries.rc@field-studies-council.org
Slapton Ley Field Centre, Slapton, Kingsbridge, Devon, TQ7 2QP, 01548 580466,
e-mail enquiries.sl@field-studies-council.org
Contact: Head Office, Preston Montford, Montford Bridge, Shrewsbury, SY4 1HW, 01743 852100; (Fax)01743 852101;
e-mail: fsc.headoffice@ field-studies-council.org www.field-studies-council.org

FLIGHTLINE
Northern Ireland's daily bird news service. Run under the auspices of the Northern Ireland Birdwatchers' Association (qv).
Contact: George Gordon, 2 Brooklyn Avenue, Bangor, Co Down BT20 5RB. 028 9146 7408; e-mail: gordon@ballyholm2.freeserve.co.uk

FORESTRY COMMISSION
The Forestry Commission of Great Britain is the government department responsible for the protection and expansion of Britain's forests and woodlands. The organisation is run from national offices in England, Wales and Scotland, working to targets set by Commissioners and Ministers in each of the three countries. Its objectives are to protect Britain's forests and resources, conserve and improve the biodiversity, landscape and cultural heritage of forests and woodlands, develop opportunities for woodland recreation and increase public understanding and community participation in forestry.
Contact: Forestry Commission, Silvan House, 231 Corstorphine Road, Edinburgh, EH12 7AT; 0131 334 0303; (Fax) 0131 334 3047; Media enquiries: 0131 314 6550. Public enquiries: 0845 367 3787.
e-mail: enquiries@forestry.gsi.gov.uk

www.forestry.gov.uk
Details of wildlife viewing sites aross the country can be found on www.forestry.gov.uk/forestry/wildwoods
Forestry Commission National Offices:
England: Great Eastern House, Tenison Road, Cambridge CB1 2DU. 01223 314546; (Fax)01223 460699, e-mail: fc.nat.off.eng@forestry.gsi.gov.uk
Scotland: Address as contact above, 0131 334 0303, (Fax)0131 314 615, e-mail: fcscotland@forestry.gsi.gov.uk
Wales: Victoria Terrace, Aberystwyth, Ceredigion SY23 2DQ. 01970 625866; (Fax)01970 626177.
Contact: Head of Communication Branch, 231 Corstorphine Road, Edinburgh, EH12 7AT. 0131 334 0303; (Fax)0131 334 4473; www.forestry.gov.uk

FRIENDS OF THE EARTH (1971; 100,000).
The largest international network of environmental groups in the world, represented in 68 countries. It is one of the leading environmental pressure groups in the UK. It has a unique network of campaigning local groups, working in 200 communities in England, Wales and Northern Ireland. It is largely funded by supporters with more than 90% of income coming from individual donations, the rest from special fundraising events, grants and trading.
Contact: 26-28 Underwood Street, London, N1 7JQ. 020 7490 1555; (Fax)020 7490 0881; e-mail: info@foe.co.uk www.foe.co.uk

GAME CONSERVANCY TRUST (1933; 22,000).
A registered charity which researches the conservation of game and other wildlife in the British countryside. More than 60 scientists are engaged in detailed work on insects, pesticides, birds (30 species inc. raptors) mammals (inc. foxes), and habitats. The results are used to advise government, landowners, farmers and conservationists on practical management techniques which will benefit game species, their habitats, and wildlife. Each June an *Annual Review* of 100 pages lists about 50 papers published in the peer-reviewed scientific press.
Contact: Deputy Director of Research, Dr Aebischer, Fordingbridge, Hampshire, SP6 1EF. 01425 652381; (Fax)01425 655848; e-mail: info@gct.org.uk
www.gct.org.uk

GAY BIRDERS CLUB (1995; 300+).
A voluntary society for lesbian, gay and bisexual birdwatchers, their friends and supporters, over the age of consent, in the UK and worldwide. The club has 3-400 members and a network of regional contacts. It organises day trips, weekends and longer events at notable birding locations in the UK and abroad; about 200+ events in a year. Members receive a quarterly newletter with details of all events. There is a Grand Get Together every 18 months. Membership £12 waged and £5 unwaged.
Contact: Memership, GeeBeeCee, BCM-Mono, London, WC1N 3XX.
e-mail: contact@gbc-online.org.uk
www.gbc-online.org.uk

GOLDEN ORIOLE GROUP (1987). Sec.
Organises censuses of breeding Golden Orioles in parts of Cambridgeshire, Norfolk and Suffolk. Maintains contact with a network of individuals in other parts of the country where Orioles may or do breed. Studies breeding biology, habitat and food requirements of the species.
Contact: Jake Allsop, 5 Bury Lane, Haddenham, Ely, Cambs CB6 3PR. 01353 740540;
e-mail: jakeallsop@aol.com
www.goldenoriolegroup.org.uk

HAWK AND OWL TRUST (1969).
Registered charity dedicated to the conservation and appreciation of all birds of prey including owls. Publishes a newsletter *Peregrine* and educational materials for all ages. The Trust achieves its major aim of creating and enhancing wild habitats for birds of prey through projects which involve practical research, creative conservation and education. Projects are often conducted in close partnership with landowners, farmers and others. Members are invited to take part in population studies, field surveys, etc. Studies of Barn and Little Owls, Hen Harrier, and Goshawk are in progress. The Trust's National Conservation and Education Centre at Newland Park, Gorelands Lane, Chalfont St Giles, Bucks, is now open to the public and offers schools and other groups cross-curricular environmental activities.
Contact: Membership administration: 11 St Mary's Close, Abbotskerswell, Newton Abbot, Devon, TQ12 5QF.
Contact: Director, Colin Shawyer, PO Box 100, Taunton TA4 2WX. 0870 990 3889;
e-mail: enquiries@hawkandowl.org
www.hawkandowl.org

IRISH RARE BIRDS COMMITTEE (1985). Assesses records of species of rare occurrence in the Republic of Ireland. Details of records accepted and rejected are incorporated in the Irish Bird Report, published annually in Irish Birds. In the case of rarities trapped for ringing, ringers in the Republic of Ireland are required to send their schedules initially to the National Parks and Wildlife Service, 51 St Stephen's Green, Dublin 2. A copy is taken before the schedules are sent to the British Trust for Ornithology.
Contact: Hon Secretary, Paul Milne, 100 Dublin Road , Sutton , Dublin 13. +353 (0)1 8325653;
e-mail: pjmilne@hotmail.com

JOINT NATURE CONSERVATION COMMITTEE (1990).
A committee of the three country agencies (English Nature, Scottish Natural Heritage, and the Countryside Council for Wales), together with independent members and representatives from Northern Ireland and the Countryside Agency. It is supported by specialist staff. Its statutory responsibilities include the establishment of common standards for monitoring, the analysis of information and research; advising Ministers on the development and implementation of policies for or affecting nature conservation; the provision of advice and the dissemination of knowledge to any persons about nature conservation; and the undertaking and commissioning of research relevant to these functions. JNCC additionally has the UK responsibility for relevant European and wider international matters. The Species Team, located at the HQ address above, is responsible for terrestrial bird conservation.
Contact: Monkstone House, City Road, Peterborough PE1 1JY. 01733 562626; (Fax)01733 555948; e-mail: comment@jncc.gov.uk
www.jncc.gov.uk

LINNEAN SOCIETY OF LONDON (1788, 2,000).
Named after Carl Linnaeus, the 18th century Swedish biologist, who created the modern system of scientific biological nomenclature, the Society promotes all aspects of pure and applied biology. It houses Linnaeus's collection of plants, insects and fishes, library and correspondence. The Society has a major reference library of some 100,000 volumes. Publishes the *Biological, Botanical and Zoological Journals*, and the *Synopses of the British Fauna.*
Contact: Executive Secretary, Adrian Thomas, Burlington House, Piccadilly, London W1J 0BF.

020 7434 4479; (Fax)020 7287 9364;
e-mail: adrian@linnean.org www.linnean.org

LITTLE OWL STUDY GROUP (2002; 37).
Formed to promote the study and conservation of Little Owls (*Athene noctua*) in Britain and to develop a population monitoring network for Little Owls. The LOSG is part of the International Little Owl Working Group, a Europe wide organisation networking Little Owl Research and Conservation. The Little Owl is declining at an alarming rate across Europe and is endangered in at least three Western European countries. To combat this, a European Species Action Plan is being developed, to put in place the necessary monitors, conservation, and education measures for its long term survival. Project *Athene* is the British leg of this SAP.(See National Projects).
Contact: Roy Leigh, C/O Biota, The Old Barn, Moseley Hall Farm, Chilford Road, Knutsford, Cheshire WA16 8RB. 0871 734 0111; (Fax)0871 734 0555;
e-mail: rsl@biota.co.uk

MANX ORNITHOLOGICAL SOCIETY see County Directory

MANX WILDLIFE TRUST see County Directory

NATIONAL BIRDS OF PREY CENTRE (1967).
Concerned with the conservation and captive breeding of all raptors. Approx 85 species on site. Birds flown daily. Open Feb-Nov.
Contact: Newent, Glos, GL18 1JJ. 0870 990 1992;
e-mail: kb@nbpc.org
www.nbpc.co.uk

NATIONAL TRUST (1895; 3.1million).
Charity depending on voluntary support of its members and the public. Largest private landowner with over 603,862 acres of land and nearly 600 miles of coast. Works for the preservation of places of historic interest or natural beauty, in England, Wales and N Ireland. The Trust's coast and countryside properties are open to the public at all times, subject only to the needs offarming, forestry and the protection of wildlife. Over a quarter of the Trust's land holding is designated SSSI or ASSI (N Ireland)and about 10% of SSSIs in England and Wales are wholly or partially owned by the Trust, as are 31 NNRs (eg Blakeney Point, FarneIslands, Wicken Fen and large parts of Strangford Lough, N Ireland). Fifteen per cent of Ramsar Sites include Trust land, as do 27%of SPAs. 71 of the 117 bird species listed in the

313

UK Red Data Book are found on Trust land.

Head of Nature Conservation: Dr H J Harvey, Estates Dept, 33 Sheep Street, Cirencester, Glos GL7 1RQ. 01285 651818.
Northern Ireland Office: Rowallane House, Saintfield, Ballynahinch, Co. Down BT24 7LH. 028 975 10721.
Contact: PO Box 39, Warrington WA5 7WD. 0870 458 4000; (Fax)020 8466 6824;
e-mail: enquiries@thenationaltrust.org.uk
www.nationaltrust.org.uk

NATIONAL TRUST FOR SCOTLAND (1931; 230,000).
An independent charity, its 90 properties open to the public are described in its annual guide.
Contact: Marketing Department, Wemyss House, 28 Charlotte Square, Edinburgh, EH2 4ET. 0131 243 9300; (Fax)0131 243 9301.
www.nts.org.uk
e-mail: information@nts.org.uk

NATURE PHOTOGRAPHERS' PORTFOLIO (1944).
A society for photographers of wildlife, especially birds. Circulates postal portfolios of prints and transparencies.
Contact: Hon Secretary, A Winspear-Cundall, 8 Gig Bridge Lane, Pershore, Worcs WR10 1NH. 01386 552103.
www.nature-photographers-portfolio.co.uk

NORTH SEA BIRD CLUB(1979; 200).
The stated aims of the Club are to: provide a recreational pursuit for people employed offshore; obtain, collate and analyse observations of all birds seen offshore; produce reports of observations, including an annual report; promote the collection of data on other wildlife offshore. Currently we hold in excess of 100,000 records of birds, cetaceans and insects reported since 1979.
Contact: Andrew Thorpe, (Recorder), Ocean Laboratory and Centre for Ecology, Aberdeen University, Newburgh, Ellon, Aberdeenshire AB41 6AA. e-mail: a.thorpe@abdn.ac.uk or nsbc@abdn.ac.uk www.abdn.ac.uk/nsbc

NORTHERN IRELAND BIRDWATCHERS' ASSOCIATION (1991; 120).
The NIBA Records Committee, established in 1997, has full responsibility for the assessment of records in N Ireland. NIBA also publishes the

Northern Ireland Bird Report.
Contact: Hon Secretary, Wilton Farrelly, 24 Cabin Hill Gardens, Knock, Belfast BT5 7AP. 028 9022 5818;
e-mail: wilton.farrelly@ntlworld.com

NORTHERN IRELAND ORNITHOLOGISTS' CLUB (1965; 150).
Operates a Tree Sparrow nestbox scheme and a winter feeding programme for Yellowhammers. Has a regular programme of lectures and field trips for members. Publishes *The Harrier* quarterly.
Contact: Jim Megarry, The Hill, 39 Ballygowan Road, COMBER, Co. Down BT23 5PG;
e-mail: maurice.hughes@nioc.fsnet.co.uk
www.nioc.fsnet.co.uk

PEOPLE'S DISPENSARY FOR SICK ANIMALS (1917).
Registered charity. Provides free veterinary treatment for sick and injured animals whose owners qualify for this charitable service.
Contact: Whitechapel Way, Priorslee, Telford, Shrops TF2 9PQ. 01952 290999; (Fax)01952 291035;
e-mail: pr@pdsa.org.uk www.pdsa.org.uk

RARE BREEDING BIRDS PANEL (1973).
An independent body funded by the JNCC and RSPB. Both bodies are represented on the panel as are BTO and ACRE. It collects all information on rare breeding birds in the United Kingdom, so that changes in status can be monitored as an aid to present-day conservation and stored for posterity. Special forms are used (obtainable free from the secretary and the website) and records should if possible be submitted via the county and regional recorders. Since 1996 the Panel also monitors breeding by scarcer non-native species and seeks records of these in the same way. Annual report is published in *British Birds*. For details of species covered by the Panel see Log Charts and the websites.
Contact: Secretary, Dr Malcolm Ogilvie, Glencairn, Bruichladdich, Isle of Islay PA49 7UN. 01496 850218; e-mail:
rbbp@indaal.demon.co.uk
www.rbbp.org.uk

ROYAL AIR FORCE ORNITHOLOGICAL SOCIETY (1965; 200).
RAFOS organises regular field meetings for members, carries out ornithological census work and mounts major expeditions annually to various UK and overseas locations. Membership is open to serving and retired members of the UK armed forces, MOD civil servants and their families. Publishes a Newsletter twice a year,

and a journal, *The Osprey,* annually jointly with the Army Ornithological Society.

Contact: Secretary, RAFOS, c/o MOD Defence Estates (Natural Environment Team), Building 21, Westdown Camp, Tilshead, SALISBURY, Wiltshire, SP3 4RS. 01452 857133; e-mail: rafos_secretary@hotmail.com www.rafos.org.uk

ROYAL NAVAL BIRDWATCHING SOCIETY
(1946; 167 full and 93 associate members and library). Covers all main ocean routes, the Society has developed a system for reporting the positions and identity of seabirds and landbirds at sea by means of standard sea report forms. Maintains an extensive world wide sea bird database. Members are encouraged to photograph birds while at sea and a library of photographs and slides is maintained. Publishes a 6 monthly Bulletin and an annual report entitled The Sea Swallow.

Contact: Hon Secretary, FS Ward Esq., 16 Cutlers Lane, Stubbington, Fareham, Hants PO14 2JN. +44 1329 665931; e-mail: francisward@btopenworld.com www.rnbws.org.uk

ROYAL PIGEON RACING ASSOCIATION
(1897; 39,000).
Exists to promote the sport of pigeon racing and controls pigeon racing within the Association. Organises liberation sites, issues rings, calculates distances between liberation sites and home lofts, and assists in the return of strays. May be able to assist in identifying owners of ringed birds caught or found.

Contact: General Manager, RPRA, The Reddings, Cheltenham, GL51 6RN. 01452 713529;
e-mail: gm@rpra.org or strays@rpra.org www.rpra.org

ROYAL SOCIETY FOR THE PREVENTION OF CRUELTY TO ANIMALS (1824; 43,690).
In addition to its animal centres, the Society also runs a woodland study centre and nature reserve at Mallydams Wood in East Sussex and specialist wildlife rehabilitation centres at West Hatch, Taunton, Somerset TA3 5RT (0870 0101847), at Station Road, East Winch, King's Lynn, Norfolk PE32 1NR (0870 9061420), and London Road, Stapeley, Nantwich, Cheshire CW5 7JW (not open to the public). Inspectors are contacted through their National Communication Centre, which can be reached via the Society's 24-hour national cruelty and advice line: 08705 555 999.

Contact: RSPCA Headquarters, Willberforce Way, Horsham, West Sussex RH13 9RS. 0870 0101181; (Fax)0870 7530048.
www.rspca.org.uk

ROYAL SOCIETY FOR THE PROTECTION OF BIRDS
UK Partner of BirdLife International, and Europe's largest voluntary wildlife conservation body. The RSPB, a registered charity, is governed by an elected body (see also RSPB Phoenix and RSPB Wildlife Explorers). Its work in the conservation of wild birds and habitats covers the acquisition and management of nature reserves; research and surveys; monitoring and responding to development proposals, land use practices and pollution which threaten wild birds and biodiversity; and the provision of an advisory service on wildlife law enforcement.

Work in the education and information field includes formal education in schools and colleges, and informal activities for children through Wildlife Explorers; publications (including *Birds*, a quarterly magazine for members, *Bird Life*, a bi-monthly magazine for RSPB Wildlife Explorers, *Wild Times* for under-8s); displays and exhibitions; the distribution of moving images about birds; and the development of membership activities through Members' Groups.

The RSPB currently manages 182 nature reserves in the UK, covering more than 313,000 acres; more than 50% of this area is owned. Sites are carefully selected, mostly as being of national or international importance to wildlife conservation. The aim is to conserve a countrywide network of reserves with all examples of the main bird communities and with due regard to the conservation of plants and other animals. Visitors are generally welcome to most reserves, subject to any restrictions necessary to protect the wildlife or habitats.

Current national projects include extensive work on agriculture, and conservation and campaigning for the conservation of the marine environment and to halt the illegal persecution of birds of prey. Increasingly, there is involvement with broader environmental concerns such as climate change and transport. The RSPB's International Dept works closely with Birdlife International and its partners in other countries and is involved with numerous projects overseas, especially in Europe and Asia.

Regional Offices:
RSPB North England, 4 Benton Terrace, Sandyford Road, Newcastle upon Tyne NE2 1QU. 0191 212 0353.
RSPB North West, Westleigh Mews, Wakefield Road, Denby Dale, Huddersfield HD8 8QD. 01484 861148.
RSPB Central England, 46 The Green, South Bar, Banbury, Oxon OX16 9AB. 01295 253330.
RSPB East Anglia, Stalham House, 65 Thorpe Road, Norwich NR1 1UD. 01603 661662.
RSPB South East, 2nd Floor, Frederick House, 42 Frederick Place, Brighton BN1 4EA. 01273 775333.
RSPB South West, Keble House, Southernhay Gardens, Exeter EX1 1NT. 01392 432691.
RSPB Scotland HQ, Dunedin House, 25 Ravelston Terrace, Edinburgh EH4 3TP. 0131 311 6500.
RSPB North Scotland, Etive House, Beechwood Park, Inverness IV2 3BW. 01463 715000.
RSPB East Scotland, 10 Albyn Terrace, Aberdeen AB1 1YP. 01224 624824.
RSPB South & West Scotland, 10 Park Quadrant, Glasgow G3 6BS. 0141 331 0993.
RSPB North Wales, Maes y Ffynnon, Penrhosgarnedd, Bangor, Gwynedd LL57 2DW. 01248 363800.
RSPB South Wales, Sutherland House, Castlebridge, Cowbridge Road East, Cardiff CF11 9AB. 029 2035 3000.
RSPB Northern Ireland, Belvoir Park Forest, Belfast BT8 7QT. 028 9049 1547
Contact: Chief Executive, Graham Wynne, The Lodge, Sandy, Beds SG19 2DL. 01767 680551; (Fax)01767 692365;
e-mail: (firstname.name)@rspb.org.uk
www.rspb.org.uk

RSPB WILDLIFE EXPLORERS and RSPB PHOENIX (formerly YOC) (1965; 152,000). Junior section of the RSPB. There are more than 100 groups run by 300 volunteers. Activities include projects, holidays, roadshows, competitions, and local events for children, families and teenagers. Publishes 2 bi-monthly magazines, *Bird Life* (aimed at 8-12 year olds) and *Wild Times* (aimed at under 8s) and 1 quarterly magazine *Wingbeat* (aimed at teenagers).
Contact: Youth Manager, Mark Boyd, RSPB Youth and Education Dept, The Lodge, Sandy, Beds SG19 2DL. 01767 680551; e-mail: explorers@rspb.org.uk and phoenix@rspb.org.uk
www.rspb.org.uk/youth

SCOTTISH BIRDS RECORDS COMMITTEE (1984).
Set up by the Scottish Ornithologists' Club to ensure that records of species not deemed rare enough to be considered by the British Birds Rarities Committee, but which are rare in Scotland, are fully assessed; also maintains the official list of Scottish birds.
Contact: Secretary, Angus Hogg, 11 Kirkmichael Road, Crosshill, Maybole, Ayrshire KA19 7RJ;
e-mail: dcgos@globalnet.co.uk
www.the-soc.org.uk

SCOTTISH NATURAL HERITAGE (1991).
Statutory body established by the Natural Heritage (Scotland) Act 1991 and responsible to Scottish Ministers. Its aim is to promote Scotland's natural Heritage, its care and improvement, its responsible enjoyment, its greater understanding and appreciation and its sustainable use.
Contact: 12 Hope Terrace, Edinburgh, EH9 2AS. 0131 447 4784; (Fax)0131 446 2277;
e-mail: enquiries@snh.gov.uk
www.snh.org.uk

SCOTTISH ORNITHOLOGISTS' CLUB (1936; 2250).
The Club has 14 branches (see County Directory). Each with a programme of winter meetings and field trips throughout the year. The SOC organises an annual weekend conference in the autumn and a joint SOC/BTO one-day birdwatchers' conference in spring. Publishes quarterly newsletter *Scottish Bird News*, the bi-annual *Scottish Birds*, the annual *Scottish Bird Report* and the *Raptor Round Up*. The SOC is developing a new resource centre in Scotland, details of which can be found on the website.
Contact: Development Manager, Bill Gardner MBE, Waterston House, Aberlady, East Lothian EH32 0PY. 01875 871 330; (Fax)01875 871 035;
e-mail: mail@the-soc.org.uk www.the-soc.org.uk

SCOTTISH SOCIETY FOR THE PREVENTION OF CRUELTY TO ANIMALS (1839).
Represents animal welfare interests to Government, local authorities and others. Educates young people to realise their responsibilities. Maintains an inspectorate to patrol and investigate and to advise owners

about the welfare of animals and birds in their care. Maintains 13 welfare centres, two of which include oiled bird cleaning centres. Bird species, including birds of prey, are rehabilitated and where possible released back into the wild.
Contact: Chief Executive, Ian Gardiner, Braehead Mains, 603 Queensferry Road, Edinburgh EH4 6EA. 0131 339 0222; (Fax)0131 339 4777;
e-mail: enquiries@scottishspca.org
www.scottishspca.org

SCOTTISH WILDLIFE TRUST (1964; 22,000). Has members' groups throughout Scotland. Aims to conserve all forms of wildlife and has over 125 reserves, many of great birdwatching interest, covering some 50,000 acres. Member of The Wildlife Trusts partnership and organises Scottish Wildlife Watch. Publishes *Scottish Wildlife* three times a year.
Contact: Chief Executive, Simon Milne, Cramond House, Off Cramond Glebe Road, Edinburgh EH4 6NS. 0131 312 7765; (Fax)0131 312 8705;
e-mail: enquiries@swt.org.uk
www.swt.org.uk

SEABIRD GROUP (1966; 350).
Concerned with conservation issues affecting seabirds. Co-ordinates census and monitoring work on breeding seabirds; has established and maintains the Seabird Colony Register in collaboration with the JNCC; organises triennial conferences on seabird biology and conservation topics. Small grants available to assist with research and survey work on seabirds. Publishes the *Seabird Group Newsletter* every four months and the journal, *Atlantic Seabirds,* quarterly in association with the Dutch Seabird Group.
Contact: Alan Leitch, 2 Burgess Terrace, Edinburgh, EH9 2BD. 0131 667 1169;
e-mail: alan.leitch@snh.gov.uk

SOCIETY OF WILDLIFE ARTISTS (1964; 76 Members, 6 Associates).
Registered charity. Annual exhibitions held in Sept/Oct at the Mall Galleries, London.
Contact: President, Andrew Stock, Federation of British Artists, 17 Carlton House Terrace, London SW1Y 5BD. 020 7930 6844;
e-mail: info@mallgalleries.com
www.swla.co.uk

SWAN SANCTUARY (THE)
Founded by Dorothy Beeson BEM. A registered charity which operates nationally, with 6 rescue centres in the UK and 1 in Ireland. Has a fully equipped swan hospital with an operating

theatre, 2 treatment rooms, x-ray facilities and a veterinary   surgeon. Present site has 3 lakes and 10 rehabilitation ponds where some 3,000 swans a year are treated. 24-hour service operated, with volunteer rescuers on hand to recover victims of oil spills, vandalism etc. A planned new site will allow visitors. Provides education and training. Reg. charity number 1002582.
Contact: Wildlife and Environmental Centre, Felix Lane, Shepperton, Middlesex TW17 8NN. 01932 240 790; e-mail: swans@swanuk.org.uk
www.swanuk.org.uk

SWAN STUDY GROUP (80).
An association of both amateur and professionals, from around the UK. Most are concerned with Mute Swans, but Bewick's and Whooper Swan biologists are also active members. The aim of the Group is to provide a forum for communication and discussion, and to help co-ordinate co-operative studies. Annual meetings are held at various locations in the UK at which speakers give presentations on their own fieldwork.
Contact: Dr Helen Chisholm, 14 Buckstone Howe, Edinburgh, EH10 6XF. 0131 445 2351;
e-mail: h.chisholm@blueyonder.co.uk

THE BRITISH LIBRARY SOUND ARCHIVE WILDLIFE SECTION (1969).
(Formerly BLOWS - British Library of Wildlife Sounds). The most comprehensive collection of bird sound recordings in existence: over 150,000 recordings of more than 8000 species of birds worldwide, available for free listening. Copies or sonograms of most recordings can be supplied for private study or research and, subject to copyright clearance, for commercial uses. Contribution of new material and enquiries on all aspects of wildlife sounds and recording techniques are welcome. Publishes *Bioacoustics* journal, CD and cassette guides to bird songs. Comprehensive catalogue available on-line at http:\\cadensa.bl.uk
Contact: Curator, Dr Joanne Nicholson, British Library, National Sound Archive, 96 Euston Road, London NW1 2DB. 020 7412 7402/3;
e-mail: nsa-wildsound@bl.uk
www.bl.uk/sound-archive

THE MAMMAL SOCIETY (1954; 2,500).
The Mammal Society is the voice for British mammals and the only organisation solely dedicated to the study and conservation of all

British mammals. They seek to raise awareness of mammals, their ecology and their conservation needs, to survey British mammals and their habitats to identify the threats they face, to promote mammal studies in the UK and overseas, to advocate conservation plans based on sound science, to provide current information on mammals through our publications, to involve people of all ages in their efforts to protect mammals, to educate people about British mammals and to monitor mammal population changes.
Contact: Enquiries, 2B Inworth Street, London SW11 3EP. 020 7350 2200; (Fax)020 7350 2211;
e-mail: enquiries@mammal.org.uk
www.mammal.org.uk

UK400 CLUB (1981).
Serves to monitor the nation's leading twitchers and their life lists, and to keep under review contentious species occurrences. Publishes a bi-monthly magazine Rare Birds and operates a website; www.uk400clubonline.co.uk.
Membership open to all.
Contact: L G R Evans, 8 Sandycroft Road, Little Chalfont, Amersham, Bucks HP6 6QL. 01494 763010; e-mail: LGREUK400@aol.com

ULSTER WILDLIFE TRUST see County Directory

WADER STUDY GROUP (1970; 600).
An association of wader enthusiasts, both amateur and professional, from all parts of the world. The Group aims to maintain contact between them, to help in the organisation of co-operative studies, and to provide a vehicle for the exchange of information. Publishes the Wader Study Group Bulletin three times a year and holds annual meetings throughout Europe.
Contact: Membership Secretary, Wader Study Group, Rod West, c/o BTO, The Nunnery, Thetford, Norfolk IP24 2PU. e-mail: rodwest@ndirect.co.uk
www.waderstudygroup.org

WALTER ROTHSCHILD ZOOLOGICAL MUSEUM
Founded by Lionel Walter (later Lord) Rothschild, the Museum displays British and exotic birds (1500 species) including many rarities and extinct species. Galleries open all year except 24-26 Dec. Adjacent to the Bird Group of the Natural History Museum - with over a million specimens and an extensive ornithological library, an internationally important centre for bird research.
Contact: Akeman Street, Tring, Herts HP23 6AP. 020 7942 6171; (Fax)020 7942 6150.

e-mail: tring-enquiries@nhm.ac.uk
www.nhm.ac.uk/visit-us/galleries/tring/

WELSH KITE TRUST (1996; 1000).
A registered charity that undertakes the conservation and annual monitoring of Red Kites in Wales. It attempts to locate all the breeding birds, to compile data on population growth, productivity, range expansion etc. The Trust liaises with landowners, acts as consultant on planning issues and with regard to filming and photography, and represents Welsh interests on the UK Kite Steering Group. Provides a limited rescue service for injured kites and eggs or chicks at risk of desertion or starvation. Publishes a newsletter Boda Wennol twice a year, sent free to subscribing Friends of the Welsh Kite and to all landowners with nesting Kites.
Contact: Tony Cross, Samaria, Nantmel, Llandrindod Wells, Powys LD1 6EN. 01597 825981;
e-mail: tony.cross@welshkitetrust.org
www.welshkitetrust.org

WELSH ORNITHOLOGICAL SOCIETY (1988; 250).
Promotes the study, conservation and enjoyment of birds throughout Wales. Runs the Welsh Records Panel which adjudicates records of scarce species in Wales. Publishes the journal Welsh Birds twice a year, along with newsletters, and organises an annual conference.
Contact: Paul Kenyon, 196 Chester Road, Hartford, Northwich CheshireCW8 1LG. 01606 77960;
e-mail: pkenyon196@aol.com
www.welshornithologicalsociety.org.uk

WETLAND TRUST
Set up to encourage conservation of wetlands and develop study of migratory birds, and to foster international relations in these fields. Destinations for recent expeditions inc. Brazil, Senegal, The Gambia, Guinea-Bissau, Nigeria, Kuwait, Thailand, Greece and Jordan. Large numbers of birds are ringed each year in Sussex and applications are invited from individuals to train in bird ringing or extend their experience.
Contact: AJ Martin, Elms Farm, Pett Lane, Icklesham, Winchelsea, E Sussex TN36 4AH. 01797 226374;
e-mail: alan@wetlandtrust.org

NATIONAL ORGANISATIONS

WILDFOWL & WETLANDS TRUST (THE)
(1946; 130,000 members and 4,700 bird
adopters).
Registered charity founded by the late Sir Peter
Scott, its mission - to conserve wetlands and
their biodiversity. WWT has nine centres with
reserves (see Arundel, Caerlaverock, Castle
Espie, Llanelli, Martin Mere, Slimbridge,
Washington, Welney, and The London Wetland
Centre in Reserves and Observatories section).
The centres are nationally or internationally
important for wintering wildfowl; they also aim
to raise awareness of and appreciation for
wetland species, the problems they face and
the conservation action needed to help them.
Programmes of walks and talks are available for
visitors with varied interests - resources and
programmes are provided for school groups.
Centres, except Caerlaverock and Welney, have
wildfowl from around the world, inc.
endangered species. Research Department
works on population dynamics, species
management plans and wetland ecology. The
Wetland Advisory Service (WAS) undertakes
contracts, and Wetland Link International
promotes the role of wetland centres for
education and public awareness.
Contact: Managing Director, Tony Richardson,
Slimbridge, Glos, GL2 7BT. 01453 891900;
(Fax)01453 890827; e-mail:
info.slimbridge@wwt.org.uk
www.wwt.org.uk

WILDLIFE SOUND RECORDING SOCIETY
(1968; 327).
Works closely with the Wildlife Section of the
National Sound Archive. Members carry out
recording work for scientific purposes as well as
for pleasure. A field weekend is held each
spring, and members organise meetings locally.
Four CD sound magazines of members'
recordings are produced for members each
year, and a journal, *Wildlife Sound*, is published
twice a year.
Contact: Hon Membership Secretary, WSRS,
Mike Iannantuoni, 36 Wenton Close,
Cottesmore, Oakham, Rutland LE15 7DR.
www.wildlife-sound.org/
e-mail: membership@wildlife-sound.org

WILDLIFE TRUSTS (THE)
A nationwide network of 46 local Wildlife Trusts
and 100 urban Wildlife Groups which work to
protect wildlife in town and country. The
Wildlife Trusts manage more than 2300 nature
reserves, undertake a wide range of other
conservation and education activities, and are
dedicated to the achievement of a UK richer in
wildlife. Publ *Natural World*. See also Wildlife
Watch.
Contact: Director-General, David Bellamy, The
Kiln, Waterside, Mather Road, Newark NG24
1WT. 0870 036 7711; (Fax)0870 036 0101;
e-mail: enquiry@wildlife-trusts.cix.co.uk
www.wildlifetrusts.org

WILDLIFE WATCH (1971; 24,000+).
The junior branch of The Wildlife Trusts (see
previous entry). It supports 1500 registered
volunteer leaders running Watch groups across
the UK. Publishes *Watchword* and *Wildlife Extra
for children and activity books for adults
working with young people*.
Contact: People and Wildlife Manager, Helen
Freeston, The Wildlife Trusts, The Kiln,
Waterside, Mather Road, Newark NG24 1WT.
0870 0367711; (Fax)00870 0360101;
www.wildlifewatch.org.uk
e-mail: watch@wildlife-trusts.cix.co.uk

WWF-UK (1961).
WWF is the world's largest independent
conservation organisation, comprising 27
national organisations. It works to conserve
endangered species, protect endangered
spaces, and address global threats to nature by
seeking long-term solutions with people in
government and industry, education and civil
society. Publishes *WWF News* (quarterly
magazine).
Contact: Chief Executive, Robert Napier, Panda
House, Weyside Park, Catteshall Lane,
Godalming, Surrey GU7 1XR. 01483 426444;
(Fax)01483 426409; www.wwf-uk.org

ZOOLOGICAL PHOTOGRAPHIC CLUB (1899).
Circulates black and white and colour prints of
zoological interest via a series of postal
portfolios.
Contact: Hon Secretary, Martin B Withers, 93
Cross Lane, Mountsorrel, Loughborough, Leics
LE12 7BX. 0116 229 6080.

ZOOLOGICAL SOCIETY OF LONDON (1826).
Carries out research, organises symposia and
holds scientific meetings. Manages the
Zoological Gardens in Regent's Park (first
opened in 1828) and Whipsnade Wild Animal
Park near Dunstable, Beds, each with extensive
collections of birds. The Society's library has a
large collection of ornithological books and
journals. Publications include the *Journal of
Zoology*, *Animal Conservation*, *Conservation
Biology* book series, *The Symposia* and *The
International Zoo Yearbook*.
Contact: Director General, Regent's Park,
London, NW1 4RY. 020 7722 3333.
www.zsl.org

NATIONAL PROJECTS

NOTICE TO BIRDWATCHERS

National ornithological projects depend for their success on the active participation of amateur birdwatchers. In return they provide birdwatchers with an excellent opportunity to contribute in a positive and worthwhile way to the scientific study of birds and their habitats, which is the vital basis of all conservation programmes. The following entries provide a description of each particular project and a note of whom to contact for further information (full address details are in the previous section).

BARN OWL MONITORING PROGRAMME
A BTO project
Volunteers monitor nest sites to record site occupancy, clutch size, brood size and breeding success. Qualified ringers may catch and ring adults and chicks and record measurements. Volunteers hold a Schedule 1 licence for Barn Owl. Contact: Dave Leech, e-mail: barnowls@bto.org

BirdTrack
Organised by BTO on behalf of BTO, RSPB and BirdWatch Ireland.
BirdTrack is a major new scheme, which developed out of Migration Watch, an Internet project to study spring migration. This year-round bird recording scheme is designed to collect large numbers of lists of birds. The idea is simple - you make a note of the birds seen at each site you visit and enter your daily observations on a simple-to-use web page. Birdwatchers can also send in other types of records including counts and casual observations. The focus of the website (www.birdtrack.net) will be spring and autumn migration, seasonal movements and the distribution of scarce species. Contact: Dawn Balmer, BTO

BREEDING BIRD SURVEY
Supported by the BTO, JNCC and the RSPB.
Begun in 1994, the BBS is designed to keep track of the changes in populations of our common breeding birds. It is dependent on volunteer birdwatchers throughout the country who can spare about five hours a year to cover a 1x1km survey square. There are just two morning visits to survey the breeding birds each year.
Survey squares are picked at random by computer to ensure that all habitats and regions are covered. Since its inception it has been a tremendous success, with more than 2,200 squares covered and more than 200 species recorded each year. Contact: Mike Raven, BTO, or your local BTO Regional Representative (see County Directory).

CONCERN FOR SWIFTS
A Concern for Swifts Group project.
Endorsed by the BTO and the RSPB, the Group monitors Swift breeding colonies, especially where building restoration and maintenance are likely to cause disturbance. Practical information can be provided to owners, architects, builders and others. The help of interested birdwatchers is always welcome. Contact: Jake Allsop, 01353 740540;

CONSTANT EFFORT SITES SCHEME
A BTO project for bird ringers, funded by a partnership of the BTO, the JNCC, The Environment and Heritage Service in Northern Ireland - National Parks & Wildlife Service (Ireland) and the ringers themselves.
Participants in the scheme monitor common songbird populations by mist-netting and ringing birds throughout the summer at more than 130 sites across Britain and Ireland. Changes in numbers of adults captured provide an index of population changes between years, while the ratio of juveniles to adults gives a measure of productivity. Between-year recaptures of birds are used to study variations in adult survival rates. Information from CES complements that from other long-term BTO surveys. Contact: Dawn Balmer, BTO.

CORMORANT ROOST SITE INVENTORY AND BREEDING COLONY REGISTER
R. Sellers and WWT.
Daytime counts carried out under the Wetland Bird Survey provide an index of the number of Cormorants wintering in Great Britain, but many birds are known to go uncounted on riverine and coastal habitats.
Dr Robin Sellers, in association with WWT, therefore established the Christmas Week Cormorant Survey which, through a network of volunteer counters, sought to monitor the numbers of Cormorants at about 70 of the most important night roosts in GB. In 1997, this project was extended to produce a comprehensive Cormorant Roost Site Inventory for Great Britain. Over 100 county bird recorders and local bird experts helped compile

the inventory, which currently lists 291 night roosts, mostly in England.

In 1990, Robin Sellers also established the Cormorant Breeding Colony Survey to monitor numbers and breeding success of Cormorants in the UK at both coastal and inland colonies. Some 1,500 pairs of Cormorants, representing perhaps 15% of the local UK population, now breed inland. New colonies are forming every year as the population inland increases annually by 19%.

The first European-wide Cormorant survey was undertaken in January 2003, organised in the UK by WWT. Anyone wishng to take part in either roost or breeding surveys should contact Colette Hall at WWT.

GARDEN BIRD FEEDING SURVEY
A BTO project.
The 2004/05 season completed 35 years of the GBFS. Each year 250 observers record the numbers and variety of garden birds fed by man in the 26 weeks between October and March. It is the longest running survey of its type in the world. Gardens are selected by region and type, from city flats, suburban semis and rural houses to outlying farms. Contact: David Glue, BTO.

BTO/CJ GARDEN BIRDWATCH
A BTO project, supported by C J WildBird Foods.
Started in January 1995, this project is a year-round survey that monitors the use that birds make of gardens. Approximately 17,000 participants from all over the UK and Ireland keep a weekly log of species using their gardens. The data collected are used to monitor regional, seasonal and year-to-year changes in the garden populations of our commoner birds. To cover costs there is an annual fee of £12. There is a quarterly colour magazine and all new joiners receive a full-colour, garden bird handbook. Results and more information are available online: www.bto.org/gbww. E-mail: gbw@bto.org
Margaret Askew, BTO, Tel: 01842 750050.

GOLDEN ORIOLE CENSUS
A Golden Oriole Group project.
With support from the RSPB, the Golden Oriole Group has undertaken a systematic annual census of breeding Golden Orioles in the Fenland Basin since 1987. In recent years national censuses have been made, funded by English Nature and the RSPB, in which some 60 volunteer recorders have participated. The Group is always interested to hear of sightings of Orioles and to receive offers of help with its

census work. Studies of breeding biology, habitat and food requirements are also carried out.
Contact: Jake Allsop, Golden Oriole Group.

GOOSE CENSUSES
A WWT project
Britain and Ireland support internationally important goose populations. During the day, many of these feed away from wetlands and are therefore not adequately censused by the Wetland Bird Survey. Additional surveys are therefore undertaken to provide estimates of population size. These primarily involve roost counts, supplemented by further counts of feeding birds.

Most populations are censused up to three times a year, typically during the autumn, midwinter, and spring. In addition, counts of the proportion of juveniles in goose flocks are undertaken to provide estimates of annual productivity. Further volunteers are always needed. In particular, counters in Scotland, Lancashire and Norfolk are sought. For more information contact: Richard Hearn, WWT, e-mail richard.hearn@wwt.org.uk.

HERONRIES CENSUS
A BTO project.
This survey started in 1928 and has been carried out under the auspices of the BTO since 1934. It represents the longest continuous series of population data for any European breeding bird. Counts are made at a sample of heronries each year, chiefly in England and Wales, to provide an index of the current population level; data from Scotland and Northern Ireland are scant and more contributions from these countries would be especially welcomed.

Herons may be hit hard during periods of severe weather but benefit by increased survival over mild winters. Their position at the top of a food chain makes them particularly vulnerable to pesticides and pollution. Contact: John Marchant, BTO.

IRISH WETLAND BIRD SURVEY (I-WeBS)
A joint project of BirdWatch Ireland, the National Parks & Wildlife Service of the Dept of Arts, Culture & the Gaeltacht, and WWT, and supported by the Heritage Council and WWF-UK.
Established in 1994, I-WeBS aims to monitor the numbers and distribution of waterfowl populations wintering in Ireland in the long term, enabling the population size and spatial and temporal trends in numbers to be identified

and described for each species.

Methods are compatible with existing schemes in the UK and Europe, and I-WeBS collaborates closely with the Wetland Bird Survey (WeBS) in the UK. Synchronised monthly counts are undertaken at wetland sites of all habitats during the winter.

Counts are straightforward and counters receive a newsletter and full report annually. Additional help is always welcome, especially during these initial years as the scheme continues to grow. Contact: Olivia Crowe, BirdWatch Ireland.

LITTLE OWLS - PROJECT *ATHENE*
Little Owl Study Group

The Little Owl is declining at an alarming rate across Europe and is endangered in at least three Western European countries. To combat this, a European Species Action Plan is being developed, to put in place the necessary monitors, conservation, and education measures for its long term survival. Project *Athene* is the British leg of this plan.

It is a two-tier monitoring programme that is open to anybody. To monitor numbers of Little Owls a playback method is employed using standardised protocol. Nest site recording provides a more in-depth information on the population dynamics of the owls. You can join the LOSG and dependant on your time and expertise, carry out Little Owl surveys in your own patch. Contact Roy Leigh: Little Owl Study Group, c/o Biota, 71-73 Ascot Court, Middlewish Road, Northwich, Cheshire CW9 7BP. 01606 333296; e-mail: RSL@biota.co.uk for further information

LOW TIDE COUNTS SCHEME
see Wetland Bird Survey

MANX CHOUGH PROJECT
A Manx registered charitable trust.

Established in 1990 to help the conservation of the Chough in the Isle of Man, leading to its protection and population increase. The main considerations are the maintenance of present nest sites, provision of suitable conditions for the reoccupation of abandoned sites and the expansion of the range of the species into new areas of the Island. Surveys and censuses are carried out. Raising public awareness of and interest in the Chough are further objects. Contact: Allen S Moore, Lyndale, Derby Road, Peel, Isle of Man IM5 1HH. 01624 843798.

MIGRATION WATCH Replaced by BirdTrack
(see above)

NEST RECORD SCHEME
A BTO Project forming part of the BTO's Integrated Population Monitoring programme carried out under contract with the JNCC.

All birdwatchers can contribute to this scheme by sending information about nesting attempts they observe into the BTO on standard Nest Record Cards or electronically via the IPMR computer package. The NRS monitors changes in the nesting success and the timing of breeding of Britain's bird species. Guidance on on how to record and visit nests safely, without disturbing breeding birds, is available in a free starter pack from the Nest Records Unit. Contact: David Leach at BTO, e-mail: nest.records@bto.org

RAPTOR AND OWL RESEARCH REGISTER
A BTO project

The Register has helped considerably over the past 29 years in encouraging and guiding research, and in the co-ordination of projects. There are currently almost 500 projects in the card index file through which the Register operates.

The owl species currently receiving most attention are Barn and Tawny. As to raptors, the most popular subjects are Kestrel, Buzzard, Sparrowhawk, Hobby and Peregrine, with researchers showing increasing interest in Red Kite, and fewer large in-depth studies of Goshawk, Osprey and harriers. Contributing is a simple process and involves all raptor enthusiasts, whether it is to describe an amateur activity or professional study. The nature of research on record varies widely – from local pellet analyses to captive breeding and rehabilitation programmes to national surveys of Peregrine, Buzzard and Golden Eagle. Birdwatchers in both Britain and abroad are encouraged to write for photocopies of cards relevant to the species or nature of their work. The effectiveness of the Register depends upon those running projects (however big or small) ensuring that their work is included. Contact: David Glue, BTO.

RED KITE RE-INTRODUCTION PROJECT
An English Nature/SNH/RSPB project supported by Forest Enterprise, Yorkshire Water and authorities in Germany and Spain

The project involves the translocation of birds from Spain, Germany and the expanding Chilterns population for release at sites in England and Scotland.

Records of any wing-tagged Red Kites in England should be reported to Ian Carter at English Nature, Northminster House, Peterborough, PEI IUA (tel 01733 455281). Scottish records should be sent to Brian Etheridge at RSPB's North Scotland Regional Office, Etive House, Beechwood Park, Inverness, IV2 3BW (tel 01463 715000).

Sightings are of particular value if the letter/number code (or colour) of wing tags can be seen or if the bird is seen flying low over (or into) woodland. Records should include an exact location, preferably with a six figure grid reference.

RETRAPPING ADULTS FOR SURVIVAL

A BTO project for bird ringers, funded by a partnership of the BTO, the JNCC, The Environment and Heritage Service in Northern Ireland - National Parks & Wildlife Service (Ireland) and the ringers themselves.
This programme started in 1998 and is an initiative of the BTO Ringing Scheme. It aims to gather re-trap information for a wide range of species, especially those of conservation concern, in a variety of breeding habitats, allowing the monitoring of survival rates.

Detailed information about survival rates from RAS will help in the understanding of changing population trends. Ringers choose a target species, decide on a study area and develop suitable catching techniques. The aim then is to catch all the breeding adults of the chosen species within the study area. This is repeated each breeding season for a minimum of five years. The results will be relayed to conservation organisations who can use the information to design effective conservation action plans. Contact: John Marchant, BTO.

RINGING SCHEME

A BTO project for bird ringers, funded by a partnership of the BTO, the JNCC, -The National Parks & Wildlife Service (Ireland) and the ringers themselves.
The purpose of the Ringing Scheme is to study survival productivity and movements by marking birds with individually numbered metal rings which carry a return address. About 2,000 trained and licensed ringers operate in Britain and Ireland, and together they mark around 750,000 birds each year.
All birdwatchers can contribute to the scheme by reporting any ringed birds they find either via the BTO website or by letter. Anyone finding a ringed bird should note the ring number, species (if known), when and where the bird was found, and what happened to it. If the bird is dead, please remove the ring, flatten it out and tape it to your letter and send it to us. If details are phoned in, please keep the ring in case there is a query. Finders who send their

name and address will be given details of where and when the bird was ringed. About 12,000 ringed birds are reported each year and an annual report is published.
Contact: Jacquie Clark, BTO.

SCARCE WOODLAND BIRD AND HABITAT SURVEY 2005 / 06

A BTO volunteer-based project, funded by the JNCC.
This project aims to assess variations in the abundance of a range of scarce and declining woodland bird species with respect to habitat and geographical region. Volunteers will be asked to map the presence of a range of species along woodland routes and to record details of the habitat. There is also a Casual Records Scheme, for which we are collecting records of eight key species from any habitat during the breeding season (see www.bto.org)
Contact: Su Gough (su.gough@bto.org).

WATERWAYS BIRD SURVEY

A BTO project
From March to July each year participants survey linear waterways (rivers and canals) to map the position and activity of riparian birds. Results show both numbers and distribution of breeding territories for each waterside species at each site WBS population trends, some dating from 1974, supplement those of the Breeding Bird Survey. Since 1998, WBS has run parallel with the Waterways Breeding Bird Survey, which uses a transect method.
WBS maps show the habitat requirements of the birds and can be used to assess the effects of waterway management. Coverage of new plots is always required, especially in poorly covered areas such as Northern Ireland, Scotland, Wales, SW England and the North East. Contact: John Marchant, BTO.

WATERWAYS BREEDING BIRD SURVEY

A BTO project, supported by the Environment Agency
WBBS uses transect methods like those of the Breeding Bird Survey to record bird populations along randomly chosen stretches of river and canal throughout the UK. Just two survey visits are needed during April-June. WBBS began in 1998 and is currently in a development phase, in which its performance is being assessed against the long-established Waterways Bird Survey. Contact BTO Regional Representative (see County Directory) to enquire if any local stretches require coverage, otherwise John Marchant at BTO HQ.

NATIONAL DIRECTORY

NATIONAL PROJECTS

WETLAND BIRD SURVEY
A joint scheme of BTO, WWT, RSPB & JNCC
The Wetland Bird Survey (WeBS) is the monitoring scheme for non-breeding waterbirds in the UK. The principal aims are:
1. to determine the population sizes of waterbirds
2. to determine trends in numbers and distribution
3. to identify important sites for waterbirds
4. to conduct research which underpins waterbird conservation.

WeBS data are used to designate important waterbird sites and protect them against adverse development, for research into the causes of declines, for establishing conservation priorities and strategies and to formulate management plans for wetland sites and waterbirds.

Once monthly, synchronised Core Counts are made at as many wetland sites as possible. Low Tide Counts are made on about 20 estuaries each winter to identify important feeding areas. Counts take just a few hours and are relatively straightforward. The 3,000 participants receive regular newsletters and a comprehensive annual report. New counters are always welcome.
Contact: Steve Holloway, at the WeBS Office, BTO HQ, E-mail WeBS@bto.org

WeBS WINTER RIVER BIRD SURVEY
A Wetland Bird Survey project (qv)
While the Wetland Bird Survey (WeBS) achieves excellent coverage of estuaries and inland still waters, rivers are poorly monitored by comparison. Consequently, WeBS undoubtedly misses a significant proportion of the UK populations of several species which use rivers, eg. Little Grebe, Mallard, Tufted Duck, Goldeneye and Goosander. A pilot survey undertaken in 2000 and 2001 indicated that the full national survey should cover at least 8,000 river sections of 500 metres length to allow an accurate estimation of population sizes for waterbirds wintering on rivers. The 2004 survey was unfortunately cancelled, as problems were encountered regarding the process of random selection of river sections for the survey. It is hoped that the January 2005 survey will be run. Many counters in addition to those involved in the pilot will be required. WWT aims to provide a dedicated Winter River Bird Survey Web site with on-line functions for registering to take part in the survey, for downloading instructions and count forms, and also for web-based data submission once the survey is completed.

Contact: WWT, 01453 891900, or e-mail: research@wwt.org.uk

WILDFOWL COLOUR RINGING
A WWT project
The Wildfowl & Wetlands Trust co-ordinates all colour ringing of swans, geese and ducks on behalf of the BTO. The use of unique coloured leg-rings enables the movements and behaviour of known individuals to be observed without recapture. The rings are usually in bright colours with engraved letters and/or digits showing as black or white, and can be read with a telescope at up to 200m. Colour-marked neck collars, and plumage dyes, have also been used on geese and swans. Any records of observations should include species, location, date, ring colour and mark, and which leg the ring was on (most rings read from the foot upwards). The main study species are Mute Swan, Bewick's Swan, Whooper Swan, Pink-footed Goose, Greylag Goose, Greenland and European White-fronted Geese, Barnacle Goose, Brent Goose, Shelduck and Wigeon. Records will be forwarded to the relevant study, and when birds are traced ringing details will be sent back to the observer. All sightings should be sent to: Research Dept (Colour-ringed Wildfowl), WWT.

WINTER GULL ROOST SURVEY (WinGS) 2003/04-2005/06
A BTO project, funded by JNCC, EN, SNH, CCW, EHSNI, Northumbrian Water.
The seventh Winter Gull Roost Survey will monitor all known major sites, as well as surveying other areas using a sampling approach. The survey aims to produce total population estimates and to identify the most important gull roost sites. Contact: Alex Banks.

WINTERING WARBLER SURVEY 2005/06
A BTO volunteer-based project.
This project aims to assess changes in wintering warbler populations in Britain, which have increased dramatically in recent decades, following a succession of mild winters. The second winter of the survey will run from 1st November 2005 to 31st March 2006 (or 15th April for Blackcap only). All records of Blackcap, Chiffchaff, Cetti's Warbler, Dartford Warbler, Firecrest and any rarer species are required, from all sites but particularly gardens, sewage works and wetlands.
Contact> Greg Conway, BTO. 01842 750050; e-mail: greg.conway@bto.org or visit www.bto.org/survey/special/wintering_warblers.htm

INTERNATIONAL DIRECTORY

Spur-winged Plover by Nick Williams

The BirdLife Partnership

BirdLife is a Partnership of non-governmental organisations (NGOs) with a special focus on conservation and birds. Each NGO Partner represents a unique geographic territory/country.

The BirdLife Network explained

Partners: Membership-based NGOs who represent BirdLife in their own territory. Vote holders and key implementing bodies for BirdLife's Strategy and Regional Programmes in their own territories.

Partners Designate: Membership-based NGOs who represent BirdLife in their own territory, in a transition stage to becoming full Partners. Non-vote holders.

Affiliates: Usually NGOs, but also individuals, foundations or governmental institutions when appropriate. Act as a BirdLife contact with the aim of developing into, or recruiting, a BirdLife Partner in their territory.

Secretariat: The co-ordinating and servicing body of BirdLife International.

Secretariat Addresses

BirdLife Cambridge Office
BirdLife International
Wellbrook Court
Girton Road
Cambridge CB3 0NA
United Kingdom
Tel. +44 1 223 277 318
Fax +44 1 223 277200
E-mail: birdlife@birdlife.org.uk
http://www.birdlife.net

BirdLife Americas Regional Office
Birdlife International
Vicente Cárdenas 120 y Japon, 3rd Floor
Quito
Ecuador
Postal address
BirdLife International
Casilla 17-17-717
Quito
Ecuador
Tel. +593 2 453 645

Fax +593 2 459 627
E-mail: birdlife@birdlife.org.ec
http://www.geocities.com/RainForest/Wetlands/6203

BirdLife Asia Regional Office
Jl. Jend. Ahmad Yani No. 11
Bogor 16161
Indonesia
Postal address
PO Box 310/Boo
Bogor 16003
Indonesia
Tel. +62 251 333 234/+62 251 371 394
Fax +62 251 357 961
E-mail: birdlife@indo.net.id
http://www.kt.rim.or.jp/~birdinfo/indonesia

BirdLife European Regional Office
Droevendaalsesteeg 3a PO Box 127, NL- 6700 AC, Wageningen
The Netherlands

Tel. +31 317 478831
Fax +31 317 478844
E-mail: birdlife@birdlife.agro.nl

European Community Office (ECO)
BirdLife International
22 rue de Toulouse
B-1040 Brussels
Belgium
Tel. +32 2280 08 30
Fax +32 2230 38 02
E-mail: bleco@ibm.net

BirdLife Middle East Regional Office
BirdLife International
c/o Royal Society for the Conservation of Nature (RSCN)
PO Box 6354
Amman 11183
Jordan
Tel: +962 6 535-5446
Fax: +962 6 534-7411
E-mail: birdlife@nol.com.jo

AFRICA

PARTNERS

Burkina Faso
Fondation des Amis de la Nature (NATURAMA), 01 B.P. 6133, Ouagadougou 01.
e-mail: naturama@fasonet.bf

Ethiopia
Ethiopian Wildlife and Natural History Society, PO Box 13303, Addis Ababa, Pub: *Agazen; Ethiopian Wildl. and Nat. Hist. News. (& Annual Report); Ethiopian Wildl. and Nat. Hist. Soc. Quarterly News (WATCH); Walia (WATCH) (Ethiopia)*.
e-mail: ewnhs@telecom.net.et
http://ewnhs.ble@telecom.net.et

Ghana
Ghana Wildlife Society, PO Box 13252, Accra, Pub: *Bongo News; NKO (The Parrots)*.
e-mail: wildsoc@ighmail.com

Kenya
Nature Kenya, PO Box 44486, 00100 GPO. Nairobi. Pub: *Bulletin of the EANHS; Journal of East African Natural; Kenya Birds*.
e-mail: office@naturekenya.org
www.naturekenya.org

Nigeria
Nigerian Conservation Foundation, PO Box 74638, Victoria Island, Lagos. Pub: *NCF Matters/News/Newsletter; Nigerian Conservation Foundation Annual Report*.
e-mail: enquiries@ncf-nigeria.org
www.africanconservation.org/ncftemp/

Seychelles
Nature Seychelles, Roche Caiman, Box 1310, Victoria, Mahe, Seychelles. Pub: *Zwazo - a BirdLife Seychelles Newsletter*.
e-mail: nature@seychelles.net
www.nature.org.sc

Sierra Leone
Conservation Society of Sierra Leone, PO BOX 1292, Freetown. Pub: *Rockfowl Link, The*.
e-mail: cssl@sierratel.sl

South Africa
BirdLife South Africa, PO Box 515, Randburg, Johannesburg 2125, South Africa, Pub: *Newsletter of BirdLife South Africa; Ostrich*.
e-mail: info@birdlife.org.za
www.birdlife.org.za

Tanzania
Wildlife Conservation Society of Tanzania, PO Box 70919, Dar es Salaam, Pub: *Miombo*.
e-mail: wcst@africaonline.co.tz

Uganda
Nature Uganda, PO Box 27034, Kampala. Pub: *Naturalist - A Newsletter of the East Africa Nat. His. Soc.*
e-mail: nature@natureuganda.org
www.natureuganda.org/

PARTNERS DESIGNATE

Tunisia
Association "Les Amis des Oiseaux", Avenue 18 Janvier 1952, Ariana Centre, App. C209, 2080 Ariana, Tunis. Pub: *Feuille de Liaison de l'AAO; Houbara, l'.* e-mail: aao.bird@planet.tn

Zimbabwe
BirdLife Zimbabwe, P O Box RV 100, Runiville, Harare, Zimbabwe. Pub: *Babbler (WATCH) (Zimbabwe); Honeyguide*.
e-mail: birds@zol.co.zw

AFFILIATES

Botswana
Birdlife Botswana, Private Bag 003 # Suite 348, Mogoditshane, Gaborone, Botswana
e-mail: blb@birdlifebotswana.org.bw
www.birdlifebotswana.org.bw

Burundi
Association Burundaise pour la Protection des Oiseaux, P O Box 7069, Bujumbura, Burundi
e-mail: aboburundi@yahoo.fr

Cameroon
Cameroon Biodiversity Conservation Society (CBCS), PO Box 3055, Messa, Yaoundé.
e-mail: gdzikouk@yahoo.fr

Egypt
Sherif Baha El Din, 3 Abdala El Katib St, Dokki, Cairo. e-mail: baha2@internetegypt.com

Rwanda
Association pour la Conservation de la Nature au Rwanda, P O Box 4290, Kigali,
e-mail: acnrwanda@yahoo.fr

Zambia
Zambian Ornithological Society, Box 33944, Lusaka 10101, Pub: *Zambian Ornithological Society Newsletter*.
e-mail: zos@zamnet.zm
www.wattledcrane.com

INTERNATIONAL DIRECTORY

FOREIGN NATIONAL ORGANISATIONS

AMERICAS

PARTNERS

Argentina
Aves Argentina / AOP, 25 de Mayo 749, 2 piso, oficina 6, 1002 Buenos Aires. Pub: *Hornero; Naturaleza & Conservacion; Nuestras Aves; Vuelo de Pajaro.*
e-mail: info@avesargentinas.org.ar
www.avesargentinas.org.ar

Belize
The Belize Audubon Society, 12 Fort Street, PO Box 1001, Belize City. Pub: *Belize Audubon Society Newsletter.*
e-mail: base@btl.net
www.belizeaudubon.org

Bolivia
Asociacion Armonia, 400 Avenida Lomas de Arena, Casilla 3566, Santa Cruz, Bolivia. Pub: *Aves en Bolivia.*
e-mail: armonia@scbbs-bo.com

Canada
Bird Studies Canada, PO Box/160, Port Rowan, Ontario N0E 1M0. Pub: *Bird Studies Canada - Annual Report; Birdwatch Canada.*
e-mail: generalinfo@bsc-eoc.org
www.bsc-eoc.org

Canada
Nature Canada, 1 Nicholas Street, Suite 606, Ottawa, Ontario, K1N 7B7. Pub: *Grass 'n Roots; IBA News Canada; Nature Canada; Nature Matters; Nature Watch News (CNF).*
e-mail: info@naturecanada.ca
www.naturecanada.ca

Ecuador
Fundación Ornithológica del Ecuador, La Tierra 203 y Av. de los Shyris, Casilla 17-17-906, Quito.
e-mail: cecia@uio.satnet.net
www.cecia.org/

Jamaica
BirdLife Jamaica, 2 Starlight Avenue, Kingston 6, Pub: *Broadsheet: BirdLife Jamaica; Important Bird Areas Programme Newsletter.*
e-mail: birdlifeja@yahoo.com
www.birdelifejamaica.com

Panama
Panama Audubon Society, Apartado 2026, Ancón, Balboa. Pub: *Toucan.*
e-mail: info@panamaaudubon.org
www.panamaaudubon.org

Venezuela
Sociedad Conservacionista Audubon de, Apartado 80.450, Caracas 1080-A, Venezuela. Pub: *Audubon (Venezuela) (formerly Boletin Audubon).*
e-mail: audubon@cantv.net

PARTNERS DESIGNATE

Mexico
CIPAMEX, Apartado Postal 22-012, D.F. 14091, Mexico. Pub: *AICA's; Cuauhtli Boletin de Cipa Mex.*
e-mail: cipamex@campus.iztacala.unam.mx
http://coro@servidor.unam.mx

Paraguay
Guyra Paraguay,, Coronel Rafael Franco 381 c/ Leandro Prieto, Casilla de Correo 1132, Asunción. Pub: *Boletin Jara Kuera.*
e-mail: guyra@guyra.org.py or
guyra@highway.com.py
www.guyra.org.py/

United States
National Audubon Society, 700 Broadway, New York, NY, 10003 -9562. Pub: *American Birds; Audubon (USA); Audubon Field Notes; Audubon Bird Conservation Newsletter.*
e-mail: audubonaction@audubon.org
www.audubon.org

Chile
Union de Ornitologis de Chile (UNORCH), Casilla 13.183, Santiago 21. Pub: *Boletin Chileno de Ornitologia; Boletin Informativo (WATCH) (Chile).*
e-mail: unorch@entelchile.net
www.geocities.com/RainForest/4372

AFFILIATES

Bahamas
Bahamas National Trust, PO Box N-4105, Nassau. Pub: *Bahamas Naturalist; Currents; Grand Bahama Update.*
e-mail: bnt@batelnet.bs
www.thebahamasnationaltrust.org/

Cuba
Centro Nacional de Áreas Protegidas (CNAP). Calle 18 a, No 1441, e/ 41 y 47, Playa, Ciudad Habana, Cuba
e-mail: cnap@snap.cu
www.snap.co.cu/

FOREIGN NATIONAL ORGANISATIONS

El Salvador
SalvaNATURA, 33 Avenida Sur #640, Colonia
Flor Blanca, San Salvador.
e-mail: salvanatura@saltel.net
www.salvanatura.org

Falkland Islands
Falklands Conservation, PO Box 26, Stanley,. or
Falklands Conservation, 1 Princes Avenue,
Finchley, London N3 2DA, UK. Pub: *Falklands
Conservation.*
e-mail: conservation@horizon.co.fk
www.falklandsconservation.com

Honduras
Sherry Thorne, c/o Cooperación Técnica, Apdo
30289 Toncontín, Tegucigalpa.
e-mail: pilar_birds@yahoo.com

Suriname
Foundation for Nature Preservation in
Suriname, Cornelis Jongbawstraat 14, PO BOX
12252, Paramaribo
e-mail: research@stinasu.sr
www.stinasu.sr

Uruguay
GUPECA, Casilla de Correo 6955, Correo
Central, Montevideo. Pub: *Achara.*
e-mail: info@avesuruguay.org.uy
www.avesuruguay.org.uy/

ASIA

PARTNERS

Japan
Wild Bird Society of Japan (WBSJ), 1/F Odakyu
Nishi Shinjuku Building, 1-47-1 Hatsudai
Shibuya-ku, Tokyo 151-061, Japan. Pub: *Strix;
Wild Birds; Wing.*
e-mail: int.center@wing-wbsj.or.jp
www.wing-wbsj.or.jp

Malaysia
Malaysian Nature Society, PO Box 10750,
50724 Kuala Lumpur. Pub: *Enggang; Suara
Enggang; Malayan Nature Journal; Malaysian
Naturalist.* www.mns.org.my
e-mail: natsoc@po.jaring.my

Philippines
Haribon Foundation, Suites 401-404 Fil-Garcia
Bldg, 140 Kalayaan Avenue cor. Mayaman St,
Diliman, Quezon CIty 1101. Pub: *Haribon
Foundation Annual Report; Haring Ibon;
Philippine Biodiversity.*

e-mail: birdlife@haribon.org.ph
www.haribon.org.ph

Singapore
Nature Society (Singapore), 510 Geylang Road,
#02-05, The Sunflower, 398466. Pub: *Nature
News; Nature Watch (Singapore).*
e-mail: nss@nss.org.sg
www.nss.org.sg

Taiwan
Wild Bird Federation Taiwan (WBFT), 1F, No. 3,
Lane 36 Jing-Long St., 116 Taipei, Taiwan,
R.O.C. Pub: *Yuhina Post.*
e-mail: wbft@bird.org.tw
www.bird.org.tw

Thailand
Bird Conservation Society of Thailand, 43 Soi
Chok Chai Ruam Mit 29, Vipahvadee-Rabgsit
Road, Sansaen-nok, Dindaeng, Bangkok 10320
Thailand. Pub: *Bird Conservation Society of
Thailand.*
e-mail: bcst@bcst.or.th
www.bcst.or.th

PARTNER DESIGNATE

India
Bombay Natural History Society, Hornbill House,
Shaheed Bhagat Singh Road, Mumbai-400 023.
Pub: *Buceros; Hornbill; Journal of the Bombay
Natural History Society.*
e-mail: bnhs@bom4.vsnl.net.in
www.bnhs.org

AFFILIATES

Hong Kong
The Hong Kong Birdwatching Society, Room
1612 Beverley Commercial Building, 87-105
Chatham Road South, Tsim Sha Tsui, Kowloon,
Hong Kong. Pub: *Hong Kong Bird Report.*
e-mail: hkbws@hkbws.org.uk
www.hkbws.org.hk

Indonesia
BirdLife Indonesia (Perhimpunan Pelestari
Burung dan Habitatnya), Jl. Dadali 32, Bogor
16161, PO. Box 310/Boo, Bogor 16003,
Indonesia.
e-mail: birdlife@burung.org
www.burung.org

Nepal
Bird Conservation Nepal, P.O.Box 12465,
Lazimpat, Kathmandu, Nepal. Pub: *Bird
Conservation Nepal (Danphe); Ibisbill.*

INTERNATIONAL DIRECTORY

e-mail: bcn@mail.com.np
www.birdlifenepal.org

Pakistan
Ornithological Society of Pakistan, PO Box 73, 109D Dera Ghazi Khan, 32200. Pub: *Pakistan Journal of Ornithology.*
e-mail: osp@mul.paknet.com.pk

Sri Lanka
Field Ornithology Group of Sri Lanka, Dept of Zoology, University of Colombo, Colombo 03. Pub: *Malkoha - Newsletter of the Field Ornithology Group of Sri Lanka.*
e-mail: fogsl@slt.lk

EUROPE

PARTNERS

Austria
BirdLife Austria, Museumplatz 1/10/8, AT-1070 Wien. Pub: *Egretta; Vogelschutz in Osterreich.*
e-mail: office@birdlife.at
www.birdlife.at/

Belgium
BirdLife Belgium (BNVR-RNOB-BNVS), Natuurpunt, Kardinaal, Mercierplein 1, 2800 Mechelen, Belgium.
e-mail: wim.vandenbossche@natuurpunt.be
www.natuurreservaten.be

Bulgaria
Bulgarian Society for the Protection of Birds (BSPB), PO Box 50, Musagenitza Complex, Block 104, Entrance A, Floor 6, BG-1111, Sofia, Bulgaria. Pub: *Neophron (& UK).*
e-mail: bspb_hq@bspb.org
www.bspb.org

Czech Republic
Czech Society for Ornithology (CSO), Hornomecholupska 34, CZ-102 00 Praha 10. Pub: *Ptaci Svet; Sylvia; Zpravy Ceske Spolecnosti Ornitologicke.*
e-mail: cso@birdlife.cz
www.birdlife.cz

Denmark
Dansk Ornitologisk Forening (DOF), Vesterbrogade 138-140, DK-1620, Copenhagen V, Denmark. Pub: *DAFIF - Dafifs Nyhedsbrev; Dansk Ornitologisk Forenings Tidsskrift; Fugle og Natur.*
e-mail: dof@dof.dk
www.dof.dk

Estonia
Estonian Ornithological Society (EOU), PO Box 227, Vesti Str. 4, EE-50002 Tartu, Estonia. Pub: *Hirundo Eesti Ornitoogiauhing.*
e-mail: eoy@eoy.ee
www.eoy.ee

Finland
BirdLife SUOMI Finland, Annankatu 29 A, PO Box 1285, FI 00101, Helsinki. Pub: *Linnuston-Suojelu; Linnut; Tiira.*
e-mail: office@birdlife.fi
www.birdlife.fi

France
Ligue pour la Protection des Oiseaux (LPO), La Corderie Royale, B.P. 90263, 17305 ROCHEFORT CEDEX, France. Pub: *Lettre Internationale; Ligue Francaise Pour La Protection des Oiseaux; Oiseau, L' (LPO); Outarde infos.*
e-mail: lpo@lpo.fr
www.lpo.fr/

Germany
Naturschutzbund Deutschland, Herbert-Rabius-Str. 26, D-53225 Bonn, Germany. Pub: *Naturschutz Heute (NABU) Naturschutzbund Deutschland.*
e-mail: nabu@nabu.de
www.nabu.de

Gibraltar
Gibraltar Ornithological and Nat. History Society, Jew's Gate, Upper Rock Nature Reserve, PO Box 843, GI. Pub: *Alectoris; Gibraltar Nature News.* e-mail: gohns@gibnet.gi
www.gibraltar.gi/gonhs

Hoopoe by Nick Williams

FOREIGN NATIONAL ORGANISATIONS

Greece
Hellenic Ornithological Society (HOS), Vas. Irakleiou 24, GR-10682 Athens, Greece. Pub: *HOS Newsletter.*
e-mail: birdlife-gr@ath.forthnet.gr
www.ornithologiki.gr

Hungary
Hungarian Orn. and Nature Cons. Society (MME), Kolto u. 21, Pf. 391, HU-1536, Budapest. Pub: *Madartani Tajekoztato; Madartavlat; Ornis Hungarica; Tuzok.*
e-mail: mme@mme.hu
www.mme.hu

Iceland
Icelandic Society for the Protection of Birds, Fuglaverndarfélag Islands, PO Box 5069, IS-125 Reykjavik, Iceland.
e-mail: fuglavernd@fuglavernd.is
www.fuglavernd.is

Ireland
BirdWatch Ireland, Rockingham House, Newcastle, Co. Wicklow, Eire. Pub: *Irish Birds; Wings (IWC Birdwatch Ireland).*
e-mail: info@birdwatchireland.org
www.birdwatchireland.ie

Israel
Society for the Protection of Nature in Israel, Hashsela 4, Tel-Aviv 66103. Pub: *SPNI News.*
e-mail: ioc@netvision.net.il
www.birds.org.il

Italy
Lega Italiana Protezione Uccelli (LIPU), Via Trento 49, IT-43100, Parma. Pub: *Ali Giovani; Ali Notizie.*
e-mail: lipusede@box1.tin.it
www.lipu.it

Latvia
Latvijas Ornitologijas Biedriba (LOB), Ak 1010, LV-1050 Riga, Latvia. Pub: *Putni Daba.*
e-mail: putni@lob.lv
www.lob.lv

Luxembourg
Letzebuerger Natur-a Vulleschutzliga (LNVL), Kraizhaff, rue de Luxembourg.L-1899 Kockelscheuer. Pub: *Regulus (WATCH); Regulus Info (& Annual Report) (WATCH); Regulus Wissenschaftliche Berichte (WATCH).*
e-mail: secretary@luxnatur.lu
www.luxnatur.lu

Malta
BirdLife Malta, 57 Marina Court, Flat 28, Triq Abate Rigord, MT-Ta' Xbiex, MSD 12, MALTA. Pub: *Bird Talk (WATCH) (Malta); Bird's Eye View (WATCH) (Malta); Il-Merill.*
e-mail: info@birdlifemalta.org
www.birdlifemalta.org

Netherlands
Vogelbescherming Nederland, PO Box 925, NL-3700 AX Zeist. Pub: *Vogelniews; Vogels.*
e-mail: info@vogelbescherming.nl
www.vogelbescherming.nl/

Norway
Norsk Ornitologisk Forening, Sandgata 30 B, N-7012 Trondheim, Norway. Pub: *Fuglearet; Fuglefauna; Var; Ringmerkaren.*
e-mail: nof@birdlife.no
www.birdlife.no

Poland
Ogólnopolskie Towarzystwo Ochrony Ptaków (OTOP), Ul. Hallera 4/2, PL-80-401 Gdansk, Poland. Pub: *Ptaki; Ptasie Ostoje.*
e-mail: office@otop.most.org.pl
www.otop.org.pl/

Portugal
Sociedade Portuguesa para o Estuda das, Aves (SPEA), Rua da Vitoria, 53-3° Esq, 1100-618, Lisboa. Pub: *Pardela.*
e-mail: spea@spea.pt
www.spea.pt

Romania
Romanian Ornithological Society (SOR), Str. Gheorghe Dima 49/2, RO-3400 Cluj. Pub: *Alcedo; Buletin AIA; Buletin de Informare Societatea Ornitologica Romana; Milvus (Romania).*
e-mail: office@sor.ro
www.sor.ro/

Slovakia
Soc. for the Prot. of Birds in Slovakia (SOVS), PO Box 71, 093 01 Vranov nad Topl'ou. Pub: *Spravodaj SOVS; Vtacie Spravy.*
e-mail: sovs@changenet.sk
www.sovs.miesto.sk

Slovenia
BirdLife Slovenia (DOPPS), Trzaska 2, PO Box 2990, SI-1000 Ljubljana, Slovenia. Pub: *Acrocephalus; Svet Ptic.*
e-mail: dopps@dopps-drustvo.si
www.ptice.org

INTERNATIONAL DIRECTORY

FOREIGN NATIONAL ORGANISATIONS

Spain
Sociedad Espanola de Ornitologia (SEO), C/
Melquiades Biencinto 34, E-28053, Madrid. Pub:
Ardeola; Areas Importantes para las Aves.
e-mail: seo@seo.org
www.seo.org

Sweden
Sveriges Ornitologiska Forening (SOF),
Ekhagsvagen 3, SE 104-05, Stockholm. Pub:
Fagelvarld; var; Ornis Svecica.
e-mail: birdlife@sofnet.org
www.sofnet.org

Switzerland
SVS/BirdLife Switzerland, Wiedingstrasse 78,
PO Box, CH-8036, Zurich, Switzerland. Pub:
*Oiwvos Ornis; Ornis Junior; Ornithologische
Beobachter; Der Ornithos; Steinadler.*
e-mail: svs@birdlife.ch
www.birdlife.ch

Turkey
Doga Dernegi, PK: 640 06445, Yeni°ehir,
Ankarae, Turkey. Pub: *Kelaynak; Kuscu Bulteni.*
e-mail: doga@dogadernegi.org
www.dogadernegi.org/

United Kingdom
Royal Society for the Protection of Birds, The
Lodge, Sandy, Bedfordshire, SG19 2DL.
e-mail: info@RSPB.org.uk
www.rspb.org.uk

PARTNERS DESIGNATE

Belarus
BirdLife Belarus (APB), PO Box 306, Minsk,
220050 Belarus. Pub: *Subbuteo - The
Belarusian Ornithological Bulletin.*
e-mail: apb@tut.by
http://apb.iatp.by/

Lithuania
Lietuvos Ornitologu Draugija (LOD),
Naugarduko St. 47-3, LT-2006, Vilnius,
Lithuania. Pub: *Baltasis Gandras.*
e-mail: lod@birdlife.lt
www.birdlife.lt

Russia
Russian Bird Conservation Union (RBCU),
Building 1, Shosse Entuziastov 60, 111123, RU-
Moscow. Pub: *Newsletter of the Russian Bird
Conservation Union.*
e-mail: mail@rbcu.ru
www.rbcu.ru/en/

Ukraine
Ukrainian Union for Bird Conservation (UTOP),
PO Box 33, Kiev, 1103, UA. Pub: *Life of Birds.*
e-mail: utop@iptelecom.net.ua
www.utop.org.ua/

AFFILIATES

Liechtenstein
Botanish-Zoologische Gesellschaft, Im Bretscha
22, FL-9494 Schaan, Liechtenstein.
e-mail: broggi@pingnet.li or renat@pingnet.li

Andorra
Associacio per a la Defensa de la Natura,
Apartado de Correus Espanyols No 96, Andora
La Vella, Principat d'Andorra. Pub: *Aiguerola.*
e-mail: and@andorra.ad
www.adn-andorra.org/

Croatia
Croatian Society for Bird and Nature Protection,
Gunduliceva 24, HR-10000 Zagreb, Croatia.
Pub: *Troglodytes.*
e-mail: jasmina@hazu.hr

Cyprus
BirdLife Cyprus, PO Box 28076, 2090 Lefkosia,
Cyprus.
e-mail: melis@cytanet.com.cy
www.birdlifecyprus.org

Faroe Islands (to Denmark)
Føroya Fuglafrødifelag (Faroese Orginithological
Society) (FOS), Postssmoga 1230, FR-110
Torshavn, Faroe Islands.
e-mail: doreteb@ngs.fo

Georgia
Georgian Centre for the Conservation of
Wildlife, PO Box 56, GE-Tbilisi 0160, Georgia.
e-mail: office@gccw.org
www.gccw.org/

MIDDLE EAST

PARTNERS

Jordan
Royal Society of the Conservation of Nature,
PO Box 6354, Jubeiha-Abu-Nusseir Circle,
Amman 11183. Pub: *Al Reem.*
e-mail: adminrscn@rscn.org.jo
www.rscn.org.jo

FOREIGN NATIONAL ORGANISATIONS

Lebanon
Society for the Protection of Nature in
Lebanon, Awad Bldg, 6th Floor, Abdel Aziz
Street, P.O.Box: 11-5665, Beirut, Lebanon.
e-mail: spnlorg@cyberia.net.lb
www.spnlb.org

PARTNER DESIGNATE

Palestine
Palestine Wildlife Society (PWLS), Beit Sahour,
PO Box 89. Pub: *Palestine Wildlife Society -
Annual Report.* www.wildlife-pal.org
e-mail: wildlife@palnet.com

AFFILIATES

Bahrain
Dr Saeed A. Mohamed, PO Box 40266, Bahrain.
e-mail: sam53@batelco.com.bh

Iran, Islamic Republic of
Dr Jamshid Mansoori, Assistant Professor,
College of Natural Resources, Tehran University,
Mojtame Sabz, Golestan Shamali, Mahestan
Ave, Shahrake Qarb, Phase 1, P.O.Box 14657,
Tehran, I.R. of Iran.
e-mail: birdlifeiran@yahoo.com

Kuwait
Kuwait Environment Protection Society, PO Box
1896, Safat 13019, Kuwait.
e-mail: rasamhory@hotmail.com
www.keps74.com

Saudi Arabia
National Commission for Wildlife Cons & Dev,
NCWDC, PO Box 61681, Riyadh 11575. Pub:
Phoenix; The.
e-mail: ncwcd@zajil.net
www.ncwcd.gov.sa/

Yemen
Yemen Society for the Protection of Wildlife
(YSPW), 29 Alger Street, PO Box 19759,
Sana'a, Yemen.
e-mail: wildlife.yemen@y.net.ye

PACIFIC

PARTNER

Australia
Birds Australia, 415 Riversdale Road, Hawthorn

East, VIC 3123, Australia. Pub: *Australia Garcilla;
Birds Australia Annual Report; Eclectus; Emu;
Wingspan (WATCH) (Australia); from
wingspan@birdsaustralia.com.au.*
e-mail: mail@birdsaustralia.com.au
www.birdsaustralia.com.au

AFFILIATES

Cook Islands
Taporoporo'anga Ipukarea Society (TIS), PO
Box 649, Rarotonga, Cook Islands.
e-mail: 2tis@oyster.net.ck

Fiji
Dr Dick Watling, c/o Environment Consultants
Fiji, P O Box 2041, Government Buildings, Suva,
Fiji.
e-mail: watling@is.com.fj
www.environmentfiji.com

French Polynesia
Société d'Ornithologie de Polynésie "Manu",
B.P. 21 098, Papeete, Tahiti.
e-mail: sop@manu.pf
www.manu.pf

Palau
Palau Conservation Society, PO BOX 1811,
Koror, PW96940. Pub: *Ngerel a Biib.*
e-mail: pcs@palaunet.com
www.palau-pcs.org/

Samoa
O le Si'osi'omaga Society Incorporated, O le
Si'osi'omaga Society Inc., P O Box 2282, Apia,
Western Samoa.
e-mail: ngo_siosiomaga@samoa.ws

New Zealand
Royal Forest & Bird Protection Society of, PO
Box 631, Wellington. Pub: *Forest & Bird; Forest
& Bird Annual Report; Forest & Bird
Conservation News.*
e-mail: office@forestandbird.org.nz
www.forestandbird.org.nz/

INTERNATIONAL DIRECTORY

INTERNATIONAL ORGANISATIONS

AFRICAN BIRD CLUB.
c/o Birdlife International as below.
e-mail (general): keithbetton@hotmail.com
(membership and sales):
e-mail: membership@africanbirdclub.org
www.africanbirdclub.org
Pub: *Bulletin of the African Bird Club.*

BIRDLIFE INTERNATIONAL.
Wellbrook Court, Girton Road, Cambridge, CB3
ONA, +44 (0)1223 277318; fax +44 (0)1223
277200,
Pub: *World Birdwatch.* www.birdlife.net

EAST AFRICA NATURAL HISTORY SOCIETY
see Kenya in preceding list.

EURING (European Union for Bird Ringing).
Euring Data Bank, Institute of Ecological
Research, PO Box 40, NL-6666 ZG Heteren,
Netherlands.
www.euring.org/index.html

**EUROPEAN WILDLIFE REHABILITATION
ASSOCIATION (EWRA).**
Les Stocker MBE, c/o Wildlife Hospital Trust,
Aston Road, Haddenham, Aylesbury, Bucks,
HP17 8AF, +44 (0)1844 292292; fax +44
(0)1844 292640,
www.sttiggywinkles.org.uk

FAUNA AND FLORA INTERNATIONAL.
Great Eastern House, Tenison Road,
Cambridge, CB1 2TT, +44 (0)1223 571000; fax
+44 (0)1223 461481, www.fauna-flora.org
e-mail: info@flora.org
Pub: *Fauna & Flora News; Oryx.* www.ffi.org.uk

LIPU-UK
**(the Italian League for the Protection of
Birds).**

David Lingard,
Fernwood,
Doddington Road,
Whisby, Lincs, LN6
9BX, +44 (0)1522
689030,

e-mail: david@lipu-uk.org www.lipu-uk.org
Pub: *The Hoopoe,* annually, *Ali Notizie,*
quarterley.

NEOTROPICAL BIRD CLUB.
As OSME below. Pub: *Cotinga.*
www.neotropicalbirdclub.org

ORIENTAL BIRD CLUB.
As OSME below. Pub: *The Forktail; Bull OBC.*
www.orientalbirdclub.org

**ORNITHOLOGICAL SOCIETY OF THE MIDDLE
EAST (OSME).**
c/o The Lodge, Sandy, Beds, SG19 2DL.
Pub: *Sandgrouse.*
www.osme.org

**TRAFFIC International (formerly Wildlife
Trade Monitoring Unit).**
219 Huntingdon Road, Cambridge, CB3 ODL,
+44 (0)1223 277427; fax +44 (0)1223
277237. Pub: *TRAFFIC Bulletin.*
e-mail: traffic@trafficint.org
www.traffic.org

**WEST AFRICAN ORNITHOLOGICAL
SOCIETY.**
R E Sharland, 1 Fisher's Heron, East Mills, Hants,
SP6 2JR. Pub: *Malimbus.*

WETLANDS INTERNATIONAL.
PO Box 471, 6700 AL Wageningen,
Netherlands, +31 317 478854; fax +31 317
478850, Pub: *Wetlands.*
e-mail: post@wetlands.org
www.wetlands.org

WORLD OWL TRUST.
The World Owl Centre,
Muncaster Castle,
Ravenglass, Cumbria,
CA18 1RQ, +44
(0)1229 717393; fax
+44 (0)1229 717107,
www.owls.org

**WORLD PHEASANT
ASSOCIATION.**
7-9 Shaftesbury St, Fordingbridge, Hants SP6
1JF. 01425 657 129; (Fax) 01425 658 053.
Pub: *WPA News.* www.pheasant.org.uk

WORLD WIDE FUND FOR NATURE.
Avenue du Mont Blanc, CH-1196 Gland,
Switzerland, +41 22 364 9111; fax +41 22 364
5358, www.panda.org

QUICK REFERENCE SECTION

Ringed Plover by Nick Williams

TIDE TABLES: USEFUL INFORMATION

BRITISH SUMMER TIME
In 2006 BST applies from 0100 on 26 March to 0100 on 29 October.

Note that all the times in the following tables are GMT.
During British Summer Time one hour should be added.

Predictions are given for the times of high water at Dover throughout the year.

The times of tides at the locations shown here may be obtained by adding or subtracting their 'tidal difference' as shown opposite (subtractions are indicated by a minus sign).

Tidal predictions for Dover have been computed by
the Proudman Oceanographic Laboratory.
Copyright reserved.

Shetland 42, 43
Orkney 44, 45

Map showing locations for which tidal differences are given on facing page.

TIDE TABLES 2005

Example 1

To calculate the time of first high water at Girvan on February 26
1. Look up the time at Dover (09 43)*
   = 09:43 am
2. Add the tidal difference for Girvan
   = 0.54
3. Therefore the time of high water at Girvan = 10:37am

Example 2

To calculate the time of second high water at Blakeney on June 18
1. Look up the time at Dover (16 17)
   = 4:17 pm
2. Add 1 hour for British Summer Time
   (17 17) = 5.17 pm
3. Subtract the tidal difference for Blakeney = - 4.07
4. Therefore the time of high water at Blakeney = 1:10 pm

*All Dover times are shown on the 24-hour clock.
Thus, 08 14 = 08.14 am; 14 58 = 2.58pm
Following the time of each high water the height of the tide is given, in metres.

(Tables beyond April 2007 are not available at the time of going to press.)

TIDAL DIFFERENCES

1	Dover	See pp 336-339		23	Morecambe	0 20
2	Dungeness		-0 12	24	Silloth	0 51
3	Selsey Bill		0 09	25	Girvan	0 54
4	Swanage (lst H.W.Springs)		-2 36	26	Lossiemouth	0 48
5	Portland		-4 23	27	Fraserburgh	1 20
6	Exmouth (Approaches)		-4 48	28	Aberdeen	2 30
7	Salcombe		-5 23	29	Montrose	3 30
8	Newlyn (Penzance)		5 59	30	Dunbar	3 42
9	Padstow		-5 47	31	Holy Island	3 58
10	Bideford		-5 17	32	Sunderland	4 38
11	Bridgwater		-4 23	33	Whitby	5 12
12	Sharpness Dock		-3 19	34	Bridlington	5 53
13	Cardiff (Penarth)		-4 16	35	Grimsby	-5 20
14	Swansea		-4 52	36	Skegness	-5 00
15	Skomer Island		-5 00	37	Blakeney	-4 07
16	Fishguard		-3 48	38	Gorleston	-2 08
17	Barmouth		-2 45	39	Aldeburgh	-0 13
18	Bardsey Island		-3 07	40	Bradwell Waterside	1 11
19	Caernarvon		-1 07	41	Herne Bay	1 28
20	Amlwch		-0 22	42	Sullom Voe	-1 34
21	Connahs Quay		0 20	43	Lerwick	0 01
22	Hilbre Island (Hoylake/West Kirby)		-0 05	44	Kirkwall	-0 26
				45	Widewall Bay	-1 30

NB. Care should be taken when making calculations at the beginning and end of British Summer Time. See worked examples above.

TIDE TABLES 2006

Time Zone **GMT**

Tidal Predictions : **HIGH WATERS 2006**

Datum of Predictions = **Chart Datum : 3.67 metres below Ordnance Datum (Newlyn)**

British Summer Time : **26th March to 29th October**

Units **METRES**

DOVER — January

DATE / DAY	Morning hr min	m	Afternoon hr min	m
1 Su	11 37	6.6	** **	**
2 M	00 06	6.7	12 27	6.6
3 Tu	00 56	6.7	13 21	6.6
4 W	01 45	6.6	14 14	6.4
5 Th	02 34	6.6	15 06	6.0
6 F	03 23	6.4	16 00	6.0
7 Sa	04 15	6.2	16 56	5.8
8 Su	05 14	6.0	17 59	5.6
9 M	06 20	5.8	19 08	5.5
10 Tu	07 29	5.7	20 17	5.6
11 W	08 37	5.8	21 16	5.8
12 Th	09 34	5.9	22 05	6.0
13 F	10 22	6.0	22 48	6.2
14 Sa	11 04	6.1	23 26	6.3
15 Su	11 42	6.2	** **	**
16 M	00 02	6.4	12 18	6.2
17 Tu	00 37	6.4	12 51	6.2
18 W	01 11	6.4	13 24	6.1
19 Th	01 43	6.3	13 55	6.0
20 F	02 14	6.2	14 26	5.8
21 Sa	02 45	6.0	15 02	5.7
22 Su	03 23	5.9	15 46	5.5
23 M	04 11	5.6	16 42	5.3
24 Tu	05 13	5.5	17 52	5.2
25 W	06 25	5.4	19 08	5.3
26 Th	07 49	5.5	20 26	5.5
27 F	08 49	5.8	21 29	5.9
28 Sa	09 49	6.1	22 22	6.3
29 Su	10 41	6.4	23 10	6.6
30 M	11 30	6.7	23 56	6.8
31 Tu	** **	**	12 19	6.8

DOVER — February

DATE / DAY	Morning hr min	m	Afternoon hr min	m
1 W	00 43	7.0	13 08	6.8
2 Th	01 26	7.0	13 53	6.6
3 F	02 09	6.9	14 37	6.4
4 Sa	02 52	6.7	15 22	6.2
5 Su	03 39	6.4	16 11	5.8
6 M	04 32	6.0	17 11	5.5
7 Tu	05 40	5.6	18 29	5.2
8 W	07 03	5.4	19 55	5.3
9 Th	08 26	5.4	21 02	5.5
10 F	09 29	5.6	21 53	5.8
11 Sa	10 17	5.8	22 34	6.1
12 Su	10 55	6.0	23 09	6.3
13 M	11 27	6.1	23 42	6.4
14 Tu	11 59	6.2	** **	**
15 W	00 15	6.5	12 29	6.3
16 Th	00 46	6.5	12 56	6.2
17 F	01 18	6.4	13 22	6.2
18 Sa	01 38	6.4	13 48	6.1
19 Su	02 03	6.2	14 17	6.0
20 M	02 34	6.1	14 55	5.8
21 Tu	03 19	5.8	15 47	5.5
22 W	04 19	5.5	17 02	5.1
23 Th	05 44	5.3	18 37	5.3
24 F	07 22	5.4	20 14	5.4
25 Sa	08 44	5.7	21 21	5.9
26 Su	09 43	6.1	22 10	6.3
27 M	10 31	6.5	22 55	6.7
28 Tu	11 17	6.8	23 38	7.0

DOVER — March

DATE / DAY	Morning hr min	m	Afternoon hr min	m
1 W	** **	**	12 01	6.9
2 Th	00 20	7.1	12 46	6.9
3 F	01 01	7.1	13 26	6.7
4 Sa	01 41	7.0	14 04	6.5
5 Su	02 20	6.7	14 45	6.2
6 M	03 04	6.3	15 33	5.8
7 Tu	03 50	5.9	16 34	5.4
8 W	05 07	5.3	17 55	5.0
9 Th	06 43	5.2	19 32	5.1
10 F	08 16	5.2	20 44	5.4
11 Sa	09 19	5.5	21 34	5.7
12 Su	10 03	5.8	22 12	6.0
13 M	10 36	6.0	22 46	6.2
14 Tu	11 03	6.1	23 18	6.4
15 W	11 33	6.3	23 48	6.5
16 Th	11 59	6.5	** **	**
17 F	00 16	6.5	12 25	6.3
18 Sa	00 40	6.4	12 49	6.3
19 Su	01 03	6.4	13 14	6.2
20 M	01 28	6.3	13 45	6.1
21 Tu	02 00	6.2	14 24	5.9
22 W	02 44	5.9	15 16	5.6
23 Th	03 50	5.5	16 36	5.4
24 F	05 31	5.3	18 02	5.8
25 Sa	07 21	5.3	20 02	* *
26 Su	08 35	5.8	21 02	6.2
27 M	09 29	6.2	21 50	6.4
28 Tu	10 14	6.6	22 32	6.8
29 W	10 56	6.8	23 13	7.0
30 Th	11 37	6.8	23 54	7.1
31 F	** **	* *	12 19	6.8

DOVER — April

DATE / DAY	Morning hr min	m	Afternoon hr min	m
1 Sa	00 33	7.0	12 57	6.7
2 Su	01 12	6.8	13 36	6.5
3 M	01 52	6.5	14 17	6.2
4 Tu	02 35	6.1	15 04	5.8
5 W	03 20	5.6	15 53	5.4
6 Th	04 42	5.1	16 23	5.0
7 F	06 20	4.9	18 58	5.0
8 Sa	07 50	5.1	20 11	5.3
9 Su	08 49	5.4	21 02	5.7
10 M	09 33	5.7	21 41	6.0
11 Tu	10 05	5.9	22 15	6.2
12 W	10 34	6.1	22 46	6.3
13 Th	11 00	6.2	23 14	6.4
14 F	11 26	6.3	23 41	6.4
15 Sa	11 52	6.3	** **	**
16 Su	00 06	6.4	12 20	6.3
17 M	00 33	6.4	12 51	6.3
18 Tu	01 05	6.3	13 28	6.2
19 W	01 43	6.1	14 11	5.9
20 Th	02 35	5.8	15 13	5.6
21 F	03 53	5.4	16 39	5.4
22 Sa	05 37	5.3	18 18	5.3
23 Su	07 08	5.5	19 38	5.7
24 M	08 14	5.9	20 35	6.1
25 Tu	09 05	6.2	21 23	6.5
26 W	09 50	6.5	22 06	6.7
27 Th	10 31	6.6	22 46	6.9
28 F	11 14	6.7	23 28	* *
29 Sa	11 54	6.8	** **	**
30 Su	00 09	6.6	12 36	6.6

TIDE TABLES 2006

Units **METRES**

Time Zone **GMT**

Tidal Predictions : **HIGH WATERS 2006**

Datum of Predictions = **Chart Datum : 3.67 metres below Ordnance Datum (Newlyn)**

British Summer Time : **26th March to 29th October**

DOVER — May

Date	Day	Morning hr min	m	Afternoon hr min	m
1	M	00 50	6.6	13 15	6.4
2	Tu	01 31	6.3	13 57	6.1
3	W	02 16	5.9	14 42	5.8
4	Th	03 09	5.5	15 37	5.5
5	F	04 15	5.2	16 43	5.2
6	Sa	05 38	5.0	18 06	5.1
7	Su	07 03	5.1	19 21	5.3
8	M	08 02	5.3	20 16	5.6
9	Tu	08 47	5.6	20 59	5.8
10	W	09 23	5.8	21 36	6.0
11	Th	09 54	6.0	22 08	6.2
12	F	10 24	6.1	22 38	6.3
13	Sa	10 55	6.2	23 09	6.3
14	Su	11 27	6.3	23 41	6.4
15	M	** **		12 04	6.3
16	Tu	00 19	6.3	12 42	6.2
17	W	01 00	6.2	13 31	6.1
18	Th	01 50	6.0	14 24	6.0
19	F	02 54	5.8	15 26	5.8
20	Sa	04 05	5.8	16 38	5.7
21	Su	05 28	5.6	17 52	5.7
22	M	06 41	5.7	19 03	5.9
23	Tu	07 43	5.9	20 00	6.1
24	W	08 35	6.1	20 51	6.3
25	Th	09 23	6.3	21 37	6.5
26	F	10 08	6.4	22 22	6.6
27	Sa	10 53	6.4	23 09	6.6
28	Su	11 38	6.4	23 54	6.5
29	M	** **		12 20	6.4
30	Tu	00 36	6.3	13 01	6.3
31	W	01 18	6.1	13 41	6.1

DOVER — June

Date	Day	Morning hr min	m	Afternoon hr min	m
1	Th	02 02	5.9	14 23	6.0
2	F	02 48	5.6	15 09	5.7
3	Sa	03 41	5.4	16 03	5.5
4	Su	04 45	5.2	17 06	5.4
5	M	05 54	5.1	18 12	5.3
6	Tu	06 57	5.2	19 14	5.4
7	W	07 49	5.4	20 04	5.6
8	Th	08 34	5.6	20 48	5.8
9	F	09 13	5.8	21 27	6.0
10	Sa	09 51	6.0	22 05	6.0
11	Su	10 31	6.1	22 45	6.2
12	M	11 13	6.3	23 28	6.3
13	Tu	11 58	6.4	** **	
14	W	00 15	6.4	12 46	6.4
15	Th	01 07	6.3	13 36	6.3
16	F	02 02	6.2	14 28	6.3
17	Sa	03 01	6.1	15 20	6.2
18	Su	04 00	5.9	16 17	6.1
19	M	05 00	5.8	17 17	6.0
20	Tu	06 02	5.8	18 20	6.0
21	W	07 04	5.8	19 21	6.1
22	Th	08 01	6.0	20 21	6.1
23	F	09 01	6.1	21 18	6.2
24	Sa	09 54	6.0	22 10	6.2
25	Su	10 41	6.2	22 57	6.3
26	M	11 26	6.3	23 42	6.2
27	Tu	** **		12 12	6.3
28	W	00 25	6.2	12 44	6.3
29	Th	01 04	6.1	13 22	6.3
30	F	01 42	5.9	14 00	6.2

DOVER — July

Date	Day	Morning hr min	m	Afternoon hr min	m
1	Sa	02 21	5.8	14 40	6.0
2	Su	03 02	5.6	15 20	5.9
3	M	03 47	5.4	16 07	5.7
4	Tu	04 38	5.3	16 58	5.5
5	W	05 35	5.2	17 58	5.4
6	Th	06 37	5.3	19 01	5.5
7	F	07 39	5.5	20 00	5.7
8	Sa	08 37	5.7	20 55	5.8
9	Su	09 29	5.8	21 44	6.0
10	M	10 15	6.1	22 32	6.2
11	Tu	11 02	6.4	23 20	6.4
12	W	11 49	6.5	** **	
13	Th	00 09	6.6	12 37	6.7
14	F	01 01	6.6	13 25	6.7
15	Sa	01 53	6.5	14 13	6.7
16	Su	02 44	6.4	14 59	6.6
17	M	03 32	6.2	15 47	6.4
18	Tu	04 22	6.0	16 41	6.2
19	W	05 26	5.7	17 50	5.9
20	Th	06 41	5.6	18 50	5.7
21	F	08 07	5.5	20 04	5.7
22	Sa	09 11	5.7	21 11	5.8
23	Su	09 46	5.9	22 05	5.9
24	M	10 31	6.1	22 50	6.1
25	Tu	11 18	6.3	23 31	6.4
26	W	11 48	6.4	** **	
27	Th	00 03	6.4	12 25	6.4
28	F	00 48	6.2	13 02	6.4
29	Sa	01 15	6.1	13 32	6.4
30	Su	01 48	6.0	14 04	6.3
31	M	02 17	5.9	14 34	6.1

DOVER — August

Date	Day	Morning hr min	m	Afternoon hr min	m
1	Tu	02 49	5.7	15 08	5.9
2	W	03 29	5.5	15 50	5.7
3	Th	04 19	5.3	16 48	5.5
4	F	05 26	5.2	18 01	5.3
5	Sa	06 47	5.2	19 22	5.3
6	Su	08 10	5.4	20 35	5.6
7	M	09 13	5.8	21 33	6.0
8	Tu	10 04	6.1	22 21	6.3
9	W	10 49	6.5	23 07	6.6
10	Th	11 33	6.8	23 54	6.8
11	F	** **		12 18	6.8
12	Sa	00 41	7.0	13 03	7.0
13	Su	01 29	6.7	13 46	6.9
14	M	02 13	6.6	14 28	6.8
15	Tu	02 56	6.3	15 12	6.5
16	W	03 44	6.0	16 04	6.1
17	Th	04 42	5.7	17 07	5.7
18	F	05 54	5.3	18 30	5.4
19	Sa	07 25	5.3	20 02	5.4
20	Su	08 42	5.5	21 12	5.6
21	M	09 36	5.8	22 03	5.9
22	Tu	10 18	6.1	22 41	6.2
23	W	10 53	6.3	23 14	6.3
24	Th	11 27	6.5	23 45	6.3
25	F	11 59	6.6	** **	
26	Sa	00 15	6.6	12 32	6.6
27	Su	00 44	6.3	13 00	6.5
28	M	01 10	6.2	13 24	6.4
29	Tu	01 34	6.1	13 48	6.3
30	W	02 00	6.0	14 14	6.1
31	Th	02 34	5.8	14 52	5.8

Time Zone **GMT**

Units **METRES**

Tidal Predictions : HIGH WATERS 2006

Datum of Predictions = Chart Datum : 3.67 metres below Ordnance Datum (Newlyn)

British Summer Time : 26th March to 29th October

DOVER — September

Date	Day	Morning hr min	m	Afternoon hr min	m
1	F	03 20	5.5	15 47	5.5
2	Sa	04 31	5.2	17 16	5.2
3	Su	06 12	5.1	19 03	5.2
4	M	07 53	5.4	20 24	5.6
5	Tu	08 58	5.9	21 20	6.1
6	W	09 47	6.3	22 07	6.5
7	Th	10 29	6.7	22 49	6.8
8	F	11 11	7.0	23 33	6.9
9	Sa	11 52	7.2	** **	*
10	Su	00 15	6.9	12 34	7.2
11	M	00 58	6.8	13 14	7.1
12	Tu	01 39	6.6	13 55	6.8
13	W	02 21	6.4	14 38	6.4
14	Th	03 04	6.1	15 32	6.0
15	F	03 49	5.7	16 41	5.4
16	Sa	04 45	5.2	18 18	5.1
17	Su	07 10	5.2	20 00	5.3
18	M	08 27	5.5	21 05	5.6
19	Tu	09 18	5.9	21 49	5.9
20	W	09 56	6.1	22 20	6.1
21	Th	10 29	6.4	22 50	6.3
22	F	11 00	6.5	23 17	6.4
23	Sa	11 31	6.6	23 45	6.4
24	Su	11 59	6.6	** **	*
25	M	00 11	6.4	12 23	6.5
26	Tu	00 33	6.4	12 46	6.4
27	W	00 58	6.3	13 08	6.3
28	Th	01 26	6.2	13 38	6.2
29	F	02 00	6.0	14 17	5.9
30	Sa	02 48	5.7	15 13	5.5

DOVER — October

Date	Day	Morning hr min	m	Afternoon hr min	m
1	Su	04 01	5.3	16 56	5.1
2	M	05 54	5.1	18 51	5.3
3	Tu	07 32	5.5	20 09	5.7
4	W	08 35	6.0	21 02	6.2
5	Th	09 24	6.5	21 46	6.6
6	F	10 04	6.8	22 26	6.8
7	Sa	10 43	7.1	23 06	6.9
8	Su	11 24	7.2	23 48	6.9
9	M	** **	*	12 05	7.2
10	Tu	00 29	6.8	12 44	7.0
11	W	01 11	6.6	13 26	6.7
12	Th	01 53	6.3	14 11	6.3
13	F	02 41	6.1	15 06	5.8
14	Sa	03 39	5.6	16 17	5.3
15	Su	04 55	5.2	17 58	5.0
16	M	06 34	5.2	19 36	5.2
17	Tu	07 53	5.4	20 35	5.5
18	W	08 45	5.8	21 18	5.8
19	Th	09 25	6.1	21 51	6.0
20	F	09 58	6.3	22 19	6.2
21	Sa	10 29	6.5	22 46	6.3
22	Su	10 59	6.5	23 13	6.4
23	M	11 26	6.5	23 38	6.4
24	Tu	11 51	6.5	** **	*
25	W	00 05	6.4	12 16	6.4
26	Th	00 34	6.3	12 44	6.3
27	F	01 07	6.2	13 19	6.1
28	Sa	01 48	6.0	14 04	5.8
29	Su	02 41	5.7	15 11	5.5
30	M	03 58	5.5	16 53	5.3
31	Tu	05 34	5.4	18 30	5.4

DOVER — November

Date	Day	Morning hr min	m	Afternoon hr min	m
1	W	07 00	5.7	19 41	5.8
2	Th	08 00	6.1	20 34	6.2
3	F	08 51	6.5	21 19	6.5
4	Sa	09 34	6.8	22 00	6.7
5	Su	10 17	7.0	22 42	6.8
6	M	10 59	7.0	23 26	6.8
7	Tu	11 41	7.0	** **	*
8	W	00 09	6.7	12 25	6.8
9	Th	00 53	6.6	13 08	6.5
10	F	01 35	6.3	13 55	6.1
11	Sa	02 21	6.0	14 47	5.7
12	Su	03 11	5.7	15 50	5.3
13	M	04 17	5.4	17 10	5.1
14	Tu	05 35	5.3	18 40	5.1
15	W	06 56	5.4	19 45	5.3
16	Th	07 56	5.6	20 33	5.6
17	F	08 42	6.1	21 12	6.0
18	Sa	09 22	6.3	21 46	6.2
19	Su	09 56	6.1	22 15	6.2
20	M	10 26	6.3	22 45	6.3
21	Tu	10 56	6.4	23 16	6.4
22	W	11 27	6.4	23 49	6.4
23	Th	** **	*	12 01	6.3
24	F	00 23	6.3	12 39	6.3
25	Sa	01 07	6.3	13 22	6.1
26	Su	01 55	6.1	14 16	5.9
27	M	02 49	6.0	15 22	5.7
28	Tu	03 53	5.8	16 38	5.6
29	W	05 04	5.8	17 52	5.8
30	Th	06 16	5.9	19 00	5.8

DOVER — December

Date	Day	Morning hr min	m	Afternoon hr min	m
1	F	07 19	6.1	19 57	6.0
2	Sa	08 16	6.3	20 49	6.2
3	Su	09 06	6.5	21 39	6.4
4	M	09 54	6.6	22 26	6.5
5	Tu	10 42	6.7	23 13	6.6
6	W	11 30	6.6	23 58	6.3
7	Th	** **	*	12 15	6.5
8	F	00 40	6.5	12 58	6.2
9	Sa	01 21	6.2	13 41	6.2
10	Su	02 03	6.0	14 26	5.8
11	M	02 48	6.0	15 16	5.7
12	Tu	03 37	5.7	16 14	5.4
13	W	04 35	5.4	17 32	5.4
14	Th	05 42	5.4	18 32	5.5
15	F	06 50	5.5	19 34	5.5
16	Sa	07 49	5.7	20 24	5.7
17	Su	08 38	5.9	21 09	5.8
18	M	09 20	6.0	21 47	5.9
19	Tu	09 58	6.2	22 00	6.1
20	W	10 35	6.3	23 00	6.2
21	Th	11 14	6.4	23 41	6.4
22	F	11 55	6.5	** **	*
23	Sa	00 23	6.2	12 40	6.4
24	Su	01 08	6.1	13 28	6.3
25	M	01 55	6.0	14 19	6.2
26	Tu	02 42	6.0	15 12	6.0
27	W	03 33	6.1	16 10	6.0
28	Th	04 31	6.1	17 10	5.8
29	F	05 36	5.9	18 13	5.7
30	Sa	06 36	5.9	19 21	5.7
31	Su	07 43	6.0	20 28	5.8

QUICK REFERENCE

DOVER — January

Date	Day	Moon	Morning hr min	m	Afternoon hr min	m
1	M		08 47	6.1	21 29	6.0
2	Tu		09 46	6.2	22 21	6.2
3	W		10 38	6.3	23 13	6.4
4	Th		11 24	**	23 48	6.5
5	F	○	**	6.5	12 06	6.3
6	Sa		00 27	6.5	12 46	6.3
7	Su		01 04	6.5	13 24	6.1
8	M		01 42	6.4	14 00	6.0
9	Tu		02 19	6.2	14 40	5.8
10	W		02 58	6.1	15 20	5.6
11	Th		03 40	5.8	16 07	5.4
12	F		04 28	5.5	17 02	5.1
13	Sa		05 27	5.3	18 09	5.0
14	Su	◡	06 37	5.2	19 24	5.1
15	M		07 48	5.3	20 30	5.3
16	Tu		08 50	5.6	21 23	5.6
17	W		09 37	5.8	22 07	6.0
18	Th		10 21	6.1	22 48	6.3
19	F	●	11 03	6.3	23 28	6.5
20	Sa		11 45	6.5	**	**
21	Su		00 12	6.7	12 30	6.6
22	M		00 56	6.8	13 17	6.6
23	Tu		01 39	6.8	14 02	6.5
24	W		02 21	6.7	14 48	6.3
25	Th		03 06	6.5	15 36	6.1
26	F		03 56	6.3	16 29	5.8
27	Sa		04 53	6.0	17 34	5.6
28	Su	∧	06 02	5.7	18 53	5.4
29	M		07 25	5.6	20 19	5.5
30	Tu		08 45	5.7	21 25	5.8
31	W		09 47	5.9	22 15	6.0

DOVER — February

Date	Day	Moon	Morning hr min	m	Afternoon hr min	m
1	Th		10 36	6.1	22 56	6.3
2	F		11 17	6.2	23 34	6.4
3	Sa	○	11 54	6.3	**	**
4	Su		00 09	6.5	12 27	6.3
5	M		00 43	6.6	13 00	6.3
6	Tu		01 17	6.5	13 31	6.1
7	W		01 48	6.3	14 00	6.0
8	Th		02 16	6.3	14 30	5.9
9	F		02 45	6.0	15 02	5.6
10	Sa		03 22	5.7	15 46	5.3
11	Su	◡	04 11	5.4	16 48	5.1
12	M		05 26	5.1	18 15	4.9
13	Tu		07 00	5.0	19 53	5.1
14	W		08 21	5.3	20 31	5.4
15	Th		09 19	5.7	21 47	5.9
16	F		10 04	6.1	22 29	6.3
17	Sa	●	10 46	6.4	23 10	6.7
18	Su		11 28	6.7	23 51	6.9
19	M		**	**	12 11	6.8
20	Tu		00 33	7.0	12 54	6.8
21	W		01 15	7.0	13 38	6.7
22	Th		01 56	6.9	14 20	6.5
23	F		02 37	6.7	15 05	6.2
24	Sa		03 25	6.3	15 57	5.8
25	Su	∧	04 24	5.8	17 04	5.4
26	M		05 41	5.4	18 36	5.2
27	Tu		07 22	5.2	20 11	5.3
28	W		08 49	5.5	21 16	5.7

DOVER — March

Date	Day	Moon	Morning hr min	m	Afternoon hr min	m
1	Th		09 47	5.8	22 03	6.0
2	F		10 29	6.0	22 39	6.2
3	Sa		11 02	6.2	23 13	6.3
4	Su	○	11 33	6.3	23 47	6.3
5	M		**	**	12 04	6.3
6	Tu		00 18	6.6	12 32	6.3
7	W		00 47	6.6	12 58	6.2
8	Th		01 12	6.4	13 22	6.0
9	F		01 35	6.3	13 46	5.9
10	Sa		01 59	6.1	14 16	5.8
11	Su		02 31	5.8	14 55	5.6
12	M	◡	03 16	5.5	15 53	5.2
13	Tu		04 34	5.1	17 28	4.9
14	W		06 26	5.0	19 18	5.1
15	Th		07 57	5.3	20 31	5.5
16	F		08 56	5.8	21 22	6.0
17	Sa		09 41	6.3	22 04	6.4
18	Su		10 24	6.5	22 45	6.8
19	M	●	11 04	6.8	23 26	7.0
20	Tu		11 47	6.9	**	**
21	W		00 06	7.1	12 29	6.9
22	Th		00 47	7.1	13 11	6.7
23	F		01 29	6.9	13 55	6.5
24	Sa		02 11	6.6	14 40	6.2
25	Su		03 02	6.2	15 34	5.8
26	M	∧	04 05	5.6	16 43	5.3
27	Tu		05 31	5.2	18 19	5.1
28	W		07 21	5.1	19 53	5.3
29	Th		08 38	5.4	20 54	5.6
30	F		09 30	5.7	21 37	5.9
31	Sa		10 07	6.0	22 14	6.2

DOVER — April

Date	Day	Moon	Morning hr min	m	Afternoon hr min	m
1	Su	○	10 36	6.1	22 46	6.4
2	M		11 06	6.2	23 19	6.5
3	Tu		11 35	6.3	23 49	6.5
4	W		**	**	12 02	6.3
5	Th		00 16	6.5	12 27	6.2
6	F		00 39	6.3	12 50	6.2
7	Sa		01 03	6.2	13 15	6.1
8	Su		01 26	6.1	13 48	5.9
9	M		02 02	5.8	14 28	5.7
10	Tu	◡	02 14	5.5	15 30	5.3
11	W		04 14	5.1	17 06	5.1
12	Th		07 06	**	18 44	5.4
13	F		07 28	5.4	19 56	5.9
14	Sa		08 27	5.9	20 49	6.3
15	Su		09 13	6.3	21 33	6.5
16	M	●	09 56	6.6	22 15	6.8
17	Tu		10 38	6.7	22 57	7.0
18	W		11 21	6.8	23 40	7.1
19	Th		**	**	12 06	6.8
20	F		00 08	7.0	12 51	6.7
21	Sa		01 08	6.7	13 36	6.5
22	Su		01 56	6.4	14 24	6.1
23	M		02 49	6.0	15 19	5.8
24	Tu	∧	03 53	5.5	16 24	5.4
25	W		05 16	5.2	17 48	5.2
26	Th		06 54	5.2	19 14	5.6
27	F		08 02	5.4	20 14	5.6
28	Sa		08 51	5.6	21 01	5.9
29	Su		09 30	5.8	21 40	6.1
30	M		10 04	6.0	22 15	6.2

SUNRISE AND SUNSET TIMES

Predictions are given for the times of sunrise and sunset on every Sunday throughout the year. For places on the same latitude as the following, add 4 minutes for each degree of longitude west (subtract if east).

These times are in GMT, except between 0100 on Mar 26 and 0100 on Oct 29, when the times are in BST **(1 hour in advance of GMT)**.

		London Rise	London Set	Manchester Rise	Manchester Set	Edinburgh Rise	Edinburgh Set
Jan	1	08 06	16 02	08 25	16 00	08 44	15 49
	8	08 04	16 11	08 23	16 09	08 40	15 59
	15	08 00	16 21	08 17	16 20	08 34	16 11
	22	07 53	16 32	08 09	16 32	08 25	16 25
	29	07 44	16 44	07 59	16 46	08 13	16 39
Feb	5	07 33	16 57	07 47	16 59	08 00	16 55
	12	07 21	17 10	07 34	17 13	07 45	17 10
	19	07 07	17 23	07 20	17 27	07 29	17 25
	26	06 53	17 35	07 04	17 41	07 12	17 40
Mar	5	06 38	17 48	06 48	17 54	06 55	17 55
	12	06 22	18 00	06 31	18 07	06 37	18 10
	19	06 06	18 12	06 14	18 20	06 18	18 24
	26	06 50	19 24	06 57	19 33	07 00	19 38
Apr	2	06 34	19 35	06 41	19 46	06 41	19 53
	9	06 19	19 47	06 24	19 59	06 23	20 07
	16	06 03	19 59	06 07	20 11	06 06	20 21
	23	05 49	20 10	05 52	20 24	05 48	20 35
	30	05 35	20 22	05 37	20 37	05 32	20 50
May	7	05 22	20 33	05 23	20 49	05 17	21 04
	14	05 11	20 44	05 11	21 01	05 03	21 17
	21	05 01	20 54	05 00	21 12	04 50	21 30
	28	04 53	21 04	04 51	21 22	04 40	21 41

Reproduced, with permission, from data supplied by HM Nautical Almanac Office, Copyright Council for the Central Laboratory for the Research Councils.

SUNRISE AND SUNSET TIMES

		London		Manchester		Edinburgh	
		Rise	Set	Rise	Set	Rise	Set
Jun	4	04 47	21 11	04 44	21 31	04 32	21 51
	11	04 44	21 17	04 40	21 37	04 28	21 58
	18	04 43	21 21	04 39	21 41	04 26	22 02
	25	04 44	21 22	04 41	21 42	04 28	22 03
July	2	04 48	21 21	04 45	21 41	04 32	22 01
	9	04 54	21 17	04 52	21 36	04 40	21 56
	16	05 02	21 11	05 00	21 29	04 49	21 48
	23	05 11	21 02	05 10	21 20	05 00	21 37
	30	05 21	20 52	05 21	21 09	05 13	21 24
Aug	6	05 32	20 40	05 33	20 56	05 26	21 10
	13	05 43	20 27	05 45	20 42	05 40	20 55
	20	05 54	20 13	05 57	20 27	05 53	20 38
	27	06 05	19 58	06 09	20 11	06 07	20 21
Sep	3	06 16	19 43	06 21	19 54	06 21	20 03
	10	06 27	19 27	06 33	19 38	06 34	19 44
	17	06 38	19 11	06 46	19 20	06 48	19 26
	24	06 50	18 55	06 58	19 03	07 02	19 07
Oct	1	07 01	18 39	07 10	18 46	07 15	18 49
	8	07 12	18 23	07 23	18 29	07 29	18 30
	15	07 24	18 08	07 36	18 13	07 44	18 13
	22	07 36	17 53	07 49	17 57	07 58	17 56
	29	06 49	16 40	07 02	16 43	07 13	16 39
Nov	5	07 01	16 27	07 15	16 29	07 28	16 24
	12	07 13	16 16	07 29	16 17	07 43	16 11
	19	07 25	16 06	07 42	16 07	07 57	15 59
	26	07 36	15 59	07 54	15 58	08 10	15 49
Dec	3	07 46	15 54	08 05	15 53	08 22	15 43
	10	07 55	15 52	08 14	15 50	08 32	15 39
	17	08 01	15 52	08 20	15 50	08 39	15 38
	24	08 05	15 55	08 24	15 53	08 43	15 41
	31	08 06	16 01	08 25	15 59	08 44	15 48

GRID REFERENCES

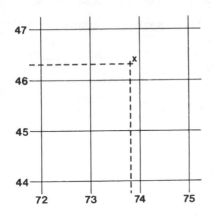

A grid reference is made up of letters and numbers. Two-letter codes are used for 100km squares on the National Grid (opposite) and single-letter codes on the Irish Grid (below).

The squares may be further subdivided into squares of 10km, 1km or 100m, allowing for increasingly specific references. On a given map the lines forming the squares are numbered in the margins, those along the top and bottom being known as 'eastings' and those along the sides as 'northings'. A reference number is made up of the relevant letter code plus two sets of figures, those representing the easting followed by the northing. According to the scale of the map they can either be read off directly or calculated by visually dividing the intervals into tenths. For most purposes three-figure eastings plus three-figure northings are adequate.

The example above, from an Ordnance Survey 'Landranger' map, illustrates how to specify a location on a map divided into lkm squares: the reference for point X is 738463. If that location lies in square SP (see map opposite), the full reference is SP738463.

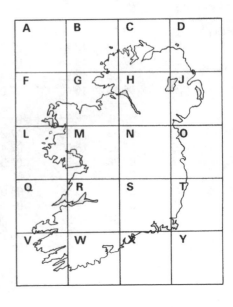

LETTER CODES FOR IRISH GRID 10km SQUARES

GRID REFERENCES

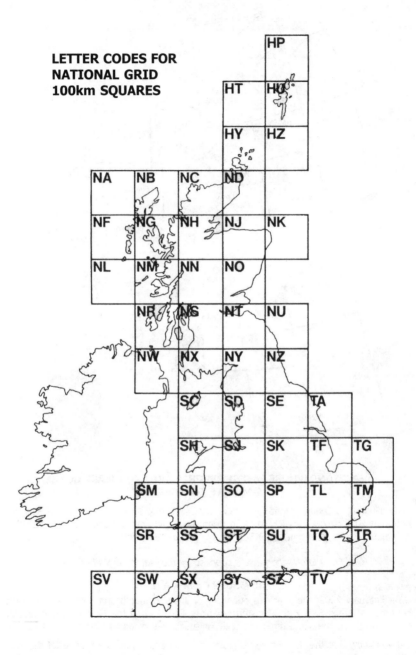

LETTER CODES FOR NATIONAL GRID 100km SQUARES

SEA AREAS

STATIONS WHOSE LATEST REPORTS ARE BROADCAST IN THE 5-MINUTE FORECASTS

Br Bridlington; C Channel Light-Vessel Automatic; F Fife Ness; G Greenwich Light-Vessel Automatic; J Jersey; L Lerwick; M Malin Head; R Ronaldsway; S Sandettie Light-Vessel Automatic; Sc Scilly Automatic; St Stornoway; T Tiree; V Valentia

From information kindly supplied by the Meteorological Office

REVISION OF SEA AREAS

On 4 February 2002, the southern boundary of areas Plymouth and Sole, and the northern boundary of areas Biscay and Finisterre were realigned along the Metarea I/II boundary at 48°27' North. At the same time, sea area Finisterre was renamed FitzRoy.

Did you know that the new FitzRoy shipping area is named after the founder of the Met Office?

THE NEW BIRDWATCHER'S CODE OF CONDUCT

Around three million adults go birdwatching every year in the UK. Following *The birdwatchers' code* is good practice, common sense and will help everybody to enjoy seeing birds.

This code puts the interests of birds first, and respects other people, whether or not they are interested in birds. It applies whenever you are watching birds in the UK or abroad. Please help everybody to enjoy birdwatching by following the code, leading by example and sensitively challenging the minority of birdwatchers who behave inappropriately.

1. The interests of the birds come first

Birds respond to people in many ways, depending on the species, location and time of year.

If birds are disturbed they may keep away from their nests, leaving chicks hungry or enabling predators to take their eggs or young. During cold weather, or when migrants have just made a long flight, repeatedly disturbing birds can mean they use up vital energy that they need for feeding.

Intentionally or recklessly disturbing some birds at or near their nest is illegal in Britain.

Whether you are particularly interested in photography, bird ringing, sound-recording or birdwatching, remember to always put the interests of the birds first.

- Avoid going too close to birds or disturbing their habitats – if a bird flies away or makes repeated alarm calls, you're too close. If it leaves, you won't get a good view of it anyway.

- Stay on roads and paths where they exist and avoid disturbing habitat used by birds.

- Think about your fieldcraft. You might disturb a bird even if you are not very close, eg a flock of wading birds on the foreshore can be disturbed from a mile away if you stand on the seawall.

- Repeatedly playing a recording of bird song or calls to encourage a bird to respond can divert a territorial bird from other important duties, such as feeding its young. Never use playback to attract a species during its breeding season.

2. Be an ambassador for birdwatching

Respond positively to questions from interested passers-by. They may not be birdwatchers yet, but good view of a bird or a helpful answer may ignite a spark of interest. Your enthusiasm could start lifetime's interest in birds and a greater appreciation of wildlife and its conservation.

Consider using local services, such as pubs, restaurants, petrol stations, and public transport. Raising awareness of the benefits to local communities of trade from visiting birdwatchers may, ultimately, help the birds themselves.

3. Know the Countryside Code, and follow it

Respect the wishes of local residents and landowners and don't enter private land without permission, unless it is open for public access on foot.

Follow the codes on access and the countryside for the place you're walking in. Irresponsible behaviour may cause a land manager to deny access to others (eg for important bird survey work). It may also disturb the bird or give birdwatching bad coverage in the media.

Access to the countryside

Legislation provides access for walkers to open country in Britain, and includes measures to protect wildlife. Note that the rules and codes are different in each part of Britain, so plan ahead and make sure you know what you can do.

4. The law

Laws protecting birds and their habitats are the result of hard campaigning by generations of birdwatchers. We must make sure that we don't allow them to fall into disrepute. In England, Scotland and Wales, it is a criminal offence to disturb, intentionally or recklessly, at or near the nest, a species listed on Schedule 1 of the Wildlife & Countryside Act 1981 (see www.rspb.org.uk/policy/wildbirdslaw for a full list). Disturbance could include playback of songs and calls. In Scotland, disturbing capercaillie and ruffs at leks is also an offence. It is a criminal offence to intentionally disturb a bird at or near the nest under the Wildlife (Northern Ireland) Order 1985.

The Government can, for particular reasons such as scientific study, issue licences to individuals that permit limited disturbance, including monitoring of nests and ringing. It is a criminal offence to destroy or damage, intentionally or recklessly, a special interest feature of a Site of Special Scientific Interest (SSSI) or to disturb the wildlife for which the site was notified.

If you witness anyone who you suspect may be illegally disturbing or destroying wildlife or habitat, phone the police immediately (ideally, with a six-figure map reference) and report it to the RSPB

5. Rare birds

Mobile phones, telephone and pager services and the internet mean you can now share your sightings instantly. If you discover a rare bird, please bear the following in mind

- Consider the potential impact of spreading the news and make an effort to inform the landowner (or, on a nature reserve, the warden) first. Think about whether the site can cope with a large number of visitors and whether sensitive species might be at risk, such as breeding terns, flocks of wading birds or rare plants. The county bird recorder or another experienced birdwatcher can often give good advice.

- On private land, always talk to the landowner first. With a little planning, access can often be arranged.

- People coming to see a rare bird can raise money for a local reserve, other wildlife project or charity. Consider organising a voluntary collection at access points to the site.

THE BIRDWATCHER'S CODE OF CONDUCT

- Rare breeding birds are at risk from egg-collectors and some birds of prey from persecution. If you discover a rare breeding species that you think is vulnerable, contact the RSPB; it has considerable experience in protecting rare breeding birds. Please also report your sighting to the county bird recorder or the Rare Breeding Birds Panel. (www.rbbp.org.uk). Also, consider telling the landowner – in most cases, this will ensure that the nest is not disturbed accidentally. If you have the opportunity to see a rare bird, enjoy it, but don't let your enthusiasm override common sense.

In addition to the guidelines above:

- park sensibly, follow instructions and consider making a donation if requested
- don't get too close so that you can take a photograph – you'll incur the wrath of everyone else watching if you scare the bird away
- be patient if the viewing is limited, talk quietly and give others a chance to see the bird too
- do not enter private areas without permission
- not everyone likes to see an 'organised flush' and it should never be done in important wildlife habitats or where there are other nesting or roosting birds nearby.

A flush should not be organised more frequently than every two hours and not within two hours of sunrise or sunset, so the bird has chance to feed and rest.

6. Make your sightings count

Add to tomorrow's knowledge of birds by sending your sightings to www.birdtrack.net This online recording scheme from the BTO, the RSPB and BirdWatch Ireland allows you to input and store all of your birdwatching records, which in turn helps to support species and site conservation. With one click, you can also have your records forwarded automatically to the relevant county recorder. County recorders and local bird clubs are the mainstay of bird recording in the UK. Your records are important for local conservation and help to build the county's ornithological history. For a list of county bird recorders, look in the County Directory of *The Yearbook,* ask at your local library, or visit www.britishbirds.co.uk/countyrecorders You can also get involved in a UK-wide bird monitoring scheme, such as the Breeding Bird Survey and the Wetland Bird Survey (see www.bto.org for details). If you've been birdwatching abroad, you can give your sightings to the BirdLife International Partner in that country by visiting www.worldbirds.org Your data could be vital in helping to protect sites and species in the country you've visited.

SCHEDULE 1 SPECIES

Under the provisions of the Wildlife and Countryside Act 1981 the following bird species
(listed in Schedule 1 - Part I of the Act) are protected by special penalties at all times.

Avocet,	Fieldfare	Osprey	Shrike, Red-backed
Bee-eater	Firecrest	Owl, Barn	Spoonbill
Bittem	Garganey	Owl, Snowy	Stilt, Black-winged
Bittern, Little	Godwit, Black-tailed	Peregrine	Stint, Temminck's
Bluethroat	Goshawk	Petrel, Leach's	Stone-curlew
Brambling	Grebe, Black-necked	Phalarope, Red-necked	Swan, Bewick's
Bunting, Cirl	Grebe, Slavonian	Plover, Kentish	Swan, Whooper
Bunting, Lapland	Greenshank	Plover, Little Ringed	Tern, Black
Bunting, Snow	Gull, Little	Quail, Common	Tern, Little
Buzzard, Honey	Gull, Mediterranean	Redstart, Black	Tern, Roseate
Chough	Harriers (all species)	Redwing	Tit, Bearded
Crake, Corn	Heron, Purple	Rosefinch, Scarlet	Tit, Crested
Crake, Spotted	Hobby	Ruff	Treecreeper, Short-toed
Crossbills (all species)	Hoopoe	Sandpiper, Green	Warbler, Cetti's
Divers (all species)	Kingfisher	Sandpiper, Purple	Warbler, Dartford
Dotterel	Kite, Red	Sandpiper, Wood	Warbler, Marsh
Duck, Long-tailed	Lark, Shore	Scaup	Warbler, Savi's
Eagle, Golden	Lark, Wood	Scoter, Common	Whimbrel
Eagle, White-tailed	Merlin	Scoter, Velvet	Wryneck
Falcon, Gyr	Oriole, Golden	Serin	

The following birds and their eggs (listed in Schedule 1 - Part II of the Act) are protected by special penalties during the close season, which is Feb 1 to Aug 31 (Feb 21 to Aug 31 below high water mark), but may be killed outside this period - Goldeneye, Greylag Goose (in Outer Hebrides, Caithness, Sutherland, and Wester Ross only), Pintail.

THE COUNTRYSIDE CODE

The new Countryside Code, launched in July 2004, followed extensive consultation with the public and stakeholders carried out through the summer of 2003. The new Code is designed to reassure land managers as new public rights of access begin, and to make the public aware of their new rights and responsibilities across the whole countryside.

Be safe – plan ahead and follow any signs
Even when going out locally, it's best to get the latest information about where and when you can go; for example, your rights to go onto some areas of open land may be restricted while work is carried out, for safety reasons or during breeding seasons. Follow advice and local signs, and be prepared for the unexpected.

Leave gates and property as you find them
Please respect the working life of the countryside, as our actions can affect people's livelihoods, our heritage, and the safety and welfare of animals and ourselves.

Protect plants and animals, and take your litter home
We have a responsibility to protect our countryside now and for future generations, so make sure you don't harm animals, birds, plants, or trees.

Keep dogs under close control
The countryside is a great place to exercise dogs, but it's every owner's duty to make sure their dog is not a danger or nuisance to farm animals, wildlife or other people.

Consider other people
Showing consideration and respect for other people makes the countryside a pleasant Environment for everyone – at home, at work and at leisure.

BIRDLINE NUMBERS
NATIONAL AND REGIONAL

Birdline name	To obtain information	To report sightings (hotlines)
National		
Bird Information Service www.birdingworld.co.uk	09068 700222	01263 741140
Flightline (Northern Ireland)	028 9146 7408	
Regional		
Northern Ireland	028 9146 7408	
Scotland*	09068 700 234	01292 611 994
Wales *	09068 700 248	01492 544 588
East Anglia www.birdnews.co.uk	09068 700 245	01603 763 388 or 08000 830 803
Midlands *	09068 700 247	01905 754 154
North East*	09068 700 246	07626 983 963
North West *	09068 700 249	01492 544 588
South East www.southeastbirdnews.co.uk	09068 700 240	07626 966 966 or 08000 377 240
South West	09068 700 241	07626 923 923

* www.uk-birding.co.uk
Charges
At the time of compilation, calls to premium line numbers cost 60p per minute.

INDEX TO BIRD RESERVES
AND OBSERVATORIES

INDEX TO RESERVES